Mathematical Buffet

Problem solving at the olympiad level

VICTOR UFNAROVSKI
JANA MADJAROVA
FRANK WIKSTROM

Studentlitteratur

The picture on the cover shows a number of equally spaced, concentric circles, along with their parallel tangent lines. Show that the intersections marked in the picture are located on a parabola.

Art. No 39120
ISBN 978-91-44-11103-2
First edition
1:2

studentlitteratur.se
Studentlitteratur AB, Lund

Printed by Eurographic Danmark A/S, Denmark 2018

CONTENTS

INTRODUCTION

Everybody knows that the only way to achieve a good result in mathematical competitions at a high level (such as the IMO — the International Mathematical Olympiad) is to solve many different problems. The only problem is to find enough problems at the right level of difficulty. Too easy problems do not teach much. Too difficult problems kill the enthusiasm, and even if the level is right, we often tend to choose the types of problems we enjoy, even though the opposite choice would probably be more useful.

The aim of this book is to resolve this difficulty and present a mathematical buffet, where everybody can find a problem to his/her taste. We have collected a large number of problems at different levels, that we have tested on our former IMO participants. Each test (called paper) consists of six problems from different areas of mathematics. Usually, but not always, the first problems are easier than the last ones and the first papers are easier than the last ones. This allows us to gradually increase the level of difficulty.

If the reader tries to take one test a week s|he will be occupied for almost five months and will definitely learn a lot. But, as in any buffet, it is not forbidden to choose only some problems or to have second helpings of the same dish.

We have made a lot of effort to explain how to think when solving the problem, as well as some possible ways to the solutions. This is why we often give several solutions, especially for the easier problems, to show a variety of approaches. For some problems, we have written additional comments and remarks, explaining the key ideas and connections.

When the food (problems) had been prepared, we realized that the reader probably would want to have access to some cutlery: a collection of appropriate definitions and statements helping both to understand the written solutions and to create one's own.

Thus we wrote a kind of dictionary, containing useful ideas and facts, which helps us to reduce the size of the solutions. As in any dictionary there are a lot of cross-references in both directions and is hardly convenient for systematic reading (page after page). Quite soon, we found that the book had taken on a life of its own, and told us to write more than we initially planned. We asked our colleagues and students and they confirmed that those demands were exactly what they wanted to see in such a book.

One of the demands was to divide the dictionary into two parts. The elementary one should contain basic ideas and facts that are normally used at the IMO level and which presumably are useful for the readers who are yet to reach this level. It is used as a source for the references to definitions and notations which are used in the remaining parts. We did not want to repeat a nice introduction part in the IMO Compendium [35] but chose to focus on problem-solving strategies instead.

The second part consists of more advanced topics, which may be more difficult at first and which sometimes are not necessary for the problems in the main part. We wanted to give some taste of deeper mathematics and we are certain that this part can be interesting for those who hardly need the elementary part. Of course, we have restricted ourselves to some selected topics, and do not pretend that our choice is perfect.

We wanted to teach our reader to see that mathematics is alive. We have used many analogies, non-formal arguments, that may look very naive, unnecessary and even irritating. But we know that for some readers those naive analogies and advice could be really useful. Our limitations of language (none of the authors have English as their first language) can increase such irritation, because sometimes we simply could not find the exact words to express the important ideas in a correct way. Hopefully the reader will be able to grasp the ideas and forgive our poor way of expressing them.

Many people helped us to improve this book. The books [20],[26], [30], [35], [41], [44], [47], [51], [52] were sources for many ideas and problems that we present. We are grateful to our colleagues Gert Almkvist, Jörgen Backelin, Ilya Bogdanov, Alexander Chapovalov, Tomas Claesson, Dag Jonsson, Kjell Elfström, Kerstin Johnsson, Peter Kumlin, Arne Meurman, Mikael Persson Sundqvist, Tomas Persson, Leonid Rabovetskiy, Amol Sasane, Jörg Schmeling, Anna Torstensson, Paul Vaderlind, Stanislav Volkov, Göran Wanby and Johan Wästlund, who all had direct or indirect influence on the content of this book. Our students solved and discussed the problems in the papers. First of all, we would like to mention Valentina Chapovalova, Malte Larsson, Lisa Lokteva, Gunnar Peng, Johan Sannemo, Mårten Wiman, David Wärn, Chen Xing and Lars Åström, but the list is much longer. We are grateful to Nazar Agakhanov, Yasuharu Asai, Valeriu Baltag, Mihai Bălună, Barbu Berceanu, Pierre Bornsztein, Peter Boyvalenkov, Radmila Bulajich, Birgit van Dalen, Gregor Dolinar, Sándor Dobos, Artūras Dubickas, Dušan Đukić, Luciano Monteiro de Castro, Claude Deschamps, Roberto Dvornicich, Steven Dunbar, Jóhanna Eggertsdóttir, Şahin Emrah, Patricia Fauring, Zuming Feng, María Gaspar, Robert Geretschläger, Radu Gologan, Hans-Dietrich Gronau, Heinrich Josef Gstöttner, Indra Haraksingh, Marteinn Harðarson, Ria Van Huffel, Yuji Ito, Sven Toft Jensen, Romualdas Kašuba, Azer Kerimov, Johannes Kleppe, Emil Kolev, Oleg Košik, Dávid Kunszenti-Kovács, Matti Lehtinen, Gordon Lessells, Charles Leytem, Qiusheng Li, Maria Losada, Alekso Malcheski, Vesna Manova-Erakovikj, Sergei Mazanik, Nguyễn Khắc Minh, Eric Müller, Givi Nadibaidze, Kai Neergård, Nikolay Nikolov, Philippe Niederkorn, Theoklitos Paragyiou, Angelo Di Pasquale, József Pelikán, Michał Pilipczuk, Maxim Pratusevich, Jürgen Prestin, Lev Radzivilovsky, Sune Reeh, Kirsten Rosenkilde, Bogdan Rublov, Naoki Sato, Nairi Sedrakyan, Lindsey Shorser, Geoff Smith, Juris Škuškovniks, Yongjin Song, Dmitry Tereshin, Anton Vassilyev, Esa Vesalainen, Igor Voronovich, John Webb, Bart Windels, Yan Loi Wong, Bin Xiong, Evangelos Zotos and many other IMO leaders and coordinators from different countries who created together this amazing IMO atmosphere without which this book would have been impossible to start (or finish).

We are also grateful to the authors of remarkable GeoGebra [60], which was used to create the pictures. Last but not least, we would like to thank our families who supported our real life while we were looking for virtual mathematical dishes.

Lund, the summer solstice of 2016.

Victor Ufnarovski
Jana Madjarova
Frank Wikström

PART I

Basic ideas and facts

CHAPTER 1

General advice

In this first part of the book, we will present some important facts and ideas that are useful for solving mathematical problems. Naturally we use ideas as well as facts when trying to solve a problem, but we would like to stress the importance of the ideas and start with those. At times, we are going to use facts and theorems that we haven't discussed in detail yet, (or won't discuss at all), but the reader should not have any difficulties understanding the important ideas. This first part is more like a dictionary than a textbook, and we have tried to make plenty of cross-references and underline the most important definitions.

As for the facts, we assume that the reader is familiar with mathematics from upper secondary school (even though this assumption may sometimes be a little off, since the syllabi in different countries cover different issues). Instead, we stress some results that may perhaps be less familiar, but nevertheless are very useful. In any case, the reader should understand that these chapters cannot replace a real and deeper study of the subjects covered in this book.

We will also assume that the reader has already reached a certain level of proficiency in problem solving. Therefore we do not discuss elementary skills, such as solving quadratic equations or finding the area of a triangle with given sides. Nevertheless we will supply a short list of basic facts that should be considered as elementary at the IMO level, that are not necessarily well-known from school.

~

It is almost impossible to explain how to solve problems — every new problem is different! Nevertheless, we can try to list some general ideas that may be useful in some situations or at least give some possible directions for the first steps towards the solution. Some of these ideas are so general that they may appear too simplistic and useless, but we will try to show how even trivial ideas can help solving real problems.

For most of the problems in this book, we do not provide traditional well-polished solutions. Instead we give less conventional solutions, which we call the **search**. Our point is that the path we take (including a few possible wrong turns) to finding a solution is more important than the solution itself. This is why we will try to explain how to think while finding a working approach to the problem. We will probably skip some details, saving time and space; if the reader understands the main idea, s|he will be able to fill in the missing pieces.

There is a dilemma. Either we know how to solve the problem (and do not need any help) or, as is usually the case, we do not. What we need is to change the problem in such a way that we can see the solution or at least how to take the next step. What are possible approaches to achieve this aim?

1.1 Tiny mathematical plays and the art of substitution

The competition has started. You have read through the formulations of the problems and you realize that you don't have a slightest idea where and how to start. After the competition is over, you are reading the official solutions and you can't help wondering: "*Why couldn't I find this myself? The solution uses nothing that is new to me and I was definitely able to solve the problem. In fact I was very close to finding the solution.*" Does this description sound familiar?

This is a really interesting question — why couldn't we solve the problem even though we feel that we knew everything necessary for the solution? Why did we choose the wrong way? And what does it even mean — the wrong way?

If we look at the process of problem-solving, we find that it consists of tiny mathematical plays that are performed over and over again in our mathematical theatre.

Some plays such as "*The quadratic equation*", "*Using induction*" and "*Homothety*" may be familiar to us and our performance on these plays is excellent. Other plays are perhaps not as familiar, such as "*Inversion*" or "*Generating functions*" and therefore we don't even attempt to use them. There is a huge difference between watching a play and acting in it. That is why solutions are relatively easy to read but are much more difficult to find.

Let us return to our competition. The main question we need to ask ourselves is, which mathematical plays are performed here? We need to act in our mathematical theatre, but first of all we need to guess: which play is hiding here? We know how the play starts and how it is expected to end, but we are unsure of the exact story line and who the actors are. Our own role is more like that of a producer's. We can try to perform several different plays that we know, but before we start acting, we need to think very carefully about the choice of actors. Being the producer, we can choose the actors and naturally we want to use the best ones available.

Again, to solve a problem we need to see what is hidden inside. We can see much clearer in a familiar environment, with convenient notation that we like and are accustomed to. Especially if the problem looks unattractive, our first question should be: *which new variables may I need here?*

Consider for example the elementary equation

$$2^{\cos 2x} + 2^{\sin^2 x} = 3.$$

Search. The equation does not look very nice. We suspect that the title of the play we need to perform sounds something like "*Quadratic equation*" but we're not too happy with the actors here. We definitely need to replace them. The equation suggests, almost directly, to use the substitution $y = \sin^2 x$, because $\cos 2x = 1 - 2\sin^2 x = 1 - 2y$. With this improved choice of actors, we get

$$2^{1-2y} + 2^y = 3,$$

which already looks much nicer. This is a good sign that we have chosen the substitution correctly. Another good sign is that the original variable has disappeared.

What is next? The same question, and a second substitution $z = 2^y$ transforms the equation to something really attractive:

$$\frac{2}{z^2} + z = 3 \quad \Leftrightarrow \quad z^3 - 3z^2 + 2 = 0.$$

So the play we were searching for was in fact "*The cubic equation*", but we already know how to solve this one. We see that $z = 1$ is a solution and can factorize the equation as

$$(z-1)(z^2 - 2z - 2) = 0.$$

It is useful to study the properties of the new variables. For example, we evidently have $0 \leq y \leq 1$ above and thus $1 \leq z \leq 2$, which immediately allows us to ignore the solutions of the equation $z^2 - 2z - 2 = 0$. □

Sometimes, as in the equation $4^x + 6^x = 9^x$ we cannot easily eliminate the old variable. Quite often a division overcomes the difficulty. Here, after dividing by 9^x ($\neq 0$), we can introduce the new variable $y = \left(\frac{2}{3}\right)^x$ and obtain a quadratic equation.

Note that if we replace 9 by 10 in the right hand side, the equation looks very similar. We can still divide by 10^x but do not get a quadratic equation. To solve $4^x + 6^x = 10^x$, we need a quite different play, namely, "*Monotonicity*": we see that the function $f(x) = \left(\frac{2}{5}\right)^x + \left(\frac{3}{5}\right)^x - 1$ is decreasing and therefore has at most one zero, which we find by inspection: $x = 1$. The division was still important.

In the equation

$$x^4 + 2x^3 + ax^2 + 2x + 1 = 0$$

there is still a quadratic equation to be found, and the main actor appears after dividing by x^2: it is $y = x + \frac{1}{x}$ because

$$x^2 + \frac{1}{x^2} = \left(x + \frac{1}{x}\right)^2 - 2 = y^2 - 2.$$

Division can often reduce the number of variables, as in

$$\frac{x+y}{x-2y} \curvearrowright \frac{s+1}{s-2},$$

where we first divide by y and then introduce $s = \frac{x}{y}$.

Sometimes we need new variables to "kill" an inconvenient condition. For example, consider the condition $abc = 1$. If we put $a = \frac{x}{y}$, $b = \frac{y}{z}$, $c = \frac{z}{x}$ we still have three variables x, y, z, but now without restrictions (except that a, b, c are non-zero). Another important example is the condition $x > 0$. If we want to avoid it, we can write $x = e^t$. Here is a classical example:

Theorem 1.1 (Weighted AM–GM inequality) *Let $x_1, \ldots, x_n$ and $\alpha_1, \ldots, \alpha_n$ be positive real numbers. If $\alpha_1 + \cdots + \alpha_n = 1$ then*

$$\alpha_1 x_1 + \alpha_2 x_2 + \cdots + \alpha_n x_n \geq x_1^{\alpha_1} x_2^{\alpha_2} \cdots x_n^{\alpha_n}$$

with equality if and only if $x_1 = \cdots = x_n$.

Proof. We follow the advice. With $x_i = e^{t_i}$ we rewrite the inequality as

$$\alpha_1 e^{t_1} + \cdots + \alpha_n e^{t_n} \geq e^{\alpha_1 t_1 + \cdots + \alpha_n t_n}.$$

Now we see that the hidden play was *Jensen's inequality* for $f(t) = e^t$ (see page 129). □

A very special case is the classical relation between arithmetic, geometric and harmonic means (abbreviated AM, GM and HM, which explains the names).

Theorem 1.2 (AM–GM–HM inequality) *For any positive real numbers* $x_1, x_2, \dots, x_n$ *we have*

$$\frac{x_1 + x_2 + \cdots + x_n}{n} \geq \sqrt[n]{x_1 x_2 \cdots x_n} \geq \frac{n}{\frac{1}{x_1} + \frac{1}{x_2} + \cdots + \frac{1}{x_n}}$$

with equality if and only if $x_1 = \cdots = x_n$.

Proof. Use $\alpha_i = \frac{1}{n}$ in theorem 1.1 to get AM–GM. Set $y_i = \frac{1}{x_i}$ to deduce GM–HM from it. □

Note that AM–GM part is valid even if we allow the variables to be non-negative (if $x_i = 0$ for some i, the geometric mean is zero).

Let us return to substitutions. If $x \geq 0$ we can consider the case $x = 0$ separately or we can use another substitution, $x = y^2$. The condition $0 \leq x \leq a$ can be expressed as $x = a\cos^2\varphi$.

Rewriting old variables as expressions in new ones may be very useful. Two typical examples are

$$A = \sqrt{1 - x^2}, \quad B = \sqrt{1 + y^2}.$$

Put $x = \sin\alpha$ and $y = \tan\beta$, demand $0 \leq \alpha, \beta \leq \frac{\pi}{2}$ and get:

$$A = \cos\alpha, \quad B = \frac{1}{\cos\beta}.$$

Expressions such as $\sqrt{x^2 + y^2}$ call for the substitution $x = R\cos\alpha$, $y = R\sin\alpha$. The main idea behind all those substitutions is the fact that $\sin^2\alpha + \cos^2\alpha = 1$. Working with trigonometry, it is useful to remember that all trigonometric functions in α can be expressed as rational functions in $t = \tan\frac{\alpha}{2}$, e.g.

$$\cos\alpha = \frac{\cos^2\frac{\alpha}{2} - \sin^2\frac{\alpha}{2}}{\cos^2\frac{\alpha}{2} + \sin^2\frac{\alpha}{2}} = \frac{1 - \tan^2\frac{\alpha}{2}}{1 + \tan^2\frac{\alpha}{2}} = \frac{1 - t^2}{1 + t^2}.$$

Consider now the following problem.

Problem 1.3 Let $a_0, a_1, \dots, a_n, \dots,$ be a sequence with $a_0 = 1$ and

$$a_{n+1} = \frac{a_n}{1+\sqrt{1+a_n^2}}.$$

a) Find a_n as a function of n.

b) Find $\lim\limits_{n\to\infty} 2^n a_n$.

Search. This looks terrible. But, recognizing the expression $\sqrt{1+a_n^2}$, it is tempting to try the standard substitution $a_n = \tan x_n$. We have:

$$\frac{a_n}{1+\sqrt{1+a_n^2}} = \frac{\tan x_n}{1+\sqrt{1+\tan^2 x_n}} = \frac{\tan x_n}{1+\frac{1}{\cos x_n}} = \frac{\sin x_n}{1+\cos x_n}.$$

This is an improvement, but we can do even better, using the half-angle formulas:

$$\frac{\sin x_n}{1+\cos x_n} = \frac{2\sin\frac{x_n}{2}\cos\frac{x_n}{2}}{2\cos^2\frac{x_n}{2}} = \tan\frac{x_n}{2}.$$

Amazing! Our problem has changed dramatically:

$$a_{n+1} = \tan x_{n+1} = \tan\frac{x_n}{2} \quad\Rightarrow\quad x_{n+1} = \frac{1}{2}x_n.$$

Because $x_0 = \frac{\pi}{4}$ we get

$$x_n = \frac{x_0}{2^n} = \frac{\pi}{4}\cdot\frac{1}{2^n}, \qquad a_n = \tan\left(\frac{\pi}{4}\cdot\frac{1}{2^n}\right),$$

and $\lim\limits_{n\to\infty} 2^n a_n = \lim\limits_{n\to\infty} 2^n \tan\left(\frac{\pi}{4}\cdot\frac{1}{2^n}\right) = \frac{\pi}{4}$.

If the last step is not evident, use a change of variables $x = \frac{\pi}{4}\cdot\frac{1}{2^n}$ and the standard limit $\lim\limits_{x\to 0}\frac{\sin x}{x} = 1$. □

A perfect way to practice the art of substitutions is to solve a lot of functional equations — see Chapter 7.

1.2 The importance of make-up

Sometimes we speak about cosmetic changes, meaning that they are not so important. But look at how important those cosmetic changes are in the theatre. A simple make-up procedure transforms our favourite actor to somebody else. He is still the same person, but looks quite different and with his new appearance we accept his role much easier.

We can do the same in our mathematical theatre. Before we start our performance, we can and often should make some cosmetic changes. They do not change the problem essentially; it is still the same problem. Nevertheless such changes may highlight some small but important details and suddenly we see the beauty of the problem and know how to proceed. Thus use some make-up early on and change the problem to make it more attractive while keeping its essence. Or, if you prefer culinary terminology, prepare the problem before eating it.

A classical example is **induction**. We have a problem where we need to prove something for all natural numbers n. Instead, we solve the improved version of the problem in which we additionally suppose that for $n-1$, it has already been proven. This additional assumption can be so strong that the solution arises almost automatically. And the price for this improvement is a (often trivial) proof for $n = 1$. A typical example is the following proof of the identity

$$1^3 + 2^3 + \cdots + (n-1)^3 + n^3 = \binom{n+1}{2}^2.$$

For $n = 1$ this is evidently true. If we suppose that the identity is valid for $n-1$, i.e.

$$1^3 + 2^3 + \cdots + (n-1)^3 = \binom{(n-1)+1}{2}^2 = \binom{n}{2}^2$$

then all that remains is to check that

$$\binom{n}{2}^2 + n^3 = \binom{n+1}{2}^2.$$

But this is routine:

$$\frac{n^2(n-1)^2}{4} + n^3 = n^2\left(\frac{(n-1)^2}{4} + n\right) = n^2\frac{(n+1)^2}{4} = \binom{n+1}{2}^2.$$

After some practise, induction proofs become quite natural, but it is still important to remember that induction needs a complete and precise formulation in order to work smoothly. Otherwise the recursion hidden in inductive proofs can be almost impossible to find and exploit.

For example, the weaker statement that the sum $1^3 + 2^3 + \cdots + n^3$ is a perfect square does not yield easily to an induction proof. We often need to reformulate the problem as strongly as possible to be able to perform the induction step. Usually the attempt to find an induction proof helps to find the reformulation. Guessing the correct formulation can be the most difficult part of the solution.

Another important method of proof is **proof by contradiction**. We want to prove a statement, but instead we suppose that the statement is false. Now we change the aim completely: we want to deduce a contradiction from this hypothesis. The advantage is that we have much more freedom — we can find a contradiction of any kind and this is sufficient. In fact, we can still prove the original statement — this would contradict our hypothesis that the statement was false. In other words, we lose nothing but gain a lot of freedom.

A classic example is the ancient proof that there exists an infinite number of primes. Indeed, suppose the opposite, that there is a finite list of primes $p_1, \dots, p_m$. Then any natural number greater than 1 is divisible by one of those primes. But the number $N = p_1 p_2 \cdots p_m + 1$ is not (the remainder when dividing by p_j is clearly 1 for every j). Here is our contradiction, and we can conclude that our hypothesis was wrong.

This proof illustrates the weakness of the method: it is non-constructive, and gives no hint as to how the prime numbers are constructed. We get our proof but gain no extra information.

The improved version of proof by contradiction is the **minimal counterexample** — we assume not only that we have a counterexample, but that we have a *minimal* one (with a suitable interpretation of minimal). Then the typical contradiction is that we construct an even smaller counterexample. *Things can only get worse* is a very optimistic statement in this approach.

We illustrate the method with another ancient result (which, by the way, was a well-kept secret in ancient times) that $\sqrt{2}$ is an irrational number.

Suppose the opposite, i.e. that $\sqrt{2}$ is rational. Then there exists a triangle with the angles 90°, 45°, 45° whose sides are rational numbers. Enlarging the triangle, we can find a similar triangle with integer sides. Now suppose that we have a minimal (for example with the respect to area) such triangle ABC with integer sides and look at it, see Figure 1.1.

An easy origami trick allows us to construct a new triangle ADE that has integer sides as well: Fold the triangle twice, first bisecting the angle at C by folding the sides AC and BC together and then forming a straight angle at E by folding AC over itself while holding on the point D. Then AED is similar to ABC (why?) and $AE = AC - CE = AC - BC$ and $AD = AB - BD =$

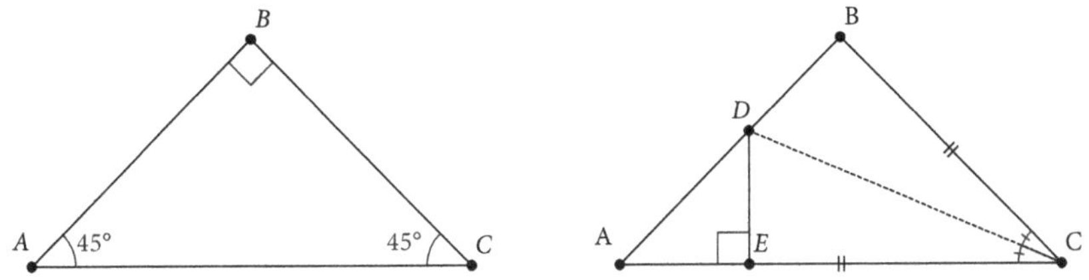

Figure 1.1 Minimal counterexample

$AB - DE = AB - AE$ are integers so AED is a smaller counter-example. A contradiction! One of the authors of this book is well-known for showing this construction at various dinner parties using paper napkins.

Note that our additional make-up procedure that reduced the question to the integer sides was essential: for triangles with rational sides, there is no minimality.

One magic word (or four, depending on the point of view) that the reader should learn is the abbreviation WLOG — *without loss of generality*. We use it when we can improve our problem by exploring some symmetries or other special properties to reduce the problem to more special cases. For example, if a triangle ABC is arbitrary, WLOG we can assume that $\angle A \le \angle B \le \angle C$ and this already gives us $\angle A \le 60°$ and $\angle C \ge 60°$. We did not change the problem, but made it slightly more convenient.

To show how this abbreviation can simplify our lives consider a very simple example — a proof of the triangle inequality $|x + y| \le |x| + |y|$. Of course, this can be done by considering different cases. But look how well our make-up works here.

First of all, WLOG we can assume that $x + y \ge 0$ (if this is not the case we simply replace x by $-x$ and y by $-y$ and this does not change the problem). Second, WLOG can we suppose that $x \ge y$ (here we use symmetry). This gives us $x \ge 0$ and now we need only to prove

$$x + y \le x + |y| \quad \Leftrightarrow \quad y \le |y|,$$

which is evident.

We finish the section with the solution of an easy problem where we want to show how efficient a preliminary preparation of the problem may be.

Problem 1.4 Some cells of an $m \times n$, $(mn \geq 2)$ board are coloured white while the others are coloured black. There is a beetle on one of the cells. Each time the beetle moves, it crawls from one cell to some other neighbouring cell (neighbouring cells share a side) and recolours this new cell in the opposite colour. Finally, the beetle flies away. Does there always exist a route such that when the beetle leaves the board, all the cells are coloured black?

Solution. Yes, such a route always exists, i.e. the beetle can always colour the whole board black. To show this, first note that there always exists a path through all the cells. If we restrict ourselves to moves in this path only, then the solution is the same as for a rectangle of size $1 \times mn$. Thus WLOG we can suppose that the size is $1 \times n$, with $n > 1$.

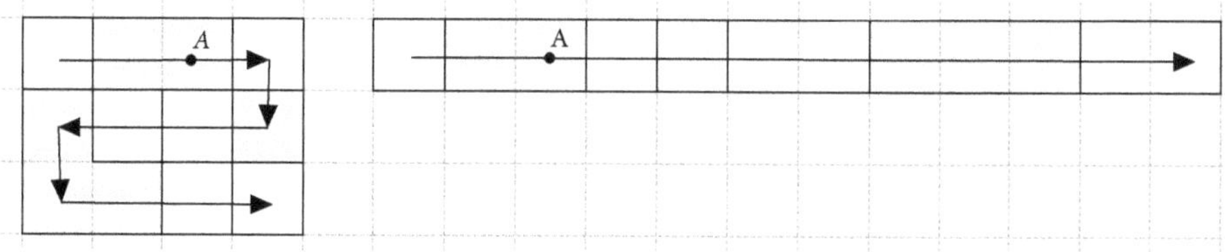

Figure 1.2 Changing the board

WLOG we can suppose that the beetle is in the very first cell (simply moving there first). WLOG we can also assume that this cell is black; otherwise we move to cell 2 and return back. Now move to the second cell and use the induction to colour the remaining $n - 1$ cells.

And that is all! The only thing we need to do is to prove the base of our induction for the size 1×2, which is elementary. □

Compare with another short solution, also based on induction.

Solution. Let A be the initial cell. Let there be some white cells (except perhaps the initial one) on the board. Consider one of them, say cell B. The beetle can crawl to B in one way or another and then come back to A using the same route. It is easy to see that as a result, cell B has been coloured black while all the other cells of the beetle's route (except A) have not changed their colour. The number of white cells (different from A) on the board has been reduced by one. Thus by induction the beetle can colour all the cells on the board (except perhaps the initial one) black and find itself back on cell A. If, at the end, cell A is coloured black, then the beetle leaves the board, otherwise the beetle crawls to the neighbouring (black) cell, say C, and colours it white, then it crawls to A and colours it black, comes back to C and colours it black and, finally, leaves the board. □

1.3 Looking for bridges

When we are solving a problem it often feels like being on one side of a river that we need to swim across to reach a particular spot on the other side. If the river is only a tiny brook we can simply jump over it as we do when solving elementary problems in school. A large brook requires some swimming skills — and learning to swim in the mathematical waves is exactly what we want to practice. But only very well-trained people can swim over a large river and reach the exact destination they want.

Normally the only thing we can do is to swim with the current in the hope of finding a bridge where we can cross the river. Bridges are made by experienced professionals, they are universal, stable and easy to use. Of course, the spot we reach on the other side is probably not exactly the one we wanted, but it will be on the other side of the river, and we will have less trouble continuing our journey by our own means.

This rough analogy is a great hint: instead of directly trying to solve a difficult problem, we look for a solid bridge that could help us. We are still swimming, but we are swimming with the current. That is what we are often doing when solving equations: we might be looking for the hidden quadratic equation, which is a very stable bridge. Consider another example.

Problem 1.5 Let $x_1, x_2, \dots, x_n$ be numbers in the interval $(0, \frac{\pi}{2})$ such that

$$\tan x_1 + \tan x_2 + \dots + \tan x_n \le n.$$

Prove that

$$\sin x_1 \cdot \sin x_2 \cdot \dots \cdot \sin x_n \le 2^{-n/2}.$$

Search. How do we find a bridge from $\tan x$ to $\sin x$? Let us swim with the current. One natural transformation is

$$\frac{\tan x_1 + \tan x_2 + \dots + \tan x_n}{n} \le 1$$

which suggests the idea of using the AM–GM inequality. Is this the bridge we are looking for? Applying the AM–GM inequality, we get

$$\tan x_1 \cdot \tan x_2 \cdot \dots \cdot \tan x_n \le 1 \quad \Leftrightarrow$$
$$\sin x_1 \cdot \sin x_2 \cdot \dots \cdot \sin x_n \le \cos x_1 \cdot \cos x_2 \cdot \dots \cdot \cos x_n,$$

and we are close to the destination point. One more multiplication and we end up with:

$$\sin^2 x_1 \cdot \sin^2 x_2 \cdot \dots \cdot \sin^2 x_n \le \frac{\sin 2x_1}{2} \cdot \frac{\sin 2x_2}{2} \cdot \dots \cdot \frac{\sin 2x_n}{2} \le 2^{-n}.$$

It was a bridge! But we could still have missed the final destination if we had tried squaring the inequality in the final step. □

Of course, someone well-trained in trigonometry can swim from $\tan x$ to $\sin x$ directly. Noting that the inequality as $\sin x \le \tan x$ is obviously not sufficient; he uses a stronger inequality $\sin x \le \frac{\sqrt{\tan x}}{\sqrt{2}}$ which, he knows, is valid for all $x \in (0, \frac{\pi}{2})$. (Why? Because $2 \sin x \cos x = \sin 2x \le 1 \Leftrightarrow 2 \sin^2 x \le \tan x$. But why does he know? Because he is well-trained.)

Using this we reduce the inequality to

$$\sqrt{\tan x_1 \cdot \tan x_2 \cdot \dots \cdot \tan x_n} \le 1$$

and finishes on the same AM–GM bridge.

Or maybe he reacts directly on the product, taking logarithms and using the *Jensen inequality* bridge instead (see page 129):

Search (Alternative solution). The inequality to be proved is equivalent to

$$\frac{\ln \sin x_1 + \ln \sin x_2 + \dots + \ln \sin x_n}{n} \le -\frac{1}{2} \ln 2.$$

Consider the function $f(t) = \ln \sin t$ on $(0, \frac{\pi}{2})$. Then $f'(t) = \cot t$ and $f''(t) = -\frac{1}{\sin^2 t}$. This means that f is a concave function on $(0, \frac{\pi}{2})$ and hence, by Jensen's inequality,

$$\frac{\ln \sin x_1 + \ln \sin x_2 + \dots + \ln \sin x_n}{n} \le \ln \sin \frac{x_1 + x_2 + \dots + x_n}{n}.$$

The function $\tan t$ is convex on the same interval, so that

$$\tan \frac{x_1 + x_2 + \dots + x_n}{n} \le \frac{\tan x_1 + \tan x_2 + \dots + \tan x_n}{n} \le 1,$$

which gives

$$\frac{x_1 + x_2 + \cdots + x_n}{n} \le \frac{\pi}{4}.$$

Hence

$$\sin \frac{x_1 + \cdots + x_n}{n} \le \frac{1}{\sqrt{2}} \quad \Rightarrow \quad \ln \sin \frac{x_1 + \cdots + x_n}{n} \le -\frac{1}{2} \ln 2.$$

□

1.4 Finding weak points

Problems, like people, have weak points. Some of them cannot resist induction, some like new dresses (or a good substitution). But like people, problems do not like to attract attention to their weaknesses and we need to look for them thoroughly. Often there is a weak point already in the formulation of the problem.

Problem 1.6 Prove that if $a_1, a_2, \ldots, a_n$ are different integers, then the polynomial

$$P(x) = (x - a_1)(x - a_2) \cdots (x - a_n) - 1$$

cannot be factorized into two polynomials of lower degree with integer coefficients.

Search. Assume the opposite, i.e., assume that $P(x) = P_1(x)P_2(x)$, where the polynomials P_1, P_2 have integer coefficients and $\deg P_1, \deg P_2 < n$. What are their weak points? Obviously a_k are very special and we test them. For each k we have $P(a_k) = P_1(a_k)P_2(a_k) = -1$. Since the numbers $P_1(a_k), P_2(a_k)$ are integers, one of them has to be equal to 1 and the other one has to be equal to -1. Hence we have $P_1(a_k) + P_2(a_k) = 0$. Definitely weak points!

Which weak points does the polynomial $Q(x) = P_1(x) + P_2(x)$ have? Like every polynomial — the weak points are the zeros, and there are at most $\deg Q(x) < n$ of them if Q is not the zero polynomial. But we have already found n zeros of Q, which shows that Q is in fact the zero polynomial, and $P_1(x) = -P_2(x)$ for all x. Consequently $P(x) = -P_1^2(x)$, which is impossible since the leading coefficient of P is 1. Hence, P cannot be factorized into two polynomials of lower degrees with integer coefficients. □

Quite often weak points are extremal points, something that is largest, shortest, widest — the winner in some abstract beauty pageant.

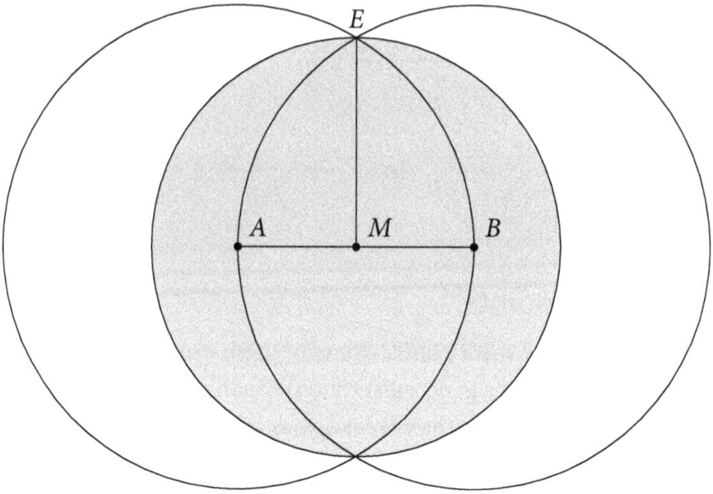

Figure 1.3 Cat and dogs

Problem 1.7 A cat from the neighbouring village keeps coming to Duncetown to annoy the dogs. Every night when all the dogs are sleeping, the cat sneaks into Duncetown, meows out loudly and runs back home. When the cat meows, all the dogs that are up to 90 m away from the cat start barking. Duncetown is a small village, and every two dogs in the village are within 100 m from each other. Is it possible for the cat to take a position such that all the dogs start barking at the same time?

Search. There are too many dogs to handle! One dog is not a problem, but two may be. Which is the worst pair or dogs? Probably A and B with the maximal distance between them.

Consider the intersection of the disks of radii $|AB|$ with centres A and B respectively (see Figure 1.3). Note that the triangle ABE is equilateral. There are no dogs outside this intersection. If the cat positions itself at the midpoint M of AB, then all the dogs will be within $\frac{\sqrt{3}}{2} \cdot |AB| \leq 50 \cdot \sqrt{3} < 90$ m from it, and they start barking at the same time. □

Note that if instead we took arbitrary points and radius 100 we could end up with a non-equilateral triangle. It was important to use the extremal points to get our solution.

1.5 On the other side of the river

Quite often the conditions of the problems are so complicated that we have no idea where to start. Instead it might be a good idea to investigate what would be a good final destination at the other side of the river. How could we finish the solution? What do we need for this? Sometimes knowing where to finish gives us a good idea of where to start.

Problem 1.8 Let α, β be two positive irrational numbers such that

$$\frac{1}{\alpha} + \frac{1}{\beta} = 1.$$

Prove that the list of the integer parts

$$[\alpha], [\beta], [2\alpha], [2\beta], \ldots, [n\alpha], [n\beta], \ldots$$

contains all positive integers and each of them appears exactly one time.

Search. Is this even true? Let us check on a calculator $\alpha = \sqrt{2} = 1.414\cdots$. Then

$$\beta = \frac{\alpha}{\alpha - 1} = \frac{\sqrt{2}}{\sqrt{2} - 1} = \frac{\sqrt{2}(\sqrt{2} + 1)}{(\sqrt{2} - 1)(\sqrt{2} + 1)} = 2 + \sqrt{2} = 3.414\cdots$$

and

$$\begin{array}{llll} \alpha = 1.414, & 2\alpha = 2.828, & 3\alpha = 4.242, & 4\alpha = 5.656, \\ 5\alpha = 7.070, & 6\alpha = 8.484, & 7\alpha = 9.898, & \ldots \\ \beta = 3.414, & 2\beta = 6.828, & 3\beta = 10.24, & \ldots \end{array}$$

Unbelievably, the sequences fit each other perfectly! But we still have no idea why it works. Let us start from the end.

What does it mean that $m = [k\alpha]$? Quite simply, that $m \le k\alpha < m + 1$ and, by the way, we can't have equality because α is irrational. Why is it impossible that $m = [l\beta]$ at the same time, thus $m < l\beta < m + 1$? Let us change the inequalities closer to our conditions. We have the inequalities

$$\frac{m}{\alpha} < k < \frac{m+1}{\alpha}, \quad \frac{m}{\beta} < l < \frac{m+1}{\beta}$$

and if we add them we get

$$m < k + l < m + 1.$$

But this is a contradiction – there are no integers between m and $m+1$! Good. Perhaps l works for α? Hardly. Evidently $\alpha > 1$ and this can be used to finish the proof of uniqueness.

What about the existence? What does it mean that $m \neq [n\alpha]$? That we miss it, or in other words that there exists k such that $k\alpha < m$ but $(k+1)\alpha > m+1$. It is the same with β and we know what to do with this. We leave this pleasure to the reader. □

Problem 1.9 Let $P(x)$ and $Q(x)$ be different polynomials such that $P(Q(x)) = Q(P(x))$ for all x. Prove that $P(P(x)) - Q(Q(x))$ is divisible by $P(x) - Q(x)$.

Search. Again, the conditions are very difficult to interpret and we have no idea how to start. Perhaps it is easier to understand how to finish? What does it mean that something is divisible by $P(x) - Q(x)$? At least this question is easier, but still difficult. Let us make it even easier: substitute. What does it mean that something is divisible by $y - z$? At last we know some answers, for example, $y^2 - z^2$ is divisible by $y - z$, or, more generally,

$$y^n - z^n = (y-z)(y^{n-1} + y^{n-2}z + \cdots + yz^{n-2} + z^{n-1})$$

is divisible by $y-z$ as well. But this also means that for any polynomial $f(x) = a_n x^n + \cdots + a_1 x + a_0$ the difference $f(y) - f(z) = a_n(y^n - z^n) + \cdots + a_1(y-z)$ is divisible by $y - z$ and probably we cannot say more than this.

Thus for any polynomial $f(x)$ we know that $f(P(x)) - f(Q(x))$ is divisible by $P(x) - Q(x)$. Good, but we have no such polynomial in our problem. But can we find one that is close to the problem? It should, of course, depend on P and Q and symmetry is expected, because P and Q have equal rights. The simplest such polynomial is $f(x) = P(x) + Q(x)$. Now $f(P(x)) - f(Q(x))$ is

$$P(P(x)) + Q(P(x)) - P(Q(x)) - Q(Q(x)) = P(P(x)) - Q(Q(x))$$

and we have found how to use the conditions! □

One can wonder if there exists non-trivial examples of such polynomials P and Q? The answer can be found in the so-called **Chebyshev polynomials** $T_n(x)$ for which $T_n(\cos x) = \cos nx$. Evidently, these polynomials satsify $T_n(T_m(x)) = T_m(T_n(x)) = T_{mn}(x)$.

Simple induction shows that such polynomials exist. Explicitly

$$T_n(x) = \frac{(x - \sqrt{x^2 - 1})^n + (x + \sqrt{x^2 - 1})^n}{2} = \sum_{k=0}^{[n/2]} \binom{n}{2k} (x^2 - 1)^k x^{n-2k}$$

but this is difficult to use. See Section 10.5 to learn more about their properties.

1.6 Complete the picture!

Mathematics has a deep aesthetic inside — correct arguments tend to be beautiful. That is why quite often an aesthetical approach helps to find a solution. If we feel that something is missing in the picture, it is often a good idea to complete it by filling in the missing detail.

Problem 1.10 Solve the equation:

$$x^3 + x^2 + x = -\frac{1}{3}.$$

Search. Aesthetically the equation $3x^3 + 3x^2 + 3x + 1 = 0$ looks better: we do not like fractions without special reasons. But how do we solve it? A standard approach to find all rational solutions in the equation $a_n x^n + \cdots + a_1 x + a_0 = 0$ with integer coefficients is to check those which have the form $x = \pm\frac{p}{q}$ where $p \mid a_0$ and $q \mid a_n$ (see Theorem 3.3 on page 101). Unfortunately this does not work here. What else? Something looks very familiar and aesthetics demands us to complete the picture that $3x^2 + 3x + 1$ is painting for us. We get:

$$3x^3 + x^3 + 3x^2 + 3x + 1 - x^3 = 2x^3 + (x + 1)^3.$$

Is this better? Oh, yes — this is a sum of two cubes and we can factorize it:

$$\begin{aligned}&\left(\sqrt[3]{2}x\right)^3 + (x + 1)^3 \\ &\quad = \left(\sqrt[3]{2}x + x + 1\right)\left(\sqrt[3]{4}x^2 - \sqrt[3]{2}x(x + 1) + (x + 1)^2\right) = 0.\end{aligned}$$

The remainder is routine. □

The same attitude can be used to handle expressions like $\sqrt{x - 2\sqrt{x - 1}}$. We want a square under the radical, i.e. something like $a^2 + 2ab + b^2$. We see only $2\sqrt{x - 1}$, but complete the picture directly:

$$\sqrt{(x - 1) - 2\sqrt{x - 1} + 1} = \sqrt{(\sqrt{x - 1} - 1)^2} = \left|\sqrt{x - 1} - 1\right|.$$

Of course, the change of variables $t = \sqrt{x-1}$ makes it easier to see this.

Speaking about the radicals, we must admit that sometimes we need more complicated and less obvious transformations, like

$$\sqrt{a \pm \sqrt{b}} = \sqrt{\frac{a + \sqrt{a^2 - b}}{2}} \pm \sqrt{\frac{a - \sqrt{a^2 - b}}{2}}$$

for $a \geq \sqrt{b}$, that are easy to prove but difficult to guess.

A classic example is completing the square. The idea is quite easy: if we see two of the three parts of $a^2 + 2ab + b^2$ we immediately add the third (and subtract it again to stay honest).

Problem 1.11 Factorize $x^4 + y^4$.

Search. The first reaction that this is a misprint; with a minus it would be perfect. But there are no misprints here and we need to do something. In any case we see two squares and should complete them to a perfect square:

$$x^4 + y^4 = x^4 + 2x^2y^2 + y^4 - 2x^2y^2 = (x^2 + y^2)^2 - 2x^2y^2.$$

But now there is a minus sign! We could factorize this if we had x^2y^2 only. Even $4x^2y^2$ would be sufficient. But 2 is not a square... Is not? It is a square — $2 = (\sqrt{2})^2$ and we finish the factorization:

$$\begin{aligned}(x^2 + y^2)^2 - 2x^2y^2 &= (x^2 + y^2)^2 - (\sqrt{2}xy)^2 \\ &= (x^2 + y^2 + \sqrt{2}xy)(x^2 + y^2 - \sqrt{2}xy).\end{aligned}$$

Perhaps not as pretty this time, which also explains why it was more difficult to find. □

Problems in geometry are perfect candidates to attack by *completing the picture.* The following simple theorem is very useful.

Theorem 1.12 *The median CM in the right-angled triangle ABC is equal to half of the hypotenuse AB.*

Proof. All we need is either to complete the triangle ABC to make the rectangle $ACBD$ or to draw a circle with its centre in M and look at the complete picture (Figure 1.4). □

In Figure 1.5 one can find some other typical examples of completion.

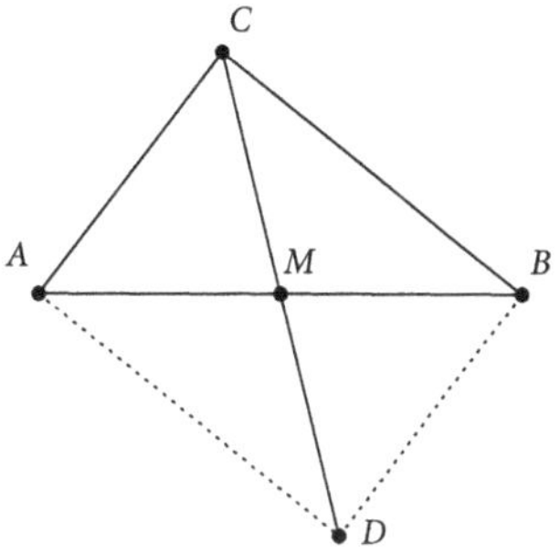

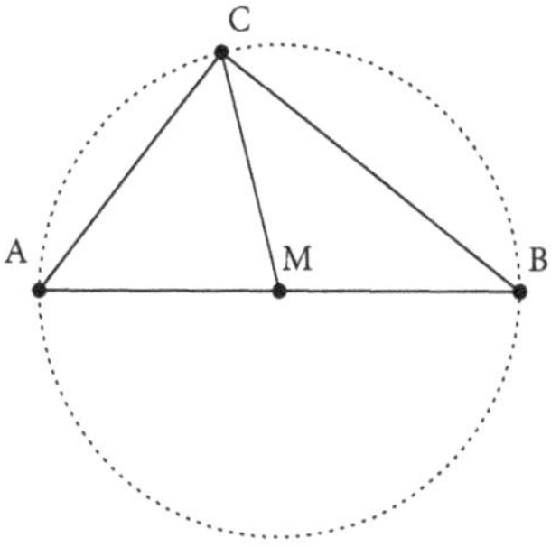

Figure 1.4 The median is half of the hypotenuse

1.7 Reversing the theorems

The triangle with the sides 3, 4, 5 is, evidently right-angled. Are you sure that it is so evident? Why? If your answer is "*By the Pythagorean theorem*", you are totally wrong. The Pythagorean theorem says that *if* a triangle is right-angled *then* $a^2 + b^2 = c^2$. Here the situation is quite the opposite. What you are using is an entirely different theorem, namely the *converse* of the Pythagorean theorem, i.e. *if* $a^2 + b^2 = c^2$ *then* the triangle is right-angled. And why is this theorem valid?

In this case, the converse follows from the cosine rule, but mistaking a theorem for its converse is so common that it is better to discuss the subject in detail.

First of all, not every theorem has a converse; there should be a conditional statement in the theorem something like "**if** Tra-ta-ta **then** Bla-bla-bla." Then in the *converse theorem* the statement sounds the opposite, namely "**if** Bla-bla-bla **then** Tra-ta-ta."

Nobody says that the converse theorem is valid; it may very well be wrong. For example the correct theorem "If $x > 0$ then $x^2 > 0$" has a very incorrect converse: "If $x^2 > 0$ then $x > 0$."

But even if the converse is correct its proof might be very different from the proof of the direct statement. Note also that if there are several conditions there might be several converse-ike statements. For example,

"If $x = 0$ and $y \neq 0$ then $x + y \neq 0$"

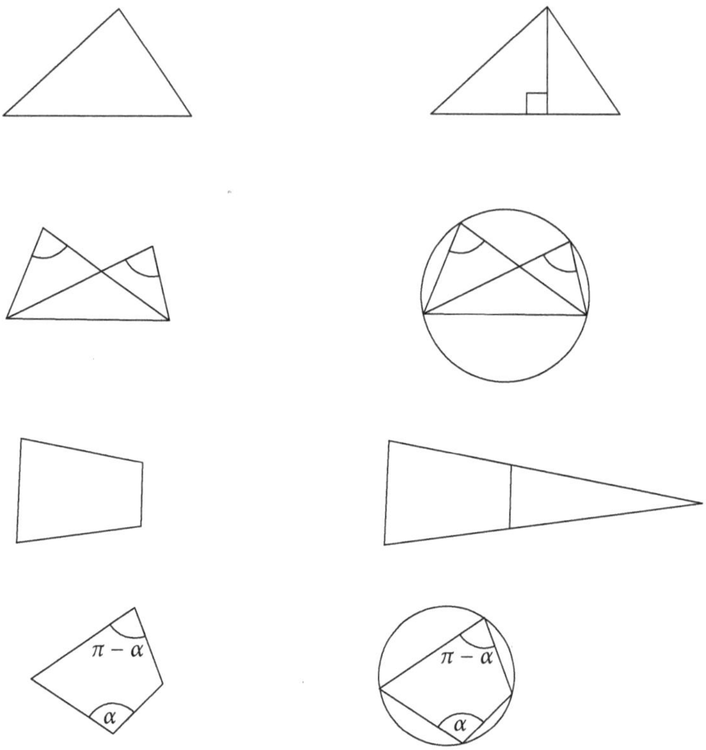

Figure 1.5 Completing pictures

has one obviously wrong converse:

"If $x + y \neq 0$ then $x = 0$ and $y \neq 0$".

But the following converse

"If $x + y \neq 0$ and $x = 0$ then $y \neq 0$"

is correct, though another converse

"If $x + y \neq 0$ and $y \neq 0$ then $x = 0$"

is a disaster.

Hopefully the reader understands how this works. There is one simple but useful principle that helps creating correct converse theorems along with their proofs.

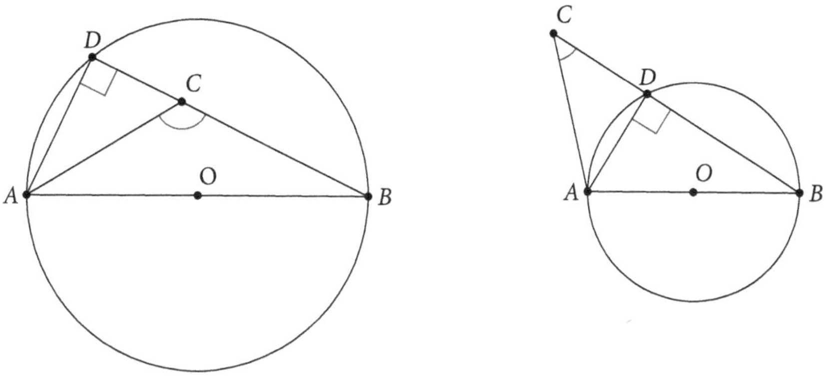

Figure 1.6 Point C outside the circle

Theorem 1.13 *Suppose that we have n different statements that sound like "If A_i then B_i." Suppose that the alternatives B_i exclude each other. Suppose also that one of the alternatives A_i always takes place. Then all the converse theorems "If B_i then A_i" are valid as well, and thus we have equivalences.*

Proof. Suppose that B_i is valid, but not A_i. Then one of A_j with $j \neq i$ should take place and, therefore B_j is valid together with B_i. But this is a contradiction that proves the statement. □

To make this less abstract consider a simple — and typical — example. We know that if we have a circle with diameter AB, and C a point on the circle different from A and B then $\angle ACB = 90°$ (**Thales' theorem**). We want to prove the converse statement, but in fact it is easier to prove something stronger.

Theorem 1.14 *Consider a circle with the diameter AB and a point C outside the line AB. Then*

$$\begin{aligned} C \text{ is on the circle} \quad &\Leftrightarrow \quad \angle ACB = 90°. \\ C \text{ is inside the circle} \quad &\Leftrightarrow \quad \angle ACB > 90°. \\ C \text{ is outside the circle} \quad &\Leftrightarrow \quad \angle ACB < 90°. \end{aligned}$$

Proof. Theorem 1.13 shows that it is sufficient to prove "⇒" instead of "⇔", because the alternatives in the right-hand sides exclude each other and one of the three alternatives in the left-hand side is always satisfied. But this is easy.

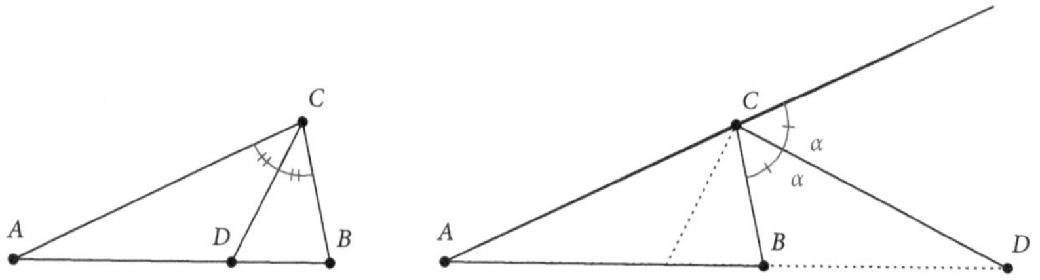

Figure 1.7 Internal and external bisector

The first implication is true. For the second, we continue BC to the intersection in D with the circle. Then (see Figure 1.6)

$$\angle ACB = \angle ADB + \angle DAC = 90° + \angle DAC > 90°.$$

Similarly we prove the third implication and we are done. □

Here is another example.

Theorem 1.15 *Consider a triangle ABC and a point D on the segment AB. Then CD is a bisector of the angle ACB if and only if $\frac{AD}{DB} = \frac{AC}{BC}$.*

Proof. If AD is a bisector then the sine rule gives:

$$\frac{AC}{AD} = \frac{\sin \angle ADC}{\sin \angle ACD} = \frac{\sin \angle BDC}{\sin \angle BCD} = \frac{BC}{BD} \quad \Rightarrow \quad \frac{AD}{DB} = \frac{AC}{BC}.$$

More interesting is why the converse is true. If we move D we get two other cases: $\angle ACD < \angle BCD \Rightarrow \frac{AD}{DB} < \frac{AC}{BC}$ and $\angle ACD > \angle BCD \Rightarrow \frac{AD}{DB} > \frac{AC}{BC}$ and it remains for us to use Theorem 1.13 to reverse all three implications. □

Every angle bisector has a sister — the bisector of the corresponding external angle, called the **external bisector**. The bisector and the external bisector are evidently orthogonal to each other. But Theorem 1.15 also has a sister.

Theorem 1.16 *Consider a triangle ABC with $AC > BC$ and a point D on the line AB and outside the segment AB. Then CD is an external bisector to the angle ACB if and only if $\frac{AD}{DB} = \frac{AC}{BC}$.*

We leave the sister proof to the reader (use that $\sin\alpha = \sin ACD$).

We recommend you to add the abbreviation **iff** (if and only if) to your collection of magic words. It not only saves time and space but also acts as a reminder for you to ask yourself: *can I replace "if" by "iff"*; is the converse statement true as well? If the answer is *"Yes"*, you may get a useful equivalence. Test this question on problem 1.8 and prove an unexpected equivalence. Check Theorem 1.12 as well.

It is especially useful to have several equivalent statements, a *bunch of equivalency*. They allow us to change the point of view on the problem quite radically. Surprisingly enough, they might also be easier to prove. For example, to show that n different statements A_i are equivalent we need only n proofs: $A_i \Rightarrow A_{i+1}$ for $i < n$ and $A_n \Rightarrow A_1$. But if we handle the equivalences pairwise, there are $\binom{n}{2} = \frac{n(n-1)}{2}$ statements and we need $n(n-1)$ proofs. Already for $n = 4$ this makes a big difference! We will try to formulate such statements when they make sense. To illustrate how different reformulations of the same statement can be, we will give two examples.

Theorem 1.17 *Let a, b, c, d be four real numbers. The following are equivalent:*

- *The number $ad - bc \neq 0$. This number is also denoted as $\begin{vmatrix} a & b \\ c & d \end{vmatrix}$ and called the **determinant** of the matrix $\begin{pmatrix} a & b \\ c & d \end{pmatrix}$.*
- $\begin{cases} ax + by = e \\ cx + dy = f \end{cases}$ *has a unique solution for each $e, f \in \mathbb{R}$.*
- $\begin{cases} ax + by = e \\ cx + dy = f \end{cases}$ *has a unique solution for some $e, f \in \mathbb{R}$.*
- *The vectors (a, b) and (c, d) are not parallel.*
- *The points $(0,0), (a,b), (c,d)$ are not collinear.*
- *For each e and f two lines $ax + by + e = 0$ and $cx + dy + f = 0$ have exactly one point in common.*
- *Every point in the plane can be written as $(au + bv, cu + dv)$ for some $u, v \in \mathbb{R}$.*
- *The change of variables $X = ax + by$, $Y = cx + dy$ is reversible (i.e. x, y can be found from X, Y).*
- *The map $f : (x, y) \mapsto (ax + by, cx + dy)$ is **injective** (i.e. maps different points to different points).*
- *The map $f : (x, y) \mapsto (ax + by, cx + dy)$ is **bijective** (i.e. has an inverse map).*

- *The complex number* $\dfrac{a+bi}{c+di}$ *is defined and is not real.*
- *(only when* $bd \neq 0$*)* $\dfrac{a}{b} \neq \dfrac{c}{d}$.
- *(only when* $bd(b+d) \neq 0$*)* $\dfrac{a}{b} + \dfrac{c}{d} \neq 2\dfrac{a+c}{b+d}$.

Proof. The key to all this is the system of equations. If the determinant is non-zero, then the unique solution can be found as

$$x = \frac{\begin{vmatrix} e & b \\ f & d \end{vmatrix}}{\begin{vmatrix} a & b \\ c & d \end{vmatrix}}, \quad y = \frac{\begin{vmatrix} a & e \\ c & f \end{vmatrix}}{\begin{vmatrix} a & b \\ c & d \end{vmatrix}}.$$

This is **Cramer's rule** and can be checked directly. If the determinant is zero then the system either has no solutions or has an infinite number of solutions (depending on the choice of e and f). We recommend the reader to prove this and all the necessary implications (there are many possible choices; use the ones you like). □

The analogous statement for several variables (and the same number of equations) is true as well but is more complicated.

Our next example is from three-dimensional geometry. We restrict ourselves to ten statements only (but the list can be made much longer).

Theorem 1.18 *Consider a tetrahedron* $ABCD$ *and the four triangles formed by its faces. The following conditions are equivalent.*

- *Opposite edges (i.e. edges with no common vertex) have equal lengths.*
- *All triangles are congruent.*
- *Each triangle has the same perimeter.*
- *Each triangle has the same area.*
- *Each triangle has the same radius of circumcircle.*
- *For each vertex, the sum of three angles sharing this vertex is equal to* 180°.
- $\angle BAD = \angle BCD = \angle ABC = \angle ADC$.
- *The segments, connecting middle points of the opposite edges are orthogonal.*
- *The inscribed and circumscribed spheres have the same centre.*
- *For the six angles between faces the sum of their cosines is equal to* 2.

Proof. The first statement obviously implies the second and from the second, the next three follow. The remaining implications are less evident.

We will discuss only one (probably, most unexpected) implication that equal areas implies equal edges, leaving the rest of the proof as a standard exercise in stereometry.

Let d be the distance between AB and CD and consider the projection of $ABCD$ onto a plane π, parallel to AB and CD. The altitudes to AB in the triangles ABC and ABD are equal (and equal to, say, h) and hence their projections are also equal (and equal to $\sqrt{h^2 - d^2}$). Then the areas of the projections of the two triangles are equal as well and as a result each of them is equal to half the area of the projection of $ABCD$. But the same is true for the projection of the triangles BCD and ACD. We conclude that all four planes form the same angle α with the plane π (since the area of the projection is the original area, multiplied by $\cos\alpha$). It follows that $h = d/\sin\alpha$. But the same is true for the altitude to CD in the triangle BCD, so the two altitudes are equal and hence $AB = CD$. The same argument shows that $AC = BD$ and $AD = BC$. □

1.8 The art of listening

Good communication with other people is based on one very easy rule: listen to them and listen to them again. Strangely enough this rule works for mathematics as well. We only need to imagine our mathematical objects to be alive and listen to what they have to say, because they know themselves much better than we do. And if we listen to them carefully, we can become their friends.

To be alive? How it is possible? It is possible — mathematics is alive!

Look at $\sin x$ and $\cos x$, they are as sister and brother, but how different they are! Look how accurate $\sin x$ is with signs: $\sin(-x) = -\sin x$; $\sin(x-y) = \sin x \cos y - \cos x \sin y$. But her brother $\cos x$ has no respect for signs at all: $\cos(-x) = \cos x$; $\cos(x - y) = \cos x \cos y + \sin x \sin y$. They are different!

Of course this is only a mnemonic rule, but such imagination helps us not only to remember but also to find new approaches. Consider one of them.

Problem 1.19 Let S be a finite non-empty set of integers greater than 1, with the following property: For any positive integer n there exists $s \in S$ such that either $(s, n) = 1$ or $(s, n) = s$. Prove that there exist $t, s \in S$ (not necessarily different) such that (t, s) is a prime number.

Search. Here is a nice family S. We do not know too much about them, but perhaps we can ask their friends? Look, here is one of them, n. He is very proud.

– "You know, I am their best friend. I have something in common with each of them and I am the first with this property."

– "Amazing! It is difficult to have so many interests. But do you have some double interest in something? "

– "Of course not. How could I be the first in this case?"

– "But who is your best friend?"

– "It is s. All his interests are mine as well."

– "You are a perfect friend! But there must be somebody who is not so close to you."

– "Yes. My relative $\frac{n}{p}$ does not like t and I have only p in common with t."

– "But in this case s and t do not have much in common either."

– "You are right. It is only p that they have in common."

Now let us translate this nice dialogue to the solution.

Solution. Let n be the least positive integer that is not relatively prime to any number from the set S. (Such a number exists since the product of all the numbers in S is not relatively prime to any number from S as $1 \notin S$.) In the prime factorization of n, every prime number must appear to the first power since otherwise n would not be the least number with the described property. The number n is not relatively prime to any number from the set S, so by the assumption in the problem there exists $s \in S$ such that $(s, n) = s$ and therefore $s \mid n$. Let p be any prime divisor of the number s. Then $p \mid n$ and according to the choice of n the number $\frac{n}{p}$ is relatively prime to a number $t \in S$. From $(t, \frac{n}{p}) = 1$ and $(t, n) > 1$ follows $p \mid t$. But as the number t is not divisible by any other prime divisor of the number n (as these prime numbers divide $\frac{n}{p}$), it follows $(t, s) = p$. □

If you are not convinced, consider another problem.

Problem 1.20 101 numbers are written on the blackboard in a line. Prove that it is possible to erase 90 of them such that the remaining 11 form a monotone sequence (thus increasing or decreasing).

Search. Again we do not know too much about the numbers, but they themselves know. "How long" — we ask one of the numbers, —"is the largest increasing subsequence that starts from you?"

– "It is of length k. And the longest decreasing one is of length l."

– "Does somebody else have the same numbers?"

– "No, I am unique. For example, those who are behind me and less than me have, naturally, a shorter decreasing sequence otherwise I could use it to make my l larger."

We do not ask more, but estimate. If all pairs k, l are different and both k, l are less than 11 then we have at most 100 such pairs. But they are different, and thus some k or l must be at least 11.

Of course, the problem can be generalized to $mn + 1$ numbers. We could prove in the same manner that either there exists a decreasing subsequence of the length $m + 1$ or an increasing subsequence of the length $n + 1$.

We will prove this in another way, allowing the numbers to order themselves. Let us ask them one after another to stay in columns in decreasing order. If there is no place at the end of any column they should start a new column. For example, if the original sequence starts as

$$4, 5, 2, 7, 1, 9, 8, 6, 3, \ldots$$

then the columns look like 4, then 4, 5, then $\begin{array}{cc} 4 & 5 \\ 2 & \end{array}$, then $\begin{array}{ccc} 4 & 5 & 7 \\ 2 & & \end{array}$ and so on until

$$\begin{array}{cccc} 4 & 5 & 7 & 9 \\ 2 & 3 & 6 & 8. \\ 1 & & & \end{array}$$

We cannot order the numbers ourselves because we do not know them, but they themselves can do it perfectly. But what is the result? Either we get a column of a length of at least $m + 1$ (and we get the decreasing subsequence and are done) or we get at least $n + 1$ columns. In the last case we ask the number k in the column $n+1$: why is it in this column, but not in the previous one? "Because, when I tried to fit here, the last number l in this column was less then me" — is the answer and we ask l why he missed the $n - 1$ column and so on, producing the subsequence $\cdots < l < k$ that is increasing. It is useful to listen to the numbers! □

To view mathematical objects as alive not only helps our intuition but also gives us a better way to remember the main idea of the proof. In the following example we want to prove that every non-constant polynomial $f(z)$ with complex coefficients has a (complex) root. But before this, let us look at an old lady who takes her dog for a daily morning walk.

She always makes n rounds around the flagpole (point) O. Because the lead is short the dog is forced to do n rounds as well. One morning, the lady was very tired and decided to make the circles smaller.

It is time to name our players. The lady's name is z^n. If $z = Re^{i\varphi}$, where R is a constant and φ changes from 0 to 2π then the lady z^n indeed makes n rounds of radius R^n around the origin O.

The dog's name is $f(z) = z^n + a_1z^{n-1} + \cdots + a_n$. The lead is therefore $a_1z^{n-1} + \cdots + a_n$ and for sufficiently large R the length of the lead is much smaller than R^n, and the dog is indeed forced to make n rounds around O as well. What happened when the lady was tired and changed R to almost zero?

If we suppose that $a_n \neq 0$ then for a very small R, the dog will mostly be concentrated near a_n and therefore makes no rounds around O at all.

Now let us change R from large to small R continuously. How could it happen that the dog decreased the number of rounds from n to zero? There is only one way to do this: for some R and some φ the dog has to go through O, in other words $f(Re^{i\varphi}) = 0$ and this is exactly what we wanted to prove. Amazingly easy! The reader should not have any difficulties in completing this to a rigorous proof (for example estimating more exactly what large and small mean and what to do if $a_0 = 0$) but will hopefully never forget the idea of the proof.

1.9 The inertia of thinking

What colour is the sky? Yes, it is blue. Which colour is grass? Naturally green. What does a cow drink? Of course water, but wasn't *milk* what first came to your mind? Why? Because there is inertia in our thinking. When we decided on one way of thinking, have stable associations such as milk and cow, it is very difficult to think differently. Try to imagine black milk. Our mind refuses the idea immediately, it is something very ugly.

This — the inertia of thinking — is one of the main obstacles when doing mathematics. The solution might be very easy, but slightly different than our

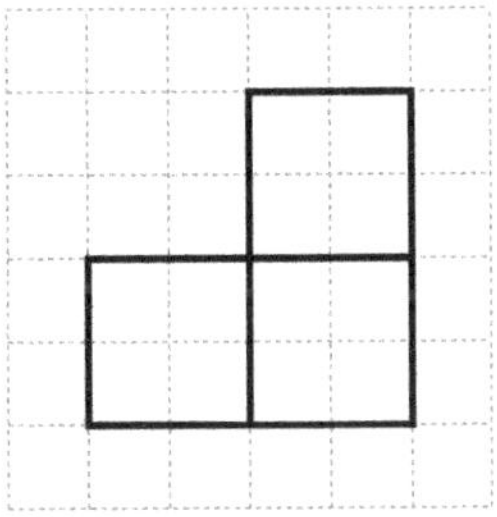

Figure 1.8 Divide this figure into four equal parts

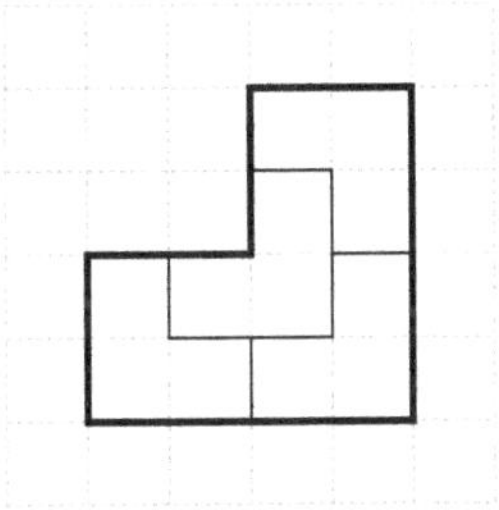

Figure 1.9 Solution of problem 1.21

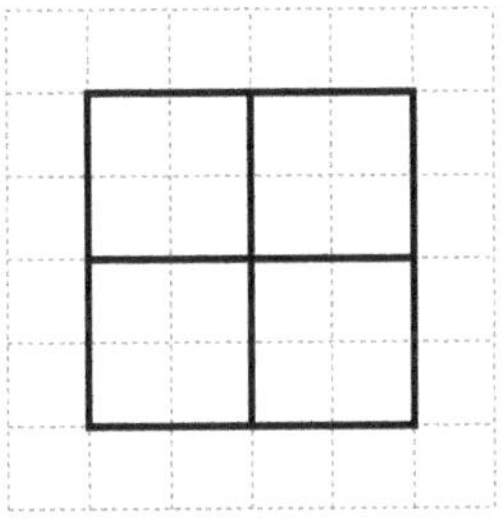

Figure 1.10 Divide this figure into five equal parts

standard way of thinking and we cannot see it. Likewise, we may begin solving the problem in one way and then it is very difficult to choose another way.

Problem 1.21 Consider Figure 1.8 consisting of three squares. Divide it into four equal parts.

Search. The total area is 12 squares and we need 3 squares for each part. After a little experimenting, we find the solution — see Figure 1.9. □

Next consider a similar, but more difficult problem. Start with a figure consisting of four squares (see Figure 1.10). Divide it into *five* equal parts.

It takes most people suprisingly long to find the solution[1]. Why? The same reason — inertia of thinking. It also probably didn't help that we pointed out that this problem was more difficult than the first one. It is hard to switch from a complicated construction to something simple!

When we know that inertia of thinking is our enemy, it is easier to fight it. The main method is to switch. To switch the approach, to switch the problem, to switch the picture, to switch yourself. It is useful to imagine how another person (your friend, teacher, the author of your favourite book or even your cat) would solve this problem. Try to be somebody else!

Once on the IMO when one of our students remembered this advice in a very difficult geometry problem, he asked himself — how would my friend (let us call him Peter) solve this? He would probably try complex numbers — he always does this.

1 simply draw four parallel lines

So our student attempted an approach using complex numbers, and found a nice property that helped him to obtain a good geometrical solution. The most interesting part of this story is that Peter himself did not solve the problem.

Let us consider another problem.

Problem 1.22 We have several bricks of the same size and a tape measure. We need to measure the longest (three dimensional) diagonal. Of course, it can be done trivially: if a, b, c are the sizes then we can measure each of them and compute $\sqrt{a^2 + b^2 + c^2}$ as the answer. The problem is that we are allowed to use only one measurement. Is it still possible?

Search. It looks impossible — we cannot measure inside the brick. On the other hand, we have no trouble finding $\sqrt{a^2 + b^2}$. Perhaps we can use the fact that we have several bricks? Can we combine them in some strange way to get the answer?

Several experiments, but no results: we can measure many different things but not this diagonal. It is a pity that the bricks are not empty inside: if they were made up of only the edges, we could measure inside the brick. But... we can create an empty brick if we take away one of them from the wall! So we take four bricks, put two of them above the others, creating a little wall and remove one of them. The empty space is our invisible empty brick in which we can measure! □

Here we see the inertia again: nobody said that the tape must touch the brick, but we restricted ourselves by this assumption. This is one of the most dangerous variants of inertia — when we impose extra conditions that are not in the problem, but which we are used to having.

Try to solve the following nice problem.

Problem 1.23 Consider a set with n elements and $n + 1$ non-empty subsets $S_1, S_2, \ldots, S_{n+1}$. Show that we can always choose two non-empty disjoint sets of indices I and J (**disjoint** means that $I \cap J = \varnothing$) such that

$$\bigcup_{i \in I} S_i = \bigcup_{j \in J} S_j.$$

Search. You started from combinatorial arguments, e.g. induction attempts, didn't you? But forget your inertia of thinking and use the following hint:

the problem is from a book in linear algebra (we suppose that the reader is familiar with the very basics of linear algebra, see Section 9.2 if this is not the case).

Linear algebra? Oh, yes — each subset S_i can be encoded as a vector v_i of length n with ones and zeros, where a one in the coordinate k means that the kth element belongs to the subset.

Thus we have $n+1$ non-zero vectors in $\mathbb{R}^n$. Of course this means that they are linearly dependent. How do we use this? Let us look at an example. If $n = 3$ and the vectors are, for example, $(1,1,0), (1,0,1), (0,1,1), (1,1,1)$ what do we get from this dependence? We have

$$(1,1,0) + (1,0,1) + (0,1,1) - 2(1,1,1) = 0$$

This can be rewritten as

$$(1,1,0) + (1,0,1) + (0,1,1) = 2(1,1,1)$$

and... this is all we need!

In general we note that we can move all the negative coefficients to the right-hand side and rewrite our dependence as

$$\sum_{i \in I} \alpha_i v_i = \sum_{j \in J} \beta_j v_j$$

with positive α_i, β_j. The corresponding subsets of indices I, J are exactly what we need. □

So do not forget always to ask yourself: *which area of mathematic does the problem come from?* It is inertia of thinking to suppose that the solution is from the same area as the problem appears to be. If this is not the case, solutions are very difficult to find, because initially we are looking in the wrong place.

Inertia of thinking also explains why it is so difficult to find a solution in two-case problems. These are the problems where to get a solution we need to consider two different cases and in each case to use a different idea. We tend to find a universal solution, but sometimes they simply do not exist and it is important to be "case-sensitive."

Problem 1.24 There is an infinite number of gangsters and every one of them tries to shoot one of his colleagues. Prove that it is possible to choose an infinite subset of gangsters not shooting one another.

Search. We consider this as a graph problem: the vertices are gangsters and we draw an edge $x \to y$ iff the gangster x tries to kill y. Thus the graph is oriented and we need to find an infinite subgraph without edges. Such translations are not necessary, but psychologically useful, because we find ourselves in a more familiar environment.

Trying to experiment with easy cases we very soon find that if there is an infinite chain: $x_1 \to x_2 \to \cdots$ then we can pick every second gangster in this chain and be done. Thus, WLOG there are no such chains which means that in every path sooner or later we come to the point where we have already been, and thus we get a cycle. Because there are no infinite chains we should obviously have an infinite number of different cycles and by taking a gangster from each cycle again we are done.

Is it this easy? We have the word "obviously" in the text. Avoid it! Usually it means either that we can prove the claim in one line (and in this case it is much better to write down this line instead) or ... that we cannot prove the claim at all. Can we prove our "obvious" statement here? Not at all; in fact the claim is wrong: suppose that each gangster except x tries to kill x (he probably deserved it). In this case we have only one cycle consisting of x and his victim y, whom he wants to eliminate.

This is always a difficult situation — when we find that our "universal" solution fails in some special case. The way out is not to repair the solution using some improvements, but to try and find another idea that is also "universal" but works in other cases.

In this problem we will consider two cases. If there exist an infinite number of gangsters who want to kill the same person, we can take them (and this is our new idea). If not then WLOG there are only a finite number of cycles and for any vertex x there are only a finite number of edges that have their endpoints in x.

Here we need the third idea, that is in fact König's lemma (you can find this lemma in Section 13.1). We adapt it as follows.

For each vertex x, we consider the number $l(x)$ — the length of the path from x to some cycle. For example, for every point y on the cycles we have $l(y) = 0$. Note that if $x \to y$ and x does not belong to a cycle then $l(x) = l(y) + 1$.

Easy induction shows that for every n, there exist only a finite number of x with $l(x) = n$: if there are infinitely many x with $l(x) = n + 1$, then infinitely many of them have edges to some y with $l(y) = n$.

Thus for each n there must exist some gangsters x with $l(x) = n$. If we take all the gangsters for which this n is odd, we get the desired set. □

To check that you understand the solution, answer the following questions: is it possible to divide all the gangsters into two peaceful infinite groups? Perhaps three?

1.10 To know and to understand

One of the most common mistakes that students make is the attempt to learn as many methods, theorems and facts as possible. It is not wrong to know a lot of useful things. What is wrong is to forget the difference between knowing and understanding. Knowledge itself is not very useful without understanding. To understand is much more than to know.

We know what prime numbers are, but do we understand them?

Even if you know the first thousand primes by heart it hardly helps to understand their nature. You understand the nature of primes much better when you study a proof of **Chebyshev's theorem** that there always exists a prime between n and $2n$.

But how can we learn to understand? First of all, by trying to solve the problem before the reading the solution. After that by repeating the proof several days later. But mostly by attempting to find another solution, another proof, because when you have several proofs you see more dimensions in the picture that is contained in the problem.

One of the best IMO participants we knew did the following. Every time when he learned some new method he tried to solve all the old problems that he had already solved by this new method. Of course, this did not work for all the problems, but he learned much better than others when this new method could be applied. The several gold medals he got at the IMO is a good argument for his approach.

To spend more time on old problems instead of learning something new? Is this really good advice? Why do we need more proofs if we already have a nice one?

Let us look at the origami proof on page 22 that $\sqrt{2}$ is irrational. It is a

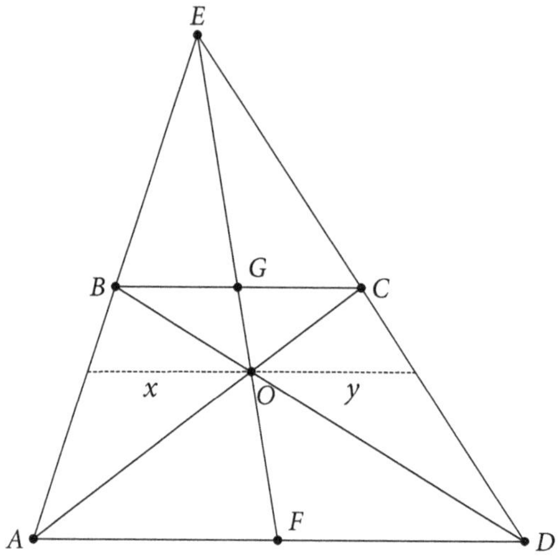

Figure 1.11 Trapezoid and median

nice proof, isn't it? But does it work for $\sqrt{3}$? Hardly. But the proof on page 78 works even for rational numbers. And of course, if we understand both of them, we are richer mathematicians.

Let us consider a classical problem where we show a variety of different proofs. Probably some of them will be not so clear to the reader, but this should pique his curiosity and the desire to learn the corresponding methods later in the text.

Problem 1.25 Let $ABCD$ be a trapezoid with $BC \parallel AD$. Let $E = AB \cap CD$ be a common point for the lines AB and CD and $O = AC \cap BD$ be a common point for the diagonals. Prove that the line EO bisects the segments AD and BC (and thus is a median in the triangle AED).

Search. WLOG we suppose that $|BC| < |AD|$. Let $F = EO \cap AD$, $G = EO \cap BC$. The aim is to find as many different proofs as possible. We start with the longest proof.

Similarity. There are a lot of similar triangles. The pairs BOG, DOF and CGO, AFO give

$$\frac{|BG|}{|DF|} = \frac{|GO|}{|OF|} = \frac{|CG|}{|AF|}.$$

On the other hand, the pairs BEG, AEF and CEG, DEF give

$$\frac{|BG|}{|AF|} = \frac{|EG|}{|EF|} = \frac{|CG|}{|DF|}.$$

Dividing these equations we get

$$\frac{|AF|}{|DF|} = \frac{|DF|}{|AF|} \quad \Rightarrow \quad |DF| = |AF|.$$

Homothety. The homothety with its centre in O that moves A to C moves D to B and therefore their midpoint F' to the midpoint G'. Thus G', O and F' are collinear. On the other hand we have a homothety with its centre in E that moves C to D and B to A. This shows that E, F', G' are collinear as well and all four points E, F', O, G' are collinear. In particular $F' = F$ and $G' = G$.

Extra line. We draw a line through O parallel to AB and want to find x (see Figure 1.11). We have, using different similarity:

$$\frac{|BC|}{x} = \frac{|AC|}{|AO|} = 1 + \frac{|CO|}{|AO|} = 1 + \frac{|BC|}{|AD|}.$$
$$\frac{|AD|}{x} = \frac{|BD|}{|BO|} = 1 + \frac{|DO|}{|BO|} = 1 + \frac{|AD|}{|BC|}.$$

Subtracting these equations we get

$$\frac{1}{x}(|BC| - |AD|) = \frac{|BC|^2 - |AC|^2}{|BC| \cdot |AD|} \quad \Rightarrow \quad \frac{1}{x} = \frac{1}{|BC|} + \frac{1}{|AD|}.$$

By symmetry we get the same result for y. Thus $x = y$ and we have a median. Note that we also get the fact that the segment $x + y = 2x$ is the harmonic mean of $|BC|$ and $|AD|$ which shows a geometrical interpretation of the AM–HM inequality.

Affine transformation. For an equilateral triangle AED the result is evident. But any triangle can be transformed to an equilateral one using an affine transformation. It remains to be noted that such transformation moves parallel lines to parallel lines and midpoints to midpoints.

Ceva's theorem. We have $1 = \dfrac{|AB| \cdot |EC| \cdot |DF|}{|BE| \cdot |CD| \cdot |AF|}$. But

$$\frac{|AB|}{|BE|} = \frac{|AE|}{|BE|} - 1 = \frac{|ED|}{|EC|} - 1 = \frac{|CD|}{|EC|}$$

and we get, after cancelling, that $1 = \frac{|DF|}{|AF|}$.

Centre of gravity. Put weight 1 at E and weights $k = \frac{|BE|}{|AB|} = \frac{|CE|}{|CD|}$ at A and D. Then the centre of gravity lies on AC and BD, thus in O. On the other hand, if F' is the midpoint of AD then it lies on EF' and therefore E, O and F' are collinear.

Complex numbers. We suppose that O is the origin and use small letters for the corresponding complex numbers. We have $a = \alpha c$ and $d = \alpha b$ for some (negative) real number α. To find e we use similarity and write

$$\frac{e-b}{c-b} = \frac{e-a}{d-a} = \frac{e-\alpha c}{\alpha b - \alpha c} = \frac{e-\alpha c}{\alpha(b-c)} \quad \Rightarrow$$

$$e - b + \frac{1}{\alpha}(e - \alpha c) = 0 \quad \Rightarrow \quad e = \frac{\alpha(b+c)}{1+\alpha} = \frac{2\alpha}{1+\alpha} \cdot \frac{b+c}{2},$$

which shows that O, E and the midpoint $\frac{B+C}{2}$ lies on the same line. If $\alpha = -1$ (where we could have a problem with the division) we have a parallelogram and E does not exist. □

Look how many different ideas we get in those proofs! Now we understand this problem completely. Do we? What about the converse statement? Formulate it and check which of the proofs that work in both directions!

1.11 To see the difference

A person lost a million dollars in one day on the stock market. He still has twenty thousand millions in his account but is much unhappier than a boy in a poor family that on the same day gets ten dollars as a present from his mother. The inequality in their wealth is still huge, but it seems that the differences are much more important than the current state. A useful lesson from this example: *always have the changes under control!*

Consider the following problem.

Problem 1.26 The sequence a_n is defined as follows:

$$a_0 = a_1 = 1; \qquad a_{n+2} = \frac{1 + a_{n+1}^2}{a_n}.$$

Prove that all terms in the sequence are integers.

Search. Is this even true? We compute the first few numbers:

$$1, 1, 2, 5, 13, 34, 89, 233, 610, \ldots$$

Amazingly the denominator always cancels. There should be some reason for this. The experienced problem-solver recognizes the Fibonacci numbers here and can try to use this observation. But if we have no such experience we can only look — what is changing, what are the differences? Subtracting the previous terms we get a new sequence:

$$0, 1, 3, 8, 21, 55, 144, \dots$$

Again, we ignore the missing Fibonacci numbers and consider the differences once again:

$$1, 2, 5, 13, 34, 89, \dots$$

We do not need much experience to see that this is the same sequence as we started with. We can state an obvious conjecture and we only need to formulate it exactly and prove it.

First of all, it is convenient to introduce the **difference operator** Δ, which for each sequence x_n creates the sequence of the differences; in other words,

$$\Delta x_n = x_{n+1} - x_n.$$

Then the first differences we have written was exactly the sequence Δa_n and the second was

$$\begin{aligned}\Delta(\Delta a_n) = \Delta(a_{n+1} - a_n) &= (a_{n+2} - a_{n+1}) - (a_{n+1} - a_n)\\ &= a_{n+2} - 2a_{n+1} + a_n.\end{aligned}$$

Our conjecture is that this is exactly a_{n+1}, or equivalently that

$$a_{n+2} = 3a_{n+1} - a_n.$$

If this is the case then all the elements are obviously integers. Thus all we need to do is to prove the conjecture and here induction looks like the most promising approach. The base case is obvious and we need only to understand why

$$\frac{1 + a_{n+1}^2}{a_n} = 3a_{n+1} - a_n.$$

By induction hypothesis we have $a_{n+1} = 3a_n - a_{n-1}$. Thus

$$\frac{1 + a_{n+1}^2}{a_n} = \frac{1 + a_{n+1}(3a_n - a_{n-1})}{a_n} = 3a_{n+1} + \frac{1 - a_{n+1}a_{n-1}}{a_n}$$

and it remains to prove that

$$\frac{1 - a_{n+1}a_{n-1}}{a_n} = -a_n \quad \Leftrightarrow \quad 1 + a_n^2 = a_{n+1}a_{n-1}.$$

But the last equality follows from the definition of a_{n+1}. □

The difference operator Δ is a useful tool. We recommend the reader to find and to prove a general formula for $\Delta^k(x_n)$. But even more useful is the lesson we learn here: it is important to study the differences. Even when there are no differences, it may be useful to present something as a difference. Here is a classical example.

Problem 1.27 Find the sum

$$\frac{1}{1 \cdot 2} + \frac{1}{2 \cdot 3} + \frac{1}{3 \cdot 4} + \cdots + \frac{1}{(n-1) \cdot n}.$$

Search. Can we present each term as a difference? This turns out to be a fruitful question, and in this problem it works perfectly because $\frac{1}{(k-1) \cdot k} = \frac{1}{k-1} - \frac{1}{k}$, we rewrite our sum as

$$\frac{1}{1} - \frac{1}{2} + \frac{1}{2} - \frac{1}{3} + \frac{1}{3} - \frac{1}{4} + \cdots + \frac{1}{n-1} - \frac{1}{n} = 1 - \frac{1}{n}.$$

It is not surprising that such types of sums are called *telescoping sums.* □

The thing that is changing is often more important that the actual objects themselves. That is why the derivative is such a strong instrument. Differences are discrete analogues of derivatives, and they are equally useful. Small changes can result in great development and it is often a good idea to investigate the result of small changes.

Let us prove two classical inequalities using this approach.

Theorem 1.28 *Let $a_1 \geq a_2 \geq \cdots \geq a_n$ and $b_1 \geq b_2 \geq \cdots \geq b_n$ be two decreasing sequences. Then*

1. ***Rearrangement inequality** If $b_1', b_2', \cdots, b_n'$ is any permutation of the second sequence then*

$$\begin{aligned} a_1b_1 + a_2b_2 + \cdots + a_nb_n &\geq a_1b_1' + a_2b_2' + \cdots + a_nb_n' \\ &\geq a_1b_n + a_2b_{n-1} + \cdots + a_nb_1. \end{aligned}$$

2. *Chebyshev's inequality*

$$n(a_1b_1 + a_2b_2 + \cdots + a_nb_n) \geq (a_1 + \cdots + a_n)(b_1 + \cdots + b_n).$$

Proof. Consider the sum $S = a_1b'_1 + a_2b'_2 + \cdots + a_nb'_n$. What happens to the sum if we exchange b'_i and b'_j? We look at the difference and find that it is equal to

$$a_ib'_i + a_jb'_j - (a_ib'_j + a_jb'_i) = (a_i - a_j)(b'_i - b'_j).$$

If $i < j$ then $a_i \geq a_j$. This means that if $b'_i \leq b'_j$, then the difference is non-positive and after the exchange, the sum is at least the same and definitely greater if the difference is negative. From this follows that to ensure that we have the largest possible sum S, we demand $b'_1 \geq b'_2 \geq \cdots \geq b'_n$ and therefore the maximal sum is $a_1b_1 + a_2b_2 + \cdots + a_nb_n$. This proves the left part of the Rearrangement inequality. The right part follows by changing the signs of b_i, because

$$-b_n \geq -b_{n-1} \geq \cdots \geq -b_1$$

(fill in the details!)

As for Chebyshev's inequality, it is sufficient to rewrite the right-hand side as the sum of n sums $S_k = a_1b_{1+k} + a_2b_{2+k} + \cdots + a_nb_{n+k}$ for $k = 0, 1, \ldots, n-1$ where the indices are taken modulo n to fit in the interval $[1, n]$, e.g. $b_{n+2} = b_2$. It remains for us to apply the Rearrangement inequality

$$a_1b_1 + a_2b_2 + \cdots + a_nb_n \geq S_k$$

and to add the results. □

It is important to understand when there is equality in either inequality. This is not so obvious. For example, if all $a_i = 0$, then every permutation of b_i gives the same sum. We need to look at the proof and can conclude that the condition is the following.

Equality at the left-hand side of the Rearrangement inequality happens only when $b'_i \geq b'_j$ for each pair i, j such that $a_i > a_j$. In particular, the sequences $b'_1, b'_2, \ldots, b'_n$ and $b_1, b_2, \ldots, b_n$ are the same if $a_1 > a_2 > \cdots > a_n$. Note that $b'_2 = b_1, b'_1 = b_2$ is still possible if $b_1 = b_2$. But if $b_1 > b_2 > \cdots > b_n$ as well, the only possibility for the equality is $b'_i = b_i$.

The conditions for equality at the right-hand side are similar. As to Chebyshev's inequality we can say more: we have the equality iff either

$a_1 = a_2 = \cdots = a_n$ or $b_1 = b_2 = \cdots = b_n$. The reason for this is that every inequality for S_k must be an equality, thus if $a_i > a_{i+1}$ we must have $b'_{i+k} = b'_{i+1+k}$ for each k.

To study the differences is often important when proving inequalities. A typical way to prove that $f(x_1, x_2, \ldots x_n) \geq 0$ is to try to change a couple of variables only and to see the difference. To see how this works, let us first study how the product xy for $x \geq y$ changes if keep $x + y$ constant. The difference is:

$$(x - a)(y + a) - xy = xa - ya - a^2 = a(x - y - a)$$

which is non-negative as long as $0 \leq a \leq x - y$. We can state this in another way: the product becomes larger when we move the numbers closer to each other (without changing their sum). This trivial remark is sufficient to to prove the **AM–GM inequality**

$$x_1 + x_2 + \cdots + x_n \geq n \sqrt[n]{x_1 x_2 \cdots x_n}$$

for $x_i \geq 0$.

Proof. Indeed, let $m = \frac{1}{n} \sum x_i$ be the arithmetic mean of the numbers in the left part of the inequality. We use induction (downward!) on the number of x_i equal to m. The base is when all $x_i = m$ and in this case the inequality is an equality. For the induction step, suppose that not all $x_i = m$. Because m is their arithmetic mean there must be one of them (say x) that is greater than m and one of them (say y) that is less than m. We replace x and y by m and $y + x - m$ increasing the product, because the numbers become closer to one another (or, if we prefer the formula, because $0 < a = x - m < x - y$). After the replacement, the number of x_i equal to m has gone up and by the induction hypothesis, the new right-hand side is less or equal than m and thus our right-hand side is strictly less than m. Note that we simultaneously proved that the equality is possible only when all x_i are equal. □

Another modification of the same approach is to replace the inequality

$$f(x_1, x_2, \ldots, x_n) \geq 0$$

by the inequality for

$$g(m, x_3, \ldots, x_n) = f(m, m, x_3, \ldots, x_n)$$

where m is some mean of x_1 and x_2. If we show that $f \geq g$ then we simplify the problem because we have reduced the number of variables. The choice of mean depends on the problem. We will consider only one example.

Problem 1.29 Prove that for non-negative numbers x, y, z we have

$$x^3 + y^3 + z^3 + 3xyz \geq x^2(y+z) + y^2(x+z) + z^2(x+y).$$

Search. Let

$$f(x,y,z) = x^3 + y^3 + z^3 + 3xyz - x^2(y+z) - y^2(x+z) - z^2(x+y).$$

According to the recipe we should study the difference

$$f(x,y,z) - f(m,m,z) = -z^2(x+y-2m) + z(3xy - x^2 - y^2 - m^2) \\ + x^3 + y^3 - x^2y - y^2x.$$

If we want this expression to be non-negative we could try the arithmetic mean $m = \frac{x+y}{2}$ to kill $-z^2$. With this choice of m, the difference becomes

$$\begin{aligned} f(x,y,z) - f(m,m,z) &= z\left(\frac{5xy}{2} - \frac{5x^2}{4} - \frac{5y^2}{4}\right) + (x^2 - y^2)(x-y) \\ &= -\frac{5z}{4}(x-y)^2 + (x+y)(x-y)^2 \\ &= (x-y)^2\left(x + y - \frac{5z}{4}\right). \end{aligned}$$

Now it is time to use cyclic symmetry. WLOG we can suppose that z has the least value which immediately implies that the difference is non-negative as we wanted.

Now it remains for us to prove that

$$g(m,z) = f(m,m,z) = z^3 - 2mz^2 + zm^2 = z(z-m)^2 \geq 0,$$

which is evident. We also found when the equality can be achieved: the two largest variables should be equal and the third is either equal to them or to zero. In fact, this is a special case of **Schur's inequality**. □

1.12 Finding the invariants

When the process in the problem is too complicated, it is important not only to follow the differences but also to find what is stable, what does not change (invariant) or is improving (semi-invariant).

Problem 1.30 Several natural numbers are written on the blackboard. In one step we can replace two numbers a, b that do not divide each other by their greatest common divisor (a,b) and their least common multiplier $[a,b]$. Prove that

- The process will finish after several steps.
- The result (the numbers written on the blackboard) is independent of the way we do the steps (but perhaps in a different order).

Search. We start with a small experiment, using 2, 3, 4. If we choose 2, 3 we get 1, 6, 4. The only choice now is to choose 4, 6 for the next step and we get 1, 2, 12 and now we are finished — we cannot choose a new pair. What if we start in another way, choosing 3, 4 at the first step? We get 2, 1, 12 immediately.

Interesting. What does not change in this process? The products seem to be the same. Can we prove this? What we are doing? If we look at the definition (see page 67), we realize that $ab = (a,b)[a,b]$, and thus we do not change the primes we have in our numbers, but only rearrange them. Good! The product does not change and we have only a finite number of ways to move the primes. Thus the process should finish.

Is this so simple? No, why can't we go back, why are there no cycles in our process? To believe in the future we need to see what is improving. What is improving in our process? The product is not changing, what about the sum? The sum seems to increase. Can we prove this? Is it true that $(a,b) + [a,b] \geq a + b$? Yes! We leave the proof to the reader (look at page 54 if you need a hint).

The second part of the problem is now almost evident: for each prime p we consider the powers p^{x_i} which appear in the numbers. This set is invariant: the powers are only exchanged in every step. At the very end if we put the remaining numbers in the increasing order, each number should divide the next. Thus the powers are arranged in increasing order as well. But there is only one way to do this. □

When we have a process, the invariants are the characteristics that are not changing, but the semi-invariants are improving during the process. By their nature they are close to each other (the invariant in one problem can be a useful semi-invariant in another one and vice versa). The invariants restrict the number of possible configurations, the semi-invariants help to prove that the process never goes in cycles. Try to solve the following problem without looking at the solution. Hint: one word is sufficient to solve the problem.

Problem 1.31 There are k black squares on an $n \times n$ board, the remaining ones are white. Each white square that has a common side with at least two black squares is recoloured black as well and this process is continued so long as it is possible to find such a white square. Prove that if at the end all the white squares become black then $k \geq n$.

Search. We test some small examples to understand the process. For example, if black squares are on the main diagonal we recolour all the white squares. Another useful example is a 3×3 board with 3 black squares in the corners. Again we can paint the whole board black, though we had a white row and a white column. Having those examples at hand we can throw away a lot of wrong approaches, but it is still not clear where the road to the solution is hidden. We suspect that the main hidden actor here is some invariant or semi-invariant, but which one? The number of components, the number of black rows? They are good semi-invariants, but do not help us very much as we see on the 3×3 example above. What are the other possible candidates? We can try to list all the possible (semi-)invariants we used before and suddenly we find a key word: perimeter! Indeed, when we recolour a white square black, we lose at least two sides from the perimeter and do not add more than two. Thus the perimeter is a perfect semi-invariant. At the end, if everything is black the perimeter is equal to $4n$. At the very beginning it is at most $4k$. Thus $4k \geq 4n$ and we are done. Why was this trivial argument so difficult to find? Inertia of thinking? □

Fixpoints, i.e. solutions of the equation $f(x) = x$ should always be considered as good candidates for invariants. See Section 10.4 for more details about fixpoints.

1.13 Do the opposite!

Sometimes a good policy to achieve a result is to do the opposite to what you are asked. Strangely enough the opposite direction can be the shortest way to the solution. For example, if we want to prove that two lines are parallel it might be easier to show that they are both orthogonal to the third line. But orthogonality is something quite opposite to being parallel.

Consider more advanced examples. The first is a system of equations.

Problem 1.32 let $a, b, c > 0$ be given real numbers. Solve the system of equations

$$\begin{cases} c - \frac{a}{x} + \frac{b}{z} = xz \\ a - \frac{b}{y} + \frac{c}{x} = xy \\ b - \frac{c}{z} + \frac{a}{y} = yz. \end{cases}$$

Search. It does not look very attractive: this is a highly nonlinear system in x, y, z. But it is linear in a, b, c! Why not do the opposite and solve it in a, b, c as variables? Like every crazy idea it may be worth a try. It is not difficult to do a standard elimination of variables and the result is amazing:

$$a = xy, \ b = yz, \ c = xz \quad \Rightarrow \quad abc = (xyz)^2 \quad \Rightarrow \quad xyz = \pm\sqrt{abc}$$

and we have no problem finding x, y, z now. □

This crazy idea works much more often than one would expect. Quite often we have a second degree equation in one of our parameters and in solving it we often get valuable information from the discriminant. Here is an extreme variant of such an approach.

Problem 1.33 Solve the equation

$$10 - 3\left(2^x - 4^{-x}\right) + 8^x - 2^{-x} - 16^x.$$

Search. Of course $y = 2^x$ looks like a good substitution, but the resulting equation has too high degree for us to be happy. We will first try something crazy instead. Note that $10 = 3^2 + 1$, and thus we can consider all this as a quadratic equation in ... the number 3! In other words, write $3 = a, 9 = a^2$ and solve the resulting equation for a. The discriminant is

$$\begin{aligned} &\left(2^x - 4^{-x}\right)^2 - 4\left(8^x - 2^{-x} - 16^x + 1\right) \\ &\quad = 4^x + 16^{-x} + 2 \cdot 2^{-x} - 4 \cdot 8^x - 4 + 4 \cdot 16^x. \end{aligned}$$

Trying to represent this as a square we rewrite it as $(2^x - 2 \cdot 4^x + 4^{-x})^2$ and our original equation can be expressed as

$$(3 - 2^x + 4^x)(3 + 4^{-x} - 4^x) = 0$$

which can now be easily solved by variable substitutions. □

Our next example is the sum $S_k = S_k(n) = 1^k + 2^k + \cdots + n^k$. If n is a parameter we would like to find a way to get S_k explicitly. Obviously, $S_0 = n$ and $S_1 = 1 + \cdots + n = \frac{n(n+1)}{2}$ and it looks quite natural to try to find S_k from the previous sums $S_{k-1}, S_{k-2}, \ldots, S_0$. But already for $k = 2$ we cannot see a direct way to do this. Because this does not work, let us do the opposite and consider S_{k+1} instead!

Note that $S_{k+1}(n+1) = S_{k+1}(n) + (n+1)^{k+1}$. On the other hand,

$$\begin{aligned} S_{k+1}(n+1) - 1 &= \sum_{j=2}^{n+1} j^{k+1} = \sum_{i=1}^{n} (i+1)^{k+1} \\ &= \sum_{i=1}^{n} \left(i^{k+1} + \binom{k+1}{1} i^k + \cdots + \binom{k+1}{k} i^1 + 1 \right) \\ &= S_{k+1}(n) + (k+1)S_k(n) + \binom{k+1}{2} S_{k-1}(n) + \cdots + n \end{aligned}$$

and we get what we wanted, because we can eliminate $S_{k+1}(n+1)$ and $S_{k+1}(n)$.

Let us check this for $k = 2$. We have $S_3(n+1) = S_3(n) + (n+1)^3$ and simultaneously

$$\begin{aligned} &S_3(n+1) - 1 = S_3(n) + 3S_2(n) + 3S_1(n) + n \quad \Rightarrow \\ &\quad 3S_2(n) = (n+1)^3 - 1 - 3\frac{n(n+1)}{2} - n = n\left(n^2 + \frac{3n}{2} + \frac{1}{2}\right) \quad \Rightarrow \\ &\quad S_2(n) = \frac{n(n+1)(2n+1)}{6}. \end{aligned}$$

It is difficult to guess such a trick, but much easier when you have seen it once.

Let us now change the subject again and consider the following question. A collection of subsets of a given set M is called an **antichain** if no one of them contains another. The question is how large can an antichain be if M has m elements? If we first consider the case $m = 2n$ then rather soon we find a good candidate for the best construction of such a collection. Namely, if we take all subsets of size n, no one contains another. We get $\binom{2n}{n}$ subsets — large enough to be a reasonable answer to our question. But how do we prove this?

Let us study something opposite: the chains! We call the collection of subsets $C_i \subset C_{i+1} \subset \cdots \subset C_j$ contained in each other a **chain** if every C_k has exactly k elements, thus the neighbours differ by one element only. We call a chain **symmetric** if $i + j = m$.

Is it possible to distribute all the 2^m subsets in M into symmetric chains? The answer is YES and here is a non-trivial algorithm to do this by induction. We start from the empty set M that has only one chain consisting of the empty set itself. If M consists of m elements (say $1, 2, \ldots, m$) and we have already constructed the chains for the set $\{1, \ldots, m-1\}$ then for each chain $C_i \subset C_{i+1} \subset \cdots \subset C_j$ with $i + j = m - 1$ we consider the two following chains:

$$C_{i+1} \subset \cdots \subset C_j$$
$$C_i \subset C_i \cup \{m\} \subset C_{i+1} \cup \{m\} \subset \cdots \subset C_j \cup \{m\};$$

(the first one, we use only if $i < j$). Because $i + 1 + j = m$, both of them are symmetric chains in M. We leave to the reader to check whether this construction indeed creates the decomposition of M into symmetric chains. We illustrate the result for small m to see how this non-trivial algorithm works.

$m = 1: \quad \{\} \subset \{1\}.$

$m = 2: \quad \{1\}; \{\} \subset \{2\} \subset \{1, 2\}.$

$m = 3: \quad \{1\} \subset \{1, 3\}; \{2\} \subset \{1, 2\}; \{\} \subset \{3\} \subset \{2, 3\} \subset \{1, 2, 3\}.$

$m = 4: \quad \{1, 3\}; \{1\} \subset \{1, 4\} \subset \{1, 3, 4\}; \{1, 2\}; \{2\} \subset \{2, 4\} \subset \{1, 2, 4\};$
$\{3\} \subset \{2, 3\} \subset \{1, 2, 3\}; \{\} \subset \{4\} \subset \{3, 4\} \subset \{2, 3, 4\} \subset \{1, 2, 3, 4\}.$

Now we can return to our original question about antichains. Note that by definition, different subsets from an antichain belong to *different* chains, and thus we find that the the number of symmetric chains is an upper bound for the size of any antichain. But because every symmetric chain in the middle contains a set with $\left[\frac{m}{2}\right]$ elements, and every such subset belongs exactly to one chain, we have found that the number of chains (and the maximum size of antichain) is exactly $\binom{m}{\left[\frac{m}{2}\right]}$. It is useful to think the opposite!

To change our thoughts in the opposite direction is important in many standard situations. We will consider several of them.

If we have a graph of the function $y = f(x)$ then to draw a graph for $y = f(x) + 1$ we simply lift the graph one unit up. What about the $y = f(x+1)$? We need to move the graph to the left — the opposite direction to what you might expect. To see what is going suppose that $f(x_0) = 0$. Then $f(x_0 - 1 + 1) = 0$ and we see clearly that the root was moved to the left.

Consider now the following map of a plane into itself:

$$f : (x, y) \mapsto (x + y, x - y).$$

What is the equation of the image of the circle $x^2 + y^2 = 1$? The "natural" answer $(x + y)^2 + (x - y)^2 = 1$ is wrong. To get the correct answer we need to consider the quite opposite map — f^{-1} (see chapter 7 if you need the definition). It is easy to check that the inverse is described as

$$f^{-1} : (x, y) \mapsto \left(\frac{x + y}{2}, \frac{x - y}{2}\right).$$

If (x, y) belongs to the image then $f^{-1}(x, y)$ belongs to the circle and the correct equation is therefore

$$\left(\frac{x + y}{2}\right)^2 + \left(\frac{x - y}{2}\right)^2 = 1 \quad \Leftrightarrow \quad (x + y)^2 + (x - y)^2 = 4$$

Thus to use f^{-1} instead of f is very important for the sets described by the equations. But when $f = f^{-1}$ (e.g. for the symmetry or the inversion) we can happily use f.

Now consider the following problem.

Problem 1.34 Show that for any natural m, there exist infinitely many terms in the Fibonacci sequence that are divisible by m.

Search. Naturally we need to work modulo m (see Section 2.2 if you do not know how to do this). So we consider the sequence $F_{n+1} = F_n + F_{n-1} \pmod{m}$ with $F_1 = F_2 = 1$.

One thing is quite obvious: because there are only finitely many choices, and every term is uniquely determined by the two previous ones, the sequence eventually becomes periodic. So, to prove that we get infinitely many terms it is sufficient to find at least one in our period. But how do we do this?

Let us change time into the opposite direction, and thus look back. Then we get the sequence $F_{n-1} = F_n - F_{n+1} \pmod{m}$ where we can allow n to be negative as well. Thus considering not the future only but the past as well we extended our sequence F_n in both directions: now n can be an arbitrary integer.

This gives two advantages. The first is that this sequence is not eventually but *purely periodic*, because we can go to the past as far as we want. Thus *every* term appears infinitely many times.

The second is that $F_0 = F_2 - F_1 = 0$ and this solves our problem. □

Both ideas: periodicity and looking into the past are very useful. Look at the paper [49] to read more about periodicity.

We finish this chapter with an IMO problem.

Problem 1.35 The positive divisors of the integer $n > 1$ are $d_1 < d_2 < \ldots < d_k$, so that $d_1 = 1, d_k = n$. Show that $d_1d_2 + d_2d_3 + \cdots + d_{k-1}d_k < n^2$.

Search. How can we connect the divisors with n? Of course, they divide n. But what is the result of the division? It is a divisor as well. So we do the opposite — we consider the set $\left\{\frac{n}{d_i}\right\}$ as the set of the divisors. Now we need to prove that

$$\frac{n^2}{d_kd_{k-1}} + \cdots + \frac{n^2}{d_2d_1} < n^2 \quad \Leftrightarrow \quad \frac{1}{d_1d_2} + \frac{1}{d_2d_3} + \cdots + \frac{1}{d_{k-1}d_k} < 1.$$

Now we see something familiar: comparing with the differences

$$\left(\frac{1}{d_1} - \frac{1}{d_2}\right) + \left(\frac{1}{d_2} - \frac{1}{d_3}\right) + \left(\frac{1}{d_{k-1}} - \frac{1}{d_k}\right) = 1 - \frac{1}{n}$$

we get the desired estimate. □

1.14 Exercises

A dictionary is not supposed to have exercises, but sometimes it is a good idea to repeat the understanding of ideas by working through some examples on your own. As a compromise we will suggest some problems with hints but without solutions. Complete solutions can be found on webpage [61].

Exercise 1.1. Solve the equations:

a) $x^4 - 4x^3 - x^2 - 8x + 4 = 0$.

b) $x(2x^2 - 1)(8x^4 - 8x^2 + 1) = \frac{1}{8}$.

Exercise 1.2. Prove that among any seven given real numbers $x_1, x_2, x_3, x_4, x_5, x_6, x_7$, there exist two, x_i, x_j, with $i \neq j$, such that

$$0 \leq \frac{x_i - x_j}{1 + x_i x_j} \leq \frac{\sqrt{3}}{3}.$$

Exercise 1.3. Let n be a natural number, $d_1, \dots, d_m$ be the list of its divisors and $a_1, \dots, a_m$ be another list, where a_i is the number of divisors of the number d_i. Let A be the sum of all a_i, and B be the sum of their cubes. Prove that $A^2 = B$.

Example: For $n = 12$ we have:

$$\begin{aligned}
&d : 1, 2, 3, 4, 6, 12;\\
&a : 1, 2, 2, 3, 4, 6;\\
&A = 1 + 2 + 2 + 3 + 4 + 6 = 18;\\
&B = 1 + 8 + 8 + 27 + 64 + 216 = 324 = 18^2.
\end{aligned}$$

Exercise 1.4. Let k be a integer and $p(x) = x^n + a_1 x^{n-1} + \cdots + a_n$ be a polynomial such that for any integer m the number $p(m)$ is an integer divisible by k. Show that k divides $n!$.

Exercise 1.5. Let

$$x_1 = 19; \quad x_2 = 97; \quad x_{n+2} = x_n - \frac{1}{x_{n+1}}.$$

Show that $x_k = 0$ for some k and find this k.

Exercise 1.6. Suppose that $a_0 = a_n = 0$ and $a_{k-1} - 2a_k + a_{k+1} \geq 0$ for all natural $k \leq n - 1$. Show that $a_i \leq 0$ for all $i \leq n$.

Exercise 1.7. Show that

$$\frac{x}{1 + x^2} + \frac{x}{2^2 + x^2} + \cdots + \frac{x}{n^2 + x^2} < \frac{\pi}{2}.$$

Exercise 1.8. Divide a segment on a plane in three equal parts using a compass only.

See also problems 15, 32, 33, 44, 59, 68, 73, 81, 108, 120 in Chapter 14.

1.15 Hints

1.1. a) Divide by x^2.

b) Put $x = \cos t$ and multiply by $\sin t$.

1.2. Put $x_i = \tan \alpha_i$.

1.3. Use induction on the number of primes in the decomposition of n and the identity on the page 21.

1.4. Consider the polynomials $p_0(x) = p(x)$ and the differences $p_{i+1}(x) = \Delta p_i(x) = p_i(x+1) - p_i(x)$. Find $p_n(x)$.

1.5. Consider $y_n = x_n x_{n+1}$ and study the differences.

1.6. Suppose that $a_{m-1} \leq 0, a_m > 0$. Study the differences.

1.7. Interpret the inequality geometrically.

1.8. Do the opposite: make it three times larger first.

CHAPTER 2

The Numbers

It is good to know general ideas, but to solve problem we also need to know some basic facts: the main theorems and definitions. We have tried to collect some of those that we consider the most important. Of course, the list of useful instruments is much longer than what we can possibly fit in this book. We are not even able to cover the necessary minimum.

We have tried to organize topics dividing them into mathematical areas, but when solving real problems, those areas often intersect and are used simultaneously. We have structured the text more like a dictionary than a textbook; and there are a lot of cross-references. Often, we use and refer to theorems or definitions that are formulated later on. The index at the end of the book should help you to find the necessary definitions.

~

We begin, naturally, with numbers.

2.1 Natural numbers and integers

Natural numbers, $\mathbb{N} = \{1, 2, 3, \ldots, n, \ldots,\}$ is one of the first abstractions that we meet in life, but even at the end of our life we still do not understand them completely. What we learn is how to use some properties of numbers. Here are a few widely used ones:

- Every non-empty subset $M \subset \mathbb{N}$ has a least element m. This element is often a good candidate for a minimal counterexample.
- Given a finite set of numbers M, there exists a natural number that does not belong to M. The same is true if we replace $\mathbb{N}$ by any infinite set of numbers, but this observation is mostly used for natural numbers, for example when we write: "Choosing n sufficiently large, ..."

- If x and y are natural numbers with $x < y$, then $x + 1 \leq y$. The same is true for integers (but of course not for rational or real numbers).
- Every natural number can be uniquely written as a product of prime numbers (with the convention that an empty product is equal to 1).

Consider the following problem.

Problem 2.1 Let n be a natural number and let a, b, c, d be different natural numbers between n^2 and $(n+1)^2$. Show that $ad \neq bc$.

Search. Suppose that, on contrary, $ad = bc$. WLOG, a is the largest of the four numbers and $b > c$, thus $a > b > c > d$. What more? We can use the fact that a divides bc. If we consider the prime factors of a, then some of them (denote their product by x) divide b and the rest (which we denote by y) divide c. Thus $a = xy$. Let z be what remains in b, i.e. $b = xz$. Similarly we decompose $c = yt$ and get $d = \frac{bc}{a} = zt$ as well.

Now it is time to look at the inequalities. Because $a = xy > b = xz$, we have $y > z \Rightarrow y \geq z + 1$. Similarly, $a > c \Rightarrow x \geq t + 1$ and together we get:

$$a = xy \geq (t+1)(z+1) = tz + t + z + 1 \geq n^2 + t + z + 1.$$

Now we see the final AM–GM bridge: because $t + z \geq 2\sqrt{tz} = 2\sqrt{d} \geq 2n$, it follows that

$$a \geq n^2 + 2n + 1 = (n+1)^2.$$

On the other hand, $a \leq (n+1)^2$, and all inequalities must in fact be equalities. In particular $t = z = n$, $x = y = n + 1$ and hence $b = c$, which is a contradiction. □

Now let us show how to play with the minimal element.

Problem 2.2 Which non-empty subsets M of natural numbers have the property that $x, y \in M, x > y \Rightarrow x - y \in M$? Which non-empty subsets I of integers have the property that $x, y \in I \Rightarrow x - y \in I$?

Search. We start with the natural numbers. First of all, M has a least element d. What are the other elements? Because we can subtract d we can always reduce other numbers. But we cannot make them less than d, and thus they should have form kd. This is a good idea, but how do we prove it? We use a minimal counterexample.

Suppose there exists some numbers in M that cannot be written as kd. Then there is a minimal such number m. Obviously, $m \neq d \Rightarrow m > d$. The number $m - d$ is also is M and is less than m. Thus $m - d = kd$, but then $m = (k+1)d$ and we get a contradiction.

Therefore all numbers in M are of the form $m = kd$. Because $kd \in M \Rightarrow (k-1)d = m - d \in M$ we can conclude that either M contains all kd for natural k *or* M is finite and is an arithmetic progression $d, 2d, \ldots, nd$ for some natural n.

For the integer case, first note that if I contains a non-zero element a, it also contains $0 = a - a$ and $-a = 0 - a$ as well. Thus $M = I \cap \mathbb{N}$ is non-empty and we can use the solution above to find M. Moreover, $a, b \in M \Rightarrow a + b = a - (-b) \in M$ which excludes the possibility that M is finite. Hence, I consists of all numbers kd for some fixed integer $d \geq 0$ and all integers k. □

Let $p_1 = 2, p_2 = 3, \ldots$ be a list of all primes. We know that this list is infinite and that every natural number a can be uniquely written as a product of powers of primes:

$$a = p_1^{a_1} p_2^{a_2} p_3^{a_3} \cdots = \prod_{i=1}^{\infty} p_i^{a_i}.$$

In this product, only finitely many a_i are non-zero, so in fact the product is finite, but for formal manipulation, it is often convenient to use infinite products. In some sense, we encode every natural number a by the infinite sequence $a_1, a_2, \ldots$ where all but finitely many elements are zero.

For example, if $a = \prod p_i^{a_i}$ and $b = \prod p_i^{b_i}$ then

- a **divides** b (denoted $a \mid b$) means that there exists a natural number c, such that $b = ac$. This is true iff $a_i \leq b_i$ for every i.
- the **greatest common divisor** $d = \prod p_i^{d_i}$ (denoted $\gcd(a, b)$ or shortly as (a, b)) is defined as the *largest* natural number d such that $d \mid a$ and $d \mid b$, and can be computed by $d_i = \min\{a_i, b_i\}$.
- the **least common multiple** $l = \prod p_i^{l_i}$ (denoted $\operatorname{lcm}(a, b)$ or shortly as $[a, b]$) is defined as the *smallest* natural number l such that $a \mid l$ and $b \mid l$, and can be computed by $l_i = \max\{a_i, b_i\}$.

It is evident that $ab = (a, b)[a, b]$, since $\max\{a_i, b_i\} + \min\{a_i, b_i\} = a_i + b_i$.

The definitions can be extended to non-zero *integers*, for example $a \mid b$ means that $|a|$ divides $|b|$. Similarly, $(a, b) = (|a|, |b|)$, and $[a, b] = [|a|, |b|]$.

As for zero, any integer divides 0, but zero divides nothing, thus when we write $a \mid b$ *we always suppose that* $a \neq 0$. We also put $(a, 0) = (0, a) = a$. But mostly, we prefer to work with natural numbers to avoid such technical complications.

The definition of the greatest common divisor and the least common multiple can also be extended to several numbers by induction, e.g.

$$\gcd(x_1, \ldots, x_n, x_{n+1}) = (\gcd(x_1, \ldots, x_n), x_{n+1}).$$

A collection of numbers are called **relatively prime** if their greatest common divisor is equal to 1. Note that the numbers $4, 6, 9$ are relatively prime, but not *pairwise* relatively prime, and thus those are different notions.

We would now like to formulate a theorem containing some elementary but important properties which are used almost automatically. We recommend to the reader to prove the theorem to be sure that s|he really understands all the details.

Theorem 2.3 *Let a, b, c be integers. Then*

- $a \mid b \Rightarrow |a| \leq |b|$ *or* $b = 0$.
- $a \mid b \Leftrightarrow (a, b) = |a|$ *and* $a \neq 0$.
- $c \mid a, c \mid b \Leftrightarrow c \mid (a, b)$.
- $c \mid a, c \mid b \Rightarrow c \mid ax + by$ *for any integers* x, y.
- $(c, ab) = (c, a)(c, b)$ *if* $(a, b) = 1$.
- $a \mid bc$ *and* $(a, b) = 1 \Rightarrow a \mid c$.
- $ac \mid bc \Leftrightarrow a \mid b$, *if* $c \neq 0$.
- $a \mid b, b \mid c \Rightarrow a \mid c$.

After these preliminaries, some problems can be solved almost without thinking.

Problem 2.4 Let x, y be integers such that $23 \mid 3x+5y$. Show that $23 \mid 5x-7y$ as well.

Solution. $23 \mid 3x+5y \Leftrightarrow 23 \mid 15x+25y \Leftrightarrow 23 \mid 15x-21y \Leftrightarrow 23 \mid 5x-7y$.□

Now it is time for one of the strongest tools, **Bézout's lemma**.

Theorem 2.5 (Bézout's lemma) *Let a and b be integers and $d = (a, b)$. Then there exist integers u and v such that $d = au + bv$. In particular, if $(a, b) = 1$ there are integer solutions to $au + bv = 1$.*

Proof. We consider only the case $(a, b) = 1$, leaving the general case to the reader. Consider the set I of all integers of the form $ax + by$. As we have seen in problem 2.2, there exists $d \in I$ such that d divides each number in I. Because $a = a \cdot 1 + b \cdot 0 \in I$ and similarly $b \in I$ then $d \mid (a, b) = 1$ and $d = 1$. But $d \in I$ and can be written as $au + bv$. □

It is important to know how to find such u and v by the help of the **extended Euclidean algorithm**. We restrict ourselves to one example, hoping that the reader will understand the idea and prove to himself that the algorithm works correctly. All we need to know is how to divide two numbers with the remainder.

Let $a = 102, b = 81$. We have:

$$\begin{array}{rcl}
102 = 1 \cdot 81 + 21 & \Leftrightarrow & 21 = 102 - 1 \cdot 81 \\
81 = 3 \cdot 21 + 18 & \Leftrightarrow & 18 = 81 - 3 \cdot 21 \\
21 = 1 \cdot 18 + 3 & \Leftrightarrow & 3 = 21 - 1 \cdot 18 \\
18 = 6 \cdot 3 + 0 & &
\end{array}$$

The left part of this computation is the usual **Euclidean algorithm** which finds that $3 = (102, 81)$. The *extended* part is on the right-hand side and using it we find, working backwards:

$$\begin{aligned}
3 &= 21 - 1 \cdot 18 && = 21 - 1 \cdot (81 - 3 \cdot 21) \\
&= 4 \cdot 21 - 1 \cdot 81 && = 4 \cdot (102 - 1 \cdot 81) - 1 \cdot 81 \\
&= 4 \cdot 102 - 5 \cdot 81,
\end{aligned}$$

so $u = 4, v = -5$.

2.2 Arithmetic in $\mathbb{Z}_m$

Let m be a natural number. We want to add an extra rule to the usual arithmetic, namely that $m = 0$. Of course this comes with a price: for example, now there are only m different numbers, because $m+1 = 1$ and $-1 = 0-1 = m-1$. For example, if $m = 5$ we obtain the following addition and multiplication tables:

+	0	1	2	3	4
0	0	1	2	3	4
1	1	2	3	4	0
2	2	3	4	0	1
3	3	4	0	1	2
4	4	0	1	2	3

·	0	1	2	3	4
0	0	0	0	0	0
1	0	1	2	3	4
2	0	2	4	1	3
3	0	3	1	4	2
4	0	4	3	2	1

.

If we want to distinguish these new "numbers" from the usual ones we can write them as $\overline{k}$. We denote the set of these numbers equipped with this modified arithmetic by $\mathbb{Z}_m$.

Thus $\mathbb{Z}_m$ consists of m different elements: $\overline{0}, \overline{1}, \overline{2}, \ldots, \overline{m-1}$ together with very simple arithmetic: use ordinary rules for addition and multiplication but at the end of the computation, take the remainder after division by m.

We may also think of $\mathbb{Z}_m$ as consisting of the usual integers **modulo** m, where we identify an integer with its remainder on division by m. In other words, a and b are equal modulo m if and only if $m \mid (a-b)$. There is a special notation for this, namely:

$$a = b \pmod{m}.$$

If you want to stress that this is not usual equality you can write $a \equiv b \pmod{m}$ instead, but usually we are not so pedantic. This notation is all we need to perform the calculations in $\mathbb{Z}_m$.

The advantage of this new arithmetic is that we replace the difficult question about divisibility by m by a much easier question about equality to zero modulo m. For example, how can we prove that the number n is divisible by 9 if and only if the sum of its digits is divisible by 9?

Suppose that n is a three digit number with digits a, b, c. Then $n = 100a + 10b + c$. Consider n modulo 9. Then $10 = 1 \pmod{9}$, $100 = 1 \pmod{9}$ and hence

$$n = 1 \cdot a + 1 \cdot b + c = a + b + c \pmod{9}$$

and we can conclude even more than we aimed for: $n - (a + b + c)$ is divisible by 9 for every n. The same proof works for an arbitrary number of digits because we have $10^k = 1^k = 1 \pmod 9$. The same test also works for checking divisibility by 3 and (with a small modification — which one?) for divisibility by 11.

Reducing a problem modulo some cleverly chosen number can often lead to an easy solution.

Problem 2.6 For which natural n is it possible to write $10^n + 3$ as a sum of two squares? Answer the same question for $10^n + 2$.

Search. For $n = 1$, it is easy to find $13 = 4 + 9$. For $n = 2$, we can check that 103 cannot be written as the sum of two squares. What is the difference between 13 and 103? For $n > 1$, the number 10^n is divisible by 4 and perhaps it is a good idea to consider the problem modulo 4. Can 3 be written as the sum of two squares in $\mathbb{Z}_4$? We have $0^2 = 2^2 = 0 \pmod 4$ and $1^2 = 3^2 = 1 \pmod 4$ and hence 3 is not the sum of two squares. Short and effective!

Working modulo 4 does not help with $10^n + 2$. But who said that we should restrict ourselves to $m = 4$ only? We can test other numbers, e.g. $m = 3$. The sum of the digits of $10^n + 2$ is equal to 3, thus $10^n + 2 = 0 \pmod 3$. What about squares? $1^2 = 2^2 = 1 \pmod 3$ and $0^2 = 0 \pmod 3$, so if $x^2 + y^2$ is divisible by 3 both x and y must be divisible by 3. But then $x^2 + y^2$ is divisible by 9 and we get the contradiction modulo 9 because $10^n + 2$ is equal to 3 modulo 9. Thus, $10^n + 2$ is *never* the sum of two squares. □

We haven't discussed division yet. For example, what is $\frac{3}{2}$ modulo 5? If we look in our multiplication table we find that it is 4, because $4 \cdot 2 = 3 \pmod 5$. Is it always possible to divide $\frac{a}{b}$ modulo m and if so, how?

Of course, we are not allowed to divide by zero, but there are other restrictions as well: if we want to divide by b the numbers b and m *must be relatively prime*. If this is the case, i.e. $(m, b) = 1$, we use our main instrument — Bézout's lemma (Theorem 2.5) — to get $mu + bv = 1$ and this shows that $bv = 1 \pmod m$. In other words, $b^{-1} = v$ and to divide by b is to multiply by v.

For example, if we want to divide by 5 modulo 23 we find, using the extended Euclidean algorithm that $1 = 2 \cdot 23 - 9 \cdot 5$ and conclude that division by 5 is multiplication by $-9 = 14 \pmod{23}$. If we look at problem 2.4 again and consider it modulo 23 we find another short automatic solution:

$$3x + 5y = 0 \pmod{23} \quad \Rightarrow \quad 5y = -3x \pmod{23}$$
$$\Rightarrow \quad y = 27x \pmod{23}.$$

Hence $5x - 7y = 5x - 189x = -184x = 23 \cdot 8x = 0 \pmod{23}$.

Returning to the general case we conclude that if $(a, m) = 1$ and $ab = ac \pmod{m}$ then we can cancel a and get the equality $b = c \pmod{m}$. If $(a, m) \neq 1$, we cannot cancel. For example $2 \cdot 2 = 0 \pmod{4}$ does not imply $2 = 0 \pmod{4}$. What we can do, and this is useful as well, is the following:

$$ab = ac \pmod{am} \quad \Rightarrow \quad b = c \pmod{m},$$

because both conditions mean that $(b - c) \mid m$. For example, we have the implication $2x = 0 \pmod{4} \Rightarrow x = 0 \pmod{2}$.

Modular arithmetic works best when $m = p$ is a prime, because then we can divide by any non-zero element. We can feel as comfortable working in $\mathbb{Z}_p$ as in $\mathbb{R}$; if fact even *more* comfortable because there is more to be said:

Theorem 2.7 *Let p be a prime number. The following is valid in $\mathbb{Z}_p$:*

- *Every linear equation $ax + b = c$ has a unique solution if $a \neq 0$.*
- ***Fermat's (little) theorem:** $a^{p-1} = 1$ if $a \neq 0$. Moreover, $a^p = a$ for each $a \in \mathbb{Z}_p$.*
- *$(a + b)^p = a^p + b^p$ for any $a, b \in \mathbb{Z}_p$.*
- *There exists c such that every non-zero element $a \in \mathbb{Z}_p$ can be written as a power $a = c^k$ for some natural number k.*

Proof. Since we have division, we can solve linear equations. To prove Fermat's theorem, consider the product of all non-zero elements $N = 1 \cdot 2 \cdot 3 \cdots (p-1)$ which is non-zero itself. Note that all elements $1a, 2a, \ldots, (p-1)a$ are different because $ka = la$ implies $k = l$ by cancellation. Therefore this set consists of all non-zero elements in $\mathbb{Z}_p$ and their product is equal to

$$N = 1a \cdot 2a \cdot 3a \cdots (p-1)a = a^{p-1} \cdot 1 \cdot 2 \cdot 3 \cdots (p-1) = a^{p-1}N.$$

Because we can cancel N, we get $1 = a^{p-1} \Rightarrow a = a^p$. For $a = 0$ the last equation is evident.

From Fermat's theorem, it follows that $(a + b)^p = a + b = a^p + b^p$.

The difficult proof of the last statement we leave as a useful exercise. *Hint*: for every b, if k is the minimal positive number such that $b^k = 1$ then $k \mid p-1$. We need to find a b for which $k = p-1$. To do this, we use the fact that the equation $x^k = 1$ has at most k solutions in $\mathbb{Z}_p$. □

Now we have a decent-sized tool box for solving many number theory problems. We start with a very useful one.

Problem 2.8 Let a, b, c, m be natural numbers such that $c^a - 1$ and $c^b - 1$ are divisible by m. Then $c^{(a,b)} - 1$ is divisible by m as well.

Search. Divisibility? We immediately rewrite the problem modulo m. We have $c^a = c^b = 1 \pmod{m}$. Without thinking, we apply our strongest instrument $d = (a,b) = au + bv$ and hope that this is the bridge we are looking for. It is:

$$c^d = c^{au+bv} = c^{au}c^{bv} = (c^a)^u(c^b)^v = 1 \pmod{m}.$$ □

We continue with another problem.

Problem 2.9 Does there exist a natural number $n > 1$ for which $n \mid 2^n - 1$?

Search. Assume that there exists such a number $n > 1$. Then $2^n = 1 \pmod{n}$. How do we proceed? It is not entirely clear, so let us try some special case, for example assume that $n = p$ is a prime. This is too easy: the condition $2^p = 1 \pmod{p}$ and Fermat's theorem $2^p = 2 \pmod{p}$ contradict each other.

This is good. At least we solved the problem for a special case, but above all we got a hint: we need to use a prime and Fermat's theorem. Which prime? We want *the best* prime p that divides n. What exactly constitutes *the best* we do not know yet, but we can decide on this later. Working modulo p, what do we get?

$$\left.\begin{aligned} 2^n &= 1 \pmod{p} \\ 2^{p-1} &= 1 \pmod{p} \end{aligned}\right\} \quad \Rightarrow \quad 2^{(n,p-1)} = 1 \pmod{p}$$

applying problem 2.8. To get a contradiction, the best choice of p will be a minimal one, because then $p-1$ and n have no common divisors, and we get $2^1 = 1 \pmod{p}$ which is impossible. □

To practice modular arithmetic, we will prove prove a couple of classical results.

Theorem 2.10 (Wilson's theorem) *A natural number p is prime if and only if $(p-1)!+1$ is divisible by p.*

Proof. If $p = ab$ with $a, b < p$ then $a \mid (p-1)!$ and $p \nmid (p-1)!+1$. If p is a prime, then every element in the product $1 \cdot 2 \cdot \dots \cdot (p-1)$ has an inverse modulo p. Two of them are inverses to themselves, namely 1 and $p-1 = -1 \pmod{p}$, and these are the only solutions to the equation $x^2 = 1$. (Since $x^2 = 1 \pmod{p} \Leftrightarrow (x-1)(x+1) = 0 \pmod{p}$, and because p is prime, $p \mid x-1$ or $p \mid x+1$.) Thus all factors except -1 cancel. □

Theorem 2.11 (Chinese remainder theorem) *Let $m_1, \dots, m_k$ be pairwise relatively prime numbers and let $c_1, \dots, c_k$ be arbitrary integers. Then there exists an integer c such that $c = c_i \pmod{m_i}$ for all $i = 1, \dots, k$. The number c is unique if we demand that $0 \le c < m = \prod m_i$, otherwise it can be replaced by $c + dm$ for any integer d.*

Proof. Uniqueness is easy: if we have two such numbers then their difference must be divisible by each m_i and therefore divisible by m, because m_i are relatively prime.

If c is a solution to the system of congruences, then $c + dm$ is as well, and we can always choose d to get a solution between 0 and $m-1$.

The difficult part is existence. We give a constructive proof using induction. The case $k = 1$ is obvious. To understand the idea, let us consider the case $k = 2$ in detail. Since $(m_1, m_2) = 1$, Bézout's lemma (Theorem 2.5) gives us u, v such that $m_1 u + m_2 v = 1$. The number $x_1 = m_2 v$ satisfies $x_1 = 1 \pmod{m_1}$ and $x_1 = 0 \pmod{m_2}$. Similarly $x_2 = m_1 u$ satisfies $x_2 = 0 \pmod{m_1}$ and $x_2 = 1 \pmod{m_2}$. Now put $c = c_1 x_1 + c_2 x_2$ and we get the solution we are looking for.

Now for the induction step, assume that we have solved the problem for k equations, and that we are given $k+1$ pieces of data. Let c be the solution corresponding to first k conditions and let $m = \prod_{i=1}^{k} m_i$. We know that $(m, m_{k+1}) = 1$ thus, as in the case $k = 2$, we can find C such that $C = c \pmod{m}$ and $C = c_{k+1} \pmod{m_{k+1}}$. But this means that $C = c = c_i \pmod{m_i}$ for $i \le k$ as well and we have finished the proof. □

2.3 Diophantine equations

Quite often we want to solve equations where the coefficients are integers and the solutions should be integers as well. Such equations are called *Diophantine* and several of them have standard solutions.

We start with the linear equation $ax + by = c$ with integer a, b, c. Let $d = \gcd(a, b)$. If d does not divide c, the equation obviously has no integer solutions. Otherwise we can cancel d, so WLOG we consider the case $(a, b) = 1$ only. There are no doubts that our main instrument — Theorem 2.5 — should be applied so let $au + bv = 1$. This immediately gives us *one* solution $x_0 = cu$, $y_0 = cv$. To find all solutions, we note that $ax - ax_0$ is divisible by b and therefore $b \mid (x - x_0) \Rightarrow x - x_0 = kb$. From $ax + by = ax_0 + by_0$, we get $akb + by = by_0$ and thus

$$x = x_0 + kb, \quad y = y_0 - ka$$

with integer k describes all the solutions.

Another classical Diophantine equation is

$$x^2 + y^2 = z^2,$$

whose solutions are called **Pythagorean triples**. Again we can restrict ourselves to the case where x, y, z are positive and have no common divisors. Let us find all the solutions in this case. First of all, considering the problem modulo 4 we immediately conclude that z has to be odd, otherwise 2 will be a common divisor. Furthermore, exactly one of x, y must be even and WLOG we can suppose that this is $x = 2t$, and thus y is odd.

Note also that $(y, z) = 1$, otherwise a prime dividing them would divide x as well. Now

$$4t^2 = z^2 - y^2 \quad \Rightarrow \quad t^2 = \frac{z+y}{2} \cdot \frac{z-y}{2}.$$

Both $\frac{z+y}{2}$ and $\frac{z-y}{2}$ are integers and relatively prime (proof: any common divisor also divides their sum and difference which are the relatively prime numbers y and z which is a contradiction). Because their product is a square t^2 each of them must be a square, say u^2 and v^2, which gives $t = uv$ and

$$x = 2uv, \quad y = u^2 - v^2, \quad z = u^2 + v^2$$

with relatively prime u and v. To get all Pythagorean triples we can multiply these solutions by any integer c and do not forget the symmetry (that y could be even as well).

2.4 Rational and irrational numbers

A rational number, as we know, is a number of the form $r = \frac{p}{q}$, where q is a natural number and p is an integer. What can be difficult about this? First of all the representation of a rational as a quotient is not unique; there are infinitely many ways to write r in this form, but normally we demand $(p, q) = 1$ for non-zero r and with this condition, there is a unique representation; we call it the *canonical form*. If you still think that all this is trivial consider the following problem.

Problem 2.12 Consider the following infinite process: In the first step, we write all integers n in their natural order as rational numbers in canonical form, i.e. $\frac{n}{1}$. If we have finished with step k then in step $k+1$ we insert between the neighbours $\frac{a}{b}$ and $\frac{c}{d}$ a new rational $\frac{a+c}{b+d}$.

Prove that every rational number appears in this process exactly once and in its canonical form.

Search. It is difficult to believe this. Consider first what is going on between 0 and 1. On the first step we have only $\frac{0}{1}$ and $\frac{1}{1}$. On step number two we insert $\frac{0+1}{1+1} = \frac{1}{2}$ and get

$$\frac{0}{1}, \frac{1}{2}, \frac{1}{1}.$$

The next three steps look like this:

$$\frac{0}{1}, \frac{1}{3}, \frac{1}{2}, \frac{2}{3}, \frac{1}{1}.$$

$$\frac{0}{1}, \frac{1}{4}, \frac{1}{3}, \frac{2}{5}, \frac{1}{2}, \frac{3}{5}, \frac{2}{3}, \frac{3}{4}, \frac{1}{1}.$$

$$\frac{0}{1}, \frac{1}{5}, \frac{1}{4}, \frac{2}{7}, \frac{1}{3}, \frac{3}{8}, \frac{2}{5}, \frac{3}{7}, \frac{1}{2}, \frac{4}{7}, \frac{3}{5}, \frac{5}{8}, \frac{2}{3}, \frac{5}{7}, \frac{3}{4}, \frac{4}{5}, \frac{1}{1}.$$

Unbelievable! Especially impressive is the fact that there are no cancellations — everything appears in canonical form. But how do we prove this? Our main tool for understanding sequences is to study the differences, and here we see something remarkable: all the differences have the form $\frac{1}{m}$ for (different) integers m. We can formulate this better: if $\frac{a}{b}$ and $\frac{c}{d}$ are neighbours then the determinant $\left|\begin{smallmatrix} a & c \\ b & d \end{smallmatrix}\right| = ad - bc$ is always -1. Can we prove at least this? It should not be difficult if we use induction. For example, one of the determinants in the next step is

$$\begin{vmatrix} a & a+c \\ b & b+d \end{vmatrix} = a(b+d) - b(a+c) = ad - bc = -1.$$

This already explains why all numbers appear in canonical form: if $p \mid a+c$ and $p \mid b+d$ then p also divides the determinant above and therefore $p = \pm 1$ and cannot be a prime. What more? Because there is no cancellation, all the denominators in the numbers that appeared on step k should be at least k, and thus the difference between the new neighbours is at most $\frac{1}{k}$. But this is good. If $r = \frac{p}{q}$, then after q steps the number should appear in the list — otherwise it can be found between two neighbours x and y and the difference $y - x = (y - r) + (r - x)$ will be too large, because the differences $y - r$ and $r - x$ are at least $\frac{1}{q}$.

We leave it to the reader to complete the details and from this deduce the proof for rational numbers outside the interval $[0, 1]$. □

You can read about the **Farey sequence** if you want to know more about this interesting construction, but we will move in another direction.

We know that every real number can be written as an infinite decimal number and the rational numbers are those that have a periodic decimal part. Some rational numbers have two different representations as decimals, for example $0.099999999\ldots = 0.100000\ldots$. The main rule is to rewrite the decimal form into a more convenient form $\frac{p}{q}$.

Problem 2.13 Divide two rational numbers:

$$\frac{3.888888888\ldots}{0.353535353\ldots}.$$

Search. We, of course, know how to divide two numbers but how do we do this here? As recommended, first we need to transform the numbers into a convenient form using the sum of an infinite geometric progression:

$$3.888888888\ldots = 3 + 0.8 + 0.08 + 0.008 + \cdots = 3 + \frac{0.8}{1 - 0.1} = 3 + \frac{8}{9} = \frac{35}{9},$$

$$0.3535353535\ldots = 0.35 + 0.0035 + 0.000035 + \cdots = \frac{0.35}{1 - 0.01} = \frac{35}{99},$$

and now the result is obviously $\frac{35}{9} \cdot \frac{99}{35} = 11$. □

A third, useful form to represent rational numbers is as a product of primes: $r = \pm p_1^{a_1} p_2^{a_2} \cdots$, where we allow negative powers. For example, if $r > 0$ and $\sqrt{r}$ is a rational number $p_1^{b_1} p_2^{b_2} \cdots$ then all $a_i = 2b_i$ must be even, thus in the canonical representation $r = \frac{p}{q}$ both p and q have to be be perfect squares because they are relatively prime. In particular, $\sqrt{n}$ is irrational whenever n is a non-square integer.

The main game with rationality is often based on the canonical representation $r = \frac{p}{q}$ because we have relatively prime numbers $(p, q) = 1$ and this is a good starting point. Another good property is that for any polynomial $p(x)$ with rational coefficients $p(r)$ is rational as well.

Problem 2.14 Find all rational numbers r such that $\sin \pi r$ is rational too.

Search. Some trivial values $0, \pm 1, \pm \frac{1}{2}$ for $\sin \pi r$ and all corresponding r are easy to find, and it is reasonable to suspect that there are no other solutions.

How do we use that $r = \frac{M}{N}$ is rational? One consequence is that $\sin rN\pi = 0$. Thus we should try to increase the angle and to understand why this leads to a contradiction. It is actually much easier to increase the angles using $\cos x$; for example we have Chebyshev polynomials to do this (see page 30).

Because $\cos x$ is a brother to $\sin x$ we ask ourselves — is the problem for $\cos x$ the same? And the answer is yes, because $\sin \pi r = \cos(\frac{\pi}{2} - \pi r)$ and $r_1 = \frac{1}{2} - r$ is rational if and only if r is rational.

So, assume $\cos x = p/q$ is rational and $q > 2$. Consequently

$$\cos 2x = 2\cos^2 x - 1 = \frac{p_1}{q_1}, \quad (p_1, q_1) = 1, \quad q_1 > q.$$

Now we have several choices. One way is to use the formula

$$2\cos x \cos{(n-1)x} = \cos nx + \cos{(n-2)x}$$

and induction. We leave this way to the reader. Another way is to use induction to get

$$\cos 2^n x = \frac{p_n}{q_n}, \quad (p_n, q_n) = 1, \quad q_1 < \cdots < q_n.$$

There are only a few choices for cosines: if $K_1 = K_2 \pmod{2N}$ then $\cos \frac{K_1 \pi}{N} = \cos \frac{K_2 \pi}{N}$, which suggests use of the pigeonhole principle: If $x = \frac{M\pi}{N}$ and $n = 2N + 1$, at least two of the numbers $\cos 2^i x$, $0 < i \le n$, have to be equal, but they all have different denominators and this contradiction shows that $q \le 2$ and we really have found all the possibilities. □

Quite often we need to find non-trivial identities to manipulate rational expressions.

Problem 2.15 Show that each rational number can be written as a sum of three cubes of rational numbers.

Search. This is a difficult problem and we want to show one possible way to solve it. We need to express x as a sum of three cubes and because we know nothing about x we can only hope that each cube is some nice expression in x and that they magically cancel each other. The first attempt is to use linear functions: is it possible that $(a_1x + a_0)^3 + (b_1x + b_0)^3 + (c_1x + c_0)^3 = x$? Expanding and looking at the constant term, we must have $a_0^3 + b_0^3 + c_0^3 = 0$ and this is the famous Fermat's equation, which only has trivial solutions, thus WLOG $b_0 = -a_0, c_0 = 0$. Looking at the coefficient for x^3, we see that a_1, b_1, c_1 also have to satisfy Fermat's equation and it is easy to conclude that there is no way to get the solution this cheaply.

The next step is to consider quadratic terms — this attempt turns out to be as unsuccessful. (See for yourself!)

We move on to cubic terms. Again Fermat's equation restricts our choices for the leading coefficients. We may assume that the third term is in fact quadratic. It turns out that it is more convenient to take the second term with minus. Thus we consider

$$f^3-g^3+h^3 = (x^3+a_2x^2+a_1x+a_0)^3-(x^3+b_2x^2+b_1x+b_0)^3+(c_2x^2+c_1x+c_0)^3.$$

After a long computation, we discover that this attempt is unsuccessful too. Before trying quadratic terms, we need to ask ourselves — do we really need to get exactly x? Of course not. For example, $ax + b$ with $a \neq 0$ would be sufficient: we could rewrite any rational number in this form. But more than this: $(ax + b)x^3$ or $(ax + b)x^6$ would be sufficient because we can divide by x^3 or x^6 and get three cubes of rational expressions. Nice! This observation gives us a lot more leeway.

Let us try to get $(ax + b)x^6$ with $a \neq 0$. To do so, we need to kill all the terms except those of degree 7 and 6. Degree 8 is easy: we need only $b_2 = a_2$. For the degree 0 we again get Fermat's equation and this time we choose $c_0 = b_0, a_0 = 0$. We now have:

$$f^3-g^3+h^3 = (x^3+a_2x^2+a_1x)^3-(x^3+a_2x^2+b_1x+b_0)^3+(c_2x^2+c_1x+b_0)^3.$$

The terms of degree 1 and 2 we get now from g and h only. In degree one we have $3b_0^2(c_1 - b_1)$, which implies $c_1 = b_1$. In degree 2 we have (using $c_1 = b_1$) the coefficient $3b_0^2(c_2 - a_2)$ which implies $c_2 = a_2$. Thus $g = x^3 + h$ and $-g^3 + h^3 = -3x^3h^2 - 3x^6h - x^9$ Therefore to kill the terms of degree < 6 we need only to kill the terms of degree < 3 in

$$(x^2 + a_2x + a_1)^3 - 3h^2 = (x^2 + a_2x + a_1)^3 - 3(a_2x^2 + b_1x + b_0)^2.$$

First of all we have $a_1^3 = 3b_0^2$. We can choose $a_1 = b_0 = 3$. After that in degree one we have $27a_2 = 18b_1$ and put $b_1 = \frac{3a_2}{2}$. The last equation in degree two will be $\frac{9}{4}(a_2 - 2)(a_2 - 6) = 0$ and if we choose $a_2 = 6$ we get

$$f = x^3 + 6x + 3, \quad g = x^3 + 6x^2 + 9x + 3, \quad h = 6x^2 + 9x + 3,$$

and

$$f^3 - g^3 + h^3 = -18x^7 - 9x^6.$$

We are almost finished: we get

$$-18x - 9 = \left(\frac{f}{x^2}\right)^3 + \left(\frac{-g}{x^2}\right)^3 + \left(\frac{h}{x^2}\right)^3.$$

If we need r we put $r = -18x + 9$, find x and get the representation. The only trouble is $x = 0$. But -9 we get directly as $-9 = (-2)^3 + (-1)^3 + 0^3$. □

It is important to know that the set of rational numbers is **dense** in the set of real numbers. This means that every real number is a limit of rational numbers. If $f(x)$ is a continuous function, we can transfer properties known for rational numbers to arbitrary real x. For example, if we know that $f(x) = ax + b$ for all rational numbers x, the same is true for all real numbers as well, because if $x = \lim x_n$ with rational x_n then

$$f(x) = f(\lim x_n) = \lim f(x_n) = \lim(ax_n + b) = ax + b.$$

Of course this is wrong if $f(x)$ is not continuous, as we can see from the Dirichlet function (see page 221).

2.5 Real numbers

Real numbers are usually the kind of number we work with. For example, if we solve a system of equations we always mean that we are looking for real solutions. If we are looking for something else (e.g. integer or complex solutions) we always say this explicitly.

What are real numbers and what are their properties? Real numbers can be understood geometrically — as points on the real line, or as infinite decimals — that is something like $a = 12.34567891011\ldots$ or $\pi = 3.1415926\ldots$ In the second case, in fact we do not know any formula for the nth digit, but there is a good geometric interpretation for π and several algorithms that allow us to find as many digits as we like. This is why we pretend that we know this number, though in reality we cannot answer trivial questions. (For example, how many zeros do we have in the first n digits of π?) We also pretend that we can add and multiply real numbers, though in reality we can hardly hope to be able to write down the decimal expansion of a^2.

The truth is that we can only handle rational numbers completely, because they are periodic. In practice, this is sufficient because rational numbers can approximate real numbers as closely as we desire. That is why we feel as we understand all real numbers as well and do not hesitate to write expressions as $2^{\sqrt{2}}$, even though it is far from obvious what we mean by this.

We hide all the difficulties in clever definitions. One of them is the *supremum*. Suppose that we have an arbitrary set M of real numbers that is bounded above. Thus there exists an upper bound: a constant C such that $m \leq C$ for every $m \in M$.

The **supremum** $\sup M$ is the *least* upper bound S, i.e. for any other upper bound B we have $S \leq B$. This least bound is (theoretically) easy to construct using the following algorithm. For simplicity we consider only the case where all the numbers in M are positive, leaving other cases to the reader's interpretation.

First we consider all integer parts for numbers in M. Because M is bounded there exists the largest one. This will (almost certainly) be the integer part of S. Then we consider all those numbers that have the same integer part and choose the largest first digit after the point. This is the first digit in S. We continue in the same way and construct all the digits of S one by one.

For example if $M = \{0.3, 0.33, 0.333, \ldots\}$, then $\sup M = 0.3333\ldots = \frac{1}{3}$. Note that this number does not belong to M. (We need to modify the approach a little bit to handle sets such as $M = \{0.9, 0.99, 0.999, \ldots\}$.)

The definition looks unpleasant at the beginning, but turns out to be very useful. For example, we can now say exactly what $2^{\sqrt{2}}$ is. If r_n is the rational approximation to $\sqrt{2}$ we get by truncating to n decimals, then $2^{\sqrt{2}} = \sup\{2^{r_n} \mid n = 1, 2, \ldots\}$.

We can make similar definitions and constructions for lower bounds, which leads us to the definition of **infimum**. For example, the infimum of $\mathbb{R}_+ = \{x \in \mathbb{R} \mid x > 0\}$ is zero. But $\mathbb{R}_+$ has no supremum.

Those notions give us an important

Theorem 2.16 *Every monotone bounded sequence a_n has a limit a.*

Proof. Take the supremum $a = \sup\{a_n \mid n = 1, 2, \ldots\}$ if the sequence is increasing or the infimum if it is decreasing. We will consider the first case only.

Note that by monotonicity, the distance $d_n = |a - a_n|$ is decreasing. If d_n does not tend to zero, then $d_n \geq c > 0$ for all n and some positive c. Then $a - a_n \geq c \Rightarrow a - c \geq a_n$ so $a - c$ is an upper bound for the sequence which is less than a, which is a contradiction. □

For example, we can *define* the number $e = 2.718281828459045\ldots$ as

$$e = \lim_{n\to\infty} 1 + \frac{1}{1!} + \frac{1}{2!} + \cdots + \frac{1}{n!}$$

and need only to show that this sequence is bounded above. Here is an evident bound:

$$B = 1 + \frac{1}{1} + \frac{1}{2} + \frac{1}{2^2} + \cdots = 1 + \frac{1}{1 - \frac{1}{2}} = 3.$$

In many books, e is defined by another limit

$$e = \lim_{n\to\infty} \left(1 + \frac{1}{n}\right)^n.$$

We leave it to the reader to prove that this sequence is increasing and bounded and that the limit is the same. It is useful to know that $\lim_{x\to 0}(1+x)^{\frac{1}{x}} = e$.

2.6 Complex numbers

The complex numbers $z = x + iy$ with $i^2 = -1$ do not appear in elementary school in most countries, and never appear in the formulations of problems at IMO. Nevertheless, they often appear in solutions, because they can help solving problems that are quite real, especially in geometry.

There are many students who hate or are afraid of Euclidian geometry, but who love algebra. For them, complex numbers provide a tool to turn a geometry problem into an algebra problem. We do not recommend this: to substitute a nice geometric condition by an awful system of equations with complex numbers is to destroy the beauty of the problem. Complex analysis is one of the most beautiful parts of mathematics, so it is not the complex numbers themselves that are ugly; it is the attempts to do everything using complex numbers, that produces ugly complex solutions. Often in the marking scheme on IMO, the solution to a geometry problem by complex numbers either gets full score or no score. In other words, solutions using complex numbers are allowed but only if they are flawless.

It is still useful to know how to create such solutions, because in emergency situations we do not look for the elegant way out; we are happy to find any door. Besides that, in the language of complex numbers some things can be explained much better. If we avoid the ugly constructions with complex numbers, we can learn to produce very nice solutions. So let us see what nice properties the complex numbers have and how to use them.

It is evident how to add, subtract or multiply complex numbers. For division, we use the conjugation rule, e.g.

$$\frac{1+2i}{3+4i} = \frac{(1+2i)(3-4i)}{(3+4i)(3-4i)} = \frac{3+6i-4i+8}{9+16} = \frac{11+2i}{25} = \frac{11}{25} + \frac{2}{25}i.$$

In most situations, the form $\frac{11+2i}{25}$ is preferable if we need to do additional work with the quotient.

The number $x = \operatorname{Re} z$ is called the *real part* of z, the number $y = \operatorname{Im} z$ is the *imaginary part* (note that the imaginary part of z is *real*!). We can identify complex numbers having zero imaginary part with real numbers. Thus the set $\mathbb{R}$ of real numbers can be viewed a subset of the set $\mathbb{C}$ of complex numbers.

If we consider a pair (x, y) as coordinates of a point Z in the plane we get a geometrical interpretation of complex numbers. For example, $0 = 0 + i0$ is the origin O. The distance to the origin $\sqrt{x^2 + y^2}$ is called the **modulus** or the **absolute value** of the complex number z and is denoted by $|z|$. If $z \neq 0$, the angle between the vector OZ and the x-axis is called the **argument** $\varphi = \arg z$. The argument is in fact only determined up to $2\pi n$, where $n \in \mathbb{Z}$, which creates some difficulties. In most situations, we can take *any* of these angles. For example, $\arg(-i) = -\frac{\pi}{2}$ and $\arg(-i) = \frac{3\pi}{2}$ both work equally well.

The complex numbers that lie on the circle $x^2 + y^2 = 1$ are exactly those numbers with $|z| = 1$. If $\varphi = \arg z$ is an argument of such a number then it is convenient to denote $z = \cos\varphi + i\sin\varphi$ as $e^{i\varphi}$. The notation is not as strange as it may seem at first. For example, it is easy to check that

$$e^{i\varphi}e^{i\psi} = e^{i(\varphi+\psi)}$$

and all other usual rules for exponents are also valid. Knowing this we can easily recover most trigonometrical formulas if we forget them. For example, if you need to rewrite $\sin 3\varphi$, simply write:

$$\begin{aligned}\sin 3\varphi &= \operatorname{Im}(e^{3i\varphi}) = \operatorname{Im}\left((e^{i\varphi})^3\right) = \operatorname{Im}\left((\cos\varphi + i\sin\varphi)^3\right)\\ &= \operatorname{Im}(\cos^3\varphi + 3i\cos^2\varphi\sin\varphi - 3\cos\varphi\sin^2\varphi - i\sin^3\varphi)\\ &= 3\cos^2\varphi\sin\varphi - \sin^3\varphi = 3\sin\varphi - 4\sin^3\varphi.\end{aligned}$$

Do not forget that $\sin x$ and $\cos x$ have exponentials as their closest relatives!

Problem 2.17 Let $a_0, a_1, \ldots, a_n$ be real numbers such that the equation

$$a_0 + a_1\cos x + \cdots + a_n\cos nx = 0$$

has $2n+1$ distinct solutions in the interval $[0, 2\pi]$. Show that

$$a_0 = a_1 = \cdots = a_n = 0.$$

Search. This looks difficult. But if we look at this as an exponential equation, the picture becomes clearer. Consider the complex number $z = e^{ix} = \cos x + i\sin x$. We have

$$\cos kx = \frac{1}{2}\left(e^{ikx} + e^{-ikx}\right) = \frac{1}{2}\left(z^k + \frac{1}{z^k}\right).$$

Using this, and multiplying through with z^{2n}, the given equation is transformed into a polynomial equation of degree at most $2n$ with $2n+1$ distinct roots. Thus the polynomial and all its coefficients are equal to zero. Hence all the coefficients a_i are equal to zero as well. □

In fact, we can define the exponential function e^z for any complex number as $e^{x+yi} = e^x e^{iy}$ and still preserve the usual rules. But it is much more important that any complex number $z = x + iy$ can be written as a $z = |z|e^{i\varphi}$. Thus we can use two different languages working with complex numbers —

Table 2.1 Translating complex expressions from one language to another

Notion	x, y	$\lvert z\rvert = R, \arg z = \varphi$	$z, \bar{z}$
x	x	$R\cos\varphi$	$\dfrac{z+\bar{z}}{2}$
y	y	$R\sin\varphi$	$\dfrac{z-\bar{z}}{2i} = \dfrac{i(\bar{z}-z)}{2}$
z	$x+iy$	$Re^{i\varphi}$	z
$-z$	$-x-iy$	$Re^{i(\pi+\varphi)}$	$-z$
$\bar{z}$	$x-iy$	$Re^{-i\varphi}$	$\bar{z}$
$\lvert z\rvert$	$\sqrt{x^2+y^2}$	R	$\sqrt{\bar{z}z}$
$\lvert z\rvert^2$	x^2+y^2	R^2	$\bar{z}z$
$\arg z$	$\begin{cases} \arctan y/x, & x>0 \\ \pi + \arctan y/x, & x<0 \\ \pi/2, & x=0, y>0 \\ -\pi/2, & x=0, y<0 \end{cases}$	φ	$\arg z$

either as real vectors (x, y) or as pairs $R = |z|$, $\varphi = \arg z$. Sometimes a third language is useful.

Recall that the **conjugate of a complex number** $z = x + iy$ is defined as $\bar{z} = x - iy$. Note that

$$\overline{z_1 + z_2} = \overline{z_1} + \overline{z_2}, \quad \overline{z_1 \cdot z_2} = \overline{z_1} \cdot \overline{z_2}, \quad \overline{z_1/z_2} = \overline{z_1}/\overline{z_2}.$$

In the third language we use a pair of complex numbers $(z, \bar{z})$. For example, $|z|^2 = \bar{z}z = z\bar{z}$ is a very useful formula. See Table 2.1 for a dictionary of translations from one language to another:

The right choice of language simplifies life. For example, the **triangle inequality** $|z_1 + z_2| \le |z_1| + |z_2|$ is obvious geometrically, but not in the coordinates. The properties $|z_1 z_2| = |z_1| \cdot |z_2|$, and $\arg(z_1 z_2) = \arg z_1 + \arg z_2$ (up to a possible difference of $2\pi n$) become evident if we write

$$z_1 z_2 = |z_1| e^{i\varphi_1} |z_2| e^{i\varphi_2} = |z_1| \cdot |z_2| \cdot e^{i(\varphi_1 + \varphi_2)}.$$

From the last equation we also see an important property of the complex plane: rotations by an angle φ around the origin is simply multiplication by

$e^{i\varphi}$. The central symmetry, in particular, is the map $z \mapsto -z$, because of *Euler's identity* $e^{i\pi} = -1$. Another map $z \mapsto \overline{z}$ describes reflection in the x-axis.

Let us see in more detail how we can interpret geometry in terms of complex numbers. First of all, the distance between two points Z_1, Z_2 is $|z_1 - z_2|$, if z_i are corresponding complex numbers. To save space, from here on we will skip the repetition of such phrases and identify the point Z with the corresponding complex number z without comment. We may interpret points as complex numbers directly, writing $Z_1 - Z_2$ instead of $z_1 - z_2$ when it is clear what we mean.

One nice property of complex numbers is that four different points Z_1, Z_2, Z_3, Z_4 in the plane (in this order) form a parallelogram if and only if

$$z_2 - z_1 = z_3 - z_4 \quad \Leftrightarrow \quad z_1 + z_3 = z_2 + z_4.$$

For example, we immediately see that the lengths are equal: $|z_2 - z_1| = |z_3 - z_4|$ and $|z_2 - z_3| = |z_1 - z_4|$. Now let us practise the art of ugly calculations.

Problem 2.18 Consider a parallelogram with sides of length a, b. Let d_1, d_2 be the lengths of the diagonals. Show that $2(a^2 + b^2) = d_1^2 + d_2^2$.

Search. The ugliest way to do this is to consider four complex numbers and to write the statement as

$$2(|z_2 - z_1|^2 + |z_2 - z_3|^2) = |z_1 - z_3|^2 + |z_4 - z_2|^2,$$

replace z_4 by $z_1 + z_3 - z_2$ and rewrite the expressions in the style of $|z_1 - z_2|^2 = (z_1 - z_2)(\overline{z_1} - \overline{z_2}) = \cdots$. Good luck, if you want to try this approach!

If we decide to use complex numbers, we should at least start with some cosmetics. Since we are free to place the coordinate system in our complex plane as we want, WLOG we can decide that one of the points is the origin 0, and the second is real a. Let the third be z with $|z| = b$. The remaining one we can choose as $z + a$, because $(z + a) - a = z - 0$ and we get a parallelogram. What we need to prove is $2(a^2 + |z|^2) = |z + a|^2 + |z - a|^2$, which should not be difficult:

$$\begin{aligned}(z + a)\overline{(z + a)} + (z - a)\overline{(z - a)} &= (z + a)(\overline{z} + a) + (z - a)(\overline{z} - a) \\ &= z\overline{z} + za + a\overline{z} + a^2 + z\overline{z} - za - a\overline{z} + a^2 = 2(|z|^2 + a^2).\end{aligned}$$

Still ugly, but possible to tolerate.

The geometric proof is not much nicer, but shorter: apply the cosine theorem to each diagonal and add the result. □

We return to our four points.

If $Z_3 \neq Z_4$ and $Z_1 \neq Z_2$ we know exactly when the line through Z_1 and Z_2 is **parallel** to the line through Z_3 and Z_4. The condition is that the number $\frac{z_1-z_2}{z_3-z_4}$ is real. For **orthogonality** we need $i \cdot \frac{z_1-z_2}{z_3-z_4}$ to be real.

For example, the line through $Z_1 = 1 + i$ and $Z_2 = 3 - 3i$ is parallel to the line between $Z_3 = 3 - i$ and $Z_4 = 2 + i$ because $\frac{-2+4i}{1-2i} = -2$ and we get more than that: the segment Z_1Z_2 is twice as long as the segment Z_3Z_4 and they, as vectors, are oppositely oriented (because of the minus sign).

Both lines are orthogonal to the line through 1 and $3 + i$, since

$$i\frac{-2+4i}{-2-i} = \frac{-4-2i}{-2-i} = 2.$$

As we have alread seen, a common task is checking that some complex number z is real. Theorem 1.17 on page 37 can be useful, but the easiest way is to show that $z = \overline{z}$. For example, to check that $i\frac{1+e^{ix}}{1-e^{ix}}$ (if defined) is real for every real x, we need to show that

$$i\,\frac{1+e^{ix}}{1-e^{ix}} = -i\,\frac{1+e^{-ix}}{1-e^{-ix}} \quad \Leftrightarrow \quad (1+e^{ix})(1-e^{-ix}) = -(1-e^{ix})(1+e^{-ix}),$$

which is trivial after expansion. Using Theorem 1.17 is about as complicated: all we need is to rewrite the number as $\frac{-\sin x + i(1+\cos x)}{1-\cos x - i\sin x}$ and to check that the determinant is equal to zero:

$$\begin{vmatrix} -\sin x & 1+\cos x \\ 1-\cos x & -\sin x \end{vmatrix} = 0.$$

The smart way is to divide the numerator and the denominator by $ie^{ix/2}$ and note that the result is $\frac{\cos(x/2)}{\sin(x/2)} = \cot(x/2)$. Another smart way is to note that 1, e^{ix} and $-e^{ix}$ all lie on the unit circle and that $e^{ix}, -e^{ix}$ are the ends of a diameter, which gives the orthogonality between lines through 1 and $\pm e^{ix}$, by Thales' theorem.

The best way to describe the **line** through two different points z_1 and z_2 is

$$z = (1-t)z_1 + tz_2$$

for all real t. For example, $t = 0$ corresponds to z_1, and the points on the segment are those for which $0 \leq t \leq 1$. The midpoint corresponds to $t = \frac{1}{2}$ and

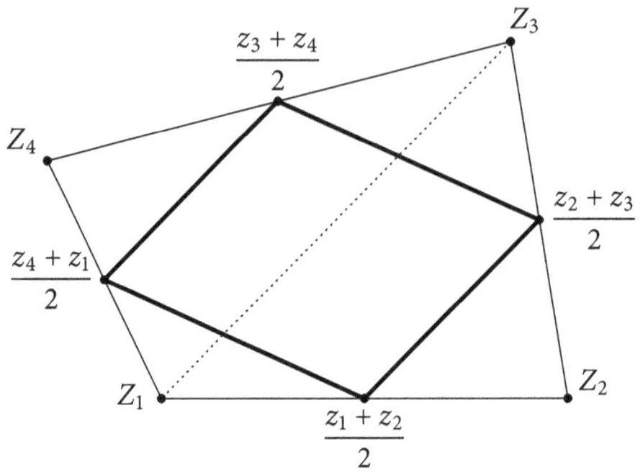

Figure 2.1 Midpoints

more generally the point that the divides the segment in the relation $m : n$ is

$$\frac{n}{m+n}z_1 + \frac{m}{m+n}z_2$$

(note — in the opposite order!)

These elementary properties are already sufficient to prove many statements.

Problem 2.19 Show that, for any quadrilateral the midpoints of its sides from a parallelogram.

Search. All the conditions are easy to express in complex numbers, so let us try this approach. If z_i are the vertices, then $\frac{z_1+z_2}{2}, \frac{z_2+z_3}{2}, \frac{z_3+z_4}{2}, \frac{z_4+z_1}{2}$ are the midpoints. Because

$$\frac{z_3+z_2}{2} - \frac{z_1+z_2}{2} = \frac{z_3-z_1}{2} = \frac{z_3+z_4}{2} - \frac{z_4+z_1}{2}$$

we are finished.

Is there a nice geometrical proof? Of course! Draw a diagonal and note that it is parallel to the middle line, connecting midpoints (see Figure 2.1).□

Now let us try something a little more serious.

Theorem 2.20 *Consider an arbitrary triangle Z_1, Z_2, Z_3. Let G be its centroid (centre of gravity, see page 4.1, O the circumcentre and H the orthocentre. Then*

- *The centroid G can be found as $\dfrac{z_1 + z_2 + z_3}{3}$. It lies on each median and divides the medians in the ratio 1 : 2.*
- *If the circumcentre O coincides with the origin, then H can be found as $z_1 + z_2 + z_3$.*
- *O, G and H lie on the same line (the* **Euler line***) and G divides OH in the ratio 1 : 2.*

Proof. We put a weight 1 at every vertex and try to find the centre of gravity. For two points Z_1, Z_2 the centre of gravity is their midpoint $M = \frac{z_1+z_2}{2}$, so we can replace Z_1 and Z_2 by M with weight 2. Thus G lies on the median from Z_3 and divides it in the ratio 1 : 2, which gives

$$G = \frac{2}{3}\frac{z_1 + z_2}{2} + \frac{1}{3}z_3 = \frac{z_1 + z_2 + z_3}{3}.$$

Evidently the same works for other medians.

Let now O be the origin, thus $|z_1| = |z_2| = |z_3| = R$. Let $H' = z_1 + z_2 + z_3$ To show that $H' = H$ it is sufficient to check orthogonality and this is straightforward. For example, to check that $H'Z_3$ is orthogonal to Z_1Z_2 we need to prove that

$$i\,\frac{z_1 + z_2 + z_3 - z_3}{z_1 - z_2} = i\,\frac{z_1 + z_2}{z_1 - z_2} = \overline{i\,\frac{z_1 + z_2}{z_1 - z_2}} = -i\,\frac{\overline{z_1} + \overline{z_2}}{\overline{z_1} - \overline{z_2}}.$$

But this is equivalent to

$$(z_1+z_2)(\overline{z_1}-\overline{z_2}) = -(z_1-z_2)(\overline{z_1}+\overline{z_2}) \quad \Leftrightarrow \quad 2z_1\overline{z_1}-2z_2\overline{z_2} \quad \Leftrightarrow \quad |z_1| = |z_2|.$$

The last part is now trivial: if we choose O as the origin we get

$$\frac{2}{3}O + \frac{1}{3}H = 0 + \frac{1}{3}(z_1 + z_2 + z_3) = G.$$

□

These results show the advantage of working with complex numbers: we get non-trivial geometric results by calculations only, and do not even need to draw a picture. But this is also the weak point: we hardly understand the nature of the result obtained and we do not gain any geometrical insights.

If the origin does not coincide with the circumcentre, we put it there by the variable substitution $W = Z - O$ and get

$$H - O = (Z_1 - O) + (Z_2 - O) + (Z_3 - O) \quad \Rightarrow \quad H = Z_1 + Z_2 + Z_3 - 2O.$$

The make-up trick with the variable substitution is used quite often and allows us to improve our working conditions.

Consider for example a regular polygon. It is easy to construct one: Take a unit circle and divide it into n equal parts. If we choose one vertex as 1 we get all of them as $1, \varepsilon, \varepsilon^2, \ldots, \varepsilon^{n-1}$, where

$$\varepsilon = e^{2\pi i/n} \quad \Rightarrow \quad \varepsilon^n = 1.$$

Moreover, if Z is an arbitrary non-zero point then the points Z, εZ, $\varepsilon^2 Z$, ..., $\varepsilon^{n-1}Z$ form a regular polygon centered at the origin and an arbitrary regular polygon we get as

$$Z_1 = C + Z, \quad Z_2 = C + \varepsilon Z, \quad Z_3 = C + \varepsilon^2 Z, \quad \ldots, \quad Z_n = C + \varepsilon^{n-1} Z$$

where C is its centre and $C + Z$ one of its vertices.

A useful identity which is frequently applied is

$$1 + \varepsilon + \varepsilon^2 + \ldots + \varepsilon^{n-1} = 0$$

and is valid because this is a geometrical progression with the sum $\frac{1-\varepsilon^n}{1-\varepsilon} = 0$. For example we can apply it to show that the centre of a regular polygon with vertices in $Z_1, \ldots, Z_n$ can be found as the arithmetic mean: $C = \frac{Z_1+\ldots+Z_n}{n}$.

Another example is the rotation around the point C. If $C = 0$ then, as we know, the rotation is the multiplication by $e^{i\varphi}$. In a general situation, if Z maps to W, then it means that $(Z - C)e^{i\varphi} = W - C$, and thus $W = (Z - C)e^{i\varphi} + C$ describes the rotation.

Complex numbers are also convenient for finding similar triangles. Consider the equation

$$\frac{z_1 - z_2}{z_3 - z_2} = \frac{w_1 - w_2}{w_3 - w_2}. \tag{2.1}$$

If we compare the absolute values, we see that the ratio between two sides of the triangles $Z_1Z_2Z_3$ and $W_1W_2W_3$ are the same. If we consider the arguments instead, we find that the angles between them are equal as well, and thus this simple equation is exactly the condition for similarity, where Z_i corresponds to W_i. Furthermore, if (2.1) is satisfied, the triangles have the same orientation:

i.e. if Z_1, Z_2, Z_3 traverses the first triangle clockwise, the same is true for W_1, W_2, W_3.

Because the reflection $z \mapsto \overline{z}$ changes the orientation, the equation

$$\frac{z_1 - z_2}{z_3 - z_2} = \frac{\overline{w_1} - \overline{w_2}}{\overline{w_3} - \overline{w_2}}$$

describes an orientation reversing similarity. It is easy to check that those two equations are the only choices for a similarity that sends Z_i to W_i.

As an example we give an easy way to check whether $Z_1Z_2Z_3$ forms an equilateral triangle. Let $\varepsilon = e^{2\pi i/3}$, $\varepsilon^3 = 1$. Then the triangle is similar to the triangle with the vertices in $1, \varepsilon, \varepsilon^2$. If the orientation is the same we get

$$\frac{z_1 - z_2}{z_3 - z_2} = \frac{1-\varepsilon}{\varepsilon^2 - \varepsilon} = \frac{-1}{\varepsilon} = -\varepsilon^2 \quad \Rightarrow \quad z_1 - (1+\varepsilon^2)z_2 + \varepsilon^2 z_3 = 0$$
$$\Rightarrow \quad z_1 + \varepsilon z_2 + \varepsilon^2 z_3 = 0.$$

If the orientation is reversed, we get the equation $z_1 + \varepsilon^2 z_2 + \varepsilon z_3 = 0$ instead.

Problem 2.21 (Napoleon's theorem.) Consider an arbitrary triangle and on its sides construct (outwards) equilateral triangles. Then the centres of those equilateral triangles themselves form an equilateral triangle. The same is true for the inward triangles.

Search. This sounds like it would be perfect for complex numbers. Consider two vertices A, B. We can complete the segment AB to an equilateral triangle by a vertex C_1 in two different ways: either $C_1 = -\varepsilon A - \varepsilon^2 B$ or $C_1 = -\varepsilon^2 A - \varepsilon B$. One of them (depending on the orientation) corresponds to the outward triangle, the other to the inward. Suppose that we have the first equation (the other case is similar). The centre of the equilateral triangle must be

$$\frac{1}{3}(A + B + C_1) = \frac{1}{3}((1-\varepsilon)A + (1-\varepsilon^2)B).$$

We get similar expressions for the other sides and we only need to check that

$$((1-\varepsilon)A + (1-\varepsilon^2)B) + \varepsilon^2((1-\varepsilon)B + (1-\varepsilon^2)C) + \varepsilon((1-\varepsilon)C + (1-\varepsilon^2)A)$$

is equal to zero which is trivial. □

In Table 2.2, we collect some useful geometric notions expressed in all three languages. Complete the table! Where is the centre of gravity, how do we find

Table 2.2 Geometric quantities expressed using complex numbers, in all three "languages". Note that the numbers a, b, c, t are real, but w, z_i are complex. Also note that there are two ways to complete z_1, z_2 to an equilateral triangle; hence the two alternatives. Draw several pictures to understand how to choose the right alternative. Empty cells means that the corresponding language is inconvenient. To save space we write $|z_i - z_j|$ in (x, y)–language to make the formulas shorter.

x, y	$\lvert z\rvert, \arg z$	$z, \overline{z}$
distance between z_1 and z_2		
$\sqrt{(x_1-x_2)^2+(y_1-y_2)^2}$	$\lvert z_1 - z_2\rvert$	$\sqrt{(z_1-z_2)(\overline{z_1}-\overline{z_2})}$
arbitrary line		
$ax+by=c$		$wz+\overline{wz}=c$
arbitrary line through $z_0 = x_0 + y_0 i$		
$a(x-x_0)=b(y-y_0)$	$\arg(z-z_0)=\alpha$ and $\arg(z-z_0)=\alpha+\pi$	$w(z-z_0)+\overline{w(z-z_0)}=0$
line through $z_0 = x_0 + y_0 i$, parametric form		
$(x,y)=(x_0,y_0)+t(a,b)$	$(x,y)=z=z_0+te^{i\alpha}$, (α constant)	$z=z_0+tw$, $(w=a+bi)$
midpoint between z_1 and z_2		
$\left(\frac{x_1+x_2}{2}, \frac{y_1+y_2}{2}\right)$		$\frac{z_1+z_2}{2}$
point dividing the segment z_1 and z_2 in relation $m:n$		
$\left(\frac{nx_1+mx_2}{m+n}, \frac{ny_1+my_2}{m+n}\right)$		$\frac{nz_1+mz_2}{n+m}$
circle with centre z_0 and radius r		
$(x-x_0)^2+(y-y_0)^2=r^2$	$\lvert z-z_0\rvert=r$	$(z_1-z_2)(\overline{z_1}-\overline{z_2})=r^2$
circle with centre z_0 and radius r, parametric form		
$\begin{cases} x=x_0+r\cos t \\ y=y_0+r\sin t \end{cases}$	$z=z_0+re^{it}$	$z=z_0+re^{it}$
the tangent to the circle $x^2+y^2=r$ in $z_0=(a,b)$		
$ax+by=r^2$	$\arg(z-z_0)=\pm\pi/2+\arg z_0$	
angle α between z_2z_1 and z_4z_4		
	$e^{i\alpha}=\frac{z_2-z_1}{z_4-z_3}\cdot\frac{\lvert z_4-z_3\rvert}{\lvert z_2-z_1\rvert}$	
z_1z_2 is orthogonal to z_3z_4		
$(x_2-x_1)(x_4-x_3)=$ $-(y_2-y_1)(y_4-y_3)$	$\arg\left(\frac{z_2-z_1}{z_4-z_3}\right)=\pm\frac{\pi}{2}$	$\overline{\left(\frac{z_2-z_1}{z_4-z_3}\right)}=-\left(\frac{z_2-z_1}{z_4-z_3}\right)$
$z_1z_2z_3$ is an equilateral triangle		
$\lvert z_3-z_1\rvert=\lvert z_2-z_1\rvert=\lvert z_3-z_2\rvert$	$z_3-z_1=e^{\pm\pi i/3}(z_2-z_1)$	$z_1+\varepsilon z_2+\varepsilon^2 z_3=0$, $(\varepsilon=e^{\pm 2\pi i/3}, \varepsilon^3=1)$

the circumcentre, the angle bisector? To practise on such easy problems is very useful if you want to learn this approach to geometry.

We will finish with a nice example.

Problem 2.22 Let c be a point outside the circle $|z| = R$. There exists two tangents from c that touch the circle in the points a and b. What is the connection between a, b and c as complex numbers?

Search. Let l be the length of the tangents. Then the orthogonality gives the equations

$$\frac{c-a}{l} = i\frac{a-0}{R}; \qquad \frac{c-b}{l} = -i\frac{b-0}{R}.$$

(the signs can be opposite but it doesn't matter). We have:

$$\frac{c-a}{c-b} = -\frac{a}{b} \qquad \Leftrightarrow \qquad c = \frac{2ab}{a+b}.$$

Thus c is the harmonic mean of a and b. □

What about the other means? The arithmetic mean $\frac{a+b}{2}$ is evidently the midpoint of the segment. As to geometric mean $\sqrt{ab}$ of two points a, b on the circle $|z| = R$ it can be interpreted as the midpoint of the arcs between them. Which of two possible arcs? In fact, both. It is important to remember that for complex numbers the square root has normally two values, so we need to be very careful choosing only one of them when doing explicit calculations.

2.7 Exercises

Exercise 2.1. Let $p(x)$ be a polynomial with integer coefficients such that $p(3) = 2$. Is it possible that $p(2003)$ is a perfect square?

Exercise 2.2. Let $p > 3$ be a prime. Show that if

$$1 + \frac{1}{2} + \cdots + \frac{1}{p-1} = \frac{m}{n}$$

then $p^2 \mid m$.

Exercise 2.3. Let $(a_n)_{n\geq 1}$ be a sequence of positive integers defined by:

(i) $a_1 > 0,\ a_2 > 0$;

(ii) a_{n+1} is the least prime divisor of $a_{n-1} + a_n$, for all $n \geq 2$.

Show that a real number x whose decimals are the digits of the numbers

$$a_1, a_2, \dots, a_n, \dots,$$

written in that order, is a rational number (digits are considered in the decimal representation).

Exercise 2.4. Let $p(x)$ be a polynomial with integer coefficients. Show that if $p(2) = 3$, $p(3) = 5$ then $p(n) \neq 0$ for every integer n.

Exercise 2.5. a) Show that if a, b, c are different rational numbers then

$$\frac{1}{(a-b)^2} + \frac{1}{(b-c)^2} + \frac{1}{(c-a)^2}$$

is a square of some rational number.

b) Show that if

$$\frac{a}{b-c} + \frac{b}{c-a} + \frac{c}{a-b} = 0$$

then

$$\frac{a}{(b-c)^2} + \frac{b}{(c-a)^2} + \frac{c}{(a-b)^2} = 0.$$

Exercise 2.6. a) Show that for every positive integer n,

$$[x] + \left[x + \frac{1}{n}\right] + \left[x + \frac{2}{n}\right] + \dots + \left[x + \frac{n-1}{n}\right] = [nx]$$

b) Show that if p and q are relatively prime odd natural numbers, then

$$\left[\frac{q}{p}\right] + \left[\frac{2q}{p}\right] + \dots + \left[\frac{(p-1)q}{p}\right] = \frac{(p-1)(q-1)}{2}.$$

Exercise 2.7. An arithmetic sequence consists of positive integers. Show that if it contains a square then it contains infinitely many squares.

Exercise 2.8. Let $p(x)$ be a non-constant polynomial with integer coefficients and positive leading coefficient. Show that for any positive integer k there exists an integer n such that all the numbers $p(n), p(n+1), \dots, p(n+k)$ are composite.

Exercise 2.9. If $\sin x = \frac{3}{5}$, show that $5^{25} \sin 25x$ is an integer not divisible by 5.

Exercise 2.10. Let A, B, C, D be different points on the unit circle. The origin O is the centre of this circle. Interpreting these points as complex numbers find the explicit expressions for the following important points:

- The point A' on the circle symmetrical to A relatively O (so AA' is a diameter).
- The point E on the circle different from A such that the lines AE and BC are orthogonal.
- The foot F of the altitude from A in the triangle ABC (i.e. the intersection of two lines above). Show that $|HF| = |FE|$, where H is the orthocentre (see page 4.1).
- The intersection P between the lines $A'H$ and BC. Show that $|HP| = |PA'|$ and $|BP| = |PC|$.
- The intersection X between the lines AC and BD.

Find (and use in other problems) obvious geometrical interpretations, e.g. that $HBA'C$ is a parallelogram and that the triangles HBE, HCE are isosceles.

Exercise 2.11. Show that $\binom{n}{p} \equiv \left[\frac{n}{p}\right] \pmod{p}$ for any prime p.

See also problems 3, 48, 56, 78, 84, 90, 93, 94 in Chapter 14.

2.8 Hints

2.1. Consider the problem (mod4) and use that $x - y$ divides $p(x) - p(y)$.

2.2. Note that $\frac{1}{k} + \frac{1}{p-k} = \frac{p}{k(p-k)}$ and we need only to show that in $\mathbf{Z}_p$ the sum

$$\sum_{k=1}^{\frac{p-1}{2}} \frac{1}{k(p-k)}$$

is equal to zero.

2.3. First get 2. If the next number is 2 or 3 we have finished. Work modulo 6. The difficult part is $a_n \equiv -1 \pmod 6$. Choose a minimal such number and show that $a_{n+2} \equiv 1 \pmod 6$.

2.4. $p(n) \not\equiv 0 \pmod 2$.

2.5. a) Set $x = \frac{1}{a-b}, y = \frac{1}{b-c}, z = \frac{1}{c-a}$.

2.6. a) Find a period of $f(x) = [nx] - \left([x] + \left[x + \frac{1}{n}\right] + \cdots + \left[x + \frac{n-1}{n}\right]\right)$.

b) Consider a rectangle $0 \le x \le p, 0 \le y \le q$ and its diagonal $y = \frac{qx}{p}$.

2.7. Let d be the difference and $a = m^2$. When does $(m + y)^2$ belong to the sequence?

2.8. Suppose that $p(x) \ge 1$ for any $x \ge m$. Put $l = p(m)\cdots p(m+k)$ and $n = m + l$. Use that $p(n+t) = (p(m+l+t) - p(m+t)) + p(m+t)$ is divisible by $p(m+t)$.

2.9. Let $5^n \sin nx = p_n$ and $5^n \cos nx = q_n$. We have $p_1 = 3, q_1 = \pm 4$ and

$$p_{n+1} = \pm 4p_n + 3q_n, q_{n+1} = -3p_n \pm 4q_n.$$

Now it is easy to check that $p_{n+4} = p_n \pmod 5$.

2.10. Let $a, b, c, d, a', e, f, h, p$ be the corresponding complex numbers. Then, clearl $a' = -a$ and $h = a + b + c$. Because $\frac{a'+h}{2} = \frac{b+c}{2}$ this should be equal to p. Show that $-\frac{bc}{a}$, which obviously belongs to the circle can be used as e. Use rules such as $\overline{a} = \frac{1}{a}$ for checking the orthogonality conditions $\overline{\frac{a-e}{b-c}} = \frac{a-e}{c-b}$. The same method uses to show that $f = \frac{h+e}{2}$: we need only to check the collinearity conditions: $\overline{\frac{f-b}{c-b}} = \frac{f-b}{c-b}$. The same approach shows $x = \frac{ac(b+d) - bd(a+c)}{ac-bd}$. This general formula is worth remembering. What happens if $ac = bd$?

2.11. Use Wilson's theorem.

CHAPTER 3

Algebra

3.1 Identities and factorization

A common wrong approach to algebraic expressions is the urge to immediately expand them. Yes, it is sometimes necessary, but most of the time we need the opposite action — a factorization — because if we succeed with this it is often easier to work with each factor separately. At least, try to collect some terms together — perhaps this can help you to see the hidden quadratic equation or other useful things such as a perfect square or a sum of cubes. To be able to factorize, we sometimes need to use the identities that are easy to prove (here we can expand!) but difficult to remember. Let us list some of the most important ones, starting from the trivial:

$$x^n - y^n = (x - y)(x^{n-1} + x^{n-2}y + \cdots + xy^{n-2} + y^{n-1}).$$

If $n = 2k + 1$ is odd, we have

$$x^n + y^n = (x + y)(x^{n-1} - x^{n-2}y + \cdots - xy^{n-2} + y^{n-1})$$

as well. There is no factorization for $x^2 + y^2$ but there is one for all other values of n. Either n is divisible by an odd prime, and we can use the factorization above or $4 \mid n$ and we complete the square as in problem 1.11 on page 32. Another useful formula is:

$$(x^2 + ax + 1)(x^2 - ax + 1) = x^4 + (2 - a^2)x^2 + 1.$$

It is important to read identities in both directions. For example, the formula for a geometric sum

$$a + ax + \cdots + ax^n = a\,\frac{1 - x^{n+1}}{1 - x}$$

for $x \neq 1$ is essentially the same as our first formula. Do not forget that for $x = 1$ we get another result, namely $(n+1)a$. The infinite version of a geometric sum,

$$\sum_{n=0}^{\infty} ax^n = \frac{a}{1-x}$$

is valid for $|x| < 1$ only (note that x may be complex as well).

Of course, everybody needs to remember the binomial expansion

$$(x+y)^n = \sum_{k=0}^{n} \binom{n}{k} x^{n-k} y^k.$$

Again, there is an infinite version

$$(1+x)^a = 1 + ax + \binom{a}{2}x^2 + \cdots + \binom{a}{k}x^k + \cdots$$

valid for $|x| < 1$ and arbitrary a, if we interpret

$$\binom{a}{k} = \frac{a(a-1)\cdots(a-k+1)}{k!}$$

as a polynomial in a of degree k. For example, if $a = n$ we get the ordinary binomial expansion, but for $a = -n$ we get

$$\binom{-n}{k} = \frac{(-n)(-n-1)\cdots(-n-k+1)}{k!} = (-1)^k \binom{n+k-1}{k}.$$

If we replace x by $-x$ we get a nice formula:

$$1 + nx + \binom{n+1}{2}x^2 + \binom{n+2}{3}x^3 + \cdots = \frac{1}{(1-x)^n}.$$

A non-trivial infinite sum which we will make use of is

$$1 + \frac{1}{2^2} + \frac{1}{3^2} + \cdots = \frac{\pi^2}{6}.$$

Here are some other identities that are worth remembering.

$$x^3 + y^3 + z^3 - 3xyz = (x+y+z)(x^2 + y^2 + z^2 - xy - xz - yz).$$

$$(a^2 + b^2)(c^2 + d^2) = (ac + bd)^2 + (ad - bc)^2.$$

$$1^2 + 2^2 + \cdots + n^2 = \frac{n(n+1)(2n+1)}{6}.$$

3.2 Polynomials in one variable

Let $f(x) = a_0x^n + a_1x^{n-1} + \cdots + a_{n-1}x + a_n$. If $a_0 \neq 0$ then $n = \deg f$. Conventionally, we set $\deg 0 = -\infty$. Evidently,

$$\deg f(x)g(x) = \deg f(x) + \deg g(x),$$
$$\deg(f(g(x)) = \deg f(x) \cdot \deg g(x),$$
$$\deg(f(x) + g(x)) \leq \max\{\deg f(x), \deg g(x)\}$$

If $a_0 = 1$, then $f(x)$ is called **monic.**

Polynomials have very similar divisibility properties as integers. If $g(x)$ is another polynomial, then there are unique polynomials $q(x)$, the *quotient*, and $r(x)$, the *remainder*, such that $f(x) = g(x)q(x) + r(x)$ with $\deg r(x) < \deg g(x)$. If $r(x) = 0$, we say that $g(x)$ **divides** $f(x)$, denoted $g(x) \mid f(x)$.

In particular $f(x) = (x-a)q(x)+f(a)$ and $f(a) = 0 \Leftrightarrow (x-a) \mid f(x)$.

Because there is a division algorithm, we also have notions of gcd and lcm. To get uniqueness, we demand the gcd and lcm to be monic, and thus a **greatest common divisor** $d(x)$ of two polynomials $f(x)$ and $g(x)$ is a monic polynomial of the highest degree that divides both $f(x)$ and $g(x)$.

The Euclidean algorithm generalizes to polynomials and can be used to find the greatest common divisor (which is often denoted simply as $(f(x), g(x))$). Note that $g(x) \mid f(x) \Leftrightarrow (f(x), g(x)) = g(x)$, if $g(x)$ is monic.

Polynomials $f(x), g(x)$ are **relatively prime** if $(f(x), g(x)) = 1$. As for integers, this is the condition for Bézout's lemma:

Theorem 3.1 (Bézout's lemma) *Let $f(x)$ and $g(x)$ be two polynomials that are relatively prime. Then there exist polynomials $u(x)$ and $v(x)$ such that*

$$u(x)f(x) + v(x)g(x) = 1.$$

(Note that multiplying by $h(x)$, we can write ***any*** *polynomial $h(x)$ as a combination of $f(x)$ and $g(x)$.)*

The Chinese remainder theorem has to be slightly reformulated.

Theorem 3.2 (Chinese remainder theorem) *Let $p_1(x), \ldots, p_k(x)$ be pairwise relatively prime polynomials and let $c_1(x), \ldots, c_k(x)$ be arbitrary polynomials with $\deg c_i(x) < m_i = \deg p_i(x)$. Then there exists (a unique!)*

polynomial $f(x)$ of degree less than $m = \sum m_i$, such that its remainder after division by $p_i(x)$ is exactly $c_i(x)$ for every $i = 1, \ldots, k$.

All these statements are valid as long as we can divide coefficients. For example, they are valid for real polynomials, for complex polynomials, for polynomials with rational coefficients, and even for polynomials with coefficients in $\mathbb{Z}_p$ when p is prime.

For polynomials over $\mathbb{Z}_m$ (if m is not a prime), the degree of the product may be less than the sum of the degrees of the factors. This never happens for polynomials with integer coefficients. It is also possible to divide polynomials over $\mathbb{Z}_m$, but only by a monic polynomial.

If we want to stress which type of numbers $\mathbb{K}$ the coefficients belong to, we write $f(x) \in \mathbb{K}[x]$.

Depending on choice of the numbers $\mathbb{K}$, there are different analogues of prime numbers — the irreducible polynomials. The polynomial $f(x)$ is **irreducible** in $\mathbb{K}[x]$ if it cannot be written as a product of two polynomials in $\mathbb{K}[x]$ of lower degree. For example, $x^2 - 2$ is irreducible in $\mathbb{Q}[x]$ or $\mathbb{Z}[x]$ but can be factorized in $\mathbb{R}[x]$ or $\mathbb{C}[x]$ as $(x - \sqrt{2})(x + \sqrt{2})$. Even in $\mathbb{Z}_2[x]$, it is not irreducible, because here $x^2 - 2 = x^2 = x \cdot x$, and in $\mathbb{Z}_7[x]$ we have $x^2 - 2 = (x + 3)(x + 4)$. But it is still irreducible over $\mathbb{Z}_3$.

There is an analogue of prime number factorization: every monic polynomial can be uniquely (up to order) written as a product of monic irreducible polynomials assuming we can divide in $\mathbb{K}$. But in $\mathbb{Z}_4[x]$ this is wrong: $(x + 2)(x + 2) = x^2 = x \cdot x$ are two different factorizations. Even worse: $(2x + 1)(2x + 1) = 1$ here. Thus the existence of division is important.

For complex polynomials the irreducible polynomials are non-zero constants and polynomials of degree one. Thus if $f(x)$ is a complex polynomial of degree n then

$$f(x) = a_0(x - x_1)(x - x_2)\cdots(x - x_n)$$

for some complex numbers x_i. Each x_i is a root of the equation $f(x) = 0$ and vice versa every root of $f(x) = 0$ is one of the x_i. Some x_i might be equal: if x_i appears k times we say that the root is of **multiplicity** k. Equivalently: $f(x) = (x - x_i)^k g(x)$ where $g(x_i) \neq 0$.

If $f(x)$ is a real polynomial then for each complex root x_i its conjugate $\overline{x_i}$ is a root as well (of the same multiplicity as x_i) and because $(x - x_i)(x - \overline{x_i}) =$

$x^2 - 2(\mathrm{Re}\, x_i)x + |x_i|^2$, every real polynomial can be factorized in real factors of degree 1 or 2. In particular, every polynomial of odd degree has a real root. The irreducible polynomials in $\mathbb{R}[x]$ are the non-zero polynomials of degree less then two and those polynomials of degree two that have no real roots. Recall that $f(x) = ax^2 + bx + c$ with $a \neq 0$ has no real roots iff the **discriminant** $b^2 - 4ac$ is negative.

For polynomials in $\mathbb{Q}[x]$ and $\mathbb{Z}_p[x]$, the situation is much more complicated. There exist irreducible polynomials of arbitrary degree n and there are no simple methods to decide whether a given polynomial is irreducible. For polynomials with integer coefficients, there are some useful partial results. First of all we know how to decide whether $f(x)$ has a linear factor with rational coefficients.

Theorem 3.3 *Let $f(x) = a_0x^n + a_1x^{n-1} + \cdots + a_{n-1}x + a_n \in \mathbb{Z}[x]$ be a polynomial of degree n. If p, q are relatively prime integers and $f\left(\frac{p}{q}\right) = 0$ then $p \mid a_n$ and $q \mid a_0$. In particular, if $f(x)$ is monic then all its rational roots are integers and can be found among the divisors of a_n.*

Proof. We have, after multiplication by q^n, that

$$a_0p^n + a_1p^{n-1}q + \cdots + a_{n-1}pq^{n-1} + a_nq^n = 0.$$

Hence p must divide a_nq^n and therefore a_n. Similarly, q divides a_0. □

Let us look at a very different approach to proving irreducibility. The idea is to consider polynomials $f(x)$ modulo some prime, and thus to work in $\mathbb{Z}_p[x]$ instead of $\mathbb{Z}[x]$. Note that the degree of the obtained polynomial $\overline{f(x)}$ may be less than the degree of the $f(x)$ if $p \mid a_0$, but is otherwise the same, for example if $f(x)$ is monic. If we can prove that $\overline{f(x)}$ is a non-constant irreducible polynomial in $\mathbb{Z}_p[x]$, it follows that $f(x)$ is irreducible in $\mathbb{Z}[x]$ as well. For example, $x^3 + 2x^2 + 7x + 13$ is irreducible in $\mathbb{Z}[x]$ because modulo 2 we have that $\overline{f(x)} = x^3 + x + 1$ is irreducible in $\mathbb{Z}_2[x]$ (if not, $\overline{f(x)}$ would have a linear factor and a root, but $\overline{f(0)} = \overline{f(1)} = 1$). Unfortunately this idea does not work that often and we need deeper ideas to prove the irreducibility of the integer polynomial $f(x)$.

Denote by $\mathrm{cont}(f(x))$ the greatest common divisor of all coefficients a_i. The following statement is known as **Gauss's Lemma**.

Theorem 3.4 *If f and g are polynomials in $\mathbb{Z}[x]$, then* $\mathrm{cont}(f(x)g(x)) = \mathrm{cont}(f(x)) \cdot \mathrm{cont}(g(x))$.

Proof. Suppose first that $\mathrm{cont}(f(x)) = \mathrm{cont}(g(x)) = 1$. We need to prove that $\mathrm{cont}(f(x)g(x)) = 1$. Suppose the opposite, that some prime p divides all the coefficients in this product. Then modulo p we have $f(x)g(x) = 0$. But $f(x)$ and $g(x)$ are non-zero modulo p and in $\mathbb{Z}_p[x]$ the degree of the product $f(x)g(x)$ is the sum of degrees and we have a contradiction.

For the general case, we write $f(x) = \mathrm{cont}(f(x))f_1(x)$, and $g(x) = \mathrm{cont}(g(x))g_1(x)$ the result follows from applying the above to $f_1(x)$ and $g_1(x)$. □

Gauss's Lemma has a very useful corollary (which is also known as Gauss's lemma).

Theorem 3.5 *A polynomial $f(x)$ with integer coefficients is irreducible in $\mathbb{Z}[x]$ if and only if it is irreducible in $\mathbb{Q}[x]$.*

Proof. Suppose that $f(x)$ is irreducible in $\mathbb{Z}[x]$ but $f(x) = f_1(x)f_2(x)$, where $f_i(x) \in \mathbb{Q}[x]$. Multiplying this equality by a large integer, we get $mf(x) = g_1(x)g_2(x)$, where $\deg g_1(x) = \deg f_i(x)$, and $g_i(x)$ have integer coefficients. Then, according to Gauss's lemma, $m \cdot \mathrm{cont}(f(x)) = \mathrm{cont}(g_1(x)) \cdot \mathrm{cont}(g_2(x))$. Thus $m = m_1m_2$, and $h_i(x) = \frac{g_i(x)}{m_i}$ have integer coefficients. We get $f(x) = h_1(x)h_2(x)$ and therefore one of $h_i(x)$ must have the same degree as $f(x)$. The same is true for $f_i(x)$ which proves that $f(x)$ is irreducible over $\mathbb{Q}[x]$. The converse implication is obvious. □

Now we can formulate a sufficient condition for $f(x)$ to be irreducible in $\mathbb{Z}[x]$.

Theorem 3.6 (Eisenstein's criterion) *Let $f(x) = a_0x^n + a_1x^{n-1} + \cdots + a_{n-1}x + a_n \in \mathbb{Z}[x]$ and p be a prime. If $p \mid a_i$ for $i > 0$ but a_0 is not divisible by p and a_n is not divisible by p^2 then $f(x)$ is irreducible in $\mathbb{Q}[x]$ (and hence in $\mathbb{Z}[x]$).*

Proof. If $f(x) = f_1(x)f_2(x)$ then $a_n = f(0) = f_1(0)f_2(0)$ and WLOG p does not divide $f_1(0)$. Modulo p we have $f(x) = a_0x^n$, and thus $f_1(x) = bx^k$ modulo p. But $f_1(0) \neq 0 \pmod{p}$ and the only choice is $k = 0$, thus $\deg f_2(x) = n$ modulo p and $\deg f_2(x) = n$ as an integer polynomial, which proves the irreducibility. □

For example, the polynomial $x^7 + 10x^3 + 5$ is irreducible over integers (take $p = 5$).

Problem 3.7 Show that for every prime p, the **cyclotomic polynomial** $f(x) = x^{p-1} + x^{p-2} + \cdots + x + 1$ is irreducible.

Search. We do not know many ways to prove irreducibility. Perhaps consider $f(x)$ modulo p? Here we get

$$f(x) = \frac{x^p - 1}{x - 1} = \frac{x^p - 1^p}{x - 1} = \frac{(x-1)^p}{x - 1} = (x-1)^{p-1} \pmod{p}.$$

Yes, we find something interesting, for example that $\binom{p-1}{k} = (-1)^k \pmod{p}$ but there is no chance to prove the irreducibility in this way.

The prime p suggest that Eisenstein's criterion could be useful. But all the coefficients are ones and we do not see how to apply it. So we are in trouble, but perhaps we can find a change of variables that helps. If we put $x = y + 1$, we have

$$\begin{aligned} f(x) &= \frac{x^p - 1}{x - 1} = \frac{(y+1)^p - 1}{y} \\ &= y^{p-1} + \binom{p}{1} y^{p-2} + \cdots + \binom{p}{p-2} y + \binom{p}{p-1}, \end{aligned}$$

which fits Eisenstein's criterion perfectly. □

The last criterion we would like to discuss is rather unexpected.

Theorem 3.8 *Let p be a prime written as $p = a_0 b^n + a_1 b^{n-1} + \cdots + a_n$ in base $b > 1$. Then the polynomial $p(x) = a_0 x^n + a_1 x_{n-1} + \cdots + a_n$ is irreducible over integers.*

For example, $113 = 10^2 + 10 + 3$ is a prime, and thus $x^2 + x + 3$ is irreducible. The proof and useful generalizations can be found in [39]. The obvious idea that a polynomial factorization $p(x) = p_1(x)p_2(x)$ implies $p = p_1(b)p_2(b)$ is not enough, because $p_i(b) = \pm 1$ might be possible.

A nice example is the polynomial $p(x) = x^4 + 6x^2 + 1$. Because all the coefficients are less than 8 and the number $p(8) = 8^4 + 6 \cdot 8^2 + 1 = 4481$ is a prime, we can conclude that $p(x)$ is irreducible. What is interesting is that modulo any prime p, the polynomial is not irreducible in $\mathbb{Z}_p[x]$ (try to prove this!)

Let us return to polynomials over $\mathbb{R}$. When are two polynomials in $\mathbb{R}[x]$ equal? If two monic polynomials of degree n agree in n different points then they are equal (because their difference has degree $< n$ and has n different roots). If they are not monic and have degree $\leq n$ we need to know that they agree in $n+1$ different points to have equality.

The following theorem is not well known, but can sometimes be useful.

Theorem 3.9 *A polynomial $p(x)$ of degree $\leq n$ is integer valued for every integer x if and only if $p(x)$ can we written as*

$$p(x) = \sum_{k=0}^{n} c_k \binom{x}{k}$$

for some integers c_k. The presentation is unique.

Proof. First of all, we have seen on page 98 that the polynomial $\binom{x}{k}$ is an integer for every $x \in \mathbb{Z}$ (not just for $x \in \mathbb{N}$). We try to construct coefficients c_m in such way that both polynomials have the same values for $x = 0, 1, \ldots, n$. We do this by induction. For $m = 0$, we take $c_0 = p(0)$. If $c_0, \ldots, c_{m-1}$ are already constructed we write the equation

$$p(m) = \sum_{k=0}^{n} c_k \binom{m}{k}$$

and use that $\binom{m}{k} = 0$ for $k > m$ and $\binom{m}{m} = 1$ to determine c_m. With $m = n$ we get two polynomials of degree $\leq n$ that coincide in $n+1$ points, so they must be equal. □

For example, $\frac{x^3-x}{3} = 2\binom{x}{3} + 2\binom{x}{2}$. If we take the products as $\binom{x}{k}\binom{y}{l}$ we can generalize the theorem to several variables.

We also know how to discover the multiple roots: If $f(x) = (x-a)^k g(x)$ with $k > 1$, then $f'(x) = (x-a)^{k-1}(kg(x) + (x-a)g'(x))$. Thus $f(x)$ has (a complex) multiple root iff $d(x) = (f(x), f'(x)) \neq 1$. Moreover, $f(x)/d(x)$ has only simple roots.

Between any two different real roots of $f(x)$, the derivative $f'(x)$ has a real root (**Rolle's Theorem**). For complex roots of $f'(x)$ we can only say that they lie in the convex hull of the complex roots of $f(x)$ (the **Gauss–Lucas Theorem**). Both these facts are less trivial.

3.3 Symmetric polynomials

A polynomial $p(x_1, x_2, \ldots, x_n)$ is called **symmetric** if switching any two variables does not change the value of the polynomial. For example $p(x, y) = x^3 + y^3 + xy$ is a symmetric polynomial, since $p(x, y) = p(y, x)$. Likewise $q(x, y, x) = x + y + z$ is symmetric, whereas $r(x, y) = x - y$ is not. Symmetric polynomials tend to show up very often, and they have many useful and beautiful properties.

One of the most important application of symmetric polynomials has to do with the connection between the roots of a polynomial and its coefficients. Let us start a little more abstractly.

Let t be a formal variable and expand

$$(t + x_1)(t + x_2)\cdots(t + x_n) = t^n + s_1 t^{n-1} + \cdots + s_n.$$

For example if $n = 3$,

$$s_1 = x_1 + x_2 + x_3, \quad s_2 = x_1x_2 + x_1x_3 + x_2x_3, \quad s_3 = x_1x_2x_3.$$

The polynomials $s_i = s_i(x_1, \ldots, x_n)$ are so-called **elementary symmetric polynomials.**

Their main property is that every symmetric polynomial $f(x_1, \ldots, x_n)$ can be uniquely expressed using s_i, for example

$$x_1^2 + x_2^2 + x_3^2 = (x_1 + x_2 + x_3)^2 - 2(x_1x_2 + x_1x_3 + x_2x_3) = s_1^2 - 2s_2.$$

Note that if x_i are roots of the equation $x^n + a_1x^{n-1} + \cdots + a_{n-1}x + a_n = 0$, then $a_i = (-1)^i s_i$. In other words, the coefficients in a monic polynomial are (up to a sign) the elementary symmetric polynomials of its roots (**Vieta's theorem**). Thus all symmetric polynomials in roots of a polynomial can be expressed in its coefficients. This observation lies at the heart of showing that general polynomial equations in degree five and higher cannot be solved in radicals.

To test how straight-forward (but boring) this approach is, consider the following question: when does a cubic polynomial $x^3 + a_1x^2 + a_2x + a_3$ have a multiple root? If x_1, x_2, x_3 are the roots then we need $D = (x_1 - x_2)(x_1 - x_3)(x_2 - x_3) = 0$. Unfortunately, this expression as a polynomial in x_1, x_2, x_3 is not symmetric: if we exchange x_1 and x_2 then D changes sign. But D^2 is symmetric and we only need to express it in s_1, s_2, s_3. This is a good opportunity to show how to do such things. The idea is quite easy – choose the first term in the lexicographical order and try to kill it using the products of s_i.

Continue in this manner until you get zero. The first term in D is $x_1^2 x_2$, and thus we need to construct a symmetric polynomial with the first term $x_1^4 x_2^2$ and $s_1^2 s_2^2$ seems like a good candidate. We have, after some long calculations:

$$\begin{aligned} D^2 - s_1^2 s_2^2 = (D - s_1^2 s_2)(D + s_1^2 s_2) \\ = -4\, x_1^4 x_2 x_3 - 4\, x_1^3 x_2^3 - 6\, x_1^3 x_2^2 x_3 - 6\, x_1^3 x_2 x_3^2 - 4\, x_1^3 x_3^3 \\ - 6\, x_1^2 x_2^3 x_3 - 21\, x_1^2 x_2^2 x_3^2 - 6\, x_1^2 x_2 x_3^3 - 4\, x_1 x_2^4 x_3 \\ - 6\, x_1 x_2^3 x_3^2 - 6\, x_1 x_2^2 x_3^3 - 4\, x_1 x_2 x_3^4 - 4\, x_2^3 x_3^3. \end{aligned}$$

Now we need to kill $x_1^4 x_2 x_3$ and use $s_1^3 s_3$:

$$\begin{aligned} D^2 - s_1^2 s_2^2 + 4 s_1^3 s_3 = -4\, x_1^3 x_2^3 + 6\, x_1^3 x_2^2 x_3 + 6\, x_1^3 x_2 x_3^2 - 4\, x_1^3 x_3^3 \\ + 6\, x_1^2 x_2^3 x_3 + 3\, x_1^2 x_2^2 x_3^2 + 6\, x_1^2 x_2 x_3^3 \\ + 6\, x_1 x_2^3 x_3^2 + 6\, x_1 x_2^2 x_3^3 - 4\, x_2^3 x_3^3. \end{aligned}$$

We continue in the same manner:

$$\begin{aligned} D^2 - s_1^2 s_2^2 + 4 s_1^3 s_3 + 4 s_2^3 = \\ 18\, x_1^3 x_2^2 x_3 + 18\, x_1^3 x_2 x_3^2 + 18\, x_1^2 x_2^3 x_3 + 27\, x_1^2 x_2^2 x_3^2 \\ + 18\, x_1^2 x_2 x_3^3 + 18\, x_1 x_2^3 x_3^2 + 18\, x_1 x_2^2 x_3^3. \end{aligned}$$

and again:

$$D^2 - s_1^2 s_2^2 + 4 s_1^3 s_3 + 4 s_2^3 - 18 s_1 s_2 s_3 = -27 x_1^2 x_2^2 x_3^2.$$

Now we see that

$$D^2 - s_1^2 s_2^2 + 4 s_1^3 s_3 + 4 s_2^3 - 18 s_1 s_2 s_3 + 27 s_3^2 = 0$$

and therefore

$$a_1^2 a_2^2 - 4 a_1^3 a_3 - 4 a_2^3 + 18 a_1 a_2 a_3 - 27 a_3^2 = 0$$

is our condition.

Quite boring work, and a rather unpredictable result that we prefer to forget immediately. It is much easier to check whether $(f(x), f'(x)) \neq 1$ in concrete examples. But there is another important application: The expression in the left-hand side of the last equation is called the **discriminant** of our cubic polynomial. If all a_i are real then the discriminant is positive if and only if the polynomial has three real roots. Indeed, if all three roots are real and different then obviously $D^2 > 0$. If two of them are complex and one (say x_3) is real then they $x_2 = \overline{x_1}$ and we see directly that $\overline{D} = -D$, which means that $D = ic$ for some real c.

This is important to remember when we translate a symmetric problem into the language of symmetric polynomials. For example, if $x_1, x_2, x_3 \geq 0$ then to get the equivalent condition we need to demand

$$s_1 \geq 0, \quad s_2 \geq 0, \quad s_3 \geq 0,$$
$$s_1^2 s_2^2 - 4s_1^3 s_3 - 4s_2^3 + 18s_1 s_2 s_3 - 27s_3^2 \geq 0,$$

which is not so pleasant and also is difficult to remember. Normally we replace the last inequality by weaker but more convenient conditions such as Maclaurin's and Schur's inequalities — see later.

Consider the following problem.

Problem 3.10 Let $f(x) = x^n + a_1 x^{n-1} + \cdots + a_{n-1}x + 1$ be a polynomial with non-negative coefficients having n real roots. Prove that $f(2) \geq 3^n$.

Search. We start by trying to understand the conditions. First of all, it is evident that we cannot have any non-negative roots because $f(x) > 0$ for non-negative x. Hence, all the roots are negative and to minimize the number of minus signs, we write the roots as $-x_i$ with $x_i > 0$. Now we have $f(x) = (x + x_1)\cdots(x + x_n)$, and thus $a_i = s_i$.

What else is there? The essential condition is that $a_n = 1$, i.e. $x_1 x_2 \cdots x_n = 1$. This should be used, but how? Let us look at $a_1 = s_1 = x_1 + x_2 + \cdots + x_n$. We can definitely use AM–GM inequality to conclude that $s_1 \geq n\sqrt{x_1 x_2 \cdots x_n} = n$. Good. What about a_2? There are $\binom{n}{2}$ terms $x_i x_j$ in s_2 and if we apply AM–GM to them we get $s_2 \geq \binom{n}{2}$. Very good. In a similar way (check!) we get $s_k \geq \binom{n}{k}$. Collecting these inequalities together, we find that for positive x

$$\begin{aligned} f(x) &= x^n + a_1 x^{n-1} + \cdots + a_{n-1}x + 1 \\ &\geq x^n + nx^{n-1} + \binom{n}{2}x^{n-2} + \cdots + \binom{n}{n-2}x^2 + \binom{n}{n-1}x + 1 \\ &= (x+1)^n. \end{aligned}$$

For $x = 2$ we get the result in the problem. □

More generally, we have **Maclaurin's inequality** which states that if $x_i \geq 0$ then

$$S_1 \geq S_2^{\frac{1}{2}} \geq \cdots S_k^{\frac{1}{k}} \cdots \geq S_n^{\frac{1}{n}},$$

where $S_k = \frac{s_k}{\binom{n}{k}}$ are the means. For example, for $n = 3$ we get:

$$\frac{s_1}{3} \geq \sqrt{\frac{s_2}{3}} \geq \sqrt[3]{s_3}.$$

It is important to use Maclaurin's inequality when we try to prove a symmetric inequality by rewriting it in terms of s_i. For example, if we want to prove

$$x_1^2 + x_2^2 + x_3^2 \geq x_1x_2 + x_1x_3 + x_2x_3,$$

then it is sufficient to consider the case $x_i \geq 0$ and one possible way is to rewrite this as

$$s_1^2 - 2s_2 \geq s_2 \quad \Leftrightarrow \quad s_1^2 \geq 3s_2$$

which follows from Maclaurin's inequality.

In two variables, Maclaurin's inequality $s_1^2 \geq 4s_2$ is equivalent to the condition in the original problem, but with a larger number of variables, more conditions are necessary to retain equivalence.

For example, as we will see later (see Theorem 3.15), **Schur's inequality**

$$x_1^r(x_1 - x_2)(x_1 - x_3) + x_2^r(x_2 - x_1)(x_2 - x_3) + x_3^r(x_3 - x_2)(x_3 - x_1) \geq 0$$

is valid for all positive r and non-negative x_i. This strong inequality is symmetric, but does not follow from Maclaurin's inequality. For example, for $r = 1$, Schur's inequality says that

$$s_1^3 + 9s_3 \geq 4s_1s_2.$$

Maclaurin's inequality can easily be deduced from lemma used in the proof of **Newton's inequality** $S_{k-1}S_{k+1} \leq S_k^2$ — see [66] for details.

3.4 Summing up

Many students are afraid of the sum and product signs. Others make the opposite mistake and use those symbols in every possible and mostly unnecessary situation, hoping that this makes their work looks more "professional". Let us try to understand how to use these symbols correctly. In most situations, we understand a sum much better when we write it as

$$a_1 + a_2 + \cdots + a_n.$$

This notation has two disadvantages, however. First, it is formally incorrect when $n = 1$, because in this case, there is no a_2 term. In practice, we ignore this formality. Another disadvantage is that the notation is too verbose. It requires us to write too much and takes too much space in the book. Both disadvantages disappear when we write the same sum as $\sum_{i=1}^{n} a_i$, and thus a good way to understand this notation is as an abbreviation of the long notation. Similarly $\prod_{i=1}^{n} a_i$ is an abbreviation for a long product. We can always rewrite the sum in the usual long form and quite often this is a good idea. But if we decide to use these symbols, we should understand how to do this correctly and to learn other advantages of this notation.

First of all, it does not matter which variable we use for the summation

$$\sum_{i=1}^{n} a_i = \sum_{j=1}^{n} a_j = \sum_{æ=1}^{n} a_{æ}$$

is the same sum as above. The only restriction is that we cannot use the variables that are "occupied". For example $\sum_{a=1}^{n} a_a$ doesn't make any sense. On the other hand we can use a change of variables and simultaneously change the lower and upper bounds. As an example

$$\sum_{k=-1}^{n-2} a_{k+2} = \sum_{k=2}^{n+1} a_{k-1}$$

denotes the same sum even though k plays a different role in each sum. Such substitutions are used quite often without any explanations and students have problems with (and hate) transformations like

$$\sum_{i=1}^{n} i^2 = \sum_{i=1}^{n+1} (i-1)^2,$$

because in fact they should be read as follows. First, because $0^2 = 0$ we have $\sum_{i=1}^{n} i^2 = \sum_{i=0}^{n} i^2$ (we change nothing adding the zero term). Second, we can make the substitution $j = i + 1 \Leftrightarrow i = j - 1$. When $i = 0$ we have $j = 1$ and when $i = n$ we have $j = n + 1$. Thus the same sum is now $\sum_{j=1}^{n+1} (j-1)^2$. At last, because the name of the variable does not matter we can change it to i back and get $\sum_{i=1}^{n+1} (i-1)^2$. But now it is not the same i that was in the beginning!

Because we are lazy, we often write our sums as $\sum_i a_i$ or simply as $\sum a_i$ and trust that the reader can recover details such as lower and upper bounds or the name of the summation variable from the context. This laziness is especially handy when we want to concentrate on the main actor in the sum — the term a_i. The laziest variant of them all (used often in tensor theory and physics) is where one skips the actual $\sum$ as well, with the convention that we sum over all the indices which appear twice (e.g. a_i^i is the same as $\sum_{i=1}^n a_i^i$). This notation (known as **Einstein's summation convention**) has some advantages but we will not use it.

We will sometimes skip the explicitly written bounds when we know them exactly. This also gives us the option to write more complicated summation conditions instead of bounds, like $\sum\limits_{0<i^2<100} a_i$, which is, in fact, the same as $\sum_{i=1}^9 a_i$.

Here are several variants of summation that are used quite often.

$$\sum_{1\le j\le i\le n} a_{ij} = \sum_{i=1}^{n}\sum_{j=1}^{i} a_{ij}.$$

Note that the upper bound of the inner sum depends on the index i in the first sum. The notation

$$\sum_{\text{sym}} x_1^{a_1}x_2^{a_2}\cdots x_n^{a_n}$$

is quite special, but convenient when we are working with symmetric polynomials. This means that we take the sum over all the possible permutations of variables $x_1,\ldots,x_n$. Because there are $n!$ such permutations we save a lot of space. For example,

$$\begin{aligned}\sum_{\text{sym}} x_1^2x_2x_3 &= x_1^2x_2x_3 + x_1^2x_3x_2 + x_2^2x_1x_3 + x_2^2x_3x_1 + x_3^2x_1x_2 + x_3^2x_2x_1\\ &= 2(x_1^2x_2x_3 + x_2^2x_1x_3 + x_3^2x_1x_2).\end{aligned}$$

Similarly

$$\sum_{\text{cyclic}} x_1^{a_1}x_2^{a_2}\cdots x_n^{a_n}$$

means that we sum over the n cyclic permutations. For example,

$$\sum_{\text{cyclic}} x_1^3x_2^2x_3 = x_1^3x_2^2x_3 + x_2^3x_3^2x_1 + x_3^3x_1^2x_2.$$

Note that this is *not* a symmetric polynomial.

We can use the same notations without indices, e.g.

$$\sum_{\text{sym}} x^2y^2z = 2(x^2y^2z + y^2z^2x + z^2x^2y) = 2\sum_{\text{cyclic}} x^2y^2z.$$

When working with prime numbers it is convenient to use the products as

$$a = \prod_{p \text{ prime}} p^{a_p},$$

where we mean that the prime p has the power a_p in the prime factorization of a. The product is taken over all primes, though in fact we need only finitely many of them. Sometimes we skip the word "prime", meaning that p is a sufficiently good notation for primes. But this may be confusing.

One of the greatest advantages of the $\sum$ symbol is that it simplifies changing the order of summation. In the simplest form this looks as follows. Suppose we have an expression $f(i, j)$ depending on two variables i, j. Then the following sums are equal:

$$\sum_{i=1}^{n}\sum_{j=1}^{m} f(i,j) = \sum_{j=1}^{m}\sum_{i=1}^{n} f(i,j),$$

because both are the sum over all points (i, j) in a rectangular $1 \leq i \leq n$, $j = 1 \leq m$.

This interchange is used much more often when the region of the summation is more complicated. For triangles it is still fairly obvious (but draw a picture in the (i, j)-plane to make sure):

$$\sum_{i=1}^{n}\sum_{j=1}^{i} f(i,j) = \sum_{j=1}^{n}\sum_{i=j}^{n} f(i,j).$$

Problem 3.11 If n and k are integers with $0 \leq k \leq n$, prove that

$$\binom{2n}{n+k} = \sum_{j} 2^{n-k-2j}\binom{n}{j}\binom{n-j}{j+k}.$$

Search. The binomial coefficient $\binom{2n}{n+k}$ can be found in the binomial expansion of $(1+x)^{2n}$ as the coefficient of x^{n+k}.

What if we try to rewrite $(1+x)^{2n}$ in another way to get something closer to the right-hand side of the identity we want to prove? Because we have $n-k$ here instead of $n+k$ perhaps it might be a good idea to look for the coefficient of x^{n-k}, using $\binom{2n}{n+k} = \binom{2n}{n-k}$. We have

$$(1+x)^{2n} = \left(x^2 + (1+2x)\right)^n = \sum_j \binom{n}{j} x^{2j}(1+2x)^{n-j}$$
$$= \sum_j \binom{n}{j} x^{2j} \left(\sum_i \binom{n-j}{i} 2^i x^i \right) = \sum_i \sum_j \binom{n}{j} x^{2j} \binom{n-j}{i} 2^i x^i$$

To find the coefficient of x^{n-k} we put $n-k = 2j+i \Rightarrow i = n-k-2j$ and the coefficient is

$$\sum_j \binom{n}{j} \binom{n-j}{n-k-2j} 2^{n-k-2j} = \sum_j \binom{n}{j} \binom{n-j}{j+k} 2^{n-k-2j},$$

which proves the identity. Note that we have skipped all the lower and upper bounds. We recommend the reader to recover them to make sure that this was really harmless.

It is always a good idea to find a combinatorial interpretation of the identity. Here we can look at the number of ways to select $n-k$ people from n couples. First of all it is exactly $\binom{2n}{n-k} = \binom{2n}{n+k}$. On the other hand, for each such selection we count the number j of complete couples selected. There are $\binom{n}{j}$ ways to choose these couples. The rest of the selected group consists of $n-k-2j$ people without their spouses. There are $\binom{n-j}{n-k-2j} = \binom{n-j}{j+k}$ ways to choose the couples from which these others are selected and 2^{n-k-2j} ways to choose one member from each such couple. Summing over j completes the count. □

3.5 Inequalities: basic properties.

Many students have difficulties with inequalities even at the very basic level. To help them, we have written this section and if the reader feels that the text is too elementary, s|he can skip forward to the next section, which is much more advanced. But for some students this section is important to read.

To understand inequalities, we start from an easy question: what does it mean that $a \geq b$ and why is this the same as $b \leq a$?

The answer is simple — *by definition* both inequalities mean that $a - b$ is a non-negative number. In the same way, $a > b$ is equivalent to $b < a$ and means that $a - b$ is positive. Note that this is a definition and all other properties of inequalities follow from this definition and the properties of numbers. For example,

$$a \geq b, c \geq d \quad \Rightarrow \quad a + c \geq b + d,$$

because

$$(a + c) - (b + d) = (a - b) + (c - d) \geq 0$$

as the sum of two non-negative numbers $a - b$ and $c - d$.

On the other hand, the difference between two non-negative numbers may have any sign. That is why

$$a \geq b, c \geq d \quad \not\Rightarrow \quad a - c \geq b - d$$

(the implication is wrong). Nevertheless the implication

$$a \geq b, c \geq d \quad \Rightarrow \quad a - d \geq b - c$$

is valid because it is again equivalent to $(a - b) + (c - d) \geq 0$. Using this approach and being careful with how we handle zero, we can easily understand which transformations are allowed. We will show some examples:

$$a \geq b \quad \not\Rightarrow \quad a^2 \geq b^2,$$

but

$$a \geq b \geq 0 \quad \Rightarrow \quad a^2 \geq b^2.$$

The reason is that $a^2 - b^2 = (a - b)(a + b)$ and thus we need $a + b \geq 0$.

The **transitivity law**:

$$a \geq b, b \geq c \quad \Rightarrow \quad a \geq c$$

follows from $a - c = (a - b) + (b - c) \geq 0$. Similarly,

$$a \geq b \quad \not\Rightarrow \quad \frac{1}{b} \geq \frac{1}{a},$$

for example because a and b can have different signs, but

$$a \geq b > 0 \quad \Rightarrow \quad \frac{1}{b} \geq \frac{1}{a}$$

because $\dfrac{1}{b} - \dfrac{1}{a} = \dfrac{a - b}{ab} \geq 0$. Finally,

$$a \geq b, c \geq d \quad \not\Rightarrow \quad ac \geq bd$$

(for example for $a = b = d = -1, c = 0$), but

$$a \geq b \geq 0, c \geq d \geq 0 \quad \Rightarrow \quad ac \geq bd.$$

To see this, write $a = b + x, c = d + y$ with $x, y \geq 0$. Then we get:

$$ac - bd = (b + x)(d + y) - bd = by + dx + xy \geq 0$$

as the sum of non-negative numbers. If the reader understood the idea s|he can probably skip Table 3.1, but we prefer to have a fairly complete list of possible and impossible inequality transformations collected in one place. We will write the transformations for non-strict inequalities only. The reader may replace the sign $\geq$ by $>$ almost everywhere.

Trigonometric inequalities are more complicated and are also rarely used. The idea is to solve the inequality on the circle and then to add the period. We restrict ourselves to one example only: $\sin x > a$ is equivalent to

$x \in \mathbb{R}$	if $a < -1$
$x \neq \frac{-\pi}{2} + \pi n, n \in \mathbb{Z}$	if $a = -1$
$\arcsin a + 2\pi n < x < \pi - \arcsin a + 2\pi n, n \in \mathbb{Z}$	if $-1 < a < 1$
no solution	if $a \geq 1$

Now let us consider some basic remarks about equivalent and non-equivalent transformations and methods for solving inequalities.

It is clear that every equation or inequality describes a set, namely the set of all combinations of variables that satisfy this equation or inequality. For example, the equation $x^2 = 1$ in one (real) variable x describes the set of two numbers: $\{1, -1\}$, but the inequality $x^2 > 0$ in the same variable describes the set of all non-zero numbers.

Note that it is important to know exactly what the variables are, since they are not necessarily present in the expression. For example, the same equation $x^2 = 1$, considered as an equation in two variables x, y describes the infinite set

$$\{(1, t), (-1, t) \mid t \in \mathbb{R}\},$$

because it can be rewritten as $x^2 + 0 \cdot y = 1$. In most situations we do not mention the set of variables, in which case they are taken to be all the explicitly written variables.

Table 3.1 Possible and impossible transformations of inequalities. In rows marked with *, there is a difference between ≥ and >.

Additive equivalences	$a \geq b$	$\Leftrightarrow$	$b \leq a$	$\Leftrightarrow$	$a \nless b$	
	$a \geq b$	$\Leftrightarrow$	$-b \geq -a$	$\Leftrightarrow$	$a - b \geq 0$	
	$a \geq b$	$\Leftrightarrow$	$a + c \geq b + c$	$\Leftrightarrow$	$a - c \geq b - c$	
but	$a \geq b$	$\not\Rightarrow$	$ac \geq bc$			
Additive implications	$a \geq b$	and	$b \geq c$	$\Rightarrow$	$a \geq c$	
	$a \geq b$	and	$c \geq d$	$\Rightarrow$	$a + c \geq b + d$	
	$a \geq b$	and	$c \geq d$	$\Rightarrow$	$a - d \geq b - c$	
but	$a \geq b$	and	$c \geq d$	$\not\Rightarrow$	$a - c \geq b - d$	
	$a \geq b$	and	$c \geq d$	$\not\Rightarrow$	$ac \geq bd$	
Multiplicative equivalences	$a \geq b > 0$	$\Leftrightarrow$	$1/b \geq 1/a > 0$	$\Leftrightarrow$	$a/b \geq 1, b > 0$	
(k odd positive integer)	$a \geq b$	$\Leftrightarrow$	$a^3 \geq b^3$	$\Leftrightarrow$	$a^k \geq b^k$	
(for $c > 0$)	$a \geq b$	$\Leftrightarrow$	$ac \geq bc$	$\Leftrightarrow$	$a/c \geq b/c$	
(for $a, b, c > 0$)	$a \geq b$	$\Leftrightarrow$	$a^c \geq b^c$	$\Leftrightarrow$	$b^{-c} \geq a^{-c}$	
but	$a \geq b$	$\not\Rightarrow$	$a^2 \geq b^2$			
	$a \geq b$	$\not\Rightarrow$	$1/b \geq 1/a$			
Multiplicative implications	$a \geq b$	and	$c \geq 0$	$\Rightarrow$	$ac \geq bc$	*
	$a > b$	and	$c \geq 0$	$\Rightarrow$	$ac \geq bc$	*
	$a \geq b \geq 0$	and	$c \geq d \geq 0$	$\Rightarrow$	$ac \geq bd$	*
	$a \geq b > 0$	and	$c \geq d > 0$	$\Rightarrow$	$a/d \geq b/c$	*
but	$a > b$	and	$c \geq d$	$\not\Rightarrow$	$ac > bd$	
Powers and logarithms ($c > 0, c \neq 1$)	$c^a \geq c^b$	$\Leftrightarrow$	$(c-1)(a-b) \geq 0$			
	$\log_c a \geq \log_c b$	$\Leftrightarrow$	$a, b > 0, (c-1)(a-b) \geq 0$			
Absolute value equivalences	$\lvert a \rvert \leq b$	$\Leftrightarrow$	$b \geq 0, a \leq b, a \geq -b$			
	$\lvert a \rvert \geq b$	$\Leftrightarrow$	$b < 0$ **or** $b \geq 0, a \geq b$ **or** $b \geq 0, a \leq -b$			
	$\lvert a \rvert \geq \lvert b \rvert$	$\Leftrightarrow$	$a^2 \geq b^2$	$\Leftrightarrow$	$(a-b)(a+b) \geq 0$	
Square root equivalences	$\sqrt{a} \leq b$	$\Leftrightarrow$	$a \geq 0, a \leq b^2, b \geq 0$			
	$\sqrt{a} \geq b$	$\Leftrightarrow$	$a \geq 0, b < 0$ **or** $a \geq b^2, b \geq 0$			
	$\sqrt{a} \geq \sqrt{b}$	$\Leftrightarrow$	$a \geq b, b \geq 0$			

Expressions such as $3 \geq 2$ are problematic. What is the set that this inequality describes? The answer is evident: the set of all possible combinations of variables. For example if we have only one variable x, then the set consists of all real numbers, but if we have two variables x, y the solution set is different:

$$\{(x, y) \mid x, y \in \mathbb{R}\}.$$

On the other hand an inequality like $2 > 3$ always describes the empty set, regardless of the set of variables!

Because inequalities and equations correspond to sets we can define operations that correspond to familiar operations on sets.

Suppose for example that we have two inequalities: $f(x) \geq 0$ and $g(x) \geq 0$. They describe two sets, say X_f and X_g. The intersection of these two sets $X_f \cap X_g$ corresponds to the system:

$$\begin{cases} f(x) \geq 0 \\ g(x) \geq 0, \end{cases}$$

where both conditions have to be satisfied. The union of these sets $X_f \cup X_g$ we denote in another way:

$$\left[\begin{array}{l} f(x) \geq 0 \\ g(x) \geq 0, \end{array}\right.$$

meaning that at least one of them should be valid, but not necessarily both. To underline that there are two alternatives, we sometimes write:

$$\left[\begin{array}{c} f(x) \geq 0 \\ \text{or} \\ g(x) \geq 0 \end{array}\right. .$$

Now, combining these notations we can create complicated combinations, describing different types of sets. If S_1 and S_2 are two such systems of equations or inequalities and X_1, X_2 are the corresponding solution sets, we say that S_1 **implies** S_2, denoted as $S_1 \Rightarrow S_2$ if $X_1 \subseteq X_2$. That is, every solution of S_1 is a solution of S_2 (but not necessarily the other way around). Similarly $S_1 \Leftarrow S_2$ means that $X_1 \supseteq X_2$.

Finally, if $X_1 = X_2$ we say that S_1 is **equivalent** to S_2 and write $S_1 \Leftrightarrow S_1$. Note that this is the same as $S_1 \Rightarrow S_2$ and $S_2 \Rightarrow S_1$ together!

As an example, we have

$$x \geq 1 \quad \Rightarrow \quad x^2 \geq 1$$

and

$$|x^2 - x| = 2 \quad \Leftrightarrow \quad \left[\begin{array}{l} \left\{\begin{array}{rcl} x^2 - x & \geq & 0 \\ x^2 - x & = & 2 \end{array}\right. \\ \left\{\begin{array}{rcl} x^2 - x & \leq & 0 \\ -(x^2 - x) & = & 2 \end{array}\right. \end{array}\right. .$$

How do we solve complicated combinations of equations? Remembering that we are working with sets we can always simplify the picture. For example,

$$\left\{\begin{array}{l} S_1 \\ \left[\begin{array}{l} S_2 \\ S_3 \end{array}\right. \end{array}\right. \quad \Leftrightarrow \quad \left[\begin{array}{l} \left\{\begin{array}{l} S_1 \\ S_2 \end{array}\right. \\ \left\{\begin{array}{l} S_1 \\ S_3 \end{array}\right. \end{array}\right. ,$$

because $X_1 \cap (X_2 \cup X_3) = (X_1 \cap X_2) \cup (X_1 \cap X_3)$. Other simple transformations are

$$\left[\begin{array}{l} S_1 \\ \left[\begin{array}{l} S_2 \\ S_3 \end{array}\right. \end{array}\right. \quad \Leftrightarrow \quad \left[\begin{array}{l} S_1 \\ S_2 \\ S_3 \end{array}\right. ; \qquad \left\{\begin{array}{l} S_1 \\ \left\{\begin{array}{l} S_2 \\ S_3 \end{array}\right. \end{array}\right. \quad \Leftrightarrow \quad \left\{\begin{array}{l} S_1 \\ S_2 \\ S_3 \end{array}\right. .$$

As a consequence it follows that any complicated combination S is equal to the union of simple systems of equations or inequalities (maybe only one of them):

$$\left[\begin{array}{l} S_1 \\ S_2 \\ \vdots \\ S_n \end{array}\right.$$

To solve a simple system, it is sufficient to solve every one of its equations or inequalities and take the intersection of the corresponding sets of solutions. So let us concentrate on a single inequality.

How do we solve an inequality? Certainly, not in the same way as we solve an equation. Let us see a typical "solution" of a simple inequality that unfortunately quite often appears in students' works. Here, we have collected many common mistakes in a single example. Suppose we are solving the inequality $\frac{1}{x} \geq x$. How many errors does the following "solution" contain?

$$\begin{aligned} \frac{1}{x} &\geq x \\ 1 &\geq x^2 \\ x^2 - 1 &\geq 0 \\ x &\geq \pm 1. \end{aligned}$$

If this had been an equation instead of an inequality, the purported solution would have been fine (with minor remarks about the case $x = 0$). But as an inequality there is an error on every line! We cannot multiply the inequality by x (since x may be negative), so the second line is wrong. We have to change the sign to $\leq$ between the second and the third line, and finally the last inequality makes no sense at all! For example, is $0 \geq \pm 1$? Impossible to answer.

So, how do we solve such an inequality correctly? It is quite simple, but we need to follow some rules. To understand them we start with equations. There are two ways to solve equations. The first, and easier way is to transform the equations, making them simpler and simpler — something like:

$$\textit{Equation-1} \quad \Rightarrow \quad \textit{Equation-2} \quad \Rightarrow \quad \cdots \quad \Rightarrow \quad \textit{Last Equation},$$

meaning that every solution to the equation *Equation-1* is also a solution to *Equation-2*, every solution to the equation *Equation-2* is also a solution to *Equation-3* and so on until we come to the *Last Equation*, which is so simple that we can read the solutions off it directly, obtaining $x_1 = a_1, x_2 = a_2, \ldots$ The only thing left to do is to check that $x_1, x_2, \ldots$ really are solutions of *Equation-1*. Those that are not, we call "false" and discard. A typical example is

$$\sqrt{x} = -x \quad \Rightarrow \quad x = x^2.$$

The second equation has two solutions: $x_1 = 0$, $x_2 = 1$, but x_2 is a false root, because $\sqrt{x}$ is always non-negative. Thus the only solution to the original equation is $x = 0$. It is also clear how the "false" solution appears. This happens when we square both sides of the original equation: each solution to the first equation is also a solution of the squared equation, but not vice versa! In other words, we have made a non-equivalent transformation.

In the second more complicated way, we try to do the following:

$$\textit{Equation-1} \quad \Leftrightarrow \quad \textit{Equation-2} \quad \Leftrightarrow \quad \cdots \quad \Leftrightarrow \quad \textit{Last Equation},$$

which means that every solution to the equation *Equation-1* is also a solution to *Equation-2*, and vice versa, every solution to the equation *Equation-2* is also a solution to *Equation-1*. Then every solution to the equation *Equation-2* is also a solution to *Equation-3* , and vice versa, every solution to the equation *Equation-3* is also a solution to *Equation-2* and so on until we come to the *Last Equation*.

In other words, we use that $\Leftrightarrow$ is $\Leftarrow$ and $\Rightarrow$. The advantage of this method is that we do not need to check the solutions obtained in the *Last Equation* — we **know** that they are solutions to the first equation. The disadvantage is that we are restricted to equivalent transformations only, and sometimes there are no simple equivalent transformation we can make. We often need to consider different cases and impose extra conditions to allow us to proceed, and this means that in fact we are forced to consider systems of equations and inequalities (and sometimes even unions of such systems!) which, for example, can look like:

$$\textit{Equation-1} \quad \Leftrightarrow \quad \left[\begin{array}{l} \textit{System-2} \\ \textit{System-3} \end{array}\right. \quad \Leftrightarrow \quad \cdots \quad \Leftrightarrow \quad \left[\begin{array}{l} x_1 = a_1 \\ \quad \vdots \\ x_n = a_n. \end{array}\right.$$

Returning to the same example with this method, we have:

$$\sqrt{x} = -x \quad \Leftrightarrow \quad \begin{cases} x \geq 0 \\ -x \geq 0 \\ x = x^2 \end{cases} \quad \Leftrightarrow \quad x = 0.$$

The condition $x \geq 0$, which says that the expression under the root should be non-negative in this example can actually be skipped (it follows from $x = x^2$), but it is easier to write it first and cancel it in the next step. In this case, keeping it around allows us to avoid solving the last equation, because $x = 0$ follows from the first two inequalities.

When we are dealing with inequalities we can use only the second approach — working with equivalence transformations, because in general it is almost impossible to check infinitely many solutions. Hence, we cannot simply copy the methods used for equations, but if we can solve the corresponding equations we can easily find the solutions of the inequalities too! This can be done using the so-called **intervals method**.

When solving an inequality, say $f(x) \geq g(x)$, we first collect all terms on one side, $f(x) - g(x) \geq 0$ and factorize the left hand side as $\frac{A(x)}{B(x)} \geq 0$. Then we find all the points where $A(x) = 0$ or $B(x) = 0$ and mark them on the real line. Naturally we cut away the points where the function is not defined, e.g. demand $f(x) \geq 0$ if we have $\sqrt{f(x)}$ somewhere. Once we have done this, the real line is divided into several intervals and we use the fact that the inequality does not change the sign in each interval. Consequently, it is sufficient to test one point from each corresponding interval to know whether the interval belongs to the set of solutions or not. In other words we solve the inequality in each interval independently by testing one point from it. (This approach assumes that all subexpressions are continuous; otherwise the left hand side might change sign without ever being zero.)

A correct solution of the inequality above would be:

$$\frac{1}{x} \geq x \quad \Leftrightarrow \quad \frac{1}{x} - x \geq 0 \quad \Leftrightarrow \quad \frac{1-x^2}{x} \geq 0.$$

We get two points $x = \pm 1$ from the denominator $x^2 = 1$ and one point: $x = 0$ from the numerator. We should also demand that $x \neq 0$, but that corresponding point we already have. Together we have four intervals. In $(1, \infty)$ there is no solution (testing for, example, $x = 1000$). The next interval $(0, 1)$ is good (we can test $x = 1/2$). The interval $(-1, 0)$ again does not contain any solutions (testing $x = -1/2$) and the final interval $(-\infty, -1)$ is again good (testing $x = -1000$). Because the inequality is non-strict we should include the solutions to the equation $x = \pm 1$ and the answer therefore is $x \in (-\infty, -1] \cup (0, 1]$ or in another form:

$$\left[\begin{array}{l} x \leq -1 \\ 0 < x \leq 1. \end{array}\right.$$

An improvement to this method is to test first only one point in the last interval and then to analyze what changes when we pass to the next interval. As an example we consider:

$$\frac{1}{x^2 - x} \geq \frac{1}{3x - 4}.$$

Note that the inequality is *not* equivalent to $x^2 - x \leq 3x - 4$, because we do not know the signs of $x^2 - x$ and $3x - 4$. The interval method works perfectly:

$$\frac{1}{x^2 - x} - \frac{1}{3x - 4} \geq 0 \quad \Leftrightarrow \quad \frac{-x^2 + 4x - 4}{(x^2 - x)(3x - 4)} \geq 0$$
$$\Leftrightarrow \quad \frac{(x-2)^2}{x(x-1)(3x-4)} \leq 0.$$

We get the points $0, 1, \frac{4}{3}, 2$ and mark them on the real line.

For large x all the expressions are positive and we put a "+" above the last interval:

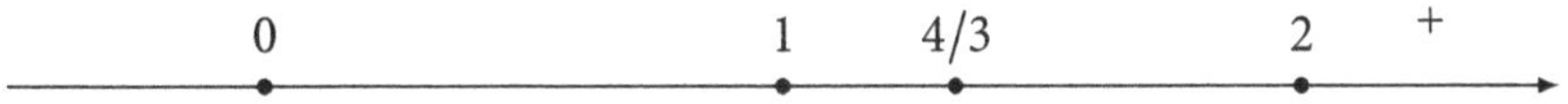

When we go to the left through 2 only the factor $(x - 2)^2$ can change sign, but it does not, and + remains.

When we pass 4/3 only the factor $(3x - 4)$ is affected and it changes a sign. We put a "–" there, and continue in the same way to end up with:

We should also remove the points $0, 1, 4/3$, because the function is not defined there.

Now we need only to look at the type of inequality we have ($\leq$) and to write the answer:

$$\left[\begin{array}{l} x < 0 \\ 1 < x < 4/3 \\ x = 2 \end{array}\right. .$$

3.6 Proving inequalities

To prove an inequality is usually much more complicated than to solve one; proving inequalities is an art. As with any art, it requires a lot of training before creating something that the artist may be proud of. Students who practice inequalities often make one of two mistakes. Some of them learn one or two inequalities (most often AM–GM and Rearrangement inequalities) and try to solve all the problems using only those primitive instruments. They may succeed in easy competitions, but fail when meet a real problem at the IMO level. Others do the opposite, they try to learn all the possible inequalities they can, and try to find which of them should fit to the current problem. Again this often works in easy one-step problems, but seldom in hard problems.

To achieve some progress in this art form, you need to know several powerful instruments, but mostly you have to learn how to transform the inequalities to be able to apply these instruments. We have already learned a few nice inequalities. Let us add a couple more. We start from a simple but useful fact.

Theorem 3.12 *Let $f(t)$ be a real polynomial $f(t) = at^2 + bt + c$ with $a > 0$ and let $D = b^2 - 4ac$ be its **discriminant**. Then for any t,*

$$f(t) \geq \frac{-D}{4a}$$

with equality iff $t = \frac{-b}{2a}$. In particular, $f(t) \geq 0$ for any real t iff $D \leq 0$.

Proof. The proof is by direct completion to a square, but we can make it simpler if we first multiply the inequality by the positive number $4a$:

$$\begin{aligned} 4af(t) &= 4a^2t^2 + 4abt + 4ac = (2at)^2 + 2 \cdot 2at \cdot b + b^2 - b^2 + 4ac \\ &= (2at + b)^2 + (4ac - b^2) = (2at + b)^2 - D. \end{aligned}$$

We see that the first term is non-negative for all values of t, being zero if and only if $t = \frac{-b}{2a}$. This proves that $4af(t) \geq -D$. The second statement follows directly from the first (but is used much more often than the first one). □

For example, the discriminant of $f(t) = 2t^2 + 3t + 5$ is $D = 3^2 - 4 \cdot 2 \cdot 5 = -31$ and thus $f(t) \geq \frac{31}{8}$ for all t.

Here is a more useful application.

Theorem 3.13 (the Cauchy–Schwarz inequality) *For any choice of real numbers $x_1, \ldots, x_n$ and $y_1, \ldots, y_n$ one has*

$$(x_1y_1 + \cdots + x_ny_n)^2 \leq (x_1^2 + \cdots + x_n^2)(y_1^2 + \cdots + y_n^2).$$

Equality occurs if and only if either all x_i are equal to zero or for some real α we have $y_i = \alpha x_i$ for each $i = 1, 2, \ldots, n$.

Proof. Let t be a real number. The sum of squares

$$\begin{aligned}(tx_1 + y_1)^2 + (tx_2 + y_2)^2 + \cdots + (tx_n + y_n)^2 \\ = t^2(x_1^2 + \cdots + x_n^2) + 2t(x_1y_1 + \cdots + x_ny_n) + (y_1^2 + \cdots + y_n)^2\end{aligned}$$

is non-negative. If all $x_i = 0$ then both statements are evident. Otherwise, according to Theorem 3.12, the discriminant

$$D = 4(x_1y_1 + \cdots + x_ny_n)^2 - 4(x_1^2 + \cdots + x_n^2)(y_1^2 + \cdots + y_n^2)$$

is non-positive and this is equivalent to the our first statement.

If $y_i = \alpha x_i$ for every i, we evidently have equality. Conversely, we need all the equalities $x_i + ty_i = 0$ for some t to get a zero discriminant, and thus we can take $\alpha = -t$. □

There are many other proofs of the Cauchy–Schwarz inequality, for example we can deduce it from the Rearrangement inequality (try!) or from Lagrange's identity (see page 136). Though the inequality looks very simple, it is more useful than you might expect. Finding the right choices of x_i and y_i is often rather difficult.

Problem 3.14 Prove that

$$(a_1 + \cdots + a_n)^2 < \frac{\pi^2}{6}\left(a_1^2 + 2^2a_2^2 + \cdots + n^2a_n^2\right).$$

Search. Where have we seen $\frac{\pi^2}{6}$ before? Yes, it is the sum $\sum_{n=1}^{\infty} \frac{1}{n^2}$. There are squares in the right-hand side as well, and maybe we can interpret this as the Cauchy–Schwarz inequality? With this idea we quickly find the answer:

$$\begin{aligned}(a_1 + a_2 + \cdots + a_n)^2 &= \left(\frac{1}{1}\cdot 1a_1 + \frac{1}{2}\cdot 2a_2 + \cdots + \frac{1}{n}\cdot na_1\right)^2 \\ &\le \left(\frac{1}{1^2} + \frac{1}{2^2} + \cdots + \frac{1}{n^2}\right)\left(a_1^2 + 2^2a_2^2 + \cdots + n^2a_n^2\right) \\ &< \frac{\pi^2}{6}\left(a_1^2 + 2^2a_2^2 + \cdots + n^2a_n^2\right).\end{aligned}$$ □

Here is another inequality that students often underestimate.

Theorem 3.15 *Let X, Y, Z, x, y, z be non-negative numbers and suppose that the triples (X, Y, Z) and (x, y, z) are similarly ordered; thus*

$$(X - Y)(x - y) \ge 0, \quad (X - Z)(x - z) \ge 0, \quad (Y - y)(Z - z) \ge 0.$$

Then

$$X(x-y)(x-z)+Y(y-x)(y-z)+Z(z-x)(z-y)\geq 0.$$

With $X=x^r$, $Y=y^r$, $Z=z^r$ *with* $r>0$ *we get* ***Schur's inequality:***

$$x^r(x-y)(x-z)+y^r(y-x)(y-z)+z^r(z-x)(z-y)\geq 0.$$

To have equality in Schur's inequality we need two of x,y,z *to be equal and the third to be* 0 *or to be equal to the other two.*

Proof. If two of the numbers x,y,z are equal then all the statements are trivial. If not, then using the symmetry in the conditions, we can WLOG suppose that $x>y>z$ and hence $X\geq Y$. Then $(z-x)(z-y)>0$ and $Z(z-x)(z-y)\geq 0$. On the other hand

$$\begin{aligned} X(x-y)(x-z)&+Y(y-x)(y-z)\\ &\geq Y(x-y)(x-z)+Y(y-x)(y-z)\\ &=Y(x-y)^2\geq 0. \end{aligned}$$

We see that for the equality we need $X=Y=Z=0$, which can't happen in Schur's inequality. □

Note that equality can be obtained in two different ways, which makes Schur's inequality rather special (and that is the reason why we have excluded the case $r=0$, even though the inequality is still valid for $r=0$). The proof looks so trivial that it is hard to believe that Schur's inequality is really deep. But look at problem 1.29 on page 55. Was it trivial? Do you recognize that this is only a special case of Schur's inequality for $r=1$?

We would now like to formulate two difficult but powerful inequalities. For them we need some special notation. Let $a=(a_1,a_2,\dots,a_n)$, $b=(b_1,b_2,\dots,b_n)$ be two decreasing finite sequences, i.e. $a_1\geq a_2\geq\cdots\geq a_n$; $b_1\geq b_2\geq\cdots\geq b_n$. The notation $a\succ b$ will be used for the following conditions:

$$\begin{aligned} a_1&\geq b_1\\ a_1+a_2&\geq b_1+b_2\\ &\vdots\\ a_1+\cdots+a_{n-1}&\geq b_1+\cdots b_{n-1}\\ a_1+\cdots+a_n&=b_1+\cdots+b_n. \end{aligned}$$

For example, $(6,2,1) \succ (4,3,2)$ and $(a_1, a_2, \dots, a_n) \succ (m, m, \dots, m)$ if m is the arithmetic mean of all a_i.

Let $x_1, x_2, \dots, x_n$ be *positive* real numbers. Their a–mean $[a]$ is defined by

$$[a] = \frac{1}{n!} \sum_{\sigma \in S_n} x_{\sigma(1)}^{a_1} \cdots x_{\sigma(n)}^{a_n}$$

where S_n is the set of all permutations of $\{1, 2, \dots, n\}$. (The sum should be taken over every permutation σ in S_n, so consists of $n!$ terms.) For example,

$$\begin{aligned}[(3,1,1)] &= \frac{1}{6}\left(x_1^3x_2x_3 + x_1^3x_3x_2 + x_2^3x_1x_3 + x_2^3x_3x_1 + x_3^3x_1x_2 + x_3^3x_2x_1\right) \\ &= \frac{1}{3}\left(x_1^3x_2x_3 + x_2^3x_1x_3 + x_3^3x_1x_2\right).\end{aligned}$$

Note that to get a symmetric sum of terms with powers $a_1, \dots, a_n$ we need only to multiply the number of the terms in the sum by $[a_1, \dots, a_n]$. For example, $\sum x_i^k = n([k, 0, \dots, 0])$ and $[(1, 0, \dots, 0)]$ is the arithmetic mean of $x_1, x_2, \dots, x_n$, whereas their geometric mean is $[(1/n, 1/n, \dots, 1/n)]$. We can easily express elementary symmetric functions:

$$s_k = \binom{n}{k} [(\underbrace{1, \dots, 1}_{k \text{ ones}}, 0, \dots, 0)],$$

for example $s_2 = 3[(1,1,0)]$.

Theorem 3.16 (Muirhead's inequality) *If $a \succ b$ then $[a] \geq [b]$. Furthermore, for $a \neq b$, equality holds iff $x_1 = \cdots = x_n$.*

For example, $x^2 + y^2 + z^2 \geq xy + xz + yz$ follows from $[(2,0,0)] \geq [(1,1,0)]$. More exactly, it follows for positive numbers, but changing the signs we can get it for non-positive numbers as well. The AM–GM inequality can be written as $[(1, 0, \dots, 0)] \geq [(1/n, 1/n, \dots, 1/n)]$ and therefore is a very special case of Muirhead's inequality.

Muirhead's inequality is quite technical. It often takes a lot of calculations, but does not demand too much thinking. The only trick that can simplify the work is the inequality $\frac{[a]+[b]}{2} \geq [\frac{a+b}{2}]$, which can be proved with the help of AM–GM (try!) Note that **Schur's inequality** can be written as $[(r+2, 0, 0)] + [(r, 1, 1)] \geq 2[(r+1, 1, 0)]$ and does *not* follow from Muirhead's inequality. But it is not forbidden to combine them, especially if we rewrite Schur's inequality in a more general form $[(r+2s,0,0)]+[(r,s,s)] \geq 2[(r+s,s,0)]$! Together, Muirhead's and Schur's inequalities form a powerful instrument for proving symmetric polynomial inequalities.

The idea is to transfer inequalities such as

$$\frac{a}{b+c}+\frac{b}{a+c}+\frac{c}{a+b} \geq \frac{3}{2}$$

for positive a, b, c into routine calculations: simply multiply everything by $2(a+b)(b+c)(a+c)$, rewrite as

$$2(a^3+b^3+c^3) \geq a^2b+a^2b+b^2a+b^2c+c^2a+c^2b$$

and note that this is the same as $[(3,0,0)] \geq [(2,1,0)]$ which follows immediately from Muirhead's inequality.

Even more complicated problems at the IMO level can be attacked in the same manner — look at [68], where one can find a proof of Muirhead's inequality as well. But we want to consider an example where this approach fails.

Problem 3.17 Prove the inequality

$$\frac{a^2+b^2}{c^2+ab}+\frac{b^2+c^2}{a^2+bc}+\frac{c^2+a^2}{b^2+ca} \geq 3$$

for all positive numbers a, b, c.

Search. We see the symmetry and attempt to avoid thinking. Multiplying everything by $(c^2+ab)(a^2+bc)(b^2+ca)$ we rewrite with some hard work the inequality as

$$\begin{aligned} & b^5c+bc^5+a^5c+ac^5+ba^5+b^5a+ \\ & \quad a^4b^2+a^2b^4+c^4a^2+c^2a^4+b^4c^2+c^4b^2+ \\ & \quad 2\left(b^2a^3c+a^3c^2b+b^2c^3a+bc^3a^2+b^3ca^2+b^3c^2a\right) \geq \\ & 3c^4ba+3b^4ac+3a^4bc+6c^2a^2b^2+3c^3a^3+3c^3b^3+3a^3b^3. \end{aligned}$$

Next we rewrite this as

$$6[(5,1,0)]+6[(4,2,0)]+12[(3,2,1)] \geq 9[(4,1,1)]+6[(2,2,2)]+9[(3,3,0)].$$

Muirhead gives us $[(4,2,0)] \geq [(3,3,0)]$; $[(3,2,1)] \geq [(2,2,2)]$ and it remains to prove (after cancelling by 3) that

$$2[(5,1,0)]+2[(3,2,1)] \geq 3[(4,1,1)]+[(3,3,0)].$$

According to our trick

$$[(5,1,0)] + [(3,2,1)] \geq 2\left[\left(4, \frac{3}{2}, \frac{1}{2}\right)\right] \geq 2[(4,1,1)].$$

so all we need is to prove

$$[(5,1,0)] + [(3,2,1)] \geq [(4,1,1)] + [(3,3,0)],$$

but here our arsenal is exhausted. Moreover, it is easy to check that other alternatives to cancel the terms fail as well.

This example show both the strong and the weak sides of Muirhead's inequality. The strong side is the possibility to prove simple polynomial inequalities by a routine procedure. The weakness is that, despite our hopes, we cannot catch *all* symmetrical inequalities in this way. Schur's inequality is the first example. By the way, do you still think that it is very simple inequality? Perhaps we can try to use the idea behind the proof of Schur's inequality to finish our problem?

Let us take a closer look at the terms we need to compare.

$$\sum_{\text{sym}} a^5b + a^3b^2c - a^4bc - a^3b^3 = \sum_{\text{sym}} a^4b(a-c) + \sum_{\text{sym}} a^3b^2(c-b).$$

The last sum over all the symmetrical permutations looks better as $\sum_{\text{sym}} b^3a^2(c-a)$. Thus we need to prove that

$$\sum_{\text{sym}} (a^4b - b^3a^2)(a-c) \geq 0 \quad \Leftrightarrow$$

$$\sum_{\text{sym}} a^2b(a^2 - b^2)(a-c) \geq 0 \quad \Leftrightarrow$$

$$\sum_{\text{sym}} a^2b(a+b)(a-b)(a-c) \geq 0.$$

This is very close to Schur's inequality or, better to say, to its proof in Theorem 3.15. We need only to rewrite this as a cyclic sum

$$\sum_{\text{cyclic}} a^2(ab + b^2 + ac + c^2)(a-b)(a-c) \geq 0.$$

Now if $A = a^2(ab + ac + b^2 + c^2)$, $B = b^2(ba + bc + a^2 + c^2)$ all we need to be able to apply Schur's theorem is to show that $(A-B)(a-b) \geq 0$. We have

$$\begin{aligned} A - B &= ab(a^2 - b^2) + c(a^3 - b^3) + c^2(a^2 - b^2) \\ &= (a-b)\big(ab(a+b) + c(a^2 + ab + b^2) + c^2(a+b)\big) \end{aligned}$$

and now $(A-B)(a-b) \geq 0$ is evident and Theorem 3.15 finishes the proof.

But this was much harder work that we anticipated, and we were lucky that the inequality we obtained at the end was good enough (this was not obvious, because cancelling the terms with the help of Muirhead's inequality we made non-equivalent transformations and could have ended up with something that wasn't good enough). Perhaps it is a good idea to compare with another proof.

From the Cauchy-Schwarz inequality we get

$$\begin{aligned} c^2 + ab &\leq \sqrt{c^2+a^2}\sqrt{c^2+b^2}, \\ a^2 + bc &\leq \sqrt{a^2+b^2}\sqrt{a^2+c^2}, \\ b^2 + ca &\leq \sqrt{b^2+c^2}\sqrt{b^2+a^2}. \end{aligned}$$

Now AM–GM inequality gives:

$$\begin{aligned} &\frac{a^2+b^2}{c^2+ab} + \frac{b^2+c^2}{a^2+bc} + \frac{c^2+a^2}{b^2+ca} \\ &\quad\geq \frac{a^2+b^2}{\sqrt{c^2+a^2}\sqrt{c^2+b^2}} + \frac{b^2+c^2}{\sqrt{a^2+b^2}\sqrt{a^2+c^2}} + \frac{c^2+a^2}{\sqrt{b^2+c^2}\sqrt{b^2+a^2}} \\ &\quad\geq 3\left(\frac{(a^2+b^2)(b^2+c^2)(c^2+a^2)}{\sqrt{c^2+a^2}\sqrt{c^2+b^2}\sqrt{a^2+b^2}\sqrt{a^2+c^2}\sqrt{b^2+c^2}\sqrt{b^2+a^2}}\right)^{1/3} = 3. \end{aligned}$$

Now we see the main disadvantage of Muirhead's inequality: having such a powerful instrument tempts us not to think. □

Let us consider another powerful tool. Recall that a function is **convex** on the interval I if the set of points

$$\{(x,y) \mid x \in I, y \geq f(x)\}$$

above its graph is convex and **concave** if $-f(x)$ is convex. Most often we check this using the second derivative: if $f''(x) \geq 0$ for all points $x \in I$ then $f(x)$ is convex on I. For example, x^2 is convex on $\mathbb{R}$ and $\sin x$ is concave on $[0, \pi]$. The function $|x|$ is convex as well, and this is a rare example where we have to use the definition ($|x|$ is not differentiable). Convex and concave functions provide us with two important inequalities.

Theorem 3.18 (Karamata's inequality) *Let f be a convex function on the interval I and all $a_i, b_i \in I$. If $a \succ b$ then*

$$\sum_{i=1}^{n} f(a_i) \geq \sum_{i=1}^{n} f(b_i)$$

(the opposite sign if f is concave). If $f(x)$ is strictly convex then there is equality iff $a_i = b_i$ for all i.

Example: for positive a, b, c:

$$\frac{1}{2a} + \frac{1}{2b} + \frac{1}{2c} \geq \frac{1}{a+b} + \frac{1}{b+c} + \frac{1}{c+a}$$

because

$$(2a, 2b, 2c) \succ (a+b, a+c, b+c)$$

if $a \geq b \geq c$ and $1/x$ is convex for $x > 0$.

Another example: for any acute triangle we have

$$(\pi/2, \pi/2, 0) \succ (\alpha, \beta, \gamma) \succ (\pi/3, \pi/3, \pi/3)$$

and $\cos x$ is concave on $[0, \pi/2]$. Hence

$$1 = 0 + 0 + 1 \leq \cos\alpha + \cos\beta + \cos\gamma \leq \frac{1}{2} + \frac{1}{2} + \frac{1}{2} = \frac{3}{2}.$$

A special case when $b_i = \dfrac{a_1 + \cdots + a_n}{n}$ is **Jensen's inequality**

$$\frac{f(a_1) + f(a_2) + \cdots + f(a_n)}{n} \geq f\left(\frac{a_1 + a_2 + \cdots + a_n}{n}\right)$$

which is used especially often. In a more general form, Jensen's inequality says that for any convex function $f(x)$ we have

$$\alpha_1 f(x_1) + \cdots + \alpha_n f(x_n) \geq f(\alpha_1 x_1 + \cdots + \alpha_n x_n)$$

for non-negative α_i such that $\sum \alpha_i = 1$. The sign is opposite if $f(x)$ is concave. We refer to [63] and [64] for the proofs.

Here are classical examples.

Problem 3.19 (IMO-2000) Let x, y, z be positive real numbers, such that $xyz = 1$. Prove that

$$\left(x - 1 + \frac{1}{y}\right)\left(y - 1 + \frac{1}{z}\right)\left(z - 1 + \frac{1}{x}\right) \leq 1.$$

Search. There are many different plays here. What if we try to start from the standard substitution $x = \frac{a}{b}, y = \frac{b}{c}, z = \frac{c}{a}$ to kill the condition $xyz = 1$? Now a, b, c are positive and we need to prove that

$$(a - b + c)(b - c + a)(c - a + b) \leq abc.$$

WLOG $a \geq b \geq c$. The first two factors are positive. If $c - a + b \leq 0$ the inequality is evident, so the problem is when the third factor is positive as well.

A sum is always easier to handle than a product and we could try to rewrite the inequality using logarithms:

$$\ln(a - b + c) + \ln(b - c + a) + \ln(c - a + b) \leq \ln a + \ln b + \ln c.$$

This looks like a Karamata-type inequality! Let us check the conditions. First, order the terms:

$$(b - c + a) \geq (a - b + c) \geq (c - a + b).$$

Next check the conditions:

$$\begin{aligned}(b - c + a) &\geq a;\\ (b - c + a) + (a - b + c) &\geq a + b;\\ (b - c + a) + (a - b + c) + (c - a + b) &= a + b + c.\end{aligned}$$

Good! Now, what about the second derivative?

$$(\ln x)'' = \left(\frac{1}{x}\right)' = -\frac{1}{x^2} < 0.$$

Perfect, we have a concave function and Karamata's inequality finishes the proof. Short and efficient! An alternative to the logarithms was to expand the product and recognize Schur's inequality from problem 1.29. □

Problem 3.20 Let $a_i > 0$. Prove that

$$\frac{a_1^3}{a_2} + \frac{a_2^3}{a_3} + \cdots + \frac{a_n^3}{a_1} \geq a_1^2 + a_2^2 + \cdots + a_n^2.$$

Search. The sum in the left hand side is cyclic, not symmetrical, so there is no hope for an approach with Muirhead. Cauchy–Schwarz or Rearrangement inequalities look more attractive, but we will temporarily forget about them and start with the recommended substitution $a_i = e^{x_i}$. We get

$$e^{3x_1-x_2} + e^{3x_2-x_3} + \cdots + e^{3x_n-x_1} \geq e^{2x_1} + e^{2x_2} + \cdots + e^{2x_n}.$$

Now it looks like Karamata's inequality. Let us check the conditions.

First we need to order the terms. Because there is no symmetry, words such as WLOG cannot help much here. So let $x_{k_1} \geq x_{k_2} \geq \cdots \geq x_{k_n}$ be the right order and

$$3x_{m_1} - x_{m_1+1} \geq 3x_{m_2} - x_{m_2+1} \geq \cdots \geq 3x_{m_n} - x_{m_n+1}$$

be the right order in another sequence. We may have different indices (taken, by the way, modulo n, so $x_{n+1} = x_1$) and this does not look very promising. But let us try.

Can we prove that $3x_{m_1} - x_{m_1+1} \geq 2x_{k_1}$? In any case, $3x_{m_1} - x_{m_1+1} \geq 3x_{k_1} - x_{k_1+1}$. But because x_{k_1} is the largest of all x_i we have

$$3x_{k_1} - x_{k_1+1} \geq 2x_{k_1} \quad \Leftrightarrow \quad x_{k_1} \geq x_{k_1+1}.$$

This gives us some hope. Take the next condition and try to use the same approach.

$$3x_{m_1} - x_{m_1+1} + 3x_{m_2} - x_{m_2+1} \geq 3x_{k_1} - x_{k_1+1} + 3x_{k_2} - x_{k_2+1}$$

because we took the largest pair on the left. But also

$$3x_{k_1} - x_{k_1+1} + 3x_{k_2} - x_{k_2+1} \geq 2x_{k_1} + 2x_{k_2} \quad \Leftrightarrow \quad x_{k_1} + x_{k_2} \geq x_{k_1+1} + x_{k_2+1},$$

because x_{k_1}, x_{k_2} is the largest pair. It works! We leave the pleasure of checking the rest of the conditions to the reader. Noting that e^x is a convex function we conclude that Karamata's inequality proves the result.

Was it possible to use the Rearrangement inequality? Almost directly! Again, suppose that $a_{k_1}^3 \geq a_{k_2}^3 \geq \cdots \geq a_{k_n}^3$ is the right order.

Then $a_{k_1}^{-1} \leq a_{k_2}^{-1} \leq \cdots \leq a_{k_n}^{-1}$ are ordered oppositely thus

$$a_1^2 + a_2^2 + \cdots + a_n^2 = a_{k_1}^3 a_{k_1}^{-1} + a_{k_2}^3 a_{k_2}^{-1} + \cdots + a_{k_n}^3 a_{k_n}^{-1}$$

is the least possible sum we can obtain in the Rearrangement inequality. □

Note that the solutions above consist of two parts: first we use some transformation to another form and only after that, we apply one of the inequalities.

In more difficult problems, we may need to do this several times and/or use different inequalities for different cases. We recommend the article [67] for further reading.

Now we would like to consider one very interesting generalization of AM–GM inequality in matrix form.

Theorem 3.21 *Let $X = (x_{ij})$ be a $m \times n$ matrix with non-negative real elements. Let $A_1, A_2, \dots, A_n$ be the arithmetic means of the columns in X and $G_1, G_2, \dots, G_m$ be the geometric means of the rows. Then*

$$A = \sqrt[n]{A_1 A_2 \cdots A_n} \ge \frac{G_1 + G_2 + \cdots + G_m}{m} = G.$$

Note that the sign of the inequality is the opposite of what you might expect! Equivalently,

$$\sqrt[n]{\prod_{j=1}^{n} \sum_{i=1}^{m} a_{ij}} \ge \sum_{i=1}^{m} \sqrt[n]{\prod_{j=1}^{n} a_{ij}}.$$

Proof. We will prove the first formula (the second one follows from the first after multiplication by m).

Suppose first that all $A_j > 0$. Using the AM–GM inequality, for any i:

$$\frac{1}{n} \sum_j \frac{a_{ij}}{A_j} \ge \sqrt[n]{\frac{a_{i1} a_{i2} \cdots a_{in}}{A_1 A_2 \cdots A_n}} = \frac{G_i}{A} \quad \Leftrightarrow \quad A \sum_{j=1}^{n} \frac{a_{ij}}{A_j} \ge n \cdot G_i.$$

Summing over i we get:

$$\begin{aligned} A \sum_{i=1}^{m} \sum_{j=1}^{n} \frac{a_{ij}}{A_j} \ge n \cdot \sum_{i=1}^{m} G_i \quad &\Leftrightarrow \quad A \sum_{j=1}^{n} \sum_{i=1}^{m} \frac{a_{ij}}{A_j} \ge n \cdot m \cdot G \\ &\Leftrightarrow \quad A \sum_{j=1}^{n} \frac{m A_j}{A_j} \ge n \cdot m \cdot G \\ &\Leftrightarrow \quad A \sum_{j=1}^{n} m \ge n \cdot m \cdot G \quad \Leftrightarrow \quad A \ge G. \end{aligned}$$

If some $A_j = 0$, then all the elements in that column are equal to zero, so all $G_k = 0$. Hence $A = G = 0$. □

We will consider several applications of this theorem. If

$$X = \begin{pmatrix} a & b \\ c & d \end{pmatrix}$$

is a 2×2 matrix, the inequality is already non-trivial (try to prove it directly!):

$$\sqrt{\frac{a+c}{2} \cdot \frac{b+d}{2}} \geq \frac{\sqrt{ab}+\sqrt{cd}}{2} \quad \Leftrightarrow \quad \sqrt{(a+c)\cdot(b+d)} \geq \sqrt{ab}+\sqrt{cd}$$

If X is an $n \times n$ circulant matrix

$$X = \begin{pmatrix} x_1 & x_2 & x_3 & \dots & x_{n-1} & x_n \\ x_2 & x_3 & x_4 & \dots & x_n & x_1 \\ x_3 & x_4 & x_5 & \dots & x_1 & x_2 \\ \dots & \dots & \dots & \dots & \dots & \dots \\ x_n & x_1 & x_2 & \dots & x_{n-2} & x_{n-1} \end{pmatrix},$$

then all the A_j and hence A are equal to the arithmetic mean $\frac{x_1+x_2+\dots+x_n}{n}$. On the other hand all the G_i (and G) are equal to the geometric mean: $\sqrt[n]{x_1 x_2 \dots x_n}$. Thus we recover the AM–GM inequality.

If X is an $n \times 2$ matrix

$$X = \begin{pmatrix} x_1^2 & y_1^2 \\ x_2^2 & y_2^2 \\ \dots & \dots \\ x_n^2 & y_n^2 \end{pmatrix}$$

we have

$$A_1 = \frac{\sum x_i^2}{n}, A_2 = \frac{\sum y_i^2}{n}, \quad G_i = \sqrt{x_i^2 y_i^2} = |x_i y_i|$$

and we get the inequality

$$\frac{\sqrt{\sum x_i^2 \cdot \sum y_i^2}}{n} \geq \frac{\sum |x_i y_i|}{n}$$

from which we again can obtain the Cauchy–Schwarz inequality. Note that it is shorter to use the second formula in the theorem (check!).

If $k \leq m$ and X is an $n \times m$ matrix with $x_{ij} = x_i^m$ for $j \leq k$ and $x_{ij} = 1$ for $j > k$:

$$X = \begin{pmatrix} x_1^m & x_1^m & \dots & x_1^m & 1 & 1 & \dots & 1 \\ x_2^m & x_2^m & \dots & x_2^m & 1 & 1 & \dots & 1 \\ \dots & \dots & \dots & \dots & \dots & \dots & \dots & \dots \\ x_n^m & x_n^m & \dots & x_n^m & 1 & 1 & \dots & 1 \end{pmatrix}$$

then $A_j = \frac{\sum x_i^m}{n}$ for $j \le k$ and $A_j = 1$ otherwise. Thus $A = \sqrt[m]{\left(\frac{\sum x_i^m}{n}\right)^k}$. On the other hand, $G_i = x_i^k$. Hence $G = \frac{\sum x_i^k}{n}$ and we get the classical **Power Mean inequality**:

$$\left(\frac{x_1^m + x_2^m + \cdots + x_n^m}{n}\right)^k \ge \left(\frac{x_1^k + x_2^k + \cdots + x_n^k}{n}\right)^m$$

which may also be rewritten as

$$\sqrt[m]{\frac{x_1^m + x_2^m + \cdots + x_n^m}{n}} \ge \sqrt[k]{\frac{x_1^k + x_2^k + \cdots + x_n^k}{n}}$$

which is valid for any real $m \ge k > 0$ with equality iff $x_1 = \cdots = x_n$.

We leave it to the reader the pleasure of formulating and finding the proof of this inequality for any real positive k, m. As a possible approach we suggest the following.

Theorem 3.22 (Hölder's inequality) *Let $x_1, x_2, \ldots, x_n$ and $y_1, y_2, \ldots, y_n$ be non-negative real numbers and p, q be positive numbers such that $\frac{1}{p} + \frac{1}{q} = 1$. Then*

$$\sum x_i y_i \le \left(\sum x_i^p\right)^{1/p} \left(\sum y_i^q\right)^{1/q}.$$

Proof. We will prove this first for rational $p = \frac{m}{k}$. Then $q = \frac{m}{l}$ where $m = k + l$. Consider the matrix X of size $n \times m$ where $x_{ij} = x_i^p$ if $j \le k$ and $x_{ij} = y_i^q$ if $j > k$:

$$X = \begin{pmatrix} x_1^p & x_1^p & \ldots & x_1^p & y_1^q & y_1^q & \ldots & y_1^q \\ x_2^p & x_2^p & \ldots & x_2^p & y_2^q & y_2^q & \ldots & y_2^q \\ \ldots & \ldots & \ldots & \ldots & \ldots & \ldots & \ldots & \ldots \\ x_n^p & x_n^p & \ldots & x_n^p & y_n^q & y_n^q & \ldots & y_n^q \end{pmatrix}$$

Then $A_j = \frac{\sum x_i^p}{n}$ for $j \le k$ and $A_j = \frac{\sum y_i^q}{n}$ for $j > k$. Thus

$$A = \left(\frac{\left(\sum x_i^p\right)^k \left(\sum y_i^q\right)^l}{n^{k+l}}\right)^{1/m} = \frac{\left(\sum x_i^p\right)^{1/p} \left(\sum y_i^q\right)^{1/q}}{n}.$$

On the other hand, $G_i = \left(x_i^{kp} y_i^{lq}\right)^{1/m} = x_i y_i$ and

$$G = \frac{\sum_i x_i y_i}{n}$$

from which the desired inequality follows.

If p is not rational and the inequality is not valid then by continuity arguments it is not valid for some rational p' close to p, which gives a contradiction. □

Note that $p = 2$ again gives the Cauchy–Schwarz inequality. We would like another generalization.

Theorem 3.23 (Minkowski's inequality) *Let p, $x_1, x_2, \dots, x_n$ and $y_1, y_2, \dots, y_n$ be non-negative real numbers. If $p \geq 1$ then*

$$\left(\sum (x_i + y_i)^p\right)^{1/p} \leq \left(\sum x_i^p\right)^{1/p} + \left(\sum y_i^p\right)^{1/p}.$$

For $0 < p < 1$, the inequality is reversed.

Proof. For $p = 1$ the inequality is obvious. We restrict ourselves to the case $p > 1$, leaving the case $p < 1$ to the reader. Let $\frac{1}{q} = 1 - \frac{1}{p} = \frac{p-1}{p}$. Note that $p, q > 0$ and $(p-1)q = p$. We have

$$\sum (x_i + y_i)^p = \sum x_i (x_i + y_i)^{p-1} + \sum y_i (x_i + y_i)^{p-1} \leq$$

(applying Hölder's inequality separately for each sum)

$$\left(\sum x_i^p\right)^{1/p} \left(\sum \left((x_i + y_i)^{p-1}\right)^q\right)^{1/q} + \left(\sum y_i^p\right)^{1/p} \left(\sum \left((x_i + y_i)^{p-1}\right)^q\right)^{1/q} =$$
$$\left(\sum x_i^p\right)^{1/p} \left(\sum (x_i + y_i)^p\right)^{1/q} + \left(\sum y_i^p\right)^{1/p} \left(\sum (x_i + y_i)^p\right)^{1/q}.$$

Therefore

$$\sum (x_i + y_i)^p \leq \left(\left(\sum x_i^p\right)^{1/p} + \left(\sum x_i^p\right)^{1/p}\right) \left(\sum (x_i + y_i)^p\right)^{1/q}.$$

Dividing by $\left(\sum (x_i + y_i)^p\right)^{1/q}$, we get

$$\left(\sum (x_i + y_i)^p\right)^{1-1/q} \leq \left(\sum x_i^p\right)^{1/p} + \left(\sum x_i^p\right)^{1/p},$$

which is equivalent to Minkowski's inequality since $1 - 1/q = 1/p$. □

3.7 Exercises

Exercise 3.1. Factorize in at least two factors (four in c).

a) $(x + y + z)^3 - x^3 - y^3 - z^3$

b) $x^{10} + x^5 + 1$

c) $x^8 + x^4 + 1$

d) $x^4(y-z) + y^4(z-x) + z^4(x-y)$

Exercise 3.2. Rewrite as sum a of squares:

$$(a_1^2 + a_2^2 + \cdots + a_n^2)(b_1^2 + b_2^2 + \cdots + b_n^2) - (a_1b_1 + \cdots + a_nb_n)^2.$$

Exercise 3.3. Let $a_1 < a_2 < \cdots < a_n$ be different real numbers.

a) Show that there exist numbers A_i such that

$$\frac{1}{(x+a_1)\cdots(x+a_n)} = \frac{A_1}{x+a_1} + \cdots + \frac{A_n}{x+a_n}.$$

b) Show that $A_1 > 0, A_2 < 0, A_3 > 0, A_4 < 0, \ldots$.

Exercise 3.4. Let a, b, c be sizes of a triangle. Show that

$$\sqrt{a+b-c} + \sqrt{b+c-a} + \sqrt{c+a-b} \le \sqrt{a} + \sqrt{b} + \sqrt{c}.$$

Exercise 3.5. Show that for any positive numbers a, b, c one has

$$a+b+c \le \frac{a^4+b^4+c^4}{abc}.$$

Exercise 3.6. Let a, b, c be positive numbers such that $abc = 1$. Show that

a) $\dfrac{1}{a^3(b+c)} + \dfrac{1}{b^3(c+a)} + \dfrac{1}{c^3(a+b)} \ge \dfrac{3}{2}$. (IMO-1995)

b) $\dfrac{1}{1+a+b} + \dfrac{1}{1+b+c} + \dfrac{1}{1+c+a} \le 1.$

Exercise 3.7. Let $a > b > 0$. Show that

$$\sqrt{ab} < \frac{a-b}{\ln a - \ln b} < \frac{a+b}{2}.$$

Exercise 3.8. Find two natural numbers a, b such that $ab(a+b)$ is not divisible by 7, but $(a+b)^7 - a^7 - b^7$ is divisible by 7^7. (IMO-1984)

Exercise 3.9. Let a, b be real numbers such that

$$3a^2b^2 + 6ab > 4a^3 + 4b^3 + 1.$$

Show that $a^2 > b$.

Exercise 3.10. Let $f(x), g(x), h(x)$ be relatively prime polynomials such that $f^n + g^n = h^n$ for some $n \geq 3$. Show that one of them is a constant.

See also problems 2, 19, 27, 40, 45, 66, 67, 80, 85, 92, 98, 102, 109 in Chapter 14.

3.8 Hints

3.1. a) Note that $x = -y, x = -z$ are two roots.

b) If $\varepsilon^3 = 1$ then ε and $\overline{\varepsilon}$ are roots.

c) Complete to a square three times.

d) $x = y, x = z$ are two roots.

3.2. Look at $(a_i b_j - a_j b_i)^2$.

3.3. a) Use induction.

b) Use the identity $\frac{a_n - a_1}{(x+a_1)\cdots(x+a_n)} = \frac{1}{(x+a_1)\cdots(x+a_{n-1})} - \frac{1}{(x+a_2)\cdots(x+a_n)}$.

3.4. Use Karamata's inequality.

3.5. Use Muirhead's inequality.

3.6. a) Try Muirhead's inequality directly or use the substitution $a = \frac{1}{x}, b = \frac{1}{y}, c = \frac{1}{z}$.

b) Set $a = x^3, b = y^3, c = z^3$.

3.7. Let $a = bx$. We need to show that

$$\sqrt{x} < \frac{x-1}{\ln x} < \frac{x+1}{2} \quad \Leftrightarrow$$

$$2\frac{x-1}{x+1} < \ln x < \frac{x-1}{\sqrt{x}}.$$

Because we have equality for $x = 1$ it is sufficient to prove inequalities for the derivatives:

$$\frac{4}{(x+1)^2} < \frac{1}{x} < \frac{x+1}{2x^{3/2}}$$

which are quite trivial for $x > 1$.

3.8. If $u = a + b, v = ab$ then $(a + b)^7 - a^7 - b^7 = 7uv(u^2 - v)^2$. Because $u^2 - v = a^2 + ab + b^2 = \frac{a^3 - b^3}{a-b}$ it is sufficient to find two numbers relatively prime to 7 such that their cubes are equal modulo 7^3.

3.9. Consider polynomial $f(x) = x^2 + 4a^3 x - 6abx + 4b^3 - 3a^2 b^2$. Because $f(1) < 0$ its discriminant is positive.

3.10. Let $n = 1$ first. Show that $\deg f$ is less that the number of different roots of the polynomial $f(x)g(x)h(x)$.

CHAPTER 4

Geometry

It is almost impossible to explain how to find mushrooms. You need to look for them. In a spot where you see nothing, an experienced mushroom-picker can find lots of delicious fungi. The same applies to geometry. The only way to learn the art of geometry is to solve a lot of different problems, to be able *to see* what is important, where to start and how. In general, the same is true for any area of mathematics, but geometry is the place where we first feel that we are in the forest where there are so many choices but we still have no idea of what to do. In this chapter we will try to give some hints that can sometimes help (and sometimes not).

4.1 The anatomy of a triangle

Medical students begin with anatomy, which seems reasonable: it is good to know the names and properties of body parts before you try to dissect or cure them. Let us follow their example and start with giving names to the most important parts of the basic geometric object — the triangle ABC. Mostly we want to introduce some standard notation. Standardization has two advantages. Firstly, it saves some space in our solutions: we can skip the needless repetitions of the same words and hope that the reader can infer what we mean. Secondly, a reader trained in the standard notation recognizes them directly in other textbooks, and the difficult process of understanding the solutions written by other people will be easier. Moreover, if the reader later wants to use the same notation in their own solutions it will be easier for others to read them, especially if s|he forgets to introduce the notation. Being original is not always a good idea. Try reading this: "Let rpO be a triangle, A be its circumradius and b be its area. Obviously $\frac{|rp|}{2} = A \sin O \ldots$" This is formally correct but quite unreadable! Do not underestimate the importance of good notation!

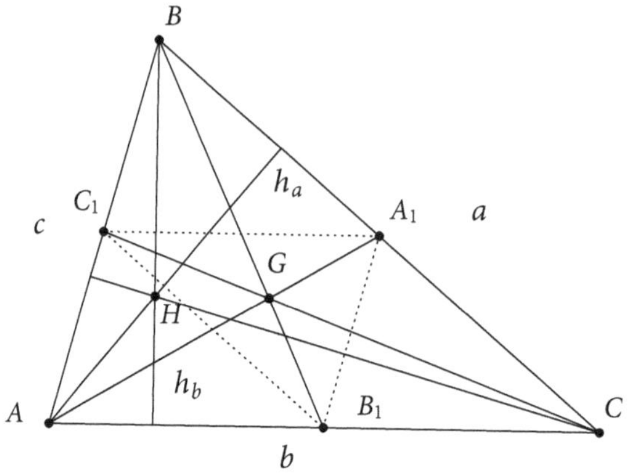

Figure 4.1 Altitudes and medians

So what is standard notation? If ABC is a triangle then its angles can be briefly denoted by A, B, C and this is convenient as long we do not plan to use the same notation for the complex numbers associated with the vertices. In the latter case we will prefer the Greek letters α, β, γ for the angles meaning that $\gamma = \angle C = \angle ACB$ et cetera. Sometimes (especially when we work with the angle bisectors) it is more convenient to use the Greek letters for the half-angles, e.g. $2\gamma = \angle ACB$, but this should be written explicitly.

Verbose notation for the lengths of the triangle's sides are $|AB|$, $|BC|$ and $|AC|$, but quite often we write only AB, BC, AC. The standard short notation for them are c, a, b (the side c is opposite to the angle C). If the triangle is right-angled, we reserve c for the hypotenuse, thus $a^2 + b^2 = c^2$ in this case. The half of the perimeter, $\frac{a+b+c}{2}$, is used so often that it gets a letter of its own, usually p (or sometimes s).

The lengths of the altitudes are usually denoted h_a, h_b, h_c meaning that the area is equal to $\frac{h_a a}{2}$. The **orthocentre** — the common point point of the altitudes — is denoted H. The **centroid** — the common point of the medians — is usually called G, because it is the centre of gravity.

The common point of the angle bisectors is always denoted I. This is the centre of the inscribed circle and r is a standard notation for its radius. Thus the area of the triangle can be calculated as pr as well (and the same formula is valid for any polygon that has an inscribed circle). Note that $\frac{1}{r} = \frac{1}{h_a} + \frac{1}{h_b} + \frac{1}{h_c}$

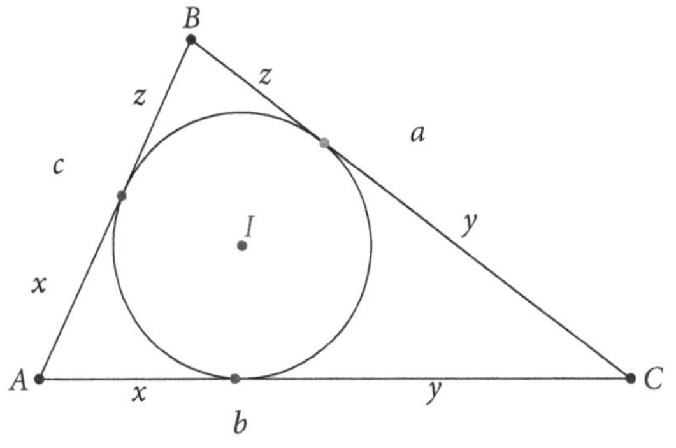

Figure 4.2 Inscribed circle

(to make this evident multiply both sides by the area). It is often necessary to calculate the length of a tangent to the inscribed circle. The following theorem is used so often that it can be cited on IMO without proof.

Theorem 4.1 *Let a, b, c be the sides of a triangle ABC. Then the length x of the tangent from the vertex A to the inscribed circle is equal to $p - a = \frac{b+c-a}{2}$. Similarly other tangents are equal to $p - b$ and $p - c$.*

Proof. It is easier to consider all the tangents, see Figure 4.2. Let x be the length of the tangents from A, y be the length of the tangents from C and z be the length of the tangents from B. Then

$$\begin{cases} y + z = a \\ x + y = b \\ x + z = c. \end{cases}$$

Summing up the equations, we get $2(x + y + z) = a + b + c$, so $x + y + z = p$. Subtracting the first equation, we get $x = p - a$. □

If we call the touch points T_A, T_B, T_C then Ceva's theorem (see page 170) shows that the cevians AT_A, BT_B, CT_C have a common point, called the **Gergonne point**. The triangle $T_A T_B T_C$ is called the Gergonne triangle. It, obviously, has the angles $\frac{\pi-A}{2}$, $\frac{\pi-B}{2}$, $\frac{\pi-C}{2}$ and I as the circumcentre. Less trivial is that the Gergonne point of ABC is the symmedianpoint of the triangle $T_A T_B T_C$ (see Section 5.1 for definitions).

Speaking about circumcircles, the letter O is reserved for the circumcentre of ABC and R for the radius of the circumcircle. The letter O reminds us that it is often a good idea to choose the origin of the coordinate system at the circumcentre because then the coordinates of H are the sums of the coordinates of the vertices A, B, C as we have seen in Theorem 2.20. The same theorem shows that O, G and H belong to the quite important Euler line. We will learn more about the Euler line in Theorem 4.22.

As suggested by the notation, r is not greater than R; in fact $2r \leq R$ with equality for the equilateral triangle only, because the distance between I and O is equal to $\sqrt{R(R-2r)}$ (another famous theorem by Euler that we recommend you to prove).

The midpoints of the sides have no standard notation, but if we denote them A_1, B_1, C_1 (where A_1 is opposite to A and is a midpoint of BC) then the middle lines A_1B_1, B_1C_1, C_1A_1 are parallel to the corresponding sides and divide the original triangle into four triangles, all of which are similar to ABC. The centre of gravity of $A_1B_1C_1$ coincides with G, but its orthocentre coincides with O (why?).

If AA_2, BB_2, CC_2 are the altitudes (A_2 lies on the line BC but not necessary on the segment BC), then the triangle $A_2B_2C_2$ is called an **orthic triangle**. If ABC is acute then its orthic triangle has the angles $\pi - 2\alpha$, $\pi - 2\beta$, $\pi - 2\gamma$ because the triangles AB_2C_2, BA_2C_2 and CA_2B_2 are similar to ABC (see Theorem 4.8). What are the angles in obtuse case?

Here is one of the most beautiful theorems in geometry.

Theorem 4.2 *The points A_1, B_1, C_1, A_2, B_2, C_2 lie on a common circle. The midpoints of the segments HA, HB, HC belong to this circle as well, and that is why it is called the* ***nine-point circle****. Its centre N is the midpoint of the segment OH and hence lies on the Euler line.*

Proof. We draw the picture first (Figure 4.3). Note that C_1B_1 is parallel to BC as a midline. Thus $A_1A_2C_1B_1$ is a trapezoid and because we expect it to be inscribed, Theorem 4.9 suggests that we should understand how we can prove that $|A_1B_1| = |A_2C_1|$.

We know that A_1B_1 is a midline, and thus $|A_1B_1|$ is half of $|AB|$. Why is the same true for $|A_2C_1|$? Asking this question is enough to see the answer: because AA_2B is a right-angled triangle and A_2C_1 is half the hypotenuse AB (by Theorem 1.12). Hence $A_1A_2C_1B_1$ is cyclic by Theorem 4.9, which means

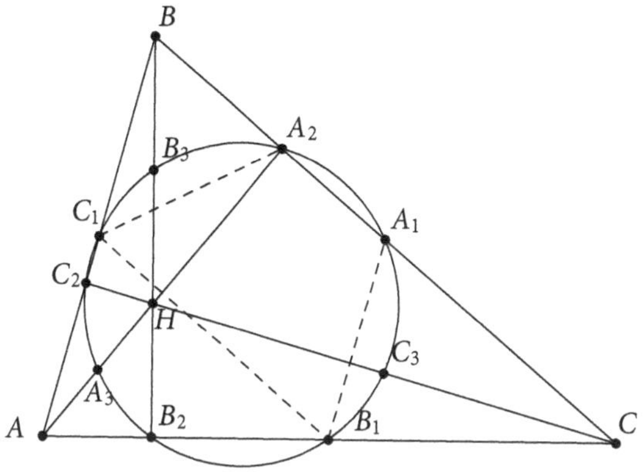

Figure 4.3 Nine-point circle

that A_2 belongs to the circumcircle of $A_1B_1C_1$. By symmetry the same is true for B_2 and C_2 and we already get six points on a common circle.

How do we get the remaining three? The picture suggests that A_1A_3 is a diameter and thus the angle $\angle A_1B_1A_3$ should be right. Can we prove at least this? Yes, it is not difficult: B_1A_3 as a midline is parallel to CH which is orthogonal to AB and therefore to A_1B_1 as well. Thus $A_3A_2A_1B_1$ is cyclic which shows that A_3 (and similarly B_3, C_3) belongs to our circle.

It remains for us to find the centre N of the nine-point circle. The easiest way for such questions is homothety (see Section 5.2). Consider the homothety with the centre H and coefficient $\frac{1}{2}$. Then A, B, C map to A_3, B_3, C_3 respectively, and therefore O maps to the circumcentre of $A_3B_3C_3$ which is N. □

Our anatomy approach suggests that we look for interesting objects inside the triangle. But in fact some of them (such as H and O) can be outside if the triangle is obtuse. The centres I_A, I_B, I_C of three external circles are always outside. Let us consider one of them. In Figure 4.4 we are mostly interested in the lengths x, y of the tangents from B and C. Obviously $x + y = a$ which implies that $|AF| + |AE| = b + c + x + y = 2p$. Because $|AF| = |AE|$ as equal tangents, both are equal to p and therefore $x = p - c$, $y = p - b$, which we recognize from the picture with the inscribed circle — but now they appear

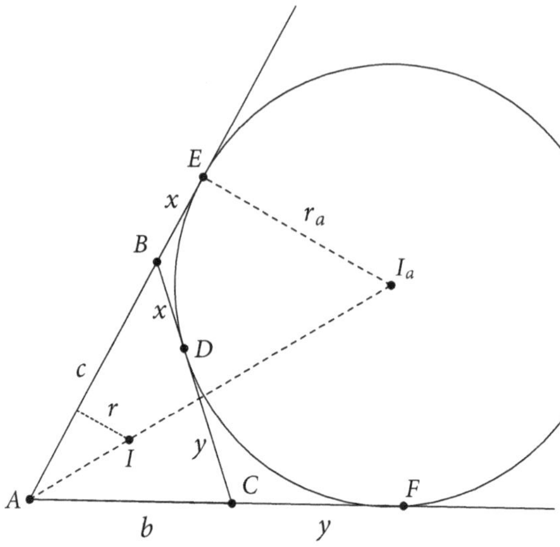

Figure 4.4 External circle

in the opposite order. The picture can help us to find r_a. By similarity:

$$\frac{r_a}{r} = \frac{p}{p-a} \quad \Rightarrow \quad r_a = \frac{pr}{p-a}$$

and it follows that the area of ABC is equal to $r_a(p-a)$ as well. Another direct consequence is that

$$\frac{1}{r_a} + \frac{1}{r_b} + \frac{1}{r_c} = \frac{1}{r}.$$

The line AD, as we see, divides the perimeter of the triangle into two equal parts. If we draw all three cevians with this property, it follows from Ceva's theorem that they have a common point, the **Nagel point**.

Right-angled triangles play very important roles and we would like to understand their anatomy in more detail.

Theorem 4.3 *Let C be a right angle in the triangle ABC. Then*

- *The midpoint D of the hypotenuse is the circumcentre and $R = \frac{c}{2}$.*
- *The median CD is the Euler line. Its length is $R = \frac{c}{2}$.*
- $r = p - c$.

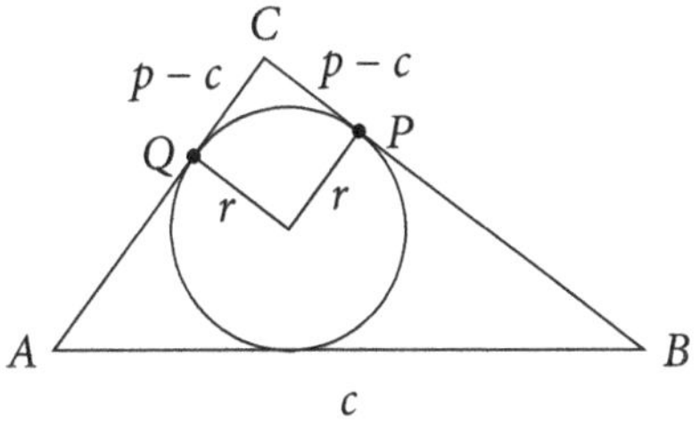

Figure 4.5 Incircle in a right triangle

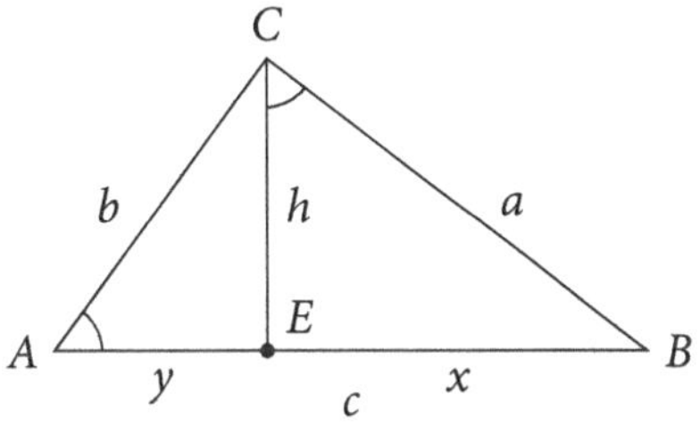

Figure 4.6 $h = \sqrt{xy}$, $a = \sqrt{xc}$

- *$h = h_c$ is the geometric mean of the part x, y in which it divides the hypotenuse c, $h = \sqrt{xy}$.*
- *If x is the projection of a on the hypotenuse then a is the geometric mean of x and c, thus $a = \sqrt{xc}$. Similarly, $b = \sqrt{yc}$.*

Proof. Because AB is a diameter of the circumcircle we get the first statement and half of the second. Evidently C is the orthocentre, so CD is the Euler line. If P, Q are the tangent points if the incircle, then $IPCQ$ is a square (we have three right angles and two equal sides). Thus Theorem 4.1 gives the third statement. For the final statements, we draw an altitude and consider similar triangles ACE and CBE (see figure 4.6). We get

$$\frac{h}{x} = \frac{b}{a} = \frac{y}{h} \quad \Rightarrow \quad h^2 = xy.$$

Likewise, considering the similarity of BCE and ABC we get:

$$\frac{a}{c} = \frac{x}{a} = \frac{h}{b} \quad \Rightarrow \quad a^2 = xc.$$

□

We see also that the centre of the nine-point circle (though they are only five in a right-angled triangle) is a midpoint of the median CD.

4.2 The degree of freedom

Mathematical plays in geometry usually have several acts, each of them being a tiny geometrical problem in itself. The difficulty is to find such problems in the original one. For this we need a reasonable plan. We should also have a large collection of such small problems memorized, and we need to be able to quickly estimate whether the plan is good enough and that the small problems are really solvable. As in architecture we should not only project an amazing beauty of our dream castle but also to think about where and how to place windows, doors and bathrooms on the inside. In other words, to make sure that every small detail in our project can be performed by the relatively primitive instruments we have access to. In this section, we would like to put aside the architectural dreams and discuss more carefully those dirty, technical parts of our geometrical construction. To say this directly — we would like to discuss how to estimate the routine calculations that we regularly need to perform while solving geometric problems.

The first question we need to ask ourselves is: how much freedom is there in our problem? How many parameters are needed to describe the desired object? For example, a circle needs only one parameter (for example its radius or its area): when we know it we know everything about the circle. If we need to know the position of the circle as well, we need two extra parameters, for example the coordinates of its centre.

A triangle needs three parameters (for example its sides) to be determined, and only two if the triangle is isosceles or right-angled. A parallelogram requires three parameters as well (e.g. two sides and an angle), but for a rhombus or a rectangle, two parameters are sufficient. We need one parameter for a square and five (e.g. its sides and one diagonal) for a quadrilateral. The trapezium demands four parameters (why?).

Suppose now that we have a problem and we have estimated its degree of freedom. The next step is to choose the corresponding number of parameters and to choose them conveniently, in such a way that we can calculate other elements in terms of the chosen ones. Some of them might be given in the problem, but most often the given parameters are not convenient. We ignore inconvenient given data, and recalculate them using the convenient parameters instead. In other words, we try to get some equations. The process sounds something like:

> "In this triangle I have 3 known parameters and know everything. In particular, I know this angle. The angle and the known side are enough to determine everything in this rhombus. Therefore I know the diagonal here."

When we have all the equations, we solve them and read the question again.

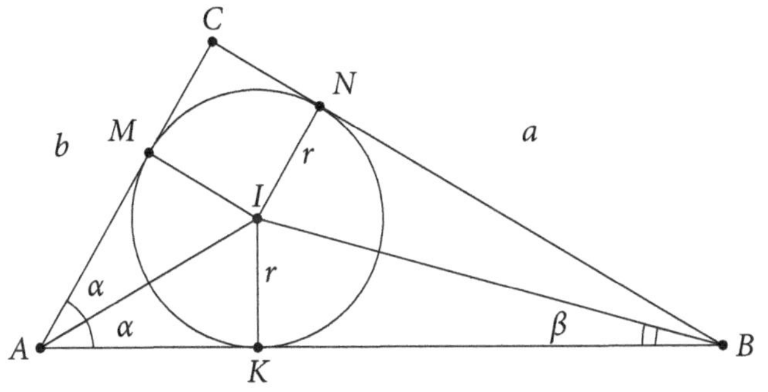

Figure 4.7 Two distances

Because the parameters were chosen conveniently we should be able to answer the question. This is not always as easy as it sounds. Sometimes we are unable to calculate something. More often, we are unable to solve the system of equations that we have obtained. For easy problems that arise as a part of larger ones, the approach works rather well, and this method is definitely an important tool. Let us test it on a simple example.

Problem 4.4 Let ABC be a right-angled triangle with $\angle ACB = 90°$. Let I be the centre of the inscribed circle. Find $|AC|$ if $|AI| = \sqrt{5}$ and $BI = \sqrt{10}$.

Search. Let us draw a picture (Figure 4.7) and estimate the degrees of freedom. We need only two parameters to define ABC and need therefore to choose convenient parameters. The two sides $a = |BC|$, $b = |AC|$ are the first parameters that come to mind but are they convenient? Thinking a little, we see that the radius $r = IK = IN = IM$ and the angle $\alpha = \angle IAK = \angle IAM$ are more flexible. Let us try using them.

In the right-angled triangle IAK we know two parameters and therefore know everything, in particular $|AI|$, which gives us the first equation. Knowing α we can find $\angle BAC$ and therefore $\angle ABC$ and its half $\beta = \angle IBK$. Therefore, we know two elements in IKB and can find $|IB|$. This is the second equation.

So let us implement this plan! The first equation can be written as $r = \sqrt{5} \sin \alpha$. Obviously $\beta = \frac{1}{2}\left(\frac{\pi}{2} - 2\alpha\right)$, and thus the second equation is $r =$

$\sqrt{10}\sin\beta = \sqrt{10}\sin\left(\frac{\pi}{4} - \alpha\right)$. Part one is done! Next we will try to find our parameters. Combining the equations we get

$$\sqrt{5}\sin\alpha = \sqrt{10}\sin\left(\frac{\pi}{4} - \alpha\right) \quad \Rightarrow \quad \sin\alpha = \cos\alpha - \sin\alpha$$
$$\Rightarrow \quad \tan\alpha = \frac{1}{2}.$$

Because $\frac{1}{\cos^2\alpha} = 1 + \tan^2\alpha$ we find directly $\cos\alpha = \frac{2}{\sqrt{5}}$ and $\sin\alpha = \frac{1}{\sqrt{5}}$, and therefore $r = 1$. Now we look at what the question was. This is easy.

$$|AB| = |AK| + |BK| = \frac{1}{\tan\alpha} + \frac{1}{\tan\left(\frac{\pi}{4} - \alpha\right)} = 2 + 3 = 5.$$

(check the details!) □

Next we will consider a slightly more advanced example.

Problem 4.5 Let a, b, c be lengths of the sides in the triangle ABC. Then $a^2 + b^2 = 5c^2$ iff the medians AD and BE are orthogonal.

Search. We start with the easier direction — suppose that the medians are orthogonal. This is only one condition, and we need two parameters to describe the situation. Let G be the centroid — the intersection point of the medians. We know that G divides the medians in the relation $1:2$ so if we choose the smaller parts as new variables $x = |GD|$, $y = |GE|$ it follows that $|AG| = 2x$, $|BG| = 2y$ so we know two values in the right-angled triangle AGB and thus we know everything about it, e.g. $|AB|$.

In the right-angled triangle AGE we know two values as well and therefore know everything about it, in particular $|AE| = \frac{b}{2}$.

Similarly we find a and our plan is ready — we only need to realize it to get our condition. First of all,

$$|AB|^2 = c^2 = (2x)^2 + (2y)^2 = 4(x^2 + y^2);$$

Similarly,

$$|AE|^2 = |AG|^2 + |GE|^2 \quad \Rightarrow \quad \left(\frac{b}{2}\right)^2 = (2x)^2 + y^2$$
$$\Rightarrow \quad b^2 = 4(4x^2 + y^2).$$

In the same way, $a^2 = 4(x^2 + 4y^2)$ and, as expected, we get

$$a^2 + b^2 = 4(5x^2 + 5y^2) = 5c^2.$$

Now we need to go the other way — deducing the orthogonality from the equation. We need three parameters. The natural candidates are x, y and $\alpha = \angle AGB$. These are sufficient to know everything about the triangles ABG and AEG. Therefore we can use the same plan as above with slightly modified calculations. Let us implement this. We get

$$|AB|^2 = c^2 = (2x)^2 + (2y)^2 - 8xy\cos\alpha = 4(x^2 + y^2 - 2xy\cos\alpha);$$

and

$$\begin{aligned} |AE|^2 &= |AG|^2 + |GE|^2 - 2|AG|\cdot|GE|\cos(\pi-\alpha) \quad \Rightarrow \\ \left(\frac{b}{2}\right)^2 &= (2x)^2 + y^2 + 4xy\cos\alpha \quad \Rightarrow \\ b^2 &= 4(4x^2 + y^2 + 4xy\cos\alpha) \end{aligned}$$

and similarly for a^2. Rewriting the condition $a^2 + b^2 = 5c^2$:

$$\begin{aligned} &4(4x^2 + y^2 + 4xy\cos\alpha) + 4(4y^2 + x^2 + 4xy\cos\alpha) \\ &\quad = 4\cdot 5(x^2 + y^2 - 2xy\cos\alpha) \end{aligned}$$

it follows that $\cos\alpha = 0$, as expected. □

Hopefully those examples are sufficient. Now for the implementation. How do we get compute everything once we know sufficiently many parameters? Mostly we use quite elementary instruments such as cosine and sine theorems (often called rules). We would also like to mention some standard calculations. For simplicity's sake, we suppose that our main parameters are the sides a, b, c of a triangle. How do we find the lengths of the altitude, the median and the angle bisector from the vertex C?

The altitude h is easy. We know from Heron's formula that the area is $p(p-a)(p-b)(p-c)$, where $p = \frac{a+b+c}{2}$ is half the perimeter. On the other hand, the area is equal to $\frac{ch_c}{2}$, so we can find the altitude h_c. On the subject of Heron's formula, we would like to mention that it is only a special case of the formula for the area of an arbitrary cyclic quadrilateral: $(p-a)(p-b)(p-c)(p-d)$, where again $p = \frac{a+b+c+d}{2}$ is half the perimeter.

We recommend you to prove and remember this formula (called sometimes Brahmagupta's formula).

As for the median m, we only need to complete the triangle to a parallelogram with AB as one diagonal and double the median as the other. Using that the sum of the squares of all the sides in a parallelogram is equal to the sum of squares of its diagonals (see Problem 2.18), we obtain $2a^2 + 2b^2 = c^2 + (2m)^2$ and we are done.

The length l of the angle bisector CD can be found with some help from the cosine theorem. Let $x = |AD|$, $y = |BD|$. Because $x + y = c$ and $\frac{x}{y} = \frac{a}{b}$ we can easily express x and y in terms of a, b, c. Let $\angle ADC = \alpha \Rightarrow \angle BCD = \pi - \alpha$. The cosine theorem gives us

$$a^2 = l^2 + x^2 - 2xl\cos\alpha; \quad b^2 = l^2 + y^2 + 2yl\cos\alpha.$$

Multiplying the first equation by y and the second by x and adding them we get:

$$a^2y + b^2x = l^2(x + y) + x^2y + y^2x.$$

Using $ay = bx$ we rewrite this equation as

$$abx + bay = l^2(x + y) + xy(x + y) \quad \Rightarrow \quad l^2 = ab - xy.$$

Here is an elegant geometric proof of the same formula. Regard the circumscribed circle. Prolong the angle bisector until it meets the circle and denote the length of the additional segment by z. We have $lz = xy$ (power of the point, see page 162). Also, by similarity

$$\frac{l + z}{b} = \frac{a}{l}.$$

Together the two equalities give $l^2 = ab - xy$.

This is the formula that is easier to remember. On page 458, the reader can find a more detailed answer and some estimates.

Do not ignore the cases where we need a fewer number of parameters than usual. For example we know that $a = 2R\sin\alpha$, and thus we need only two parameters to find the circumradius. We also know that to find a median from a right-angled triangle we need only to know the length of the hypotenuse — see Theorem 1.12. Other examples are: the length of the mid-line (we need only the length of the parallel side), the distance from the centroid to the vertex (we need only a median), the area (two parameters such as p and r or a, h_a are sufficient), and the distance $|OI| = \sqrt{R(R-2r)}$.

4.3 How to remember trigonometry

"I do not remember", said a student in an oral exam, "but I can always look it up if I need it." The teacher was not pleased. "This doesn't seem to be true", he said, "because you sure need it now."

Is it really possible to remember, say, all those trigonometrical formulas? Is it even necessary? The answer is YES and NO. We do not need to remember all the formulas, but we must remember some of them and know how to deduce the rest. The logic is much easier to remember and we need only our brain to obtain the formulas when we need them.

On page 84, we have already seen an example this approach. Let us try it again to deduce two basic formulas. Because $e^{i(x+y)} = e^{ix}e^{iy}$ is equivalent to

$$\cos(x+y) + i\sin(x+y) = (\cos x + i\sin x)(\cos y + i\sin y),$$

we immediately get, comparing the real and the imaginary parts, that

$$\cos(x+y) = \cos x\cos y - \sin x\sin y \tag{4.1}$$

$$\sin(x+y) = \sin x\cos y + \cos x\sin y. \tag{4.2}$$

These two formulas are in fact better to remember. The other variations with different signs, we do not remember but deduce them when necessary, e.g.

$$\begin{aligned}\cos(x-y) &= \cos x\cos(-y) - \sin x\sin(-y)\\ &= \cos x\cos y + \sin x\sin y\end{aligned} \tag{4.3}$$

(because we remember that cosine is not accurate with the signs).

Setting $x = y$ we get,

$$\cos 2x = \cos^2 x - \sin^2 x; \quad \sin 2x = 2\sin x\cos x,$$

which we can hardly forget. Neither can we forget that

$$1 = \cos^2 x + \sin^2 x$$

(which we get by putting $x = -y$, but in fact this is the Pythagorean theorem). It is important that we can combine these three formulas. For example, we get useful factorizations:

$$(\cos x \pm \sin x)^2 = 1 \pm \sin 2x$$

and

$$(\cos x + \sin x)(\cos x - \sin x) = \cos 2x.$$

Another important conclusion is that all the squares $\sin^2 x$, $\cos^2 x$, $\tan^2 x$, $\cot^2 x$ and $\cos 2x$ can be easily expressed in one or other (and this is all we need to remember). For example, we easily deduce

$$\begin{aligned}
1 + \cos 2x &= 2\cos^2 x \\
1 - \cos 2x &= 2\sin^2 x \\
1 + \tan^2 x &= \frac{1}{\cos^2 x}.
\end{aligned}$$

This also gives a good hint: **use larger angles if you need smaller powers.** Here is a slightly more advanced example confirming this approach.

$$\begin{aligned}
\sin^6 x + \cos^6 x &= (\sin^2 x + \cos^2 x)(\sin^4 x - \sin^2 x\cos^2 x + \cos^4 x) \\
&= (\sin^4 x + 2\sin^2 x\cos^2 x + \cos^4 x) - 3\sin^2 x\cos^2 x \\
&= (\sin^2 x + \cos^2 x)^2 - \frac{3}{4}\sin^2 2x \\
&= 1 - \frac{3}{8}(1 - \cos 4x) = \frac{5 + 3\cos 4x}{8}.
\end{aligned}$$

One good property of the squares is that we have similar formulas for solving the equations:

$$\begin{aligned}
\sin^2 x = a \quad &\Rightarrow \quad x = \pm\arcsin\sqrt{a} + \pi n,\ n \in \mathbb{Z}, \\
\cos^2 x = a \quad &\Rightarrow \quad x = \pm\arccos\sqrt{a} + \pi n,\ n \in \mathbb{Z}, \\
\tan^2 x = a \quad &\Rightarrow \quad x = \pm\arctan\sqrt{a} + \pi n,\ n \in \mathbb{Z},
\end{aligned}$$

where $0 \le a \le 1$ in the first two formulas and $a \ge 0$ in the last.

Quite often we need express the product of two functions as a sum or vice versa. Again we do not try to remember the formulas but deduce them. For example, adding (4.1) and (4.3) we get

$$\cos(x + y) + \cos(x - y) = 2\cos x \cos y,$$

which gives the expression of the product of two cosines. Subtracting them we find the product of two sines:

$$\cos(x-y)-\cos(x+y)=2\sin x\sin y.$$

In a similar way we get the product of a sine and a cosine:

$$\sin(x+y)+\sin(x-y)=2\sin x\cos y.$$

If we put here $A = x + y$, $B = x - y$ we can rewrite the formulas the other way around:

$$\cos A+\cos B=2\cos\frac{A+B}{2}\cos\frac{A-B}{2}$$
$$\sin A+\sin B=2\sin\frac{A+B}{2}\cos\frac{A-B}{2}$$

Deduce the formula for the differences! Of course after such training, it is much easier to remember the formulas as well: we remember approximately how they look and choose the right signs by testing known examples.

What about the tangent? Everything follows from the sine and cosine by division, for example:

$$\tan(x+y)=\frac{\sin(x+y)}{\cos(x+y)}=\frac{\sin x\cos y+\cos x\sin y}{\cos x\cos y-\sin x\sin y}=\frac{\tan x+\tan y}{1-\tan x\tan y}.$$

We get also $\tan 2x=\frac{2\tan x}{1-\tan^2 x}$ which is useful and worth remembering.

As we noticed on page 19 the tangent of half of the angle is a universal substitution. Note also that we always have to be careful with the sign, when we compute the sine or the cosine of the half of the angle, but there is less trouble with the tangent:

$$\tan\frac{x}{2}=\frac{\sin\frac{x}{2}}{\cos\frac{x}{2}}=\frac{2\sin\frac{x}{2}\cos\frac{x}{2}}{2\cos^2\frac{x}{2}}=\frac{\sin x}{1+\cos x}.$$

Let us check if this is sufficient for IMO-1962 problem.

Problem 4.6 Solve the equation

$$\cos^2 x+\cos^2 2x+\cos^2 3x=1.$$

Search. A natural attempt is to find a common factor. For this we need first to eliminate the constant. The best way to do this is to increase the angles and reduce the degrees. Let us try this directly.

$$2\cos^2 x-1+2\cos^2 2x-1+2\cos^2 3x=0\Leftrightarrow\cos 2x+\cos 4x+2\cos^2 3x=0.$$

Table 4.1 Values of the trigonometric functions for the most important angles

n	α	$\sqrt{n}$	$\sin\alpha$	$\cos\alpha$	$\tan\alpha$
0	0°	0	$\frac{\sqrt{0}}{2} = 0$	1	0
1	30°	1	$\frac{\sqrt{1}}{2} = \frac{1}{2}$	$\frac{\sqrt{3}}{2}$	$\frac{1}{\sqrt{3}}$
2	45°	$\sqrt{2}$	$\frac{\sqrt{2}}{2}$	$\frac{\sqrt{2}}{2}$	1
3	60°	$\sqrt{3}$	$\frac{\sqrt{3}}{2}$	$\frac{1}{2}$	$\sqrt{3}$
4	90°	2	$\frac{\sqrt{4}}{2} = 1$	0	–

Of course we need to care which angles should be chosen, but so long all looks perfect because we already see the next transformation which gives us the desired common factor:

$$2\cos 3x\cos x + 2\cos^2 3x = 0 \Leftrightarrow \cos 3x(\cos x + \cos 3x) = 0$$

and we get the final step

$$2\cos 3x\cos 2x\cos x = 0.$$

It is hard to believe today that this was an IMO-problem. □

Now let us look at some concrete angles. The most natural way to remember them is geometry, for example that $\sin 30^\circ = \frac{1}{2}$ can be seen in half an equilateral triangle. There is an interesting mnemonic for writing down the most important values. Let us fill the table with the important angles: $0, 30^\circ, 45^\circ, 60^\circ, 90^\circ$ — we have no trouble remembering them. Numbering the rows from zero, write down the root of the number $\sqrt{n}$ in the next column. Dividing these numbers by two we get all the sines in the next column! Reversing the column with the sines, we get the cosines. Division gives, naturally, tangents and we are done! Two minutes — and we recovered all the values!

One easy question: when do we use degrees and when do we use radians? When we want to express our respect to the ancient Greeks (and Babylonians) and work with geometry, use degrees. In all other cases, use radians! The reason is that some of the most important formulas such as $|\sin x| \le |x|$, and the series expansions of $\sin x$ or $e^{ix} = \cos x + i \sin x$ are expressed in radians and look much uglier in degrees. We do not really need degrees, but we respect the geometry created by the Greeks and follow their notations in geometry.

Besides the standard angles above, at the IMO level it is good to also recognize the angles $15°, 75°, 18°, 36°, 72°$. The first two are easy, e.g.:

$$\begin{aligned}\tan(45° \pm 30°) &= \frac{\tan 45° \pm \tan 30°}{1 \mp \tan 45° \cdot \tan 30°}\\ &= \frac{1 \pm \frac{1}{\sqrt{3}}}{1 \mp \frac{1}{\sqrt{3}}} = \frac{\sqrt{3} \pm 1}{\sqrt{3} \mp 1} = \frac{(\sqrt{3} \pm 1)^2}{2} = 2 \pm \sqrt{3}.\end{aligned}$$

For the others, first we need to create a cubic equation. Let $\alpha = 18°$ and $t = \sin\alpha$. Then (note that $\cos\alpha \ne 0$)

$$\begin{aligned}\sin 4\alpha = \sin 72° = \cos\alpha \quad &\Rightarrow \quad 2\sin 2\alpha \cos 2\alpha = \cos\alpha\\ &\Rightarrow \quad 4\sin\alpha(1 - 2\sin^2\alpha) = 1\\ &\Rightarrow \quad 4t(1 - 2t^2) = 1\\ &\Rightarrow \quad 8t^3 - 4t + 1 = 0.\end{aligned}$$

We can easily guess one root $t = \frac{1}{2}$ which we are not interested in. After a polynomial division by $2t - 1$ we get

$$4t^2 + 2t - 1 = 0 \quad \Rightarrow \quad t = \sin 18° = \cos 72° = \frac{\sqrt{5} - 1}{4}.$$

After that we can easily find

$$\cos 36° = 1 - 2t^2 = \frac{\sqrt{5} + 1}{4}$$

and we get access to all the angles.

Angles such as $\pm\alpha + k\pi$, $\pm\alpha + \frac{k\pi}{2}$, can be readily transformed by the rules (4.1)–(4.3), but there is also an easy trick worth remembering. We need to find a function and a sign for the result. The function is easy: it is the same when we have integral multiples of π and the opposite when we have half-integral multiples. As for the sign, the rule is the following: Draw the unit circle and look where the the obtained angle lies when α is between 0 and $\frac{\pi}{2}$. If your original function is positive for these values, choose plus, otherwise minus. For example,

$$\sin(\pi - \alpha) = \sin\alpha; \quad \cos(\pi - \alpha) = -\cos\alpha.$$

We did not change the function, the sine is positive but the cosine is negative in the interval between $\frac{\pi}{2}$ and π. Similarly,

$$\sin\left(\frac{\pi}{2} - \alpha\right) = \cos\alpha; \quad \cos\left(\frac{\pi}{2} - \alpha\right) = \sin\alpha; \quad \tan\left(\frac{\pi}{2} - \alpha\right) = \cot\alpha,$$

and thus we can always replace a function by its opposite. Geometrically, it seems obvious if we consider a right-angled triangle, but the formulas are valid for any α (not just $0 < \alpha < \frac{\pi}{2}$).

The unit circle is also useful for solving equations. For example,

$$\sin\alpha = \sin\beta \quad \Leftrightarrow \quad \alpha = \beta + 2\pi n \quad \text{or} \quad \alpha = \pi - \beta + 2\pi n, n \in \mathbb{Z}.$$

Indeed, on the circle angles with the same y-coordinate are weither equal or their sum is is equal to π. The circle determines the angle up to $2\pi n$. Similarly,

$$\begin{aligned} \cos\alpha = \cos\beta \quad &\Leftrightarrow \quad \alpha = \pm\beta + 2\pi n \\ \tan\alpha = \tan\beta \quad &\Leftrightarrow \quad \alpha = \beta + \pi n \end{aligned}$$

and we do not need to remember all these — only draw the picture when necessary.

On the subject of triangles we would like to mention a couple of interesting connections.

Problem 4.7 Suppose that $0 \leq \alpha, \beta, \gamma \leq \frac{\pi}{2}$. Show that $\alpha + \beta + \gamma = \pi$ iff

$$\cos^2\alpha + \cos^2\beta + \cos^2\gamma + 2\cos\alpha\cos\beta\cos\gamma = 1.$$

If no angle is equal to $\frac{\pi}{2}$ then show that this is also equivalent to the condition

$$\tan\alpha + \tan\beta + \tan\gamma = \tan\alpha\tan\beta\tan\gamma.$$

Note that both identities are valid for any triangle with the angles α, β, γ.

Search. We always look for common factors. In the tangent identity we have one immediately:

$$\tan\alpha + \tan\beta + \tan\gamma(1 - \tan\alpha\tan\beta) = 0$$

and we see something that looks familiar — the terms in the expansion of $\tan(\alpha+\beta)$. Once we recognize this, it is easy to rewrite the expression as

$$\tan(\alpha+\beta)+\tan\gamma=0 \quad\Leftrightarrow\quad \tan(\alpha+\beta)=\tan(\pi-\gamma).$$

With the conditions on our angles, there is only one possibility: $\alpha+\beta=\pi-\gamma$ and we are done.

The first identity is slightly more sophisticated. First we need to eliminate 1. The natural way to do this is to use the double angles. We multiply the identity by 2 and rewrite it as

$$\cos 2\alpha+\cos 2\beta+2\cos^2\gamma+4\cos\alpha\cos\beta\cos\gamma=0.$$

To get common factors, we rewrite a sum as a product and a product as a sum:

$$\begin{aligned}
&2\cos(\alpha+\beta)\cos(\alpha-\beta)+2\cos^2\gamma+2(\cos(\alpha+\beta)+(\cos(\alpha-\beta))\cos\gamma=0\\
&\Leftrightarrow\quad \cos(\alpha+\beta)(\cos(\alpha-\beta)+\cos\gamma)+(\cos\gamma+(\cos(\alpha-\beta))\cos\gamma=0\\
&\Leftrightarrow\quad (\cos(\alpha+\beta)+\cos\gamma)(\cos(\alpha-\beta)+\cos\gamma)=0
\end{aligned}$$

and now it should be easy. If the first factor is zero then

$$\cos(\alpha+\beta)=\cos(\pi-\gamma)$$

and again with the restrictions on the angles, the only possibility is $\alpha+\beta=\pi-\gamma$. If the second factor is equal to zero then our conditions imply

$$\cos(\alpha-\beta)=\cos\gamma=0 \quad\Leftrightarrow\quad \gamma=\alpha=\frac{\pi}{2},\quad \beta=0.$$

In any case we get $\alpha+\beta+\gamma=\pi$. □

4.4 Angle chasing

If you do not know where to start, try finding a cyclic quadrilateral. The idea of this very good advice is to start angle chasing. This means that we try to find all possible dependencies between angles hoping that some of them can lead us to something useful (or at least a new chase). As in any chasing, we need some weapon. Let us look at our arsenal.

We start with **similarity**. Two similar polygons give us equal angles. Congruency is, naturally, a special case of this powerful weapon. For example, if two triangles have the same sides then their angles are equal as well. A slightly less trivial fact is that if corresponding sides are all in the same proportion then the triangles are similar. If only one ratio is equal, say $|AB| : |CB| = |A'B'| : |C'B'|$ but the angles between the corresponding sides are equal ($\angle ABC = \angle A'B'C'$) again we have similarity.

In 90 percent of the cases we do the opposite: claim that that triangles are similar because they have two equal angles. Thus similarity is mostly a result of the angle chasing. When we get a similarity of ABC and $A'B'C'$ it is often a good idea to write down the relations between the corresponding lengths:

$$\frac{|A'B'|}{|AB|} = \frac{|B'C'|}{|BC|} = \frac{|A'C'|}{|AC|}$$

and study more carefully which (of any) part of this that may be useful and how to rewrite it. If we are looking for another similarity we probably need to exchange a pair of terms. But more often we need to rewrite one of equalities (e.g. the first) in another form such as

$$|A'B'| = \frac{|B'C'|}{|BC|}|AB|.$$

because this gives us a comparison between two different lengths that can appear somewhere else.

The following theorem is trivial but rather useful.

Theorem 4.8 *Let ABC be an acute triangle and let AD and CE be its altitudes. Then the triangles ABC and BDE are similar.*

Proof. We do not need a picture for this. First, ADB and CEB are similar because they are right triangles with a common angle B. We follow the recommendation and write:

$$\frac{|AD|}{|CE|} = \frac{|BD|}{|BE|} = \frac{|AB|}{|BC|}.$$

Looking at this, we choose those elements that we are interested in. All of them are in the last equality and if we note that the triangles DBE and ABC have a common angle B then this equality is exactly what we need to obtain the desired similarity.

Was the condition that ABC is acute essential? □

Our next weapon is the **external angle.**

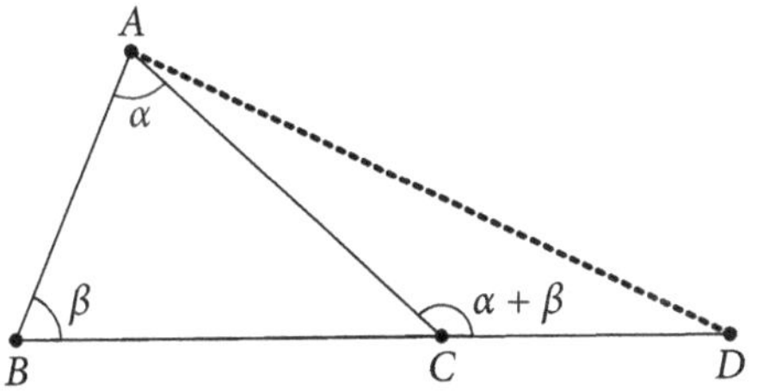

Figure 4.8 External angle

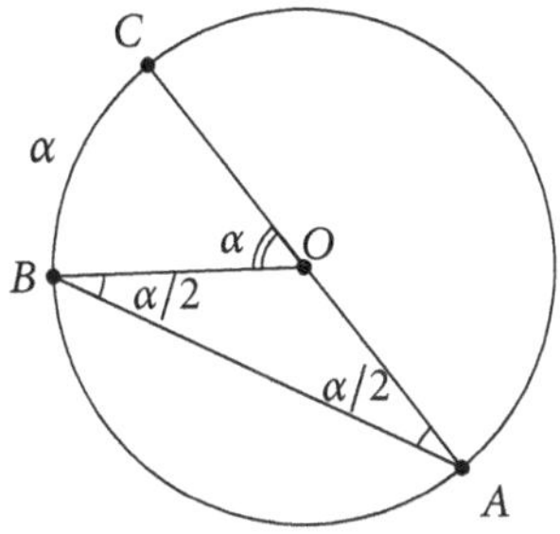

Figure 4.9 The arc and the angle

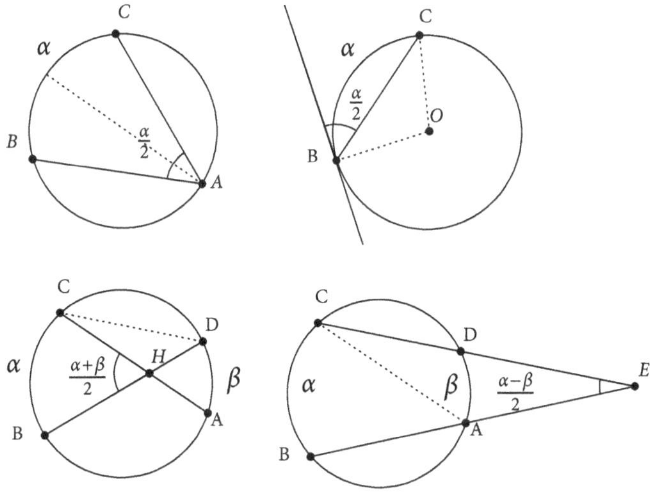

Figure 4.10 The arcs and the angles

If ABD is a triangle and C is a point between B and D, then the external angle $\angle ACD$ is the sum of two internal angles in ABC: $\angle ACD = \angle CAB + \angle CBA$. This simple observation is often quicker than summing up angles in a triangle (from which it obviously follows).

But the main arsenal for angle chasing is the **circle**. There are some pictures that you should be able recognize immediately, and we want to show them even if some of them are completely trivial. It is convenient to imagine that the circle has radius 1. Then there is a good connection between the angles and the lengths of the arcs. Everything is based on Figure 4.9. First we see that the

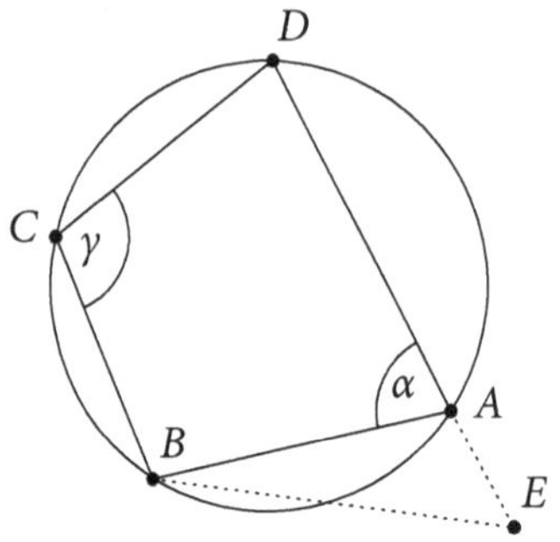

Figure 4.11 Cyclic quadrilateral

central angle BOC is exactly equal to the length α of the arc BC. Because it is an external angle to AOB we get $\angle BAO = \frac{\alpha}{2}$. The next pictures (Figure 4.10) shows, knowing the arcs, how to find the angles between any two interesting lines. All we need is to draw an additional line. If we draw it through the centre O in the first picture, we get $\angle BAC = \angle BAO + \angle CAO = \frac{\alpha'}{2} + \frac{\alpha''}{2} = \frac{\alpha}{2}$. The reader should check that the result is the same even when O is outside the triangle ABC. Most important is, of course, that the angle $\angle BAC$ is always the same — half of the arc BC for any point A as long the point A is chosen on the complementary arc. If we move A to the same arc we get, of course, half of the complementary arc: $\pi - \frac{\alpha}{2}$. An evident, but useful special case is that BC is a diameter iff $\angle BAC = \frac{\pi}{2}$.

The line through the centre in the next picture shows that the **angle between the tangent and and a chord** can be obtained as $\frac{\pi}{2} - \angle CBO = \frac{\pi}{2} - \frac{\pi-\alpha}{2}$ and again is equal to $\frac{\alpha}{2}$. This is a very useful fact! The two last pictures show how to find angles knowing two arcs. The reader should not have any problems understanding them — this is the first exercise in angle chasing.

Another important remark is the following. If $a = |BC|$ we can move A on the circle without changing the angle (or changing to $\pi - \alpha$). In particular, we can make a triangle ABC right-angled which gives $a = 2R\sin\alpha$ and this is true for any triangle! Note how the sine rule follows from this.

Next we would like to consider **cyclic quadrilaterals**, (those that may be inscribed in a circle) in more detail.

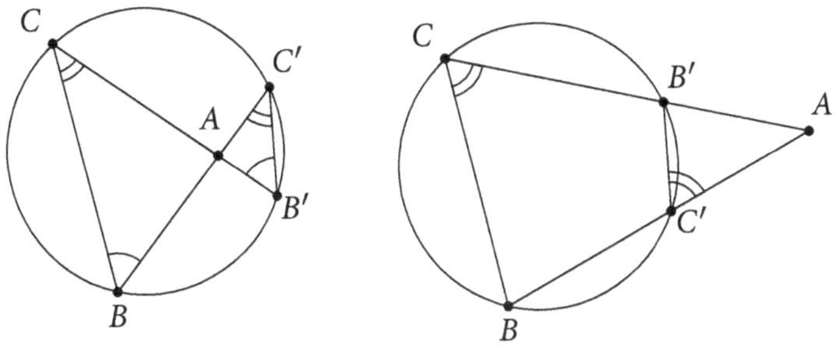

Figure 4.12 Similar triangles

Theorem 4.9 *A quadrilateral $ABCD$ is cyclic iff $\angle DAB + \angle DCB = \pi \Leftrightarrow \angle ACB = \angle ADB$.*

If $ABCD$ is cyclic then $|AB| = |CD| \Leftrightarrow AD \parallel BC$. Thus a trapezoid is cyclic iff it is isosceles.

A parallelogram is cyclic iff it is a rectangle.

Proof. Consider Figure 4.11. If $ABCD$ is cyclic then C and D are on the same arc for AB which gives equal angles $\angle ACB$ and $\angle ADB$. On the other hand A and C lie on the complementary arcs for BD which gives $\alpha + \gamma = \pi$. To show the converse note that if E is outside the circle then $\angle AEB < \alpha$, and thus the sum $\angle DEB + \angle DCB < \pi$. Similarly the sum is larger than π if E is inside the circle. Thus the sum of the angles is equal to π iff E is on the circle. The converse to the other statement is proved similarly.

If $ABCD$ is cyclic and $|AB| = |CD|$ then (because equal chords correspond to equal angles) we have $\angle BDA = \angle CBD \Leftrightarrow AD \parallel BC$.

If $AD \parallel BC$ then $\alpha + \angle ABC = \pi$ which means that $ABCD$ is cyclic iff $\angle ABC = \angle DCB$ which proves the last two statements. □

Now we have enough weapons to begin the chasing.

Theorem 4.10 *Suppose that a circle k contains points B and C but does not contain A. Let B' and C' be the points where the the lines AC and AB intersect k. Then the triangles ABC and $AB'C'$ are similar.*

Proof. If A is inside the circle (see Figure 4.12), we evidently have two pairs of equal angles. If A is outside we have them as well, but to see it note that, for example, $\angle BCA = \pi - \angle BC'B' = \angle AC'B'$. □

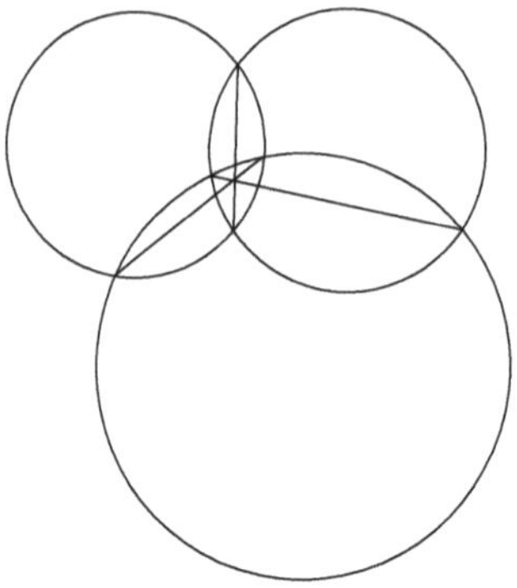

Figure 4.13 Three concurent chords

From the similarity in Theorem 4.10, it follows that the product $|AB| \cdot |AC'|$ is independent of the choice of B and this number (up to sign) is called the **power of the point** A. We choose the sign to be positive if A is outside the circle and negative if it is inside. If A is on the circle we define the power as zero. This is convenient because it allows us to calculate the power of the point A as $d^2 - r^2$, where r is the radius of the circle and d is the distance between A and the centre O. To see this, we choose B and C' in such a way that the line through them contains O. Then $|AB| \cdot |AC'| = (d + r)|d - r|$.

If A is outside the circle, then we can move B to be closer to C' and in the limit, we see that the power of the point A is equal to the square of the length of the tangent from A. Prove this directly using similarity!

If there are two different circles then the locus of points which have the same power relative to both circles is a line (find the proof!) called **radical axis** of the circles. If the circles intersect, then this line simply goes through their common points (because they have zero power) and is the common tangent if the circles are tangent to each other. Think about how to construct the radical axis when the circles have no common points.

Let us consider the following nice problem.

Problem 4.11 Show that if every pair of three different circles has a common chord then the lines containing those three chords are concurrent.

Proof. It looks like a very difficult problem, but the solution is easy. Consider the intersection of two lines. Because each of them is a radical axis, this point

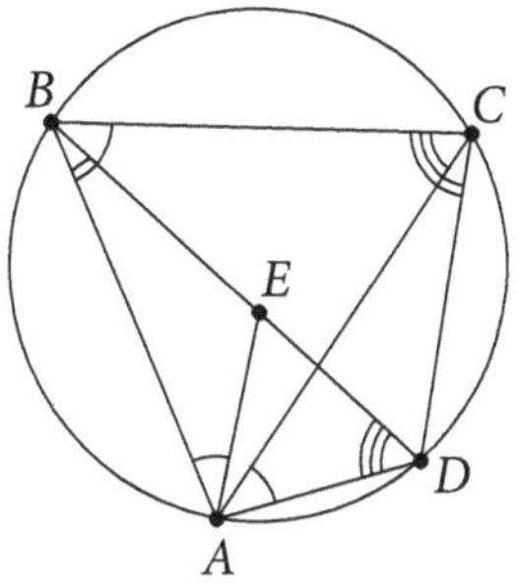

Figure 4.14 Looking for similar triangles

has the same power relatively to each of the three circles and therefore belongs to the third radical axis as well.

Here is another beautiful proof: Consider three spheres with the same centres and radii as the circles. Any two of them have a common circle whose projection on the plane is a chord. This circle intersects the third sphere at two points whose projection is the common point for all three chords.

The difference between the proofs is that the first works even in the case when the common point is outside the circles. □

Let us practise more angle chasing. Consider an inscribed quadrilateral $ABCD$ and draw two diagonals. We can already recognize two pairs of similar triangles, but we would like more. Choose a point E on the diagonal BD such that $\angle BAE = \angle DAC$ (see Figure 4.14). How many new similar triangles can you find now? One pair is quite evident: ABE and ACD have two equal angles. What more? Note that $\angle BAC = \angle EAD$. Because $\angle BCA = \angle EDA$ as well, we deduce that the triangles ABC and AED are similar. Now we need only to follow the recipe and write the equalities, hoping to get something interesting. We have:

$$\frac{|AB|}{|AC|} = \frac{|BE|}{|CD|} = \frac{|AE|}{|AD|}; \qquad \frac{|AB|}{|AE|} = \frac{|BC|}{|ED|} = \frac{|AC|}{|AD|}.$$

We will try, as recommended, to find the lengths and it follows from the first

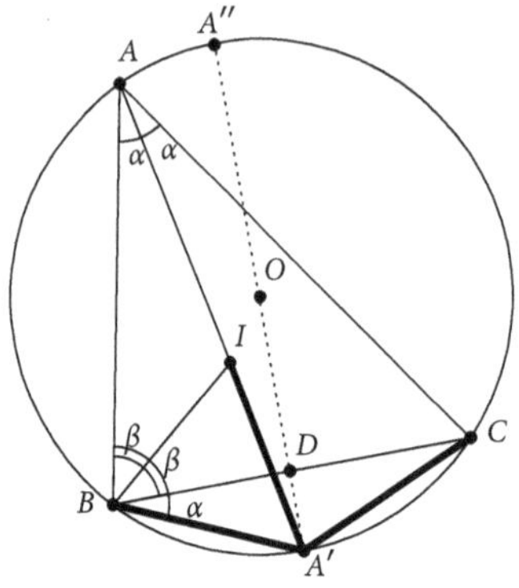

Figure 4.15 Superlemma

and the last equations that

$$\left.\begin{aligned} |BE| &= \frac{|AB|\cdot|CD|}{|AC|} \\ |ED| &= \frac{|AD|\cdot|BC|}{|AC|} \end{aligned}\right\} \quad \Rightarrow \quad |BD| = \frac{|AB|\cdot|CD| + |AD|\cdot|BC|}{|AC|}.$$

Hence the product of the diagonals is equal to the sum of the products of the opposite sides. An amazing result for so little effort! We formulate this in a more general setting, leaving the proof of the generalization to the reader.

Theorem 4.12 (Ptolemy's theorem) *A quadrilateral is cyclic if and only if the product of the diagonals is equal to the sum of the products of the opposite sides. For a non-cyclic quadrilateral ABCD, the* **Ptolemy inequality** *is valid:*

$$|AB|\cdot|CD| + |AD|\cdot|BC| > |AC|\cdot|BD|.$$

Ptolemy's theorem is a quite strong instrument and is used in many difficult problems.

The following lemma is so useful that our students named it:

Theorem 4.13 (Superlemma) *Let the bisector of the angle $\angle BAC$ intersect the circumcircle of the triangle ABC at a point A′ (different from A), we name it the* **superpoint***. Let D be the midpoint of the side BC and I be the in-centre. Then*

- *The points D, A' and circumcentre O are collinear. In particular, DA' is orthogonal to BC.*
- *If A'' is the second intersection point of the circumcircle and the line DA', then the points A, A'' and the excentres I_B, I_C are collinear, thus AA'' is the external bisector.*
- $|BA'| = |IA'| = |CA'| = |I_A A'|$.

Proof. Consider Figure 4.15 and do a little angle chasing. Obviously, $|BA'| = |CA'|$ because they correspond to the same angle α. Thus $BA'C$ is isosceles and $A'D$ is the median and the altitude, which proves the first statement. Because $A''A'$ is a diameter, the line AA'' is orthogonal to the bisector AA' and therefore is the bisector of the external angle, which proves the second statement.

Because I is a common point of the angle bisectors, we have $\angle BIA' = \beta + \alpha$ as an external angle. But $\angle IBA' = \beta + \alpha$ as well and we get that $IA'B$ is isosceles which proves the first equalities in the last statement. The reader should not have difficulties to prove the remaining equality, but we want to get much more. □

Theorem 4.14 *Consider the triangle ABC, the three lines connecting its excentres I_A, I_B and I_C, its circumcircle and the three superpoints A', B', C'. Then*

- *The in-centre I is the orthocenter of the triangle $I_A I_B I_C$ and the segments $I_A A$, $I_B B$, $I_C C$ are its altitudes.*
- *The circumcircle is the nine-point circle of the triangle $I_A I_B I_C$.*
- *The corresponding nine points are A, B, C, superpoints A', B', C' and the points A'', B'', C'' from the superlemma.*
- *In particular, A', B', C' are midpoints of the segments II_A, II_B, II_C and A'', B'', C'' are midpoints of the segments $I_B I_C$, $I_A I_C$, $I_A I_B$.*
- *The superpoint A' is the circumcentre of the cyclic quadrilateral IBI_AC.*
- *A homothety with centre in I and coefficient 2 sends the triangle $A'B'C'$ to the triangle $I_A I_B I_C$.*

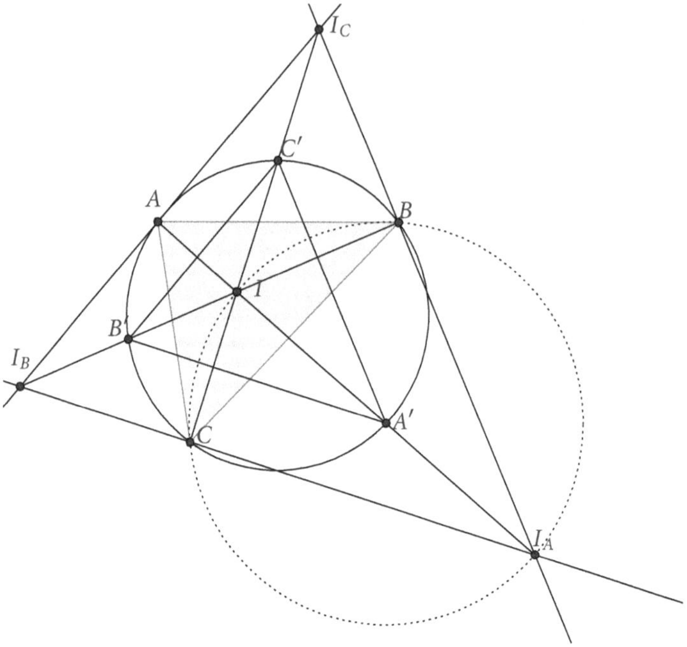

Figure 4.16 Superpoints and excircles

Proof. Because the bisectors are orthogonal to external bisectors, the first statement follows directly. This means that A, B and C viewed as foots for the altitudes belong to the nine-point circle of the triangle $I_A I_B I_C$ which proves the second statement. Theorem 4.2 implies two next statements. In particularly $|I_A A'| = |IA'|$ and this finishes the proof of the superlemma, from which follows that IBI_AC is cyclic. The homothety statement is obvious as well. □

The nice picture in Figure 4.16 is definitely worth remembering. So, when you have an angle bisector in your problem it might be a good idea to consider a circumcircle and study the superpoint — the point where the bisector intersects it. Here is a situation when the superpoint arises naturally.

Problem 4.15 Two circles are tangent to each other internally at the point A. The tangent at point D to the smaller circle intersects the large circle at the points B and C. Prove that AD is the angle bisector in the triangle ABC.

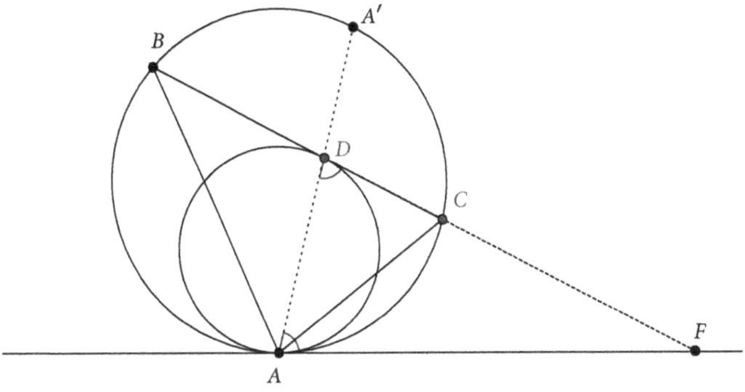

Figure 4.17 Superpoint.

Search. As recommended, we extend the bisector to meet the superpoint A' — see Figure 4.17. All we need is to prove that the corresponding arcs BA' and $A'C$ are equal. We need to use in some way the difficult condition that the circles are tangent to each other. This means that they have a common tangent. We draw it and extend the other tangent to intersect it at some point F. Now we can start the angle chasing. First see we that $\angle FDA = \angle FAD$. Translating this equality to the language of arcs, we see that the sum of the lengths of the arcs BA' and AC is the same as the length of the arc ACA' from which our result follows immediately. □

Now let us consider superpoints as complex numbers. As usual we can choose the circumcentre to be 0 and for simplicity suppose that circumradius is 1. Let $A = e^{2i\alpha} = x^2, B = e^{2i\beta} = y^2, C = e^{2i\gamma} = z^2$. The product $yz = e^{\beta+\gamma}$ lies exactly in the middle of the arc BC and must be either the superpoint A' or A''. Which of them depends of the choice of angles. For example, we can choose them so that yz is on the same arc as A so it will be A''. Then $A' = -yz$ will be the superpoint. We can choose the third angle in a such way that $B' = -xz$ as well. Then $I = -xy - xz - yz$ and automatically $C' = -xy$ will be the third superpoint. To see this, note that the number

$$\overline{Z} = \overline{\frac{A-I}{A-A'}} = \overline{\frac{x^2+xy+xz+yz}{x^2+yz}} =$$

$$\frac{\frac{1}{x^2}+\frac{1}{xy}+\frac{1}{xz}+\frac{1}{yz}}{\frac{1}{x^2}+\frac{1}{yz}}=\frac{yz+xy+yz+x^2}{yz+x^2}=Z$$

is real, which means that the points A, I and A' are collinear. Similarly B, I and B' are collinear, which shows that I is, as stated, the intersection of two bisectors. We leave to the reader to check that for example $I_b = xy + yz - xz$. Note that it would be impossible to choose the angles to get I as $xy + xz + yz$ (why?).

4.5 Centre of gravity and barycentric coordinates

The idea to use the centre of gravity as in Theorem 2.20 is rather deep and can be used in a lot of problems. We can use it efficiently when we want to find the ratio in which a point divides a given segment — all we need to do is to place the weights in such a way that the centre of gravity appears at the desired point.

We start from the basic remark that if two positive weights x and y are placed at A and B, then the centre of gravity of those two points is on the line AB at the point P such that $x|AP| = y|BP|$, thus P divides AB in the ratio $y : x$ and $P = \frac{x}{x+y}A + \frac{y}{x+y}B$ if we interpret the points as complex numbers. This formula suggests that there is no need to restrict ourselves to positive weights only. For example, if $y = 0$ we get point A and for $x > 0$, $0 > y > -x$ we find the point P to the left of A. What if $x > 0$, $y = -x$? We get a point at infinity and for some problems this has a meaning. But we restrict ourselves to the case $x + y > 0$ which is sufficient to describe all the points on the line AB.

To work in the plane we need three points A, B and C not all on the same line. Again we can place weights x, y, z at these points and find their centre of gravity P (see Figure 4.18).

If we replace the weights x and z by the weight $x + z$ at their centre of gravity $E = \frac{x}{x+z}A + \frac{z}{x+z}C$, we do not change the common centre of gravity P. But now we can find it explicitly as

$$\begin{aligned} P &= \frac{x+z}{x+y+z}E + \frac{y}{x+y+z}B \\ &= \frac{x}{x+y+z}A + \frac{y}{x+y+z}B + \frac{z}{x+y+z}C. \end{aligned}$$

Thus the point P is uniquely determined by the weights. We say that $(x : y : z)$

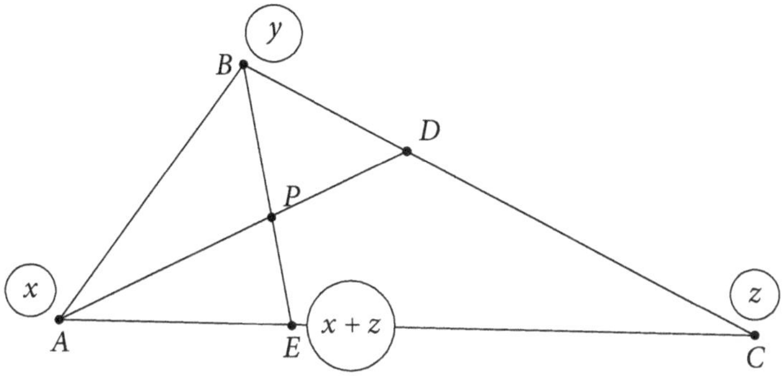

Figure 4.18 Barycentric coordinates

are **barycentric coordinates** of the point P. How unique are they? If we replace the weights by the proportional cx, cy, cz, it is clear that we do not change the centre of gravity, and thus $(x : y : z) = (cx : cy : cz)$ for $c \neq 0$. That is why we write colons instead of commas – we want to emphasize that the barycentric coordinates are defined up to multiplication by a non-zero constant.

If we want the barycentric coordinates to be unique, we can demand $x + y + z = 1$. Under this restriction, each point in the plane has unique barycentric coordinates and they are quite convenient. For example, all points inside the triangle have positive x, y, z. The line AB corresponds to the points with a zero z-coordinate. To see this note that $A = (1 : 0 : 0)$, $B = (0 : 1 : 0)$ and any point $C = (1-t)A + tB$ on the line has the coordinates $(1-t : t : 0)$. Choosing the signs of x, y, z we can describe all 7 parts in which three lines AB, AC, BC divide the plane (check!).

Here we use the fact that if $P_i = (x_i : y_i : z_i)$ with $x_i + y_i + z_i = 1$ and $\alpha + \beta = 1$, then the point $\alpha P_1 + \beta P_2$ has barycentric coordinates $(\alpha x_1 + \beta x_2 : \alpha y_1 + \beta y_2 : \alpha z_1 + \beta z_2)$. Thus we have quite reasonable mathematics. Note also that if $A = 1, B = i, C = 0$ then $P = x + iy$ has $(x : y : 1 - x - y)$ as barycentric coordinates, and we can consider ordinary coordinates as a very special case of barycentric ones.

For those who are familiar with linear algebra we would like to mention the conditions when three points P_i are collinear. If $(x_i : y_i : z_i)$ are their barycentric coordinates then a necessary and sufficient condition for

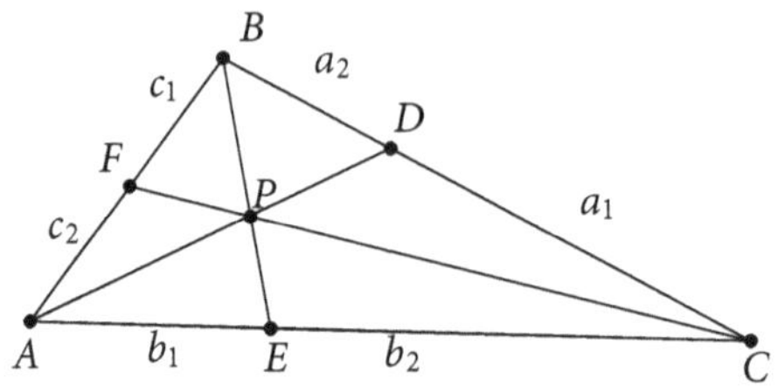

Figure 4.19 Ceva's theorem: $a_1b_1c_1 = a_2b_2c_2$

colinearity is that the determinant of their coordinates is equal to zero:

$$\begin{vmatrix} x_1 & y_1 & z_1 \\ x_2 & y_2 & z_2 \\ x_3 & y_3 & z_3 \end{vmatrix} = 0.$$

Now it is time for some applications. First of all we have complete control over the relationships between the lengths. For example, in Figure 4.18 we have $BD : DC = z : y$ and $BP : PE = (x + z) : y$, which is far from being evident. As a consequence we also have control over all the possible areas. If we suppose for simplicity that the area of ABC is 1 then we find that the area of ABE is $\frac{z}{x+z}$ and the area of ABP is

$$\frac{x+z}{y+x+z} \cdot \frac{z}{x+z} = \frac{z}{x+y+z}.$$

In particular if $x+y+z = 1$ then we have another interpretation of the barycentric coordinates: x, y, z are the areas of three triangles BPC, APC, APB.

Next we would like to show how to use barycentric coordinates to prove theorems. The idea is to choose the weights to obtain the centre of gravity at the point of interest. Here is a good example.

Theorem 4.16 (Ceva's theorem) *Let D, E, F be points on the segments BC, AC, DE respectively. The lines AD, BE, CF have a common point P inside the triangle iff $a_1b_1c_1 = a_2b_2c_2$ (see Figure 4.19).*

More generally, if D, E, F are on the lines BC, AC, DE and we interpret points as complex numbers, then

$$(C-D)(A-E)(B-F) = -(B-D)(C-E)(A-F)$$

iff the lines AD, BE, CF either have a common point or are all parallel.

Proof. We will prove the first part of the theorem. Let P be the intersection point of AD and BE. We need to choose weights to get the centre of gravity at P. First we choose the weight a_1 at B and the weight a_2 in C. This places their centre of gravity at D and the common centre of gravity is now on the line AD. Next we need to choose the weight x at A. If $xb_1 = a_2b_2$ then we can replace A and C by E and therefore the common centre of gravity of all three weights is on the segment BE as well and therefore at P. Now if we want CF to pass through P, we need the centre of gravity of the weights in A and B to be at F, and therefore

$$xc_2 = a_1c_1 \quad \Leftrightarrow \quad a_1b_1c_1 = a_2b_2c_2.$$

We leave the proof and the interpretation of the second part of the theorem to the reader. We could use the lengths of the corresponding segments instead of complex numbers. Using appropriately chosen signs before each length depending on which part of the line we find the corresponding points and negative barycentric coordinates, we can get a similar proof. Parallel lines appear when the barycentric coordinates $(x : y : z)$ satisfy $x + y + z = 0$, in which case P is at infinity. In some sense three parallel lines also have a common point, but this common point is at infinity.

Usually Ceva's theorem is formulated for vectors, i.e.

$$\frac{CD}{BD} \cdot \frac{AE}{CE} \cdot \frac{BF}{AF} = -1,$$

where we can divide the vectors because they are parallel (but the number obtained after the division can be negative if the vectors point in opposite directions). We usually prefer to use complex numbers to avoid division by zero. □

It is not surprising that the segment connecting the vertex to the point on the line containing the opposite side of the triangle is usually called a **cevian**. It follows immediately from Ceva's theorem that medians, angle bisectors and altitudes each have a common point, called the **centroid** G, the **incentre** I and the **orthocentre** H respectively. Obviously the centroid G has barycentric coordinates $(1 : 1 : 1)$ and therefore divides each median in the ratio $1 : 2$. It is useful to know the barycentric coordinates of other important points as well and we collect these in Table 4.2.

Table 4.2 Barycentric coordinates for the most important points in a triangle. Here a, b, c are the side lengths and $p = \frac{a+b+c}{2}$ is the semiperimeter. To save space, we denote $s_a = b^2 + c^2 - a^2$, $s_b = a^2 + c^2 - b^2$, $s_c = a^2 + b^2 - c^2$.

Point	Notation	Barycentric coordinates
Centroid	G	$(1:1:1)$
Incentre	I	$(a:b:c)$
Excentre	I_A	$(-a:b:c)$
Excentre	I_B	$(a:-b:c)$
Excentre	I_C	$(a:b:-c)$
Orthocentre	H	$(s_bs_c:s_cs_a:s_as_b)$
Circumcentre	O	$(a^2s_a:b^2s_b:c^2s_c)$
Nagel point	N	$(p-a:p-b:p-c)$
Gergonne point	Ge	$((p-b)(p-c):(p-a)(p-c):(p-a)(p-b))$
Symmedian point	L	$(a^2:b^2:c^2)$

Now some comments. The coordinates of the incentre and the excentres follow from Theorems 1.15 and 1.16. The coordinates of the orthocentre we can easily check by the cosine theorem. For example,

$$\cos B = \frac{a^2+c^2-b^2}{2ac} = \frac{s_b}{2ac}$$

and if AD is the altitude then

$$BD:CD = c\cos B : b\cos C = c\cdot\frac{s_b}{2ac} : b\cdot\frac{s_c}{2ab} = s_b : s_c.$$

To get the circumcentre O, we use normalized coordinates for G and H and that $s_as_b + s_as_c = s_a \cdot 2a^2$. Then by Theorem 2.20 we have:

$$\begin{aligned} O &= \frac{3G}{2} - \frac{H}{2} = \frac{1}{2}(1:1:1) - \frac{1}{2(s_bs_c+s_cs_a+s_as_b)}(s_bs_c:s_cs_a:s_as_b) \\ &= \frac{1}{2(s_bs_c+s_cs_a+s_as_b)}(s_as_b+s_as_c:s_as_b+s_bs_c:s_cs_a+s_cs_b) \\ &= (a^2s_a:b^2s_b:c^2s_c). \end{aligned}$$

For the remaining points, the reader can find the definitions and sufficient information to verify the coordinates in Sections 4.1 and 5.1.

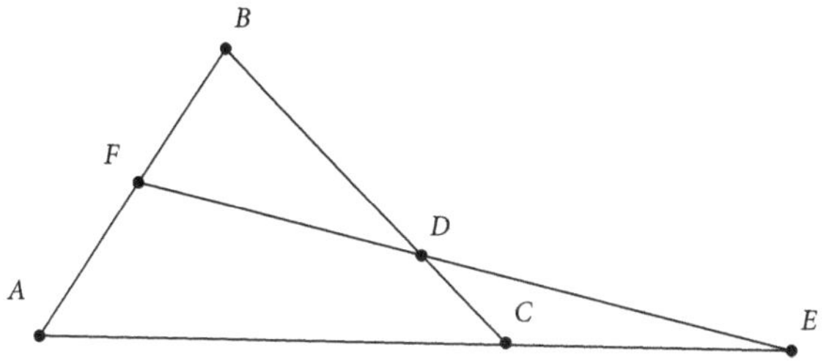

Figure 4.20 Menelaus' theorem

Note that replacing one coordinate by zero we get the coordinates of the point on the corresponding side. For example, if AA_1 is the altitude then the coordinates of A_1 are $(0 : s_c : s_b)$ and $(a^2 s_a : 0 : c^2 s_c)$ is the point where BO intersects AC.

Next we would like to prove another classical theorem.

Theorem 4.17 (Menelaus) *Consider three points D, E, F on the lines BC, AC and AB respectively and interpret them as complex numbers. The points are collinear iff*

$$(C - D)(A - E)(B - F) = (B - D)(C - E)(A - F).$$

Equivalently (if the points are different from the vertices), in vector form:

$$\frac{CD}{BD} \cdot \frac{AE}{CE} \cdot \frac{BF}{AF} = 1.$$

Proof. If D, E and F are collinear, either all three points are outside the triangle or exactly two of them are on the sides. The calculations are similar and we will consider only the case where one point, say E, is outside (see Figure 4.20). The orientation of the vectors shows that two of the fractions are negative and one (namely $\frac{AE}{CE}$) is positive. Thus we only to check the lengths. We find the barycentric coordinates. Obviously,

$$F = (|BF| : |AF| : 0), \quad D = (0 : |CD| : |BD|).$$

To find E note that

$$C = \frac{|CE|}{|AE|}A + \frac{|AC|}{|AE|}E \quad \Rightarrow \quad E = -\frac{|CE|}{|AE|}A + \frac{|AE|}{|AC|}C.$$

Thus $E = (-|CE| : 0 : |AE|)$ and we understand the negative weights better.

Now it is sufficient to use the determinant to get the result:

$$0 = \begin{vmatrix} |BF| & |AF| & 0 \\ 0 & |CD| & |BD| \\ -|CE| & 0 & |AE| \end{vmatrix} = |BF| \cdot |CD| \cdot |AE| - |AF| \cdot |BD| \cdot |CE|$$

and this is the equation for the lengths of the vectors we are looking for.

This was quite technical and not a very nice proof, but we wanted to show the technique to practise using barycentric coordinates. We recommend to the reader to find a simple geometrical proof as well.

We leave it to the reader to fill in some missing details for the converse statement. □

Consider now the following problem.

Problem 4.18 Show that a triangle is isosceles if and only if its centroid G, its incentre I, and its circumcentre O are collinear.

Search. In one direction this is evident. In the opposite, barycentric coordinates provides a direct technical approach. Let a, b, c be the lengths. Then using the coordinates from the table and expanding the determinant we get that

$$0 = \begin{vmatrix} 1 & 1 & 1 \\ a & b & c \\ a^2(b^2+c^2-a^2) & b^2(a^2+c^2-b^2) & c^2(a^2+b^2-c^2) \end{vmatrix}$$

$$\begin{aligned} &= a^4b - a^4c + a^3b^2 - a^3c^2 - a^2b^3 + a^2c^3 \\ &\qquad - ab^4 + ac^4 + b^4c + b^3c^2 - b^2c^3 - bc^4. \end{aligned}$$

Now the challenge is to factorize this expression. A good idea is to consider the right hand side as a polynomial in one variable only, say a. Obviously $a = b$, $a = c$ are roots of the polynomial, so it should be divisible by $(a-b)(a-c) = a^2 - (b+c)a + bc$. Polynomial division gives us:

$$(a-b)(a-c)(a^2b - a^2c + 2ab^2 - 2ac^2 + b^3 + b^2c - bc^2 - c^3) = 0.$$

The last factor should be divisible by $b - c$ as well. Dividing again (this time as a polynomial in b) we get

$$\begin{aligned}(a-b)(a-c)(b-c)&(b^2+2bc+2ab+a^2+c^2+2ac)\\ &=(a-b)(a-c)(b-c)(a+b+c)^2=0,\end{aligned}$$

from which the result follows.

Again quite technical, but this was a really hard problem, which is not easy to solve purely geometrically. Note that we could replace O by H because G, O and H are collinear, but this hardly makes the problem easier. □

4.6 When are the points collinear?

Quite often the geometrical problem is to prove that three different points A, B, C are **collinear**, which means that they lie on the same line. Usually the first impression is that we have no idea how to prove this, because the conditions look completely irrelevant. It may be a good idea to discuss how we in principle can prove such statements: what are the possible ways and instruments? There are only a handful of reasonable approaches and in this section we would like to consider the most useful ones. First we will briefly describe the ideas and after that try them on different problems. For simplicity, we suppose that we want to prove that B lies on the segment AC.

New point. We find another point P such that A, B, P are collinear and B, C, P are collinear. Another variant: we find a new point Q on the line AC and prove that it has the same properties as B.

Angles. We take another point D and prove that $\angle DBA + \angle DBC = \pi$ using other angles and active angle chasing via similar triangles and different circles. Alternatively we draw a line DE containing B and prove that $\angle DBA = \angle EBC$.

Coordinates. We calculate coordinates of our points and prove that the vectors AB and BC are parallel, e.g. by showing that the determinant is equal to zero:

$$\begin{vmatrix} x_A - x_C & y_A - y_C \\ x_B - x_C & y_B - y_C \end{vmatrix} = 0.$$

Complex numbers. We interpret A, B, C as complex numbers and show that $z = \frac{A-C}{B-C}$ is a real number (probably by verifying that $z = \overline{z}$).

Barycentric coordinates. We find the coordinates (or interpret points as complex numbers) and show that $\alpha A + \beta C = B$ for some real α, β with $\alpha + \beta = 1$.

Centre of gravity. We place weights at various points in such a manner that B is the center of gravity of the whole collection and such that we can split the weights in two parts, having A and C as their respective centres of gravity.

Known lines. We show that points belong to well-known lines, e.g. bisectors, radical axes, the Euler line, the Simson line.

Locus. We show that all three points have some common property and that the locus of points with this property is a line.

Homothety or inversion We show that some homothety (or inversion) with its centre in B maps A to C.

Third dimension. We consider the problem in three-dimensional space and show that the points belong to two different planes and therefore lie on their intersection.

Triangle inequality. We show that $|AB| + |BC| = |AC|$.

Projective geometry. We use classical theorems such as Desargues', Pappus's or Pascal's — see Section 12.4.

For example, we have used the approach with barycentric coordinates when we studied the Euler line in Theorem 2.20. Consider now a couple of classical problems.

Problem 4.19 Let M and N be midpoints of the sides AB and CD in the quadrilateral $ABCD$. Let E and F be midpoints of the diagonals AC and BD and L be a midpoint of the segment EF. Show that M, N and L are collinear.

Search. There are so many midpoints that the idea of the centre of gravity is almost obvious. Put equal weights in A, B, C, D. Then the centre of gravity of A and B is at M and the centre of gravity of C and D is at N so the centre of gravity is on the line MN. On the other hand, the centres of gravity for pair of weights of A, C and B, C lies at E and F respectively, and thus the common centre of gravity is at L and we are done. But, of course, we could solve the problem with the help of complex numbers or barycentric coordinates by showing that $L = \frac{M+N}{2}$. Easy problems have many solutions! □

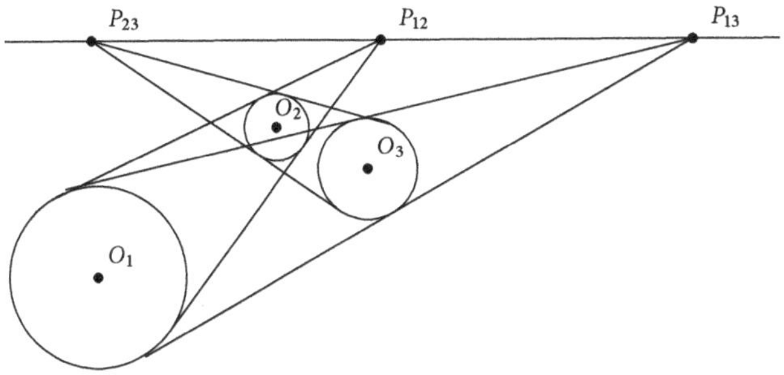

Figure 4.21 Three homotheties, Problem 4.20

Problem 4.20 Consider three circles K_i with different diameters and centres. For each pair of circles consider the two common external tangents and their intersection points P_{ij}. Prove that these three points are collinear.

Search. In fact we do not need a picture here. We see three different homotheties h_{ij} with centres at P_{ij} that send K_i to K_j and try to use them. If we perform first h_{12} and after that h_{23} we obviously send K_1 to K_3. Because the composition of two homotheties with different coefficients is again a homothety (see Section 5.2) we obviously get h_{13}. Good! Now let us study the movement of P_{13}. It moves after h_{12} to some point P and moves back after h_{23}, because it is stable under h_{13}. What does this mean? P_{12}, P_{13} and P are collinear and P_{23}, P, P_{13} are collinear and all four points are collinear!

Here is another, three-dimensional approach. Consider three homothetic cones where the circles are bases. Three vertices of these cones define a plane and all points P_{ij} obviously belong to this plane, because they lie on the lines connecting the vertices. The intersection with the original plane gives us the line we are looking for. □

Problem 4.21 Let O be the centre of the circle inscribed in a quadrilateral $ABCD$. Let E and F be midpoints of the diagonals AC and BD. Show that E, F and O are collinear.

Search. What are reasonable plans? It is probably possible to calculate the distances $|OE|$, $|OF|$ and $|EF|$ in coordinates or complex numbers and to use

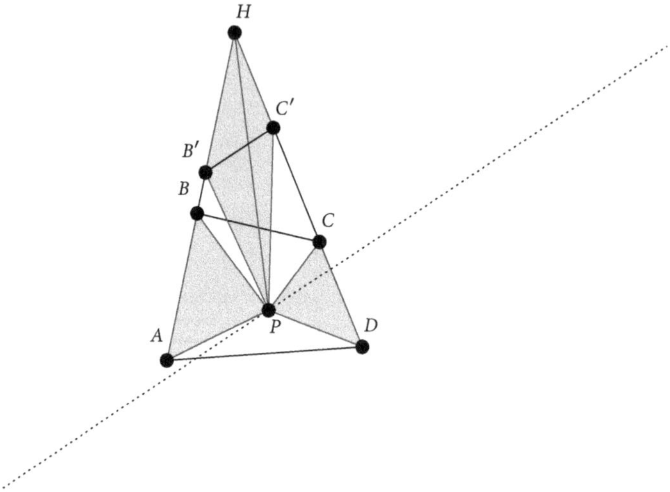

Figure 4.22 Sum of two areas

the triangle inequality, but this does not look very attractive. The centre of gravity approach looks more promising. If we put equal weights (say a) at A and C and equal weights b at B and D we need only to choose a and b in such a way that the common centre of gravity is at O. This still looks like it will take a lot of effort. Perhaps we can use some common properties for the points?

The midpoint reminds us about the median and the medians like areas. Perhaps there are some area properties? Aha, here it is: for all three points P, the sum of the areas of APB and CPD is exactly half of the area of $ABCD$ (for the midpoints we use the fact that the medians divide the area of a triangle into two equal parts; for O that the radii are the altitudes in the corresponding triangles and that $|AB| + |CD| = |AC| + |BD|$).

Now we only need to prove that for any quadrilateral $ABCD$, the locus of the points that have the the same sum of areas of APB and CPD is the line. But is this true? Oh, no! If we consider a square all the points inside the square have this property. Too bad. But perhaps this is only an exception? Not quite, the same problem occurs with any rhombus. On the other hand a rhombus is an exception for our problem as well: in this case E, F and D coincide, since the diagonals are the angle bisectors.

So perhaps this idea works for quadrilaterals that are not parallelograms? Suppose that, for example, AB and CD are not parallel. We need to exploit this fact and the reasonable way to do this is to consider the intersection H of these two lines (see Figure 4.22). How can it help? Can we make some cosmetic improvements? Interesting — we can move the segment AB on the line AB without changing the area. The same applies for CD. So why not move them to H? In other words consider points B', C' on the lines AB and CD such that $|HB'| = |AB|$ and $|HC'| = |DC|$ and now the question is the following: for which points P do we have that the sum of the areas of HPB' and HPC' is the same constant K?

At least this is true for the points on the segment $B'C'$; in this case K is the area of the triangle $HB'C'$. Good! But what if we change the constant? This is easy — we draw a line parallel to $B'C'$ and this is exactly the line we are looking for! It was not trivial, but it was useful! □

Now we would like to consider two important lines we have already mentioned above.

Theorem 4.22 *(i) The following points in the triangle ABC are collinear: the centroid (centre of gravity) G, the orthocenter H, the circumcentre O, the centre N of the nine-point circle, the* **de Longchamps point** *L (which is defined as the orthocentre of the anticomplementary triangle $A'B'C'$ that we get by drawing lines through the vertices A, B, C parallel to the opposite sides). If ABC is an equilateral triangle, then all these points coincide. Otherwise they are different and the corresponding line is called the* **Euler line**. *More exactly:*

$$G = \frac{2}{3}O + \frac{1}{3}H; \quad N = \frac{1}{2}O + \frac{1}{2}H; \quad O = \frac{1}{2}L + \frac{1}{2}H.$$

(ii) The Euler lines of the triangles ABH, BCH, CAH and ABC have a common point (which is N).

(iii) The incentre I of the inscribed circle belongs to the Euler line only when the triangle is isosceles.

(iv) The Euler lines of the triangles ABI, BCI, CAI have a common point S that also belongs to the Euler line of ABC. This point is sometimes called the **Schiffler point**.

Proof. **(i)** The theorem is a convenient collection of the facts proved in the different places. Theorems 2.20 and 4.2 are basic statements. Because the

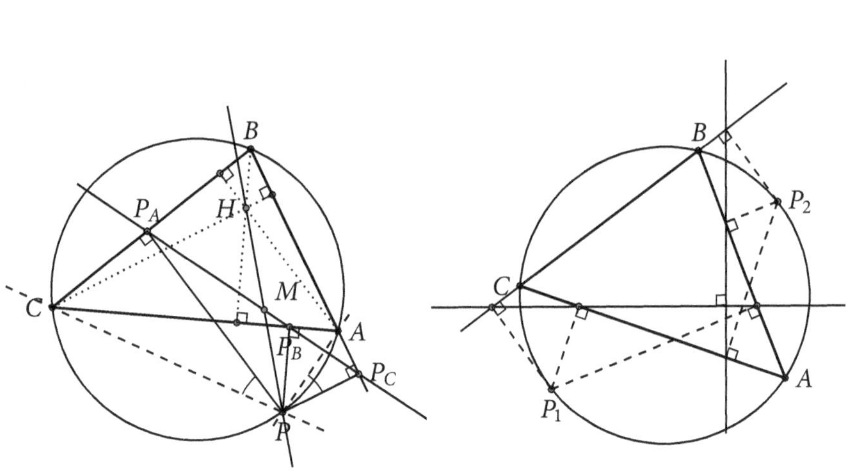

Figure 4.23 The Simson line (left). Simson lines associated with diametrically opposed points are orthogonal (right).

homothety with the centre at G and the coefficient -2 maps ABC to $A'B'C'$, it maps H to L. Thus L lies on the Euler line and

$$\begin{aligned} |HG| : |GL| = 2 \quad &\Rightarrow \quad G = \frac{2}{3}H + \frac{1}{3}L \\ &\Rightarrow \quad \frac{2}{3}O + \frac{1}{3}H = \frac{2}{3}H + \frac{1}{3}L \\ &\Rightarrow \quad O = \frac{1}{2}L + \frac{1}{2}H. \end{aligned}$$

Because ABH, BCH, CAH and ABC have a common nine-point circle and its centre lies on their Euler lines we get the proof of **(ii)**.

In Theorem 4.25 we have a proof of **(iii)**. The last statement is more complicated and we leave it as a good but difficult exercise. □

Theorem 4.23 *Let ABC be an arbitrary triangle and P be an arbitrary point. The orthogonal projections of P on the lines AB, BC and CD are collinear if and only if P belongs to the circumcircle. The corresponding line is called the* ***Simson line*** *of the point P.*

The Simson line bisects the segment PH (where H is the orthocentre) and the corresponding midpoint belongs to the nine-point circle.

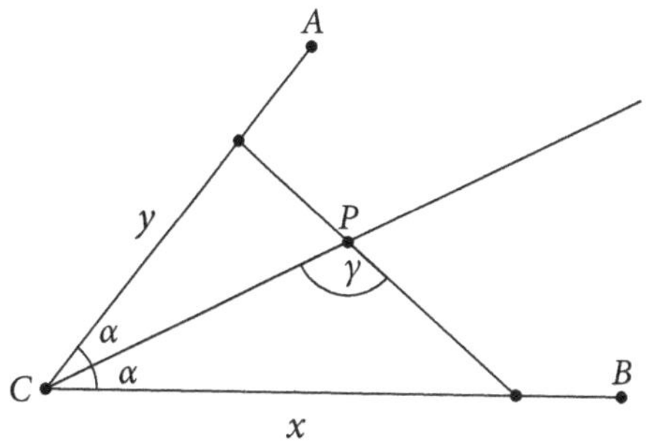

Figure 4.24 A point on the bisector

If P_1, P_2 are two opposite points on the circumcircle, then their Simson lines are orthogonal to each other and and meet on the nine-point circle.

Proof. We will prove only the main statement leaving the other two as a useful exercise. Suppose first that P belongs to the circumcircle, e.g. to the arc AC. Let P_A be the projection on BC, P_B the projection on AC and P_C the projection on AB. We start angle chasing — there is a cyclic quadrilateral PP_ABP_C as a good starting point. This gives us

$$\angle P_APP_C = \pi - \angle ABC = \angle APC \quad \Rightarrow \quad \angle P_CPA = \angle P_APC = \alpha$$

and only one of the points P_A and P_C (say P_A) lies on the side of the triangle. Continuing chasing we find in the cyclic quadrilateral PP_CAP_B that $\angle P_CP_BA = \alpha$ as well. Similarly in PP_BP_AC we find that $P_AP_BC = \alpha$. But this is possible only when these angles are vertical, which proves that P_C, P_B and P_A are collinear.

Reversing this proof (starting from the vertical angles and finishing with $\angle P_APP_C = \pi - \angle ABC$) we get the converse theorem. Check this! □

The Simson line is quite a useful instrument in many IMO problems.

Sometimes there are non-linear equations describing collinearity.

Theorem 4.24 *Let P be a point on the angle bisector of $\angle ACB$. If the line through P cuts the segments of the length x and y from the rays CA and CB, then the sum $\frac{1}{x} + \frac{1}{y}$ is a constant independent on the choice of line.*

In particular, the equation

$$\frac{1}{x} + \frac{1}{y} = \frac{a+b+c}{ab}$$

is satisfied if and only if the line passes through the incentre I and

$$\frac{1}{x} + \frac{1}{y} = \frac{a+b-c}{ab}$$

iff the line passes through the ex-centre I_C.

Proof. Consider Figure 4.24. By the sine rule we have

$$\frac{|CP|}{x} + \frac{|CP|}{y} = \frac{\sin(\gamma+\alpha)}{\sin\gamma} + \frac{\sin(\gamma-\alpha)}{\sin\gamma} = 2\cos\alpha,$$

which proves the first statement.

When the result is so nice there should be some geometrical reason for this. Can we find a nice proof? A golden rule in geometry that we ignored in our picture is **to draw an altitude**! But if we draw two altitudes of the same length h from the point P and look at the areas of triangles we find immediately that

$$\frac{hx}{2} + \frac{hy}{2} = \frac{xy\sin 2\alpha}{2} \quad \Rightarrow \quad \frac{1}{y} + \frac{1}{x} = \frac{2\sin\alpha\cos\alpha}{h} = \frac{2\cos\alpha}{|CP|}.$$

Definitely a nicer proof!

If $P = I$ is the in-centre then we can choose the line that passes through B to find the constant. Then $x = a$ and, because the line is a bisector, $y = \frac{a}{a+c}b$ and we get

$$\frac{1}{x} + \frac{1}{y} = \frac{1}{a} + \frac{a+c}{ab} = \frac{a+b+c}{ab}.$$

To get the result for the ex-circle we use homothety with the coefficient $\frac{p-a}{p} = \frac{a+b-c}{a+b+c}$. □

4.7 Equal sizes in a triangle

Isosceles and equilateral triangles appear so often that it would be nice to have a collection of conditions that are sufficient to deduce these properties. Besides some obvious conditions we would like to present a few non-trivial ones. The theorems below are rather long and do not look attractive. You should probably consider them as compact references. Try looking through the conditions (and proofs!) without any serious attempt to memorize them. You will get a picture in your mind that is easy to recover when you need it. Later, solving other problems, you will be able to recognize some of the conditions below. Return to these theorems and you will find them much more useful and attractive.

We will start with isosceles triangles.

Theorem 4.25 *Let I be the in-centre, H the orthocentre, G the centroid (centre of gravity) and O the centre of the inscribed circle for a triangle ABC. The following conditions are equivalent:*

(i) *ABC is isosceles (the lengths of two of its sides are equal).*

(ii) *Two of the angles in the triangle are equal.*

(iii) *An altitude or an angle bisector divides the area of the triangle into equal parts.*

(iv) *An altitude or a median bisect the corresponding angle.*

(v) *An altitude or a median or an angle bisector divides the perimeter of the triangle into equal parts.*

(vi) *Two of the medians have the same length.*

(vii) *Two altitudes have the same length.*

(viii) *Two angle bisectors have the same length.*

(ix) *Two angle bisectors divide each other in equal ratios.*

(x) *Some line divides the triangle into equal triangles.*

(xi) *The median and the altitude from the same vertex have the same length.*

(xii) *The median and the angle bisector from the same vertex have the same length.*

(xiii) *The altitude and the angle bisector from the same vertex have the same length.*

(xiv) *One of the points I, H, O belongs to the line containing one of the medians.*

(xv) *One of the points G, H, O belongs to the line containing one of the angle bisectors.*

(xvi) *One of the points I, G, O belongs to the line containing one of the altitudes.*

(xvii) *The three points I, H, G lie on the line.*

(xviii) *A line parallel to one of the sides cuts a cyclic quadrilateral.*

If the triangle is acute, there are more conditions.

(xix) *Two altitudes divide each other in equal ratios.*

(xx) *For some point F on the segment AC the triangles AFB and CFB are similar.*

(xxi) *Three of the points A, B, C, I, H, G lie on the line.*

Proof. We recommend you to prove all the equivalences as a good exercise in congruency of triangles, but some comments may be useful. Draw the corresponding pictures if you need to understand them better!

Both *(i)* and *(ii)* are equivalent to the congruency of the triangle ABC to itself (e.g. if $\angle A = \angle C$ then the congruency is between ABC and CBA). It is obvious that in this case the altitude, median and angle bisector from the same vertex coincide. Thus all the remaining statements follow almost immediately and we need only to find the ways back.

The angle bisector always lies between the median and altitude (prove this!). Thus if two of them have the same length all of them coincide. Thus conditions *(xi)–(xiii)* imply *(iv)*.

Because it is the **median** that divides the area into equal parts from *(iii)*, *(iv)* and *(v)* we easily find that the corresponding lines divide a triangle into congruent triangles and we get *(x)*.

The line in *(x)* should pass through one of the vertices (say B). Let F be the intersection of this line with AC. The angle $\angle AFB$ is external to the triangle BFC and therefore can be equal to the angle $\angle BFC$ only. Equal angles correspond to equal lengths of the sides in the congruent triangles and we get *(i)*. Thus all the conditions *(i)–(v)* and *(x)–(xiii)* are equivalent.

If AA_1 and CC_1 are equal medians then the triangles AGC_1 and CGA_1 are congruent (they have two equal sides and an equal angle between them), and thus $|BA| = 2|AC_1| = 2|CA_1| = |BC|$. Thus *(vi)* $\Rightarrow$ *(i)*. Considering the area of the triangle we find that *(vii)* $\Rightarrow$ *(i)* as well.

Let AA_2 and CC_2 be angle bisectors from *(ix)*. Then by Theorem 1.15

$$\frac{|AC_2|}{|AC|} = \frac{|C_2I|}{|IC|} = \frac{|IA_2|}{|AI|} = \frac{|CA_2|}{|AC|} \quad \Rightarrow \quad |AC_2| = |CA_2| = x.$$

The same theorem gives

$$\frac{|AB|}{|AC|} = \frac{|BC| - x}{x}; \quad \frac{|BC|}{|AC|} = \frac{|AB| - x}{x} \quad \Rightarrow \quad \frac{|AB|}{|BC|} = \frac{|BC| - x}{|AB| - x}$$
$$\Rightarrow \quad |AB| = |BC|$$

and we get *(i)* again.

The conditions *(xiv)*–*(xvi)* without the point O are directly translated to *(iii)* or *(iv)*. If O is one of the the points (and B is the vertex) we connect O with A and C to get the isosceles triangle AOC. If BO is the angle bisector we obviously get *(ii)*. If M is the midpoint of AC and B, O, and M are on the same line in the case when BO is the median or the altitude. This obviously implies *(iii)*.

An inscribed trapezoid is isosceles, and thus *(xviii)* $\Rightarrow$ *(ii)*.

Consider now some less trivial implications. For one of them, *(xvii)* $\Rightarrow$ *(i)*, we use the notion of the Euler line and problem 4.18.

If the triangle is acute then *(xxi)* $\Rightarrow$ *(xvii)*, because if we have two vertices then only H can be collinear with them and in this case the triangle is right-angled. To show the implication *(xx)* $\Rightarrow$ *(ii)* we can use the same arguments as in the case *(x)* above. Even here we need to exclude the case of right-angled triangles only.

If the altitudes in *(xix)* intersect each other at Q then Q is inside the triangle and it is acute. Their endpoints form a cyclic quadrilateral and considering the power of the point Q we immediately conclude that their lengths should be equal. Thus we get *(vii)*.

At last consider the implication *(viii)* $\Rightarrow$ *(i)*. Try first to find your own proof before reading! It is much more difficult that one can expect.

In fact, it is much easier to prove the following: If $|AB| > |BC|$ then for the corresponding angle bisectors we have the inequality $|AA_3| > |CC_3|$ (see Figure 4.25).

To prove this suppose that $\angle BAC = 2\alpha < \angle BCA = 2\beta$. Note first that

$$\alpha + \beta = \frac{1}{2}(\angle BAC + \angle BCA) < \frac{\pi}{2}.$$

Let P be a point on AA_3 such that $\angle PCC_3 = \alpha$. Note that

$$|AP| = 2R\sin(\alpha + \beta) > 2R\sin(\alpha + \alpha) = |CC_3|$$

because both angles here are acute. Thus $|AA_3| > |AP| > |CC_3|$. □

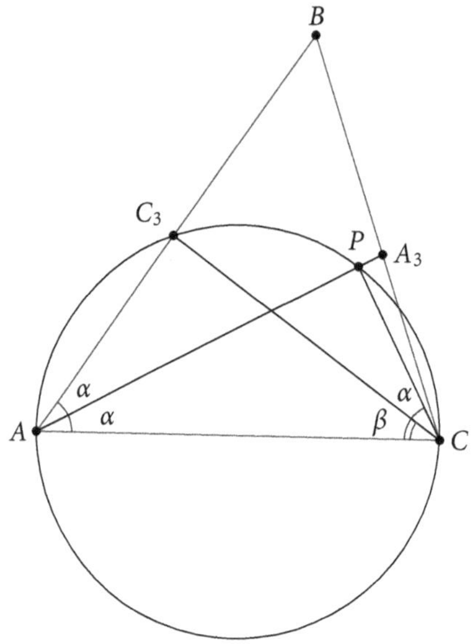

Figure 4.25 $\alpha < \beta \Rightarrow AA_3 > CC_3$

Now consider the conditions for an equilateral triangle. We use the same notation as above.

Theorem 4.26 *The following conditions for the triangle ABC are equivalent.*

(i) ABC is equilateral, and thus all of its sides are equal.

(ii) All its angles are equal.

(iii) The triangle is isosceles and one of the angles is $60°$.

(iv) At least two of the points I, G, H, O coincide.

(v) $A + \varepsilon B + \varepsilon^2 C = 0$, *if we interpret A, B and C as complex numbers and* $\varepsilon = e^{2\pi i/3}$ *or* $\varepsilon = e^{-2\pi i/3}$.

(vi) All medians have the same lengths.

(vii) All altitudes have the same lengths.

(viii) All angle bisectors have the same lengths.

(ix) Two angle bisectors divide each other in ratio $1:2$.

(x) Two altitudes divide each other in ratio $1:2$.

(xi) *For any point inside the triangle the sum of the distances to its three sides is the same.*

(xii) *If r is the radius of the inscribed circle and R is the radius of the circumcircle then $R = 2r$.*

(xiii) *All points P inside the circumcircle satisfy the inequality*

$$\frac{1}{|PA|} + \frac{1}{|PB|} + \frac{1}{|PC|} > 5.$$

(xiv) *A triangle obtained from ABC after some rotation of 120° has its sides parallel to the sides of the triangle ABC.*

Proof. As usual the proof of the equivalence is a good exercise. Here are some hints. The equivalence between *(i)–(iv)*, *(vi)–(viii)* are evident or follows from the previous theorem. The condition *(v)* was proved in Section 2.6. Another convenient condition with complex numbers is $A - B = (C - B)e^{\pm i\pi/3}$.

To see that *(ix)* and *(x)* are equivalent to *(iv)*, it is sufficient to note that if two cevians divide each other in the ratio $1 : 2$ and their intersection P is inside the triangle only if P is the centroid G. To show this consider the barycentric coordinates $(x : y : z)$ of P. We can choose $z = 1$ and suppose that the cevians from the vertices corresponding coordinates x and y are given. These conditions mean that $x + y = 2$ or $2(x + y) = 1$ (it was not given *how* P divides the cevians) and $x + 1 = 2y$ or $2(x + 1) = y$. Because $x, y > 0$ (P is inside!) the only possible solution is $x = y = 1$ and we get the coordinates of G.

To get *(i)* $\Rightarrow$ *(xi)*, divide the triangle into three and calculate the area as the sum of three areas. If two sides are not equal the altitudes from the corresponding vertices are not equal and using continuous arguments we find some points inside the triangle close to the vertices with different sums of the distances.

For *(xii)* we use the Euler theorem that $|OI|^2 = R(R-2r)$, For *(xiii)* we find a point on the circumcircle where we have an equality. Trying the complex numbers in *(xiv)* your learn them better, but there is an easier argument. □

Because triangles and parallelograms are relatives, we would like to present some equivalent conditions for being a parallelogram.

Theorem 4.27 *The following conditions for a quadrilateral $ABCD$ are equivalent:*

(i) *$ABCD$ is a parallelogram: $AB \parallel CD$ and $BC \parallel AD$.*

(ii) *$\angle A = \angle C$ and $\angle B = \angle D$*

(iii) *$\angle A + \angle B = \angle A + \angle C = \pi$.*

(iv) *The triangles ABC and CDA are similar (exactly in this order, $A \leftrightarrow C$, $B \leftrightarrow D$, $C \leftrightarrow A$).*

(v) *The triangles ABC and CDA are congruent (exactly in this order).*

(vi) *Two of opposite sides are parallel and have the same length.*

(vii) *Each pair of the opposite sides have the same length: $|AB| = |CD|$, $|BC| = |AD|$.*

(viii) *The diagonals AC and BD bisect each other.*

(ix) *The angle bisectors of the opposite angles are parallel (for each pair).*

(x) *The angle bisector of $\angle B$ is orthogonal to angle bisectors of $\angle A$ and $\angle C$.*

(xi) *The diagonals divide the area into four equal parts.*

(xii) *The distances from the opposite vertices to the corresponding diagonals are equal.*

(xiii) *The vectors $\overrightarrow{AB}$ and $\overrightarrow{DC}$ are equal.*

(xiv) *If A, B, C, D are interpreted as complex numbers then $A + C = B + D$.*

The proof we leave to the reader — here there are no difficulties at all. The same applies to the following theorem.

Theorem 4.28 *The following conditions for a quadrilateral $ABCD$ are equivalent:*

(i) *$ABCD$ is a rhombus: $|AB| = |BC| = |CD| = |AD|$.*

(ii) *The diagonals are the bisectors of the angles and are orthogonal to each other.*

(iii) *$ABCD$ is a parallelogram and one of the diagonals bisects one of the angles.*

(iv) *$ABCD$ is a parallelogram and the diagonals are orthogonal.*

(v) *All the triangles in which the diagonals divide the quadrilateral are congruent.*

(vi) *ABCD is a parallelogram and has an inscribed circle.*

4.8 Exercises

Exercise 4.1. Given a triangle ABC, denote the orthocentre by H, the circumcentre by O, and the midpoint of the side BC of the triangle by A'. Prove that $AH = 2OA'$.

Exercise 4.2. Let $ABCD$ be a cyclic quadrilateral and let H_A, H_B, H_C and H_D be the orthocentres of the triangles BDC, CDA, DAB and ABC, respectively. Prove that the quadrilaterals $ABCD$ and $H_AH_BH_CH_D$ are congruent.

Exercise 4.3. Let $ABCD$ be a square, and let E be a point on its diagonal BD. Suppose that O_1 is the centre of the circumcircle of $\triangle ABE$ and O_2 is the centre of the circumcircle of $\triangle ADE$. Show that AO_1EO_2 is a square.

Exercise 4.4. Let ABC be an equilateral triangle and let D be an internal point on the side BC. A circle, tangent to BC at D, intersects AB internally at M and N, and AC internally at P and Q. Show that

$$BD + AM + AN = CD + AP + AQ.$$

Exercise 4.5. Prove that if two triangles are inscribed in the same circle, then neither incircle is contained strictly inside the other.

Exercise 4.6. The convex hexagon $ABCDEF$ is such that $\angle B + \angle D + \angle F = 360^\circ$ and $AB \cdot CD \cdot EF = BC \cdot DE \cdot FA$. Prove that $BC \cdot AE \cdot FD = CA \cdot EF \cdot DB$.

Exercise 4.7. For given segment AB describe the set of all possible incentres of triangles ABC.

Exercise 4.8. Denote by θ the angle between the equal legs in an isosceles triangle. Given that the lengths of the sides are $\sqrt{\sin\theta}$, $\sqrt{\sin\theta}$ and $\sin\theta$. Show that the area of the triangle is greater than $\frac{1}{\pi}$.

Exercise 4.9. Prove the identity for the angles in a triangle:

$$\sin\alpha + \sin\beta + \sin\gamma = 4\cos\frac{\alpha}{2}\cos\frac{\beta}{2}\cos\frac{\gamma}{2}$$

See also problems 17, 20, 22, 47, 52, 57, 64, 99, 105, 107, 112, 116 in Chapter 14.

4.9 Hints

4.1. Use the Euler line.

4.2. Use angle chasing to show that ADH_AH_D is a parallelogram.

4.3. Use that $AO_1 = O_1E$ and $EO_2 = O_2A$ and that $\angle ABE = \angle ABD = \angle ADB = \angle ADE = 45°$.

4.4. Use the power of the point.

4.5. Use that $|OI_i|^2 = R(R - 2r_i)$ and apply triangle inequality.

4.6. Translate the problem to the language of complex numbers.

4.7. What are possible values for the angle AIB?

4.8. $\frac{8}{25} > \frac{1}{\pi}$.

4.9. Transform all products to the sums.

CHAPTER 5

Geometric transformations

One part of geometry whose importance is often underestimated are geometric transformations. The main idea is to create, using such transformations, new objects that are related to the original ones, but are *different* and sometimes more convenient. It is not so easy to find an appropriate transformation, but when the reader succeeds creating their first proof using geometric transformations, they will definitely enjoy it and will be eager to learn more. The aim of this chapter is to help them to create this first proof. We will consider the complex number interpretations of our transformations, because sometimes they are easier to understand.

5.1 Reflection

We will begin with an easy question. Look at Figure 5.1 and try to figure out how to continue the sequence. If it was easy for you ask your friends — you will be surprised how difficult is to see the evident for many of us. It looks so trivial: each picture is symmetric so consider the left part of each of them and the pattern become obvious. Why was this so difficult? The answer is

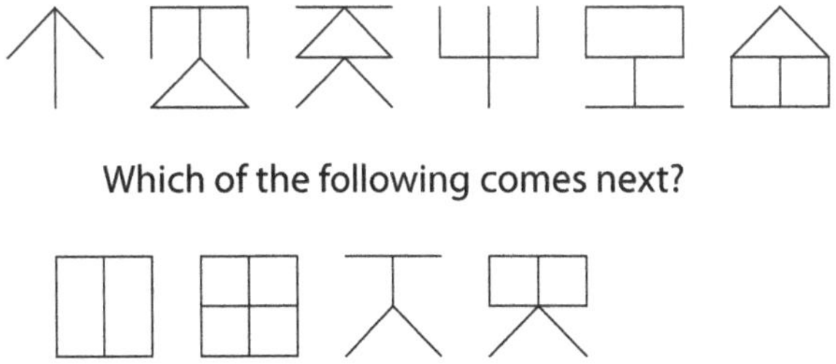

Figure 5.1 Consider the symbols in the first row. Which of the symbols in the second row should come next?

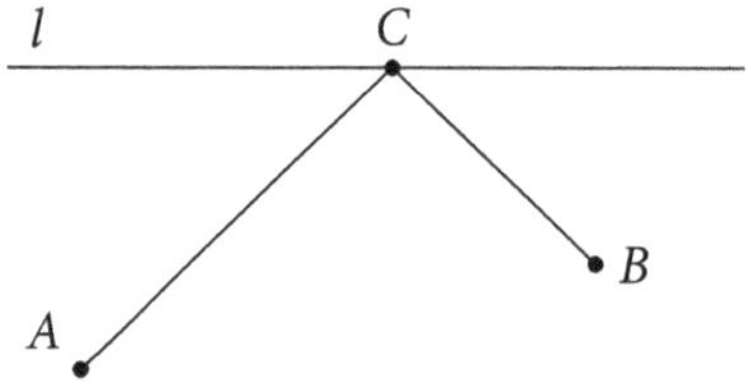

Figure 5.2 Minimize $|AC| + |BC|$

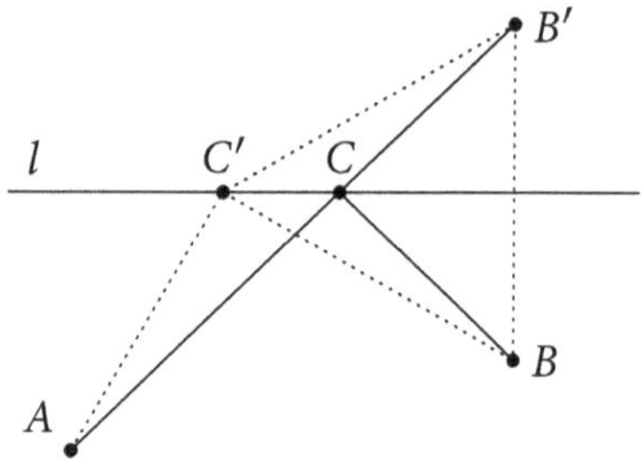

Figure 5.3 Reflection

that the reflection, which creates the symmetric picture from its left part is much deeper thing that we normally think. We simply underestimate it.

Now consider a classic example.

Problem 5.1 Let A and B be two points on the same side of the line l (see Figure 5.2). How should we choose the point C on l such that $|AC| + |BC|$ is minimal?

Search. How do we find C? To write an expression for the distance expressed in coordinates looks like the least attractive option. Perhaps some physics can help? We remember that when light finds the shortest path, there are some conditions about equal angles.

Probably this is the answer, but how do we find such a C and how do we explain this mathematically? Why does light reflect in a mirror in such a way? Mirror…Let's look inside the mirror and make a reflection (see Figure 5.3)! Now all becomes evident. If B' is the mirror image of B then C is obviously the intersection of AB' and l. Indeed: if C' is another point on the line then

$$|AC'| + |BC'| = |AC'| + |B'C'| > |AB'| = |AC| + |B'C| = |AC| + |BC|$$

and we are done. □

Though this looks like a trivial solution, it is not! Consider a slightly more advanced problem.

Problem 5.2 Through a point P exterior to a given circle passes a secant and a tangent to the circle. The secant intersects the circle at A and B, and the tangent touches the circle at C, on the same side of the diameter through P as

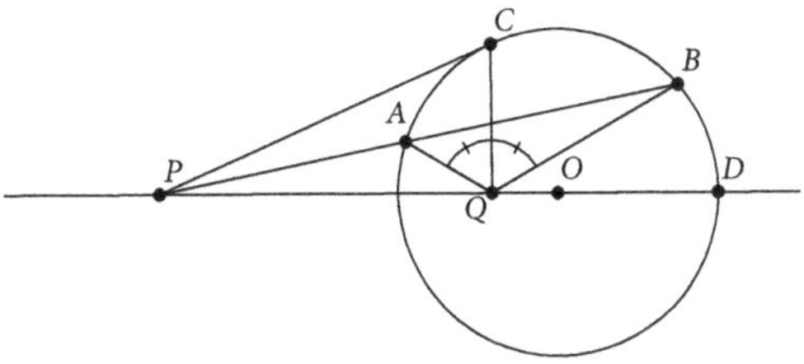

Figure 5.4 Tangent and secant

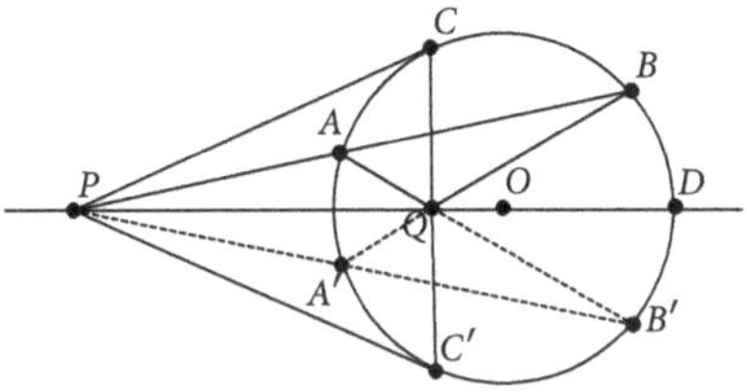

Figure 5.5 Tangent and secant reflected

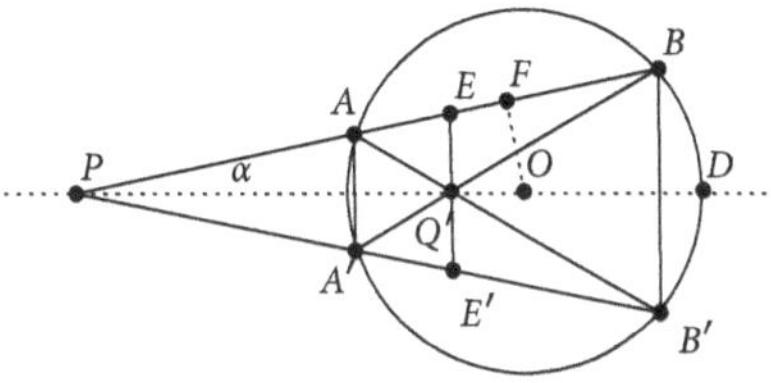

Figure 5.6 Q' is the same for all A

AB. The orthogonal projection of C onto the diameter is Q. Prove that QC bisects $\angle AQB$.

Search. This looks like an easy exercise but the problem puts up a bigger fight than expected. In fact, this is a good sign — when we solve the problem we learn something important. But how do we start here? How do use the fact that we have a tangent? Either we need angle chasing or we need to user a power of the point P or both. The latter looks more promising because the point A can be chosen arbitrarily. But we still do not know how to use this idea. How can we finish the proof? If CQ is a bisector then we have $\angle PQA = \angle BQD$ as well. This reminds us of the previous problem 5.1, and the idea of using reflection as a first step becomes natural.

Let us look at the completed drawing (Figure 5.5). Now we see the picture much better: all we need is to prove that A', Q and B are collinear. A natural way to do this is to consider the point Q': the intersection of AB' and $A'B$, which by symmetry is on the line PD and it remains to prove that it coincides with Q. Because A is arbitrary we need only to prove that Q' is always the same (the tangent will be the limit case). So we have reduced the problem to a much easier one: to show that Q' is independent of the choice of A (see Figure 5.6). This we can calculate directly. $|PQ'| = |EQ'| \cot\alpha$. We recognize another familiar picture and because $|EE'|$ is a harmonic mean of $|AA'|$ and $|BB'|$ (see page 49 or prove it directly) we find that $|EQ'| = \frac{|AA'|\cdot|BB'|}{|AA'|+|BB'|}$. But $\frac{|AA'|}{|PA|} = \frac{|BB'|}{|PB|} = 2\sin\alpha$ so we need only to show that

$$|PQ'| = \frac{|AA'|\cdot|BB'|}{|AA'|+|BB'|}\cot\alpha = 2\frac{|PA|\cdot|PB|}{|PA|+|PB|}\cos\alpha$$

is independent of A. By the power of the point this is true for the numerator, so our last task is to get (using the midpoint F on AB) that

$$\frac{|PA|+|PB|}{\cos\alpha} = \frac{2|PA|+|AB|}{\cos\alpha} = 2\frac{|PA|+|AF|}{\cos\alpha} = 2\frac{|PF|}{\cos\alpha} = 2|PO|$$

is independent on A as well. It is neither the shortest nor the prettiest solution, but it was exactly the way one of the authors used to solve the problem. □

Angle bisectors are the lines that are most often used as reflection lines. The reason is evident: the angle maps to itself, so after reflection we have something very close to our original problem. Naturally, we mostly use an angle in a triangle. If the sides of the triangle are a, b, c then we know that the bisector CD of the angle $\angle C$ divides c in the ratio $b : a$. Suppose that CE is another line through the same vertex, which divides c in the ratio $x : y$. If CE' is the reflection of this line over the bisector, it is easy to check (do this, using e.g. the sine rule!) that this line divides c in the ratio $\frac{b^2}{a^2}\cdot\frac{y}{x}$. If CE is a median, we obtain a very useful line called the **symmedian**, which divides c in the ratio $b^2 : a^2$ (see Figure 5.7). Note that the important part of the line — the segment CF is also often called the symmedian.

The formula above also shows that the reflections of three concurrent cevians are concurrent cevians as well. The common point L of the three symmedians has its own name (the **Lemoine** or the **symmedian point**) and evidently has the barycentric coordinates $(a^2 : b^2 : c^2)$. To show that a symmedian is indeed a useful instrument, we will formulate a trivial but useful theorem.

Theorem 5.3 *Consider a triangle ABC and points A_1, B_1 different from the vertices on the segments BC and AC (Figure 5.8). The following are equivalent.*

- *The quadrilateral ABA_1B_1 is cyclic.*

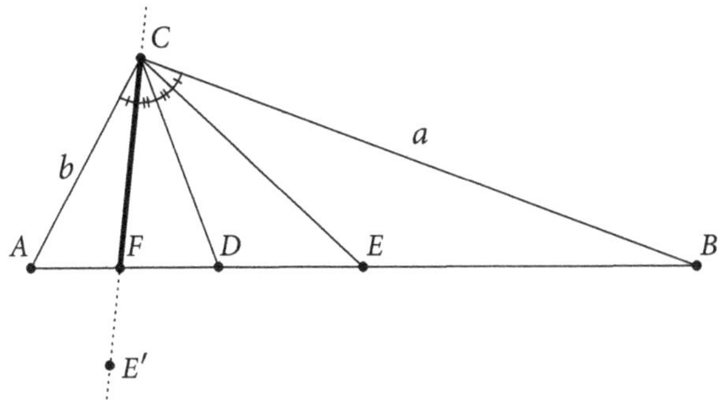

Figure 5.7 CE — median, CE' — symmedian, $|AF| : |FB| = b^2 : a^2$

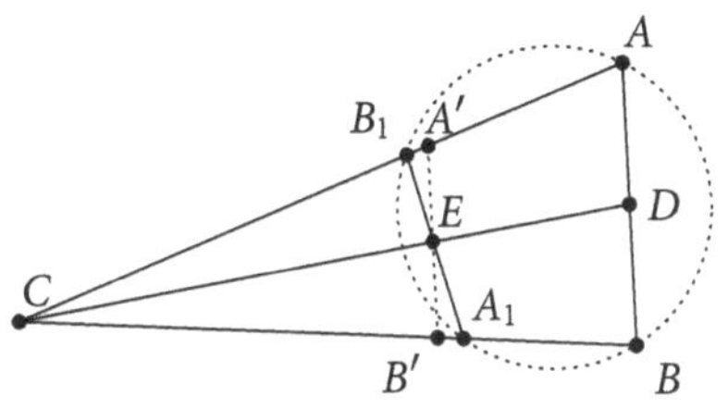

Figure 5.8 CD– median, CE– symmedian

- *A_1B_1 is **antiparallel** to AB, which means that the triangles ABC and A_1B_1C are similar (in this order, so A corresponds to A_1 and B corresponds to B_1).*
- $|CA| \cdot |CB_1| = |CB| \cdot |CA_1|$.
- *The line that bisects segment AB (median) is a symmedian in AB_1C_1.*
- *The line that bisects segment A_1B_1 (median) is a symmedian in ABC.*

Proof. Most equivalences are already considered when we discussed the power of the point in Theorem 4.10. The new ones are the symmedian statements, which follow directly if we use reflection over the angle bisector, which transfers the antiparallel line A_1B_1 to a parallel line $A'B'$. For the converse, we need the slightly less trivial statement that if a line through C bisects two segments AB and $A'B'$ (A' belongs to the segment AC and B' to the segment BC) then the segments are parallel. We leave these details to the reader. □

If we move A_1 towards B we get another useful theorem giving equivalent conditions to the condition that CB is a tangent to the circumcircle of the triangle ABB_1 at the point B. We challenge the reader to formulate and prove the theorem.

Here is an example where a symmedian appears naturally.

Problem 5.4 The tangents at points C and C' to the circumcircle of the triangle CBC' intersect at the point P. Prove that PB is a symmedian in this triangle.

Search. As most problems concerning symmedians, the solution is harder to find than expected. Here we can use Figure 5.5 (which explains the unusual notation). We see that BQ is a median and need only to prove that the angle bisector of $\angle CBC'$ is simultaneously the angle bisector of $\angle ABA'$. But this follows from the symmetry, because the arc AC is equal to the arc $A'C'$ thus $\angle CBA = \angle A'BC'$. All the difficulties were in the proof above that the picture is correct. Thus it is worth remembering.

Note that there exists another case, when B lies on the opposite side to CC'. We leave this to the reader. □

Now we would like to discuss some computational aspects. In the language of complex numbers, there is an important reflection: $z \mapsto \bar{z}$. To find a reflection over a line through the origin, it is sufficient to know a vector v parallel to this line. Then the arbitrary vector u reflects to $u - 2\frac{u \cdot v}{|v|^2}v$. If v has the coordinates $X = \binom{a}{b}$, then the corresponding map can be described as a matrix multiplication:

$$\begin{pmatrix} x \\ y \end{pmatrix} \mapsto A \begin{pmatrix} x \\ y \end{pmatrix}$$

where

$$A = I - 2\frac{XX^T}{X^TX} = \begin{pmatrix} 1 & 0 \\ 0 & 1 \end{pmatrix} - \frac{2}{a^2+b^2}\begin{pmatrix} a^2 & ab \\ ab & b^2 \end{pmatrix}$$

5.2 Homothety

A **homothety** is a map defined by a point O — the centre of the homothety and a non-zero real number k (ratio). Each point P maps to the point P' such that P, P' and O are collinear and $OP' = k \cdot OP$. Here, we view OP

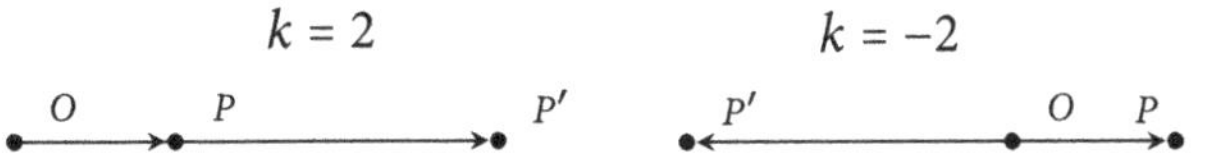

Figure 5.9 Homothety maps P to P'.

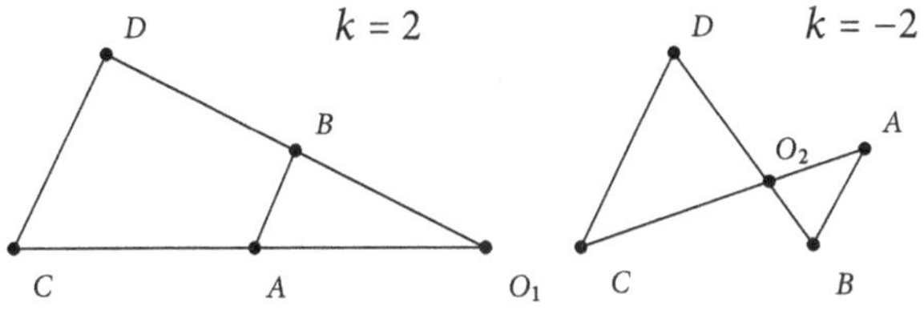

Figure 5.10 Two homotheties mapping AB to CD.

and OP' as vectors. Thus if $k > 0$ then the points P, P' are on the same side of O. If $k < 0$ then O is between P and P' — compare the cases $k = 2$ and $k = -2$ in Figure 5.9.

If we interpret points as complex numbers and O is chosen as 0 then the map is simply multiplication by k. This allows easy proofs of the following results (though we encourage the reader to find geometrical proofs as well).

Theorem 5.5 *Any homothety maps parallel lines to parallel lines, circles to circles and does not change angles or ratios of lengths. More generally it maps any figure to a similar figure whose area is multiplied by k^2. In particular midpoints map to midpoints, medians to medians, etc.*

Vice versa, for any two parallel segments AB and CD of different lengths there exist two homotheties that send AB to CD (only one homothety with $k = -1$ if they have the same length but are different).

Similarly for any two circles with different radii, there exist two homotheties that send one circle to the other (and only one if they are different and have the same radius).

Proof. Consider for example a circle with the equation $|z - z_0| = R$. Then $z' = kz$ satisfies the equation $|z' - kz_0| = |k|R$ and we see that its centre z_0 maps exactly to kz_0 while the radius is multiplied by $|k|$. Similarly one can prove other statements.

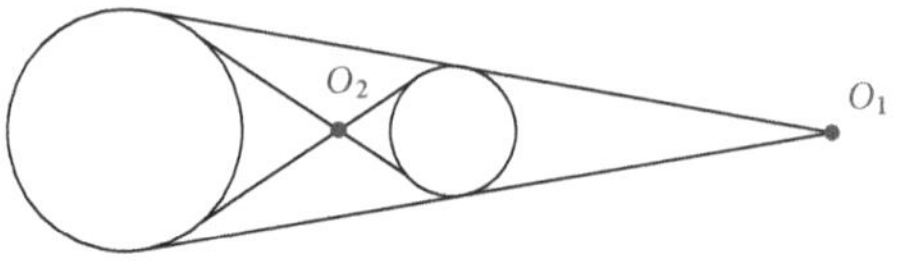

Figure 5.11 Two homotheties mapping the circles.

More interesting are the converses. For the segments there are two cases: either A maps to C and B to D or A maps to D and B to C. Connecting the corresponding points we find the possible centres O_1, O_2 — see Figure 5.10.

Generally, if we know P, its image P' and $k < 0$ then we can find the centre O using barycentric coordinates. Indeed

$$O = \frac{|k|}{|k|+1}P + \frac{1}{|k|+1}P' = \frac{k}{k-1}P + \frac{1}{1-k}P'.$$

The reader should check why the same formula works even for $k > 0$ when $k \neq 1$.

This allows us to find two homotheties that send a circle with centre P and radius r to another circle with centre P' and radius R. Indeed, knowing the radii we find $k = \pm\frac{R}{r}$ and we can use the formula above. □

It the circles are outside each other we can construct the homotheties geometrically using the common tangents: external for $k > 0$ and internal for $k < 0$ — see Figure 5.11. In the last case the circles should not have any intersections. Obviously, if two circles tangent each other then their common point is a very useful homothety centre.

The nice properties of homothety can be applied to many problems. Sending two similar figures to each other we can find many interesting connections and similarities. We have already seen such examples in the proof of Theorems 4.2 and 4.22 or in the solution to the problem on page 48. Consider another one, where we use both symmetry and homothety.

Problem 5.6 The point P lies inside the triangle ABC and is such that $\angle BPC$ is a right angle and $\angle BAP = \angle BCP$. If M, N denote the midpoints of the segments AC, BC, respectively, prove that $\angle PMN$ is a right angle.

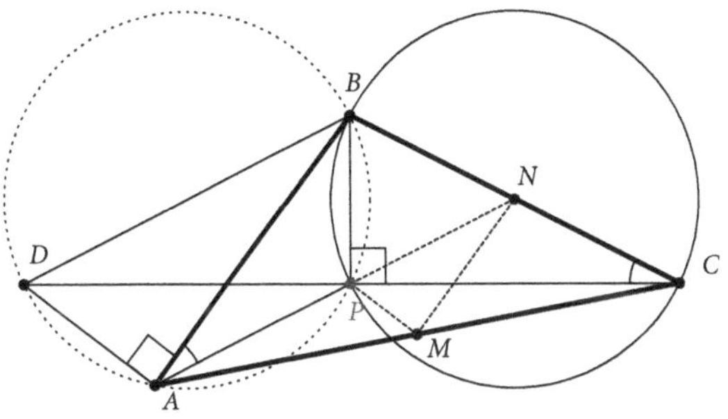

Figure 5.12 ABD and MNP are homothetic.

Solution. Draw a circle k on BC as a diameter. It is then given that the point P lies inside the triangle and on the circle k. Draw a second circle k_1 through the points B and P, and of the same radius as k. Since $\angle BAP = \angle BCP$, it follows that the point A lies on k_1. Let D be the (second) point of intersection of the line CP and k_1. Since $BP \perp CD$, BD is a diameter of k_1, and thus $\angle BAD = 90°$. Also, by congruence, $CP = DP$. It now follows that the triangles BAD and NMP are homothetic, with centre C and ratio $\frac{1}{2}$, and thus that $\angle PMN = \angle DAB = 90°$. □

Suppose that we perform two homotheties: first with centre O_1 and ratio k_1 and then with centre O_2 and ratio k_2. Is the composition a homothety as well? The answer is almost *yes*.

We describe the first homothety using the language of complex numbers. Because the centre is not necessarily at the origin, we need to make a change of variables, $w = z - o_1$, and to rewrite the map $w \mapsto k_1 w$ in the original variables to get $z \mapsto k_1(z - o_1) + o_1$. Rewriting the second homothety in a similar fashion as $z \mapsto k_2(z - o_2) + o_2$ we find that their composition is

$$\begin{aligned} z \mapsto f(z) &= k_2(k_1(z - o_1) + o_1 - o_2) + o_2 \\ &= k_1 k_2 z - k_1 k_2(o_1 - o_2) + o_2. \end{aligned}$$

So, if this is a homothety then the ratio is evidently $k = k_1 k_2$. What is the centre? The centre is a fixed point of the mapping and we can find it by solving the equation $f(z) = z$. When $k \neq 1$ there is solution and the composition is a homothety. But when $k = 1$ there is a solution only when $o_1 = o_2$. Otherwise the map $z \mapsto f(z)$ will be a parallel translation (another very useful geometrical transformation). See page 177 for an example of how this (not so obvious) result can be used.

In Section 2.6, we used another powerful transformation — rotation. Rotations can be combined with a homotheties. When their centres coincide with the origin, there is a very natural interpretation of the composition — it is simply multiplication by a constant complex number. Geometrically this is far from being evident, but it is important that using such maps, we can send any segment to another one preserving the angles and similarities while having control over how the lengths are changing. Such advanced approaches are not used very often, however; there are almost always simpler ways.

5.3 Exercises

Exercise 5.1. Let f be a geometric congruence in the plane that is a composition (in arbitrary order) of a translation and a rotation (where the angle of rotation is not a multiple of 2π). Show that there exists a unique point P such that $|PA| = |Pf(A)|$ for each point A.

Exercise 5.2. a) Show that if an altitude in a triangle is its symmedian, then the triangle is either right-angled or isosceles.

b) Show that the Lemoine point of a right-angled triangle is the midpoint of its altitude on the hypotenuse.

Exercise 5.3. Let P be the intersection point of the line AC and the tangent in B to the circumcircle of the triangle ABC. Let PD be another tangent. Show that the line BD is a symmedian in ABC.

Exercise 5.4. The circles ω_1 and ω_2 intersect at A and B. Their common (outer) tangents intersect at O. A line l through O intersects ω_1 at P and some other point, and ω_2 at Q and some other point, where P and Q are closer to O than the other points of intersection. Denote $M = AP \cap BQ$, $N = AQ \cap BP$, and let C be a point on l such that $CM = CN$. Prove that the following does not depend on the choice of l:

- the measure of $\angle AMB$;
- the length of the segment MN;
- the length of the segment CM.

Exercise 5.5. Let ABC be a triangle, let B_1 be the midpoint of the side AB and C_1 the midpoint of the side AC. Let P be the point of intersection, other than A, of the circumscribed circles around the triangles ABC_1 and AB_1C. Let P_1 be the point of intersection, other than A, of the line AP with the circumscribed circle around the triangle AB_1C_1. Prove that $2AP_1 = 3AP$.

See also problems 18, 21, 29, 39 in Chapter 14.

5.4 Hints

5.1. Study the fixed point.

5.2. Use that the altitude is antiparallel to a cathetus.

5.3. Show that BD and tangents in A and C are concurrent.

5.4. O is perfect as a homothety centre. Other homotheties are useful as well.

5.5. Use another transformation - the inversion (read section 12.3 if necessary).

CHAPTER 6

Combinatorics

6.1 Enumerating the objects

In some sense, combinatorics begins with enumeration. The first techniques we learn are how to enumerate different sets of objects: We start with a finite set N with n elements.

- There are 2^n different subsets of N (because we have two choices for each element: to be or not to be an element in the subset).
- There are $n!$ permutations of the elements in N (the first object can be placed in n different ways, the second in $n - 1$ and so on).
- There exist m^n functions $f : N \to M$ from N to a set M with with m elements. (Each function value can be chosen in m different ways, and there are n of them.) By taking $M = \{0, 1\}$, we can identify a function $f : N \to M$ with a subset of N (how?) and recover the first formula above.

We also learn intuitively plausible multiplicative and additive principles: when to add and when to multiply while enumerating (try to formulate them rigorously — keywords are *independent* and *mutually exclusive* choices).

Next we learn how many ways there are to choose k objects from N. The answer is $\binom{n}{k}$ and we open up ourselves to the incredible ocean of binomial coefficients.

Before we dive into this ocean, we would like to point out that there are in fact *four, different* possible answers to the question "in how many ways can we choose k objects from N". Indeed, the question is ill-formulated. There are two additional pieces of information that we did not specify: a) whether the order in which we choose the objects is essential, and b) whether an object may be chosen several times, i.e if repetition is allowed or not. The complete answer to the question is summarized in Table 6.1.

When the conditions are not specified we mean that we do not allow repetition and the order is irrelevant (and this is a typical choice). We will always

Table 6.1 Number of ways to enumerate objects. In how may ways can we choose k objects from a set of n objects?

	Repetitions are allowed	Repetitions are not allowed
Order is relevant	n^k	$\frac{n!}{(n-k)!}$
Order is not relevant	$\binom{n+k-1}{k}$	$\binom{n}{k}$

follow this agreement. Now we would like to prove the least obvious formula in the table (all others follow from the multiplicative principle — check!).

Theorem 6.1 *The equation* $x_1 + \cdots + x_n = k$ *has* $\binom{n+k-1}{k}$ *solutions in non-negative integers* x_i.

Proof. First we note that every solution corresponds to a particular way of choosing k objects from N where repetition is allowed and the order of the choice is irrelevant: We interpret x_i as the number of times we have chosen the ith object.

Next, we give a different interpretation. Write k stars in a row and divide them into n groups (possibly empty), using $|$ as separators. We claim that every such distribution of stars and bars corresponds to a solution to the equation, and vice versa, each solution can be interpreted in this language. For example, the grouping $| * * | * | | | *$ corresponds to the solution $0+2+1+0+0+1=4$ to the equation $x_1 + x_2 + x_3 + x_4 + x_5 + x_6 = 4$.

Now comes the magical trick. The group consists of k stars and $n-1$ bars, $k+n-1$ objects in total. To create such a group we take $k+n-1$ stars, choose k of them and replace them by bars. This can be done in precisely $\binom{n+k-1}{k}$ different ways. □

The importance of this proof is not only the result but the path leading up to it. We found *two different* interpretations of the numbers we were looking for, and this is a common technique in combinatorics — trying to find alternative interpretations to rephrase the problem.

It is time to collect (and even prove) some well-known facts about binomial coefficients $\binom{n}{k}$.

First of all,

$$\binom{n}{k} = \frac{n!}{k!(n-k)!} = \frac{n(n-1)(n-2)\cdots(n-k+1)}{k!},$$

where the second formula is more convenient, at least for computing by hand. The rule to remember is that $\binom{n}{k}$ is a polynomial (in n) of degree k, so there are exactly k factors in the numerator. For example,

$$\binom{n}{4} = \frac{n(n-1)(n-2)(n-3)}{4!}, \quad \binom{n}{1} = n, \quad \binom{n}{0} = 1.$$

Note that $\binom{n}{k} = 0$ for $k > n$ or $k < 0$ (there is nothing to choose!).

The most important properties of the binomial coefficients are the following:

Theorem 6.2

$$\binom{n}{k} = \binom{n}{n-k};$$

$$\binom{n}{k} = \binom{n-1}{k-1} + \binom{n-1}{k}, \qquad \textit{for } n > 0;$$

$$\sum_{k=0}^{n} \binom{n}{k} a^k b^{n-k} = (a+b)^n, \qquad \textbf{\textit{the binomial theorem.}}$$

Proof. Choosing k objects from a set of n objects is the same as choosing the $n - k$ that remain. This proves the first formula.

For the second formula, imagine that one of the n objects is specially marked. Any subset of k chosen objects either contains the marked object or it does not. There are $\binom{n-1}{k}$ subsets *not* containing the marked object, and $\binom{n-1}{k-1}$ subsets that contain the marked object (if we pick the marked object, we should only pick $k - 1$ unmarked ones). This is the interpretation of the second formula.

For the final formula, expanding the product $(a + b)(a + b)\cdots(a + b)$, there are $\binom{n}{k}$ ways to form the term $a^k b^{n-k}$: we have to choose a from k factors (and b from the remaining factors $n - k$ factors). □

For example

$$\binom{n}{n-2} = \binom{n}{2} = \frac{n(n-1)}{2}.$$

The second formula in the theorem can be rewritten as the well-known Pascal triangle:

$$
\begin{array}{ccccccccc}
 & & & & 1 & & & & \\
 & & & 1 & & 1 & & & \\
 & & 1 & & 2 & & 1 & & \\
 & 1 & & 3 & & 3 & & 1 & \\
1 & & 4 & & 6 & & 4 & & 1
\end{array}
$$

where each number is the sum of the two above it. In general, a row looks like

$$\binom{n}{0} \quad \binom{n}{1} \quad \binom{n}{2} \quad \cdots \quad \binom{n}{n-1} \quad \binom{n}{n}.$$

Let us write down a few more advanced formulas.

Theorem 6.3 *let n, m be non-negative integers. Then*

$$\binom{n}{0}+\binom{n}{1}+\binom{n}{2}+\binom{n}{3}+\cdots+\binom{n}{n}=2^n;$$
$$\binom{n}{0}-\binom{n}{1}+\binom{n}{2}+\cdots+(-1)^n\binom{n}{n}=0;$$
$$\binom{n}{0}+\binom{n+1}{1}+\cdots+\binom{n+m}{m}=\binom{n+m+1}{m};$$
$$\binom{n}{0}^2+\binom{n}{1}^2+\binom{n}{2}^2+\cdots+\binom{n}{n}^2=\binom{2n}{n}.$$

Proof. Each of these formulas can be proven in different ways. We show only one for each, encouraging the reader to find others.

In the first formula, there are two different ways to compute the number of subsets of a set with n elements. In the left hand side is the number of subsets with 0 elements plus the number of subsets with 1 elements and so on. This sum must the total number of subsets, which we already know is 2^n.

Putting $a = 1$, $b = -1$ in the Binomial Theorem 6.2, we get the second formula.

Induction on m allows us to rewrite the left-hand side in the third formula as $\binom{n+m}{m-1}+\binom{n+m}{m}$ and get $\binom{n+m+1}{m}$ according to Theorem 6.2.

Consider the expansions of

$$(1+t)^n(1+t)^m=(1+t)^{m+n}$$

and compare the coefficient of $t^n = t^k t^{n-k}$ in the two expansions. We get the more general forumla

$$\sum_{k=0}^{n} \binom{n}{k}\binom{m}{n-k} = \binom{n+m}{n}$$

and specializing to $m = n$, we recover the formula we were aiming for, again with some help of Theorem 6.2. □

6.2 Elements of graph theory

Graph theory is not part of standard school mathematics, but is very useful to know when solving olympiad problems. We would like to discuss some basics of graph theory, mostly the definitions, notations and a few simple, important theorems.

A **graph** G consists of two sets — the set of **vertices** V and the set of **edges** E. An edge e is simply a pair of *different* vertices $v_1, v_2 \in V$ which we interpret as a connection between two vertices. The terminology is sometimes a little different, but we will assume that in our graphs, for each pair of vertices v_1 and v_2 there is *at most* one edge connecting them. Also, we do not distinguish between the edges v_1, v_2 and v_2, v_1.

If we consider *directed edges*, where v_1, v_2 is different from v_2, v_1, we call the graph **oriented.** If we allow multiple edges between pairs of vertices we speak about **multigraphs** (oriented or not). In (oriented) multigraphs, we allow loops — edges starting and ending at the same vertex.

In this book, we will mostly deal with graphs.

To visualize a graph, it is convenient to make a picture, drawing vertices as points and edges as segments or curves connecting them (with a small arrow if the graph is oriented). Such segments may intersect in other points that are not vertices, and we ignore those extraneous intersections. Here some important examples (see Figure 6.1 for illustrations):

The **complete graph** K_n is the graph with n vertices in which each pair of distinct vertices is connected by an edge. It obviously has $\binom{n}{2}$ edges.

The **cycle graph** (or simply a **cycle**) C_n has vertices $v_1, \ldots, v_n$ and edges only between v_i and v_{i+1} for all $i < n$ and between v_n and v_1. We can visualize C_n as a regular polygon; C_3 is a triangle and C_4 a square.

In a **bipartite graph**, the set of vertices can be divided into two disjoint sets V_1 and V_2 such that every edge has one end in V_1 and the other in V_2. A

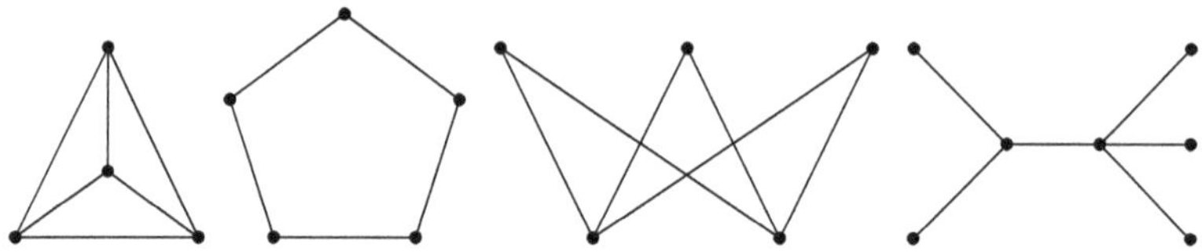

(a) K_4 (b) C_5 (c) $K_{2,3}$ (d) A tree

Figure 6.1 Some examples of graphs

typical example is formed by a set of clubs D_i and their members x_j. If we draw an edge for every membership $x_j \in D_i$ we get a bipartite graph, where V_1 is the set of the clubs and V_2 the set of all their members.

A complete bipartite graph is a bipartite graph such that for every $v_1 \in V_1$ and $v_2 \in V_2$ there is a connecting edge. If $|V_1| = m, |V_2| = n$ then the graph is denoted $K_{m,n}$ and has mn edges.

A graph $G_1 = (V_1, E_1)$ is a **subgraph** of $G_2 = (V_2, E_2)$ if $V_1 \subseteq V_2$ and $E_1 \subseteq E_2$.

We should also say which graphs we consider to be the same. A graph $G_1 = (V_1, E_1)$ is **isomorphic** to a graph $G_2 = (V_2, E_2)$ if there exists a bijection between vertices $f : V_1 \to V_2$ such that $\{v_1, v_2\} \in E_1 \Leftrightarrow \{f(v_1), f(v_2)\} \in E_2$.

A **path of length** n in a graph is a sequence of vertices $v_0, \ldots, v_n$ such that there exists an edge between v_{i-1} and v_i for every $i = 1, \ldots, n$. If all the vertices are different we call it a **simple path** and can identify it with a subgraph.

A graph is **connected**, if for any two vertices v and w there is a path from v to w.

A path with $v_0 = v_n$ is called a **cycle**, and a **simple cycle** if no other vertex is repeated. A simple cycle can be identified with a subgraph C_n.

A connected graph that does not contain any cycles is called a **tree**.

If a graph is not connected it is simply the disjoint union of its connected subgraphs (usually called **components**).

After so many definitions, we need to start using them. Let us begin with something simple.

Theorem 6.4 *Suppose that a graph G has n vertices. Then*

- *If G has at least n edges it contains a cycle.*
- *If G is connected then G is a tree iff it has* $n - 1$ *edges.*

Proof. We will prove the first statement using induction on n. The base $n = 3$ is obvious: the triangle C_3 is the only possible graph for $n \leq 3$. If there is a vertex with 0 or 1 adjacent edges, we throw it away (together with the potential edge) and use the induction hypothesis.

Otherwise, every vertex has at least two adjacent edges. Choose one of them and start moving from vertex to vertex (along existing edges). Since each vertex has at least two edges, we can keep going, and sooner or later we will return to a vertex that we have already visited, thus finding a cycle.

To prove the second statement, first note that a tree cannot have more than $n - 1$ edges (otherwise the graph would contain a cycle, from what we just proved). On the other hand, since the graph is connected it must have at least $n - 1$ edges, because if we consider paths from a fixed vertex to all the other vertices, all of those $n - 1$ paths must have different terminal edges. Consequently, a tree has precisely $n - 1$ edges.

Conversely, if the graph contains a cycle we can delete one edge, and the remaining graph will still be connected. Continuing deleting edges, we finally obtain a tree which we know has $n - 1$ edges. This argument shows that if there are $n - 1$ edges in a connected graph then there are no cycles. □

For each vertex $v \in V$ we define its **degree** $d_G(v)$ as the number of edges adjacent to v. If $e = |E|$ is the number of the edges, then $2e = \sum_{v \in V} d_G(v)$ because every edge is counted twice in the sum. Albeit trivial, this equation is nevertheless very useful. For example, it immediately follows that every tree with $|V| > 1$ has at least one vertex of degree one. If we remove it and the corresponding edge, we again get a tree and this allows us to do induction proofs.

An important class of graphs consist of the **planar graphs** — those that can be drawn in the plane without any intersecting edges (except at the vertices). For example K_3, K_4 and $K_{2,3}$ are planar (even though the drawing of $K_{2,3}$ in figure 6.1 does not show this). On the other hand, it can be shown that K_5 and $K_{3,3}$ are not planar. A planar graph divides the plane into f regions and if $n = |V|$ is the number of the vertices, then $f + n - e = 2$ (**Euler's formula**). Euler's formula is a strong tool that can be applied to polyhedra as well, with f being the number of faces (convince yourself of this!).

An **Eulerian path** is a path that visits every edge exactly once. A graph has an Eulerian path if and only if 1) the graph is connected, and 2) the degree of every vertex (possibly with two exceptions) is even. The Eulerian path can be taken as a *cycle* iff all degrees are even. The proof is a straight-forward exercise on induction, and is left to the reader.

A **Hamiltonian path** visits every vertex exactly once, with the analogous definition for a **Hamiltonian cycle.** Graphs having Hamiltonian cycles are called **Hamiltonian graphs.** There is no simple characterisation of Hamiltonian graphs, but there are some necessary and/or sufficient criteria. The first one was proved by Ore.

Theorem 6.5 (Ore) *Suppose that $n = |V| \geq 3$. If $d_G(x) + d_G(y) \geq n$ for all non-adjacent $x, y \in V$, then the graph is Hamiltonian. If $d_G(x)+d_G(y) \geq n-1$ for all $x \neq y$ then G has a Hamiltonian path.*

Ore's condition is sufficient but far from being necessary. For example, if $n > 1$, the n-dimensional cube is Hamiltonian. First, recall that the **n-dimensional cube** I^n can be represented by the set of vectors $v = (\varepsilon_1, \ldots, \varepsilon_n)$, where every $\varepsilon_i \in \{0, 1\}$. In particular, the n-dimensional cube has 2^n vertices. There is an edge between two vertices iff the corresponding vectors differ in exactly one position. For example, in the 3-dimensional cube, there are edges from $v = (1, 1, 0)$ to three other vertices, namely: $(0, 1, 0), (1, 0, 0)$ and $(1, 1, 1)$. Hence, the degree of each vertex is n and we cannot apply Ore's theorem if $n > 2$. But we can still construct a Hamiltonian cycle by induction starting from $n = 2$. Note that I^2 is just the ordinary square which is a cycle in itself:

$$(0,0) \to (0,1) \to (1,1) \to (1,0) \to (0,0).$$

Suppose that $v_0 \to v_1 \to \cdots \to v_{2^n} \to v_0$ is a Hamiltonian cycle in I^n. For $v = (\varepsilon_1, \ldots, \varepsilon_n)$ we denote by (v, ε) the vertex $(\varepsilon_1, \ldots, \varepsilon_n, \varepsilon)$ in I^{n+1}. Then

$$\begin{aligned}(v_0, 0) &\to (v_1, 0) \to \cdots \to (v_{2^n}, 0) \\ &\to (v_{2^n}, 1) \to (v_{2^n-1}, 1) \to \cdots \to (v_0, 1) \to (v_0, 0)\end{aligned}$$

is a Hamiltonian cycle in I^{n+1}.

A simple, but useful necessary condition for a graph to be Hamiltonian is the following: if we throw away k vertices and the corresponding adjacent edges in a Hamiltonian graph, then the new graph has at most k components.

Problem 6.6 For which m and n is the complete bipartite graph $K_{m,n}$ Hamiltonian?

Search. If $m = n > 1$ we find a Hamiltonian cycle by inspection:

$$v_1 \to u_1 \to v_2 \to u_2 \to \cdots \to v_n \to u_n \to v_1.$$

This appears to be the only solution and this is easy to confirm. If $m < n$ then throwing away all m vertices in V_1 we also throw away all the edges and end up with $n > m$ components. □

The **complement** $\overline{G}$ of a graph G is the graph with the same set of the vertices, but the edge between v_1 and v_2 is present in $\overline{G}$ if and only if it is absent in G. For example, the complement of K_n is a graph without any edges. In general, a set of vertices is called **independent** (in G) if there are no connecting edges between them. Here is a good estimate for the size of an independent set.

Problem 6.7 Show that G contains at least

$$\sum_{v \in V} \frac{1}{d_G(v) + 1}$$

independent vertices.

Search. Induction on the size of V looks like a reasonable approach. If we have a single vertex then it is independent and the sum is equal to one as well, which establishes the base for induction.

Next, we try to remove one vertex v_0 to organize the induction step. It is reasonable to choose a vertex of minimal degree k (this amounts to a so-called *greedy* approach, which works surprisingly often). If $v_1, \ldots, v_k$ are the neighbours of v_0 there is no reason to keep them, so we remove them as well (together with the adjacent edges). In the remaining graph $H = (U, F)$ by induction we can find an independent set of size $\sum_{u \in U} \frac{1}{d_U(u)+1}$. Adding v_0 we increase the size of the independent set by one. What is the difference between the sums? If we show that it is at most one we are done. But the difference is no greater than

$$\sum_{i=0}^{k} \frac{1}{d_G(v_i) + 1} \le \sum_{i=0}^{k} \frac{1}{d_G(v_0) + 1} = (k+1)\frac{1}{k+1} = 1.$$

□

6.3 The pigeonhole principle

If you put $kn + 1$ rabbits into n boxes then at least one of the boxes must contain at least $k + 1$ rabbits. This is the classic formulation of the **pigeonhole principle.** The proof is easy; if not, then there can be no more than $n \times k$ rabbits which is a contradiction.

When we write "*by the pigeonhole principle*" we indicate that we know this proof and prefer not to repeat it again. For example, by the pigeonhole principle, in any set of $n + 1$ integers there must be (at least) two such that their difference is divisible by n. What we mean is that if we put the numbers in the same residue class modulo n in the same box, then there are only n different boxes. In one of the boxes we find two numbers (which are our rabbits) and their difference is divisible by n.

The pigeonhole principle is often used more generally, something like "if we a lot of things to distribute, then the distribution can't be sparse everywhere". For example, consider the following statements:

- At least one of the angles in triangle must be at least 60°;
- If a union of a finite number of sets is infinite, then at least one of the sets is infinite.
- If we mark each real number r by some integer $n(r)$ then at least one integer is used infinitely many times.

All of these statements contain the same idea: if not then by counting we get a contradiction. (Maybe the last one is more advanced; we need to know that the set of reals is essentially larger than the set of integers – see problem 10.8 .)

To illustrate the idea, consider the following problem.

Problem 6.8 Consider a quadrilateral $ABCD$ and four discs having its sides as diameters. Show that the discs cover the quadrilateral.

Search. What does this problem have in common with the pigeonhole principle? This is the wrong question. The right question is — when does a disc cover a given point P? Here we know the answer: see Theorem 1.14. Now connect the point P inside the quadrilateral with the vertices. By the pigeonhole principle, one of the angles APB, BPC, CPD, DPA must be at least 90°. □

Let us return to the classic pigeonhole principle. We can rarely apply it directly.

Sometimes the problem is to understand what the rabbits are, sometimes what the boxes are (and sometimes both). Consider the following classic problem.

Problem 6.9 Show that from any set of n integers, we can choose a non-empty subset (possibly with a single element) whose sum is divisible by n.

Search. The boxes are evidently residue classes modulo n. It is not as evident what the rabbits should be. The sums look like good candidates and we have as many as $2^n - 1$ ways to create a sum. That is too many! Let us imagine how we would like to finish the proof. We would want to say something like this: "and by the pigeonhole principle two of our sums are in the same box, and thus their difference is divisible by n." The problem is that this difference is not necessary a sum itself; for example if the numbers are a_i then the difference $(a_1 + a_2) - (a_3 + a_4)$ is probably not a sum. This gives us an idea of we want from our sum-rabbits: the difference of two sums should be a sum as well, and this is something we can ensure: the sums

$$a_1, a_1 + a_2, a_1 + a_2 + a_3, \ldots, a_1 + a_2 + \cdots + a_n$$

have this property. Now the problem is that we have too few rabbits: one fewer than we need. So it may very well happen that we get exactly one rabbit in each box. But that is not as bad as it sounds; if that happens, then one of the sums, corresponding to the residue zero is already divisible by n. On the other hand, if the zero box is empty, there are only $n - 1$ remaining boxes for our n rabbits and everything works out perfectly.

It turned out that even the boxes were not as evident as we thought. Another way to save the argument would have been to add 0 to our list of sums to get $n + 1$ perfect rabbits for our n original boxes. □

In the next problem the main difficulty is to find the correct boxes. Try to solve it before reading the solution.

Problem 6.10 Consider $n + 1$ positive different integers none of which is greater than $2n$. Show that among them we can choose

(i) two relatively prime numbers;

(ii) two numbers such that one divides another;

(iii) three (not necessary different) numbers such that one of them is the sum of two others.

Show that if we have only n numbers not greater than $2n$ then either all the numbers are greater than $\frac{2n}{3}$ or the least common multiple of two of them is not greater than $2n$.

Search. *(i)* We want to have boxes such that any two numbers in a box are relatively prime. The obvious choice is the n pairs $\{1,2\},\{3,4\},\{5,6\},\ldots,\{2n-1,2n\}$.

(ii) Now we need such boxes that any two numbers in a box divide each other. After some experiments with small n we easily find the solution: $\{1,2,4,8,\ldots\},\{3,3\cdot 2,3\cdot 4,\ldots\},\{5,5\cdot 2,5\cdot 4,\ldots\},\ldots,\{2n-1\}$, i.e. each box consists of the numbers $k2^m$ where k is a fixed odd number and $m \in \mathbb{N}$ varies.

Obviously statement *(ii)* is correct even if the collection of numbers may contain repeated elements.

(iii) Let a_1 be the minimal number. Then the set $\{a_i - a_1 \mid i > 1\}$ contains n numbers and taken together with the original set there $2n+1$ numbers, none of them greater than $2n$. Hence two of them must be equal and $a_i - a_1 = a_j \Rightarrow a_i = a_j + a_1$.

Consider now the last statement. We do not see how to apply the pigeonhole principle directly. But suppose that we have one number a less than $\frac{2n}{3}$. How can we use it? If we replace it by $2a$ and $3a$ we get $n+1$ numbers and can apply statement *(ii)* to them. The rest is evident and we leave it to the reader. □

Consider some continuous applications. The following theorem is attributed to Kronecker.

Theorem 6.11 *Let α be irrational. Then for every $\varepsilon > 0$ there is an integer m such that $m\alpha$ differs from an integer by less than ε. Moreover, the set of numbers $\{m\alpha + n \mid m, n \in \mathbb{Z}\}$ is dense in the real line.*

Proof. Subtracting the integer part of $m\alpha$ we can restrict ourselves to numbers in $[0,1]$. First we want to show that we can find such a number arbitrarily close to zero, from which our first statement follows.

Consider the fractional parts $k\alpha - [k\alpha]$ for all natural numbers k. All these numbers are different (because α is irrational) and belong to $[0,1]$.

If we divide $[0,1]$ into n equal parts we find, with some help from the pigeonhole principle, two different integers k_1 and k_2 such that $k_1\alpha - [k_1\alpha]$

and $k_2\alpha - [k_2\alpha]$ belong to the same sub-interval. If we denote by β the fractional part of the difference $(k_1 - k_2)\alpha$ we therefore have $|\beta| < \frac{1}{n}$. Because n can be arbitrarily large we are arbitrarily close to zero.

Now if want to be close to an arbitrary $0 < r < 1$ we find first $p \geq 0$ such that $p\beta > r$. Then $|p\beta - r| < |p\beta - (p-1)\beta| = |\beta|$ and it remains to note that $p\beta$ can be written in the form $m\alpha + n$.

Here is another non-trivial statement. Let us call a point in the plane where both coordinates are integers, an *integer point.*

Theorem 6.12 *Let S be any figure of area less than 1. Then we can translate S in such a way that in the new position it will not contain any integer points.*

Proof. We again use the modulo 1 idea used in the previous proof. Consider a given coordinate system, cut the plane into unit squares and identify all of these squares. Since the area of S is less than 1, at least one point P in the unit square will not be covered by points in S (the area covered by S will be *less then or equal* to the area of S, since some points may be covered several times).

Now move the origin of the coordinate system to P, which is equivalent to translating S, and we are done. □

Speaking about areas and integer points, we would be remiss not to mention the following, very useful combinatorial result.

Theorem 6.13 (Pick's formula) *Let P be a polygon and suppose that all its vertices are integer points. Then the area of P is equal to* $a + \frac{b}{2} - 1$, *where a is the number of interior integer points and b is the number of boundary integer points.*

Proof. If P is a rectangle of size $m \times n$ with the sides parallel to the axes we have $a = (m-1)(n-1)$, $b = 2(m+n)$ and the formula indeed gives the area mn. If we draw a diagonal we check the formula for a right-angled triangle with the sides parallel to the axes. We inscribe an arbitrary triangle in a rectangle and subtract the areas of the remaining right-angled triangles. Finally, the arbitrary polygon is divided into triangles and we use induction. We leave it to the reader the details of this proof. □

One trivial consequence of the theorem is that there is no equilateral triangle with integer vertices. Indeed, by Pick's formula, such a triangle would have a rational area. On the other hand, the area is equal to $a^2\frac{\sqrt{3}}{4}$ with integer a^2 which is a contradiction.

6.4 Colouring the problem

Colouring is a useful combinatorial tool. Sometimes it is a good idea to colour different part of objects in different colours to highlight their differences. A classical example is the following.

Problem 6.14 Consider an 8×8 board and cut off two cells that lie on the same diagonal. Show that the remaining board is impossible to cut into 2×1 pieces (dominos).

Search. This almost looks like a chessboard. The only thing missing is the colouring. If we colour our board like a normal chessboard, we find that two our missing cells have the same colour and the solution is now evident, because each domino consists of one white and one black cell. □

The colouring in the next problem is less obvious.

Problem 6.15 A rectangle is "nice" if at least one of its side has integer length. Prove that every rectangle that can be cut into nice rectangles is nice itself.

Search. We need to understand what is nice in being "nice". Consider again a chessboard colouring, but where all the cells have size $\frac{1}{2} \times \frac{1}{2}$. Then in every nice rectangle, exactly half the area is white. When we prove the opposite the problem will be solved. We leave this to the reader. □

6.5 Exercises

Exercise 6.1. Show that there exists a positive integer n such that the decimal expansion of 2^n begins with $2016\cdots$.

Exercise 6.2. Prove that $\sum_{k=0}^{n}\binom{n}{k}^3 = \sum_{k=0}^{n}\binom{n}{k}^2\binom{2k}{n}$.

Exercise 6.3. A magician shows the following trick. He leaves the room and one of the spectators places n coins in a line (with either side up) and points at one of the coins. The magician's assistant turns one coin over and leaves the room. The magician comes back into the room and immediately picks up the coin selected by the spectator. For which n can the Magician perform this trick?

Exercise 6.4. Prove that the sum of the digits of the number 2^n (in its decimal representation) is unbounded as $n \to \infty$. Is the following statement true: for each number A there exists $N = N(A)$ such that for all $n > N(A)$ the sum of the digits of the number 2^n is greater than A?

Exercise 6.5. Let S be a finite set of points in the plane such that no three of them are collinear. For each convex polygon P with vertices from the set S, let $a(P)$ be the number of vertices of P, and let $b(P)$ be the number of points in S outside P. Prove that for all $x \in \mathbb{R}$

$$\sum_P x^{a(P)}(1-x)^{b(P)} = 1,$$

where the sum is taken over all convex polygons with vertices in S. (A line segment, a point and the empty set are considered to be convex polygons with 2, 1 and 0 vertices respectively.)

Exercise 6.6. There are n ladies, and each of them knows about some scandal that is not known to any of the others. They communicate by telephone, and whenever two ladies make a call, they pass on to each other every scandal that they know at the time. How many calls are needed before all the ladies know all the scandals?

Exercise 6.7. Given m natural numbers x_i not greater than n and n natural numbers y_j not greater than m prove that it always possible to choose two non-empty sets of indices I and J such that

$$\sum_{i \in I} x_i = \sum_{j \in J} y_j.$$

See also problems 8, 28, 34, 41, 50, 60, 65, 96, 104, 113, 118 in Chapter 14.

6.6 Hints

6.1. lg 2 is irrational.

6.2. Use identities

$$\sum_{i+j=s} \binom{x}{i}\binom{y}{j} = \binom{x+y}{s}$$

$$\binom{x}{y}\binom{y}{z} = \binom{x}{x-z}\binom{x-z}{x-y} = \binom{x}{x-z}\binom{x-z}{y-z}$$

to prove more general result:

$$\sum_k \binom{n}{k}\binom{m}{k+l}\binom{2k+l}{n} = \sum_i \binom{n}{i}\binom{m}{i}\binom{m}{n+l-i}.$$

6.3. Use induction to show that it is possible when n is a power of two. Use the pigeonhole principle to show that this is impossible for other values n.

6.4. Prove a more general result for g^n, where g is an even natural number not divisible by 10.

Define a sequence a_n as follows: $a_0 = 0$ and a_{k+1} is the first natural number such that $2^{a_{k+1}} > 10^{a_k}$, i.e $a_1 = 1, a_2 = 4, a_3 = 14, \ldots$.

Prove that for $n > a_k$ the sum of the digits g^n is greater than k (which solves both variants of the problem).

6.5. Let $c(P) = n - (P) - b(P), y = 1 - x$. View

$$\sum_P x^{a(P)}y^{b(P)} = \sum_P x^{a(P)}y^{b(P)}(x+y)^{c(P)} = \sum_P \sum_{i=0}^{c(P)} \binom{c(P)}{i} x^{a(P)+i}y^{b(P)+c(P)-i}.$$

as a homogeneous polynomial of degree n in two independent variables x, y. Interpret the coefficient of $x^r y^{n-r}$ combinatorially. The desired result follows from:

$$\sum_P x^{a(P)}y^{b(P)} = \sum_{r=0}^{n} \binom{n}{r} x^r y^{n-r} = (x+y)^n = 1.$$

6.6. The answer $2n - 4$ for $n \geq 4$ is easy to guess. See [30] for the difficult proof.

6.7. Let $a_i = x_1 + \cdots + x_i, b_j = y_1 + \cdots + y_j$. If $a_m \leq b_n$ find the smallest $k = f(i)$ with $b_k \geq a_i$ and study the differences $c_i = b_k - a_i$.

CHAPTER 7

Functions and functional equations

This chapter contains general advice on functional equations, but first we would like to give exact definitions of functions and their properties, because many students' understanding of functions is intuitive and emotional rather than based on formal definitions. Mathematics is an exact science and its understanding begins from the understanding of the definitions.

7.1 Functions and permutations

We begin with the definition of a function $f : X \to Y$. Note that the **function** consists of three parts: the set X, often called **domain**, the set Y, sometimes called **codomain**, and the rule that for every $x \in X$ defines a unique element $y \in Y$ which is denoted by $f(x)$. Thus $x \overset{f}{\mapsto} y$. If we change any of the three parts, we get a *different* function. For example, the following four functions are all different, even though the rule is the same:

$$
\begin{aligned}
f_1 &: \mathbb{R} \to \mathbb{R}, && x \mapsto x^2;\\
f_2 &: \mathbb{R} \to \mathbb{R}^+, && x \mapsto x^2;\\
f_3 &: \mathbb{R}^+ \to \mathbb{R}, && x \mapsto x^2;\\
f_4 &: \mathbb{R}^+ \to \mathbb{R}^+, && x \mapsto x^2.
\end{aligned}
$$

Here $\mathbb{R}^+$ denotes the set of non-negative real numbers.

Mathematicians are lazy and we do not like to describe our functions completely. We often write only the rule, supposing that the reader can easily recognize the sets X and Y from the context. Conventionally, for a real-valued function f, we take X to be the set of all points x for which the rule $f(x)$ makes sense and we take $Y = \mathbb{R}$. So, for the function $\sqrt{x}$ we mean

$$g_1 : \mathbb{R}^+ \to \mathbb{R}, \quad x \mapsto \sqrt{x}$$

but not

$$g_2 : \mathbb{R}^+ \to \mathbb{R}^+, \quad x \mapsto \sqrt{x}.$$

In fact, we are so lazy that we do something formally incorrect: we say "the function $f(x)$", meaning the rule $x \overset{f}{\mapsto} f(x)$. For example, we say "function $\sin x$", even though a more correct name would be "the function sin", because $\sin x$ is the value of this function at some point x, not the actual function. And what we mean in fact is the function

$$\sin : \mathbb{R} \to \mathbb{R}, \quad x \mapsto \sin(x).$$

After all this it is not surprising that we write $\sin x$ instead for $\sin(x)$ — the reason is the same: this is convenient for people who are as lazy as us. It is also not surprising that we prefer to write "the function x^2" when we refer to the function f_1 above. What *is* surprising is that we wonder why the notion of function is so difficult for our students.

To frighten our students even more we use some foreign, strange sounding words. We say that the function $f(x)$ is **surjective** meaning that for each y the equation $f(x) = y$ has *at least* one solution x. When for each y the same equation has *at most* one solution we call the function **injective**. In other words, a function f is injective iff: $x_1 \neq x_2 \Rightarrow f(x_1) \neq f(x_2)$. For example, the functions f_2, f_4 and g_2 above are surjective, whereas the functions f_3, f_4, g_1, g_2 are injective. The function f_1 is neither surjective nor injective.

When the equation $f(x) = y$ has *exactly one* solution for each $y \in Y$, i.e. when the function is both surjective and injective, we say that f is **bijective**. In this (and only this) case, we can create an inverse function f^{-1}, that maps in the opposite direction $f^{-1} : Y \to X$ with the rule sending y to the unique solution of the equation $f(x) = y$. For example f_4 and g_2 are bijective functions and are inverses to each other: $f_4^{-1} = g_2$, $g_2^{-1} = f_4$. Note that if f is bijective, then f^{-1} is also bijective and $(f^{-1})^{-1} = f$.

Some authors use English terminology and call the injective functions *one-to-one*. The problem is that other authors use the same name for bijective functions, so be very careful if you use the word *one-to-one*!

If we can add, multiply or divide in Y we can do the same with the functions $f : X \to Y$, for example $(f + g)(x) = f(x) + g(x)$. Even more importantly, we have another operation — **composition**. If $f : X \to Y$, and $g : Y \to Z$, we can put the arrows together creating new function $h : X \to Z$,

$h(x) = g(f(x))$. We can see it as

$$x \mapsto y = f(x) \mapsto z = g(y).$$

The notation is $h = g \circ f$. Note the order, it is f that is applied first!

For example, $\sin x^2$ is the composition $\sin \circ f_1$. Note that function composition is not *commutative*. In general $f \circ g \neq g \circ f$. (Putting your clothes on after you have showered is not the same thing as showering with your clothes on!)

For the bijective function f we have

$$f^{-1} \circ f = \mathrm{id}_X; \quad f \circ f^{-1} = \mathrm{id}_Y,$$

where the identity function $\mathrm{id}_X : X \to X$ changes nothing: $\mathrm{id}_X(x) = x$.

An interesting property of a surjective function f is that we may cancel from the right:

$$g_1 \circ f = g_2 \circ f \quad \Rightarrow \quad g_1 = g_2$$

for any functions g_i allowing the composition. Conversely, a function f that admits right-cancellation for all g_1 and g_2 must be surjective. Similarly, injective functions f can be cancelled from the left:

$$f \circ g_1 = f \circ g_2 \quad \Rightarrow \quad g_1 = g_2.$$

The notation f^2 is often used for the composition $f \circ f$ by itself, but the trouble is that the same notation is also used for the square of f. For example, we read $\sin^2 x$ as the square $(\sin x)^2$ and not as $\sin(\sin(x))$. Fortunately, we rarely notice that such expressions are ambiguous. If you want to avoid any risk of ambiguity, an alternative notation is $f^{\circ 2}$.

There is an *incredible* number of functions. In a certain sense, there are many more functions than real numbers, in fact there are as many functions $f : \mathbb{R} \to \mathbb{R}$ as there are subsets of $\mathbb{R}$. To explore this idea would lead us to the intriguing area of *cardinal arithmetic*, but that would take us too far astray.

The functions that we most often use: the elementary functions such as rational or trigonometric functions, form a very small part of the world of functions. Let us consider a few more exotic creatures.

The **Dirichlet function** $f(x)$ is defined quite simply: $f(x) = 1$ if x is a rational number and $f(x) = 0$ otherwise. But try to draw its graph! Impossible. The Dirichlet function is discontinuous at each point! But its composition $f \circ f = 1$ is a constant function.

If we instead take **Thomae's function**

$$f(x) = \begin{cases} \dfrac{1}{q}, & \text{if } x = \dfrac{p}{q} \text{ is a rational number with } (p,q) = 1 \\ 0, & \text{if } x \text{ is irrational,} \end{cases}$$

we get an even more interesting function that is continuous at all irrational numbers and discontinuous at all rational numbers.

Another generalization is the **characteristic function** $\chi_M : X \to \mathbb{R}$ which is defined for any subset M of given set X by the rule $\chi_M(x) = 1$ if $x \in M$ and $\chi_M(x) = 0$ otherwise. For example, the Dirichlet function is the characteristic function for the rational numbers. Some useful properties are:

$$\chi_{M \cap N} = \chi_M \cdot \chi_N; \quad \chi_{M \cup N} = \chi_M + \chi_N - \chi_M \cdot \chi_N.$$

The function $f(x) = [x]$ calculates for each x its integer part (i.e. the largest integer that is not greater than x) and has a very nice graph, which looks like a staircase. The difference $\{x\} = x - [x]$ calculates the fractional part of x. The function $\{x\}$ has a nice graph as well (draw it!). This function is periodic with period 1.

What is the difference between a map and a function? There is no difference, but we usually speak about functions when the domain and codomain consist of numbers, and maps in more abstract settings. For example by a linear function we mean $f(x) = ax + b$, but a linear map is a map between two vector spaces.

When $f : N \to N$ is bijective and N is finite we call it neither a function nor a map but a **permutation**. In fact we sometimes use this term for infinite sets as well but this is not as common. If the set N has n elements there are $n!$ different permutations. The set of all those permutations is normally denoted S_n. Conventionally, we use Greek letters, most often σ and τ to denote permutations and we often identify N with the set $\{1, 2, \ldots, n\}$. If $\sigma \in S_n$, we can define its **sign** as a strange product:

$$\operatorname{sgn} \sigma = \prod_{1 \le i < j \le n} \frac{\sigma(i) - \sigma(j)}{i - j}.$$

The number $\operatorname{sgn} \sigma$ is always ± 1 and is sometimes denoted by $(-1)^\sigma$. Its importance lies in the fact that $\operatorname{sgn}(\sigma \circ \tau) = \operatorname{sgn} \sigma \cdot \operatorname{sgn} \tau$.

If we call a permutation that exchanges only two elements (and leave the remaining ones unchanged) a **transposition**, then every permutation can be written as a composition of transpositions (in other words: we can always permute objects in an arbitrary order using only two hands and exchanging two objects at a time). We can do this in many different ways, but it turns out that the number of the transposition will always be even if $\operatorname{sgn}\sigma = 1$ (that is why such permutations are called **even**). Similarly the permutations for which $\operatorname{sgn}\sigma = 1$ are called **odd** and they always require an odd number of transpositions.

If this is the first time the reader sees these definitions s|he will hardly be excited: everything looks too abstract. But in fact, the theory of permutations is a surprisingly useful tool. Consider the following problem.

Problem 7.1 The Rubic cube was rotated 999 times (rotations 180° are considered as two rotations). Show that it is not in the same position as in the beginning.

Proof. All we need to show is that a rotation is an odd permutation of the corners. Because the composition of an odd number of odd permutations is an odd permutation, the result cannot be the identity (which is an even permutation). Let us check one of six possible rotations, where we rotate the upper part. If we denote the corners from 1 to 8 (and the four first ones are on top) then the permutation σ looks like

$$\sigma(1) = 2; \quad \sigma(2) = 3; \quad \sigma(3) = 4; \quad \sigma(4) = 1;$$
$$\sigma(5) = 5; \quad \sigma(6) = 6; \quad \sigma(7) = 7; \quad \sigma(8) = 8.$$

It is easy to check that $\operatorname{sgn}\sigma = -1$ and we are done. □

Of course, one can say much more and do the calculations better, but we only wanted to give a taste of this approach. Try solving this problem in another way!

We return to our real-valued functions. Normally we can say more about them if we know some additional properties. One of the most important classes of functions are **continuous functions** $f : X \to \mathbb{R}$. The formal definition is that $\lim f(x_n) = f(\lim x_n)$ for every convergent sequence $x_n \in X$ with $\lim x_n \in X$, but this (or other equivalent definitions) require understanding of the theory of limits. Intuitively, continuity means that small changes of x give rise to small changes of $f(x)$ but in practice, we use that every differentiable function is continuous where defined. For example, $1/x$ is a continuous function as long we do not try to define it at $x = 0$. But any such attempt will evidently fail. When $X = [a, b] = \{x \mid a \le x \le b\}$ is a closed, bounded interval, we can say a lot more about continuous functions.

Theorem 7.2 *Let $f : [a, b] \to \mathbb{R}$ be a continuous function. Then*

- *f is bounded and attains its maximum and minimum values. In other words, there exist $x_m, x_M \in [a, b]$ such that for any $x \in X$ we have:*

$$m = f(x_m) \le f(x) \le f(x_M) = M.$$

- *f attains all values between m and M, thus for any $m \le v \le M$ there exists $x \in [a, b]$ such that $f(x) = v$. In particular, f attains all the values between $f(a)$ and $f(b)$. This fact is usually called* ***the intermediate value theorem.***
- *If $f(a)f(b) \le 0$ then there exists c such that $f(c) = 0$.*
- *If all values $f(x)$ belong to $[a, b]$, then there exists a fixed point x such that $f(x) = x$.*

Proof. We refer to any good textbook in calculus for the proof of the first two statements. The third statement follows directly from the second, but is used so often that we have formulated it separately. The last statement follows from the third if we apply it to the continuous function $g(x) = f(x) - x$: evidently $g(a) \le a - a = 0$ and $g(b) \ge b - b = 0$. □

The first statement is wrong for an open interval $(a, b) = \{x \mid a < x < b\}$. The example $f(x) = 1/x$ on $(0, 1)$ shows that a continuous function can be unbounded on an open interval, and $f(x) = x$ on the same interval shows that even if the function is bounded it may have no maximum or minimum value..

The last statement has an analogue in higher dimensions: (**Brouwer's fixed-point theorem**) — see Section 10.4.

Speaking about the maximum and minimum we would like to clarify how to use derivatives to find them. A typical bad approach is that a student writes the equation $f'(x) = 0$, finds a solution x_0 and claims that this is a maximum (or minimum) point. Much more than that is required! Let us discuss this carefully. First recall the definitions.

- A function $f : X \to \mathbb{R}$ has a (global) maximum at the point x_0 if $f(x_0) \ge f(x)$ for every $x \in X$. In a similar way we define a global minimum. Note that the definition depends on the set X. For example, the function

$f(x) = x^2$ defined on the (closed) interval $1 \le x \le 2$ has a maximum at the point $x = 2$. If we instead consider x^2 on the interval $-1 \le x \le 1$, this will be another function which has a maximum at two points: $x = -1$ and $x = 1$. On the open interval $-1 < x < 1$ we have yet another function with no maximum at all.

- The function $f(x)$ has a local maximum at the point x_0 if there exists a (small) open interval $I = (a, b) = \{x \mid a < x < b\}$ containing x_0 such that (the new) function $f(x)$ considered on $I \cap X$ has a global maximum at the point x_0. The difference between a global and local maximum is like the difference between Miss Sweden and Miss World: she might be the same person, but is probably not.
- In the same way we define local minima. The word *extremum* is the common name for a local minimum or maximum.

Note that none of the above has anything to do with derivatives. What can we say about differentiable functions?

- If the domain X is an open interval (a, b) (where $a = -\infty$ and $b = +\infty$ are allowed) and $f(x)$ is differentiable on X, then $f'(x_0) = 0$ at every extremum. But the extremum points, the global minimum or maximum may be absent even if the equation $f'(x) = 0$ has solutions. The existence of extrema must also be proven.
- If $f(x)$ is differentiable at x_0, $f'(x_0) = 0$ and $f''(x_0) > 0$ then f has a local minimum at x_0. Similarly, if $f'(x_0) = 0$ and $f''(x_0) < 0$ then f has a local maximum at x_0. These conditions are sufficient but not necessary! No conclusions can be made about the global minimum or maximum.
- If X is a closed interval $[a, b]$ (no infinities here!), $f(x)$ is differentiable on (a, b) and continuous on $[a, b]$ then $f'(x_0) = 0$ in every extremum point inside (a, b). Besides the points where f' vanishes, f may also have an extremum at $x = a$ and $x = b$ (even if the derivative at the endpoints is non-zero). In contrast with the open intervals, here Theorem 7.2 guarantees the existence of the global minimum and maximum.

This is all we can say, and the derivative is a good tool to find points where an extremum can occur. To find a global maximum or minimum, we need a little more work (for example studying the signs of the derivative). The best case is when we are dealing with a differentiable function on a closed interval

$[a, b]$. In this case, we need only to compute $f(a)$, $f(b)$ and $f(x_i)$ for all x_i inside the interval where $f'(x_i) = 0$. The largest of these values is the global maximum, the smallest — the global minimum.

Another good case is the following: If $f(x)$ is differentiable in (a, b) and the equation $f'(x) = 0$ has only *one* solution x_0 in (a, b), then f has its global minimum at x_0 if $f''(x_0) > 0$ or its global maximum if $f''(x_0) < 0$. Try to motivate this.

Note that, if f fails to be differentiable at some points (as in the case $f(x) = |x|$) we also need to check those points, because f can have an extremum there as well.

An interesting case is a periodic continuous function $f(x)$. If T is a period then it is sufficient to consider the function on the closed interval $[0, T]$. But because the interval is closed, there must be a global minimum and maximum, and these can only occur at points where the derivative is zero (or where the derivative doesn't exist). Why can we ignore the endpoints of $[0, T]$?

Add the function $f(x) = x^{2k} \sin(\frac{1}{x})$ to your list of exotic functions. If we additionally put $f(0) = 0$, we get a function that you can differentiate exactly k times, but no more. Even more curiously, the kth derivative isn't even continuous at $x = 0$.

We will need a little more terminology. A function $f : \mathbb{R} \to \mathbb{R}$ is called **increasing** if $x \leq y$ implies $f(x) \leq f(y)$. Similarly, f is **decreasing** if $x \leq y \Rightarrow f(x) \geq f(y)$. A function that is either increasing or decreasing is called **monotone**.

7.2 Functional equations

Functional equations are very popular in competitions. The reason for their popularity is that their difficulty can vary and can demand knowledge from any part of mathematics. On the other hand, though solutions can be quite elementary, performing a little mathematical research is often required to find it.

The equation may look strange, but the main difficulty is that the task feels unfamiliar: you are looking not for a number x but for the function itself; it is the rule f that is unknown! To understand this, consider a very simple problem.

Problem 7.3 Find all monotone functions $f : \mathbb{R} \to \mathbb{R}$ such that $(f(x))^2 = 1$.

Search. We skip an obvious trap — the temptation to answer directly that $f(x) = 1$ — and instead factor the equation:

$$(f(x))^2 - 1 = 0 \quad \Leftrightarrow \quad (f(x) - 1)(f(x) + 1) = 0.$$

It follows that $f(x)$ is a constant: either $f(x) = 1$ or $f(x) = -1$, which should be the obvious answer. But something is very suspicious: we have never used the condition that f is monotone. Are we missing something?

This question is sufficient to find a gap: we have indeed obtained that $f(x) = \pm 1$ for each x, but nobody has said that the chosen sign is the same for *all* x. If we can choose the sign of $f(x)$ for each x independently, we get a huge number of possible functions. For example, we can choose -1 when x is rational and $+1$ when x is irrational in line with the Dirichlet function. Moreover, if M is any subset of $\mathbb{R}$ then using its characteristic function $\chi_M(x)$ we get all possible solutions to our equation as $f(x) = 2\chi_M(x) - 1$.

Now suppose that $f(x)$ is monotone. WLOG we can suppose that f is increasing (otherwise we replace it by $-f$). Suppose that $f(x)$ is not a constant function. Then $f(a) = 1$ for some a and we get $f(x) = 1$ for $x \geq a$ as well. Similarly, $f(b) = -1$ for some b and $f(x) = -1$ for all $x < b$ as well. Now we need to find a breakpoint. As usual with real numbers we avoid the minimum or maximum, but prefer to use the supremum or the infimum. Let $c = \sup\{x \mid f(x) = -1\}$. It obviously exists, and $f(x)$ is negative for $x < c$ and positive for $x > c$ (otherwise it is not a supremum). As for its value in c we see that both alternatives are possible. To summarize we can say that either $f(x)$ is a constant function 1 or -1 (to stress that $f(x)$ is a constant we write $f(x) \equiv 1$ or $f(x) \equiv -1$) or for each constant c we have four solutions:

$$f_c^+(x) = \begin{cases} 1, & \text{if } x \geq c \\ -1, & \text{if } x < c, \end{cases} \qquad f_c^-(x) = \begin{cases} 1, & \text{if } x > c \\ -1, & \text{if } x \leq c \end{cases}$$

and $-f_c^+(x), -f_c^-(x)$. □

So the solution to a functional equation might consist of a huge number of functions, including some very exotic ones. So how can we find them? The idea is to use the freedom to choose x and other variables involved in the equation. We try to substitute them by some expressions (this may be very complicated, e.g. involving f as well) to get new equations. The aim is to combine those equations and to receive some extra information about f:

its zeros, fixed points, values in some special points. Perhaps we can prove that the function is surjective or injective or monotone. Every new piece of information we obtain restricts the possible choices of f and if we are lucky enough we get down to a reasonable number of choices which we can check directly by plugging them into the equation.

A typical start is to study $f(0)$ or such a that $f(a) = 0$. Using them we can get simpler equations and continue our research.

We shouldn't assume that $f(x)$ is a polynomial or other nice function, but if we have no other ideas, we can try these special cases first — perhaps we can at least find a possible solution to the functional equation (quite often the solutions are not very complicated functions). If we succeed in finding a good candidate $f_0(x)$, we can try to consider $g(x) = f(x) - f_0(x)$ and to rewrite the equation in terms of $g(x)$. The new equation may be easier to solve. For example, if f_0 is the only solution then we need to prove that $g(x) \equiv 0$ which might be easier. Another similar approach is the substitution $f(x) = h(x)f_0(x)$.

Let us consider a more complicated example.

Problem 7.4 Find all real functions $f(x)$ such that

$$f(f(x-y)) = f(x)f(y) - f(x) + f(y) - xy,$$

for all real x, y.

Search. We try to determine $c = f(0)$ first. Taking $x = y$ gives $f(c) = (f(x))^2 - x^2$ for all $x \in \mathbb{R}$, and hence $(f(x))^2 - x^2 = (f(0))^2 - 0^2 = c^2 = f(c)$ for all $x \in \mathbb{R}$. Next we choose $y = 0$ and get $f(f(x)) = cf(x) - f(x) + c$ for all $x \in \mathbb{R}$.

Hence for all t that can be written as $f(x)$ (i.e. for all t in the image of f) we have $f(t) = (c-1)t + c$. If $f(x)$ is surjective we could conclude that $f(x)$ is a linear function and check it directly. But without knowing that f is surjective, such a conclusion is impossible (which students often forget).

Nevertheless for t in the image of f we can combine our knowledge:

$$\begin{aligned} c^2 = f^2(t) - t^2 &= (c-1)^2t^2 + 2c(c-1)t + c^2 - t^2 \\ &= ct\big((c-2)t + 2(c-1)\big) + c^2, \end{aligned}$$

which gives $t((c-2)t + 2(c-1)) = 0$ for all t in the image.

Here we draw an important conclusion, namely that the image of f consists of at most two values, namely $t = 0$ and (for $c \neq 2$) $t = -\frac{2(c-1)}{c-2}$.

We continue to investigate the value of c. Assume first that $c \neq 0$.

Hence c is the only non-zero value attained by f and since $f(c) = c^2$ is such a value, we get $f(c) = c^2 = c$. Since $c \neq 0$ it follows that $c = 1$ and hence $-\frac{2(c-1)}{c-2} = 0$, a contradiction to the existence of a non-zero value.

Hence $c = 0$ and for all t in the image we have $f(t) = -t$.

We have now established that $(f(x))^2 = x^2$ for all $x \in \mathbb{R}$. Assume that there exists $x_0 \in \mathbb{R}$, $x_0 \neq 0$ such that $f(x_0) = x_0$. But such a point would have to be in the image of f and for all such points we have $f(t) = -t$, which is a contradiction. Hence

$$f(x) = -x \qquad \forall x \in \mathbb{R}$$

and it is easy to check that this function satisfies the equation. □

This was a typical solution. Note the importance of the last statement (so many students lose points and sometimes IMO medals forgetting to verify their candidate solutions!) Another place to lose points is to conclude from $(f(x))^2 = x^2$ that either $f(x) = x$ or $f(x) = -x$. Again, this is true for each separate x but it should be proven that the same sign can be chosen simultaneously. Such difficulties often occur while solving functional equations. We will consider one more problem.

Problem 7.5 Determine all functions $f : \mathbb{R} \to \mathbb{R}$ such that $f(1) = 1$, and

$$f(xy + f(x)) = xf(y) + f(x),$$

for all $x, y \in \mathbb{R}$.

Search. Inserting $x = 1$ gives $f(y+1) = f(y) + 1$; if we now choose $y = 0$, we get $f(1) = f(0) + 1$, i.e. $f(0) = 0$. It follows by induction that $f(n) = n$ for all $n \in \mathbb{Z}$; we shall show that $f(x) = x$ is indeed the only solution (it is quite obvious that it is a solution). Choose $y = 0$ to conclude that $f(f(x)) = f(x)$ for all $x \in \mathbb{R}$. Let us now successively choose $y = f(x)$ and $x = f(y)$; We get

$$\begin{aligned} f(xf(x) + f(x)) &= f((x+1)f(x)) \\ &= xf(f(x)) + f(x) = xf(x) + f(x) \end{aligned}$$

and

$$\begin{aligned} f(yf(y)+f(f(y))) &= f((y+1)f(y)) \\ &= (f(y))^2 + f(f(y)) = (f(y))^2 + f(y) \end{aligned}$$

for all $x, y \in \mathbb{R}$. Combining the two last equalities (replacing y by x in the second) gives

$$f(x)(x - f(x)) = 0 \qquad \forall x \in \mathbb{R}.$$

This means that for each $x \in \mathbb{R}$ either $f(x) = 0$, or $f(x) = x$. Let x_0 be such that $f(x_0) = 0$. The functional equation yields

$$f(x_0 \cdot 1 + f(x_0)) = f(x_0 + 0) = 0 = x_0 f(1) + f(x_0) = x_0,$$

and it follows that $x_0 = 0 = f(x_0)$. Hence the only solution to our equation is $f(x) \equiv x$. □

It is very important to know which additional properties the function $f(x)$ may have, because this can significantly simplify the problem. For example, we will see on page 263 that **Cauchy's equation** $f(x+y) = f(x) + f(y)$ has highly non-trivial solutions. But if we know that $f(x)$ is continuous there are much fewer choices.

Indeed, if we put $x = y = 0$, we find $f(0) = 0$. By induction it follows that $f(nx) = nf(x)$ for all real x and all natural n. Condequently, the same equality holds for all integers n, since $f(-x) + f(x) = f(0) = 0$ and hence $f(-nx) = f(n(-x)) = nf(-x) = -nf(x)$. It also follows that

$$f(x) = f(m\frac{x}{m}) = mf(\frac{x}{m})$$

and hence that $f(rx) = rf(x)$ for all real x and all rational r. If we now put $f(1) = a$, we get $f(r) = ar$ for all rational r. Since the set of rational numbers is dense in the set of real numbers, by continuity we get $f(x) = ax$.

We will finish this section with an advanced example.

Problem 7.6 Find all bounded continuous functions $f : \mathbb{R} \to \mathbb{R}$ such that

$$(f(x))^2 - (f(y))^2 = f(x+y)f(x-y),$$

for all $x, y \in \mathbb{R}$.

Search. The first step is to try to guess the answer. Looking through the list of elementary bounded functions we find that $f(x) = \sin x$ is a solution to the functional equation, and this does not bode well. Trigonometric functions are much more difficult to handle than, say, polynomials.

It is easy to check that all functions of the form $f(x) = a \sin bx$, $a, b \in \mathbb{R}$, solve the functional equation as well, and we shall show that there are no other bounded continuous solutions.

The most difficult part is to establish that every solution is *periodic*. We will start with something easier, namely showing that all solutions are *odd*.

So, let f be a bounded continuous nontrivial solution of the equation. Set $x = y = 0$; we get $0 = (f(0))^2$, so that $f(0) = 0$. For $x = 0$, $y \in \mathbb{R}$ we have $-(f(y))^2 = f(y)f(-y)$, so that either $f(y) = 0$, or $f(-y) = -f(y)$. In case $f(y) = 0$ the equation gives $(f(-y))^2 - 0 = f(0)f(2y) = 0$, so $f(-y) = 0$ as well. In any case, f has to be an odd function. Good.

This means we can restrict our considerations to positive values of x. To find a candidate for a period we try to find another zero of the function than zero. Assume the opposite, that $f(x) \neq 0$ for all $x > 0$.

It is clear that if f is a solution, then so is the function $af(bx)$, $a, b \in \mathbb{R}$. Thus we may as well assume that $f(x) > 0$ for $x > 0$.

Let $x > y > 0$. Then $x \pm y > 0$ and we get $(f(x))^2 - (f(y))^2 > 0$, which, together with $f(x), f(y) > 0$, implies that f is strictly increasing. This is an important observation. Either we find another solution to our equation, or a contradiction. We need to use the fact that f is bounded.

Since f is bounded and increasing, we can guarantee that the limit $\lim_{x\to\infty} f(x)$ exists and is finite. Let $\lim_{x\to\infty} f(x) = L$ and let x tend to infinity in the functional equation with y fixed positive. We get (using continuity) $L^2 - (f(y))^2 = L^2$. Thus $f(y) = 0$, which contradicts the assumption $f(x) \neq 0$ for $x > 0$.

Hence there exists some $\bar{x} > 0$ such that $f(\bar{x}) = 0$. By induction it is easy to see that $f(n\bar{x}) = 0$ for all $n \in \mathbb{N}$. We have come much closer to finding the period and it seems that a minimal such $\bar{x} > 0$ would be a good candidate. First we need to be sure that such a minimal value really exists. Thus we shall show that there exists a smallest number among the positive zeroes of f.

Assume the opposite, i.e. assume that there exists a sequence $\{x_n\}$ such that $x_n \to 0$ as $n \to 0$, and $f(x_n) = 0$ for all $n \in \mathbb{N}$. Then we have too many zeros. More precisely, since the set of numbers $\{kx_n\}$, $k, n \in \mathbb{N}$, is dense in the set of positive real numbers (why?), by continuity it follows that f is identically equal to 0, which is a contradiction, since we assumed that f is a nontrivial solution. Done.

Let x_0 be the smallest positive zero of f ($x_0 = \inf\{x > 0 \mid f(x) = 0\}$). Again, using the scaling: if $f(x)$ is a solution, then so is $af(bx)$, WLOG we can assume that $x_0 = \pi$, that $f(x) > 0$ in the interval $(0, \pi)$, and that $\max_{x\in(0,\pi)} f(x) = 1$.

Our next goal will be to show that f is periodic with period 2π. Let us begin by taking $x \in (0, \pi)$. Then

$$(f(k\pi \pm x))^2 - (f(x))^2 = f(k\pi)f(k\pi \pm 2x) = 0,$$

for all $k \in \mathbb{N}$, and it follows that $|f(k\pi \pm x)| = f(x) > 0$. Since $\pi - x \in (0, \pi)$, we get $f(\pi - x) = f(x) > 0$. Furthermore, from the equation

$$f(k\pi + x)f(k\pi - x) = 0 - (f(x))^2,$$

keeping in mind that f is odd, we get $f(k\pi + x) = -f(k\pi - x)$, (for $k \in \mathbb{Z}$) and, in particular, $f(\pi + x) = -f(x) = -f(\pi - x) < 0$.

From the identities $f((k+1)\pi+x) = -f((k+1)\pi-x) = -f(k\pi+(\pi-x))$, inductively we get $f(k\pi + x) = (-1)^k f(x)$, in particular $f(2\pi + x) = f(x)$, so far only for $x \in (0, \pi)$.

To prove the same identities for an arbitrary x, it is enough to note that for each $x \in \mathbb{R}$, x not a multiple of π, there exist $x_1 \in (0, \pi)$ and $k \in \mathbb{Z}$ such that $x = k\pi + x_1$. Thus f is 2π-periodic (the alternating signs imply that the period is not smaller than 2π).

Next we will show that f attains its maximum at $\frac{\pi}{2}$. Again, from the functional equation,

$$\begin{aligned}(f(\tfrac{\pi}{2}))^2 - (f(x))^2 &= f(\tfrac{\pi}{2} + x)f(\tfrac{\pi}{2} - x)\\ &= f(\tfrac{\pi}{2} - x)f(\pi - (\tfrac{\pi}{2} - x)) = f(\tfrac{\pi}{2} - x)^2.\end{aligned}$$

Thus $f(\frac{\pi}{2}) \geq |f(x)|$ for all real x, and, according to our previous assumptions WLOG, $f(\frac{\pi}{2}) = 1 = \sin\frac{\pi}{2}$.

We shall now show that $f(x) = \sin x$ for all real x. By what we have established so far we have $(f(\frac{\pi}{2}))^2 - (f(\frac{\pi}{4}))^2 = f(\frac{3\pi}{4})f(\frac{\pi}{4}) = (f(\frac{\pi}{4}))^2$, and it follows that $f(\frac{\pi}{4}) = \sin\frac{\pi}{4}$. Using induction, it is straight-forward to prove that $f(\frac{\pi}{2^n}) = \sin\frac{\pi}{2^n}$, and $f(\frac{k\pi}{2^n}) = \sin\frac{k\pi}{2^n}$, $k \in \mathbb{N}$, $x > 0$ (we leave the details to the reader). Since the numbers $\{\frac{k\pi}{2^n}\}$ are dense among the positive numbers, by continuity $f(x) = \sin x$ for all positive x. Both functions are odd, which implies that $f(x) = \sin x$ for all real x.

By what we have proved, and the comments at the beginning of the proof, all bounded continuous solutions of the functional equation are $f(x) = a \sin bx$, $a, b \in \mathbb{R}$. □

7.3 Exercises

Exercise 7.1. Let $f(x) = ax^2 + bx + c$. Show that if the equation $f(x) = x$ has no real solutions then the equation $f(f(x)) = x$ has no real solutions as well.

Exercise 7.2. Show that if $(a + b + c)c < 0$ then $b^2 > 4ac$.

Exercise 7.3. Determine all functions $f : \mathbb{R} \setminus \{-1, 1\} \to \mathbb{R}$ which satisfy the equation

$$f\left(\frac{x-3}{x+1}\right) + f\left(\frac{3+x}{1-x}\right) = x.$$

Exercise 7.4. Find all continuous functions $f : \mathbb{R} \to \mathbb{R}$ such that

$$f(x+y) + f(x-y) = 2f(x) + 6xy^2, \quad \forall x, y \in \mathbb{R}.$$

Exercise 7.5. Find all functions f, $f : \mathbb{N} \to \mathbb{N}$, satisfying the equality

$$f(m - n + f(n)) = f(m) + f(n),$$

for all $m, n \in \mathbb{N}$.

See also problems 36, 46, 58, 71, 87, 91, 110 in Chapter 14.

7.4 Hints

7.1. If $f(x) > x$ for all x then $f(f(x)) > f(x) > x$.

7.2. Show that $f(x) = ax^2 + bx + c$ has a root.

7.3. Set $t = \frac{x-3}{x+1}$ first and $t = \frac{3+x}{1-x}$ after that. Add.

7.4. Set $u = x + y, v = x - y$. WLOG $f(0) = 0$.

7.5. Study $F(n) = f(n) - n$.

PART II

Advanced tools

CHAPTER 8

Number Theory

In this second part of the book, we would like to collect some more advanced methods and instruments that can be interesting for an experienced problem solver. Some parts are not necessary at the IMO level and are included only because we consider them to be beautiful parts of mathematics that suggest different ways of thinking.

~

8.1 Order and primitive elements

If m is not prime there is still an analogue of Fermat's little theorem in $\mathbb{Z}_m$.

Theorem 8.1 (Euler's theorem) *If* $(a, m) = 1$ *then*

$$a^{\varphi(m)} = 1,$$

where $\varphi(m)$ *is* ***Euler's (totient) function*** *which is defined as the number of invertible elements in* $\mathbb{Z}_m$ *and can be obtained from the prime factorization* $m = p_1^{a_1} \cdots p_k^{a_k}$ *as*

$$\varphi(m) = m\left(1 - \frac{1}{p_1}\right)\left(1 - \frac{1}{p_1}\right)\cdots\left(1 - \frac{1}{p_k}\right).$$

Proof. The proof is the same as for Fermat's theorem (see Theorem 2.7) if we take N as the product of all $\varphi(m)$ invertible elements. To get the formula for $\varphi(m)$, we first show that $\varphi(m_1 m_2) = \varphi(m_1)\varphi(m_2)$ if $(m_1, m_2) = 1$ and then calculate explicitly $\varphi(p^k) = p^k - p^{k-1} = p^k\left(1 - \frac{1}{p}\right)$ for each prime p. We leave the details to the reader. □

For example, $\varphi(24) = 24\left(1 - \frac{1}{2}\right)\left(1 - \frac{1}{3}\right) = 8$. Thus $a^8 = 1 \pmod{24}$ if $(a, 24) = 1$. Since $(a, 24) = 1$ iff $(a, 6) = 1$ we also get $a^2 = 1 \pmod{6}$, but in fact the strongest result is $a^2 \equiv 1 \pmod{24}$ because in the two factors $(a-1)(a+1)$ at least one is divisible by 3, at least one is divisible by 4 and both are even numbers. Thus Euler's theorem (and its special case Fermat's theorem) does not always give the best possible result.

If $(a, m) = 1$, there is a minimal positive n such that $a^n \equiv 1 \pmod{m}$. This number is usually called the **order** of a (modulo m) and is denoted as $\operatorname{ord}_m(a)$.

Minimality implies that for any k such that $a^k \equiv 1 \pmod{m}$ the order n must divide k. Indeed, writing $k = dn + r$ with $0 \le r < n$ it follows that $a^r \equiv 1 \pmod{m}$ and therefore $r = 0$. In particular, $n \mid \varphi(m)$.

The elements with order exactly equal to $\varphi(m)$ are called **primitive (roots)**. Primitive roots exist if and only if $m = 2, 4, p^k, 2p^k$ where p is an odd prime and k is a natural number. If x is a primitive root, then each invertible element in $\mathbb{Z}_m$ can be written as some power x^a which is a good starting point for multiplicative problems.

To show that x is primitive it is sufficient to check that $x^{\varphi(m)/q} \not\equiv 1 \pmod{m}$ for each prime divisor of $\varphi(m)$. For example, 10 is a primitive root modulo 17 because $10^8 \equiv 2^8 5^8 \equiv 16^2 25^4 \equiv 8^4 \equiv 16^3 \equiv -1 \not\equiv 1 \pmod{17}$.

When $m = p$ is a prime, the order of any element divides $p-1$, a primitive element x always exists and x^a is primitive iff $(a, p-1) = 1$.

Orders and primitive elements are powerful instruments for solving many number theory problems. Here is an example.

Problem 8.2 Let p be a prime and n be a natural number. Which remainder can the sum

$$S = 1^n + 2^n + \cdots + (p-1)^n$$

have modulo p?

Search. Fermat's theorem suggests that we should begin with investigating the case when $p-1$ divides n. Then each summand is equal to 1 modulo p and consequenty, $S \equiv p-1 \pmod{p}$.

If $p-1 \nmid n$ then a primitive root x is useful. Note that $y = x^n \not\equiv 1 \pmod{p}$. Then

$$yS = x^n + (2x)^n + \cdots + (x(p-1))^n.$$

Because all the elements $x, 2x, \ldots, x(p-1)$ are different it is still the same sum, but written in another order, and therefore $yS = S$ which gives us $S = 0$. □

8.2 Quadratic equations in $\mathbb{Z}_p$ and the Legendre symbol

Theorem 8.3 *Let p be a prime number. The following are equivalent:*

- $p = 1 \pmod 4$ *or* $p = 2$.
- *The equation* $x^2 = -1$ *has a solution in* $\mathbb{Z}_p$.
- p *can be written as a sum of two squares.*

Proof. $p = 2$ is trivial, so consider an odd prime p. If $p = 4n + 1$, we divide all $4n$ non-zero elements in $\mathbb{Z}_p$ into $2n$ pairs $\{x_i, -x_i\}$. Thus the product of all those elements is

$$\prod_{i=1}^{2n} x_i \cdot (-x_i) = (-1)^{2n} \prod_{i=1}^{2n} x_i^2 = x^2.$$

On the other hand x^2 is the product of all the non-zero elements, and consequently $x^2 = (p-1)! = -1 \pmod p$ according to Wilson's theorem.

Suppose now that we have an x satisfying $x^2 = -1 \pmod p$. We want to write p as a sum of two squares. Let $k = [\sqrt{p}]$. Obviously,

$$k < \sqrt{p} < k+1 \quad \Rightarrow \quad k^2 < p < (k+1)^2.$$

Consider now the $(k+1)^2$ elements in $\mathbb{Z}_p$ of the form $i + jx$, with $0 \leq i, j \leq k$. Because there are more elements than there exist in $\mathbb{Z}_p$ (at least) two of them must be equal:

$$\begin{aligned} i + jx = i' + j'x \pmod p \quad &\Rightarrow \quad (i - i') = x(j' - j) \pmod p \\ &\Rightarrow \quad (i-i')^2 = x^2(j'-j)^2 = -(j'-j)^2 \pmod p. \end{aligned}$$

Now consider the non-negative numbers $a = |i - i'|$, $b = |j' - j|$. Then, by the above, $a^2 + b^2$ is divisible by p. On the other hand

$$0 < a^2 + b^2 \leq k^2 + k^2 < 2p,$$

so $a^2 + b^2 = p$, as we wanted.

Modulo 4 the equation $x^2 + y^2 = 3$ has no solutions, and thus if an odd p is a sum of two squares then $p = 1 \pmod 4$. □

How do we solve a general the quadratic equation $ax^2 + bx + c = 0$ in $\mathbb{Z}_p$? As usual: if $p \neq 2$ then

$$x = \frac{-b \pm \sqrt{b^2 - 4ac}}{2a}.$$

If $p = 2$ we cannot divide but can check directly whether 0 or 1 are solutions.

The main problem is, nevertheless, to decide if a given number a is a square modulo p. It is not a problem when $a = 0$ or $a = 1$. But already for $a = -1$ it is not so evident, as we have seen above. Fortunately there exists a straight-forward, but non-trivial way to decide this. Because

$$\begin{aligned} x^2 = y^2 \pmod{p} \quad &\Leftrightarrow \quad (x-y)(x+y) = 0 \pmod{p} \\ &\Leftrightarrow \quad x = \pm y \pmod{p} \end{aligned}$$

exactly half of the non-zero elements in $\mathbb{Z}_p$ are squares. We define the **Legendre symbol** $\left(\frac{a}{p}\right)$ to be 1 if a is a non-zero square modulo p and to be -1 if it is not. For completeness we define it as zero if $a = 0 \pmod{p}$. Note that the Legendre symbol is an integer, but it is mostly used as an element in $\mathbb{Z}_p$. Our first statement is that for odd primes p we have

$$\left(\frac{a}{p}\right) = a^{\frac{p-1}{2}} \pmod{p}.$$

Indeed, consider a non-zero a ($a = 0$ being trivial). If c is a primitive element then $a = c^k$. If k is even then a is a square and we can write exactly half of the non-zero elements in this way. Therefore $a = c^k$ is *not* a square when k is odd.

If $b = a^{\frac{p-1}{2}}$ then $b^2 = 1$ by Fermat's theorem. Thus $b = \pm 1$ (if $a \neq 0$). For even k we obviously get $b = 1$, but because the order of c is exactly $p - 1$, it follows that $b \neq 1$ for odd k, thus $b = -1$ in this case.

Now we will give another interpretation. If a, b are integers and $(a, p) = 1$ then the map $x \mapsto ax + b$ is a permutation π of the set of non-zero elements in $\mathbb{Z}_p$. The exciting and unexpected fact is that the sign of this permutation (see page 222) coincides with $\left(\frac{a}{p}\right)$:

$$\begin{aligned} \operatorname{sgn}(\pi) &= \prod_{0<i<j<p} \frac{\pi(i) - \pi(j)}{i-j} = \prod_{0<i<j<p} \frac{ai - aj}{i - j} \\ &= \prod_{0<i<j<p} \frac{a(i-j)}{i-j} = a^{\binom{p}{2}} = (a^p)^{\frac{p-1}{2}} = a^{\frac{p-1}{2}} = \left(\frac{a}{p}\right) \pmod{p}. \end{aligned}$$

Based on this observation, we can prove the following properties.

Theorem 8.4 *Let p, q be odd primes and a, b arbitrary integers. Then*

$$\left(\frac{a}{p}\right)\left(\frac{b}{p}\right)=\left(\frac{ab}{p}\right);$$

$$\left(\frac{-1}{p}\right)=(-1)^{\frac{p-1}{2}};\qquad \left(\frac{2}{p}\right)=(-1)^{\frac{p^2-1}{8}};$$

$$\left(\frac{q}{p}\right)=\left(\frac{p}{q}\right)(-1)^{\frac{p-1}{2}\cdot\frac{q-1}{2}}\qquad \textit{(quadratic reciprocity law).}$$

For example: is 11 a square modulo 101? We will check directly:

$$\left(\frac{11}{101}\right)=\left(\frac{101}{11}\right)(-1)^{\frac{11-1}{2}\cdot\frac{101-1}{2}}=\left(\frac{2}{11}\right)=(-1)^{\frac{11^2-1}{8}}=-1$$

and the answer is *no*.

There exist many beautiful proofs of the quadratic reciprocity law. We will give a sketch of a combinatorial one, because it allows us to connect some different subjects and methods.

Consider the set S consisting of all pairs (i,j) with $i\in\mathbb{Z}_p$, $j\in\mathbb{Z}_q$. We can define two permutations

$$\pi_p:(i,j)\mapsto(i,j+pi),\qquad \pi_q:(i,j)\mapsto(i+qj,j).$$

Since these permutations do not change one of the coordinates we can find their signs by looking at the action on the other coordinate:

$$\operatorname{sgn}(\pi_p)=\left(\left(\frac{p}{q}\right)\right)^p=\left(\frac{p}{q}\right);\quad \operatorname{sgn}(\pi_q)=\left(\left(\frac{q}{p}\right)\right)^q=\left(\frac{q}{p}\right).$$

The product of these two numbers gives us the sign of the permutation $\sigma=\pi_p\pi_q^{-1}$ which moves a pair $(i+jq,j)$ to $(i,j+pi)$ (through the pair (i,j)).

Now we would like to interpret our set S differently, as $\mathbb{Z}_{pq}$. Each number x in $\mathbb{Z}_{pq}$ can be uniquely written as $i+jq$ with $0\le i<p$, $0\le j<q$. According to the Chinese remainder theorem, the map $i+jq\mapsto j+ip$ is a permutation and this is the same permutation σ as above. But now we can calculate its sign in a different way. It is equal to $(-1)^k$, where k is the number of pairs (x,x') such that $x=i+jq>x'=i'+j'q$ but $j+pi<j'+pi'$. Obviously this is equivalent to the conditions $j>j'$; $i<i'$, and thus $k=\binom{q}{2}\binom{p}{2}$ which gives us the desired expression $(-1)^k=(-1)^{\frac{p-1}{2}\cdot\frac{q-1}{2}}$.

We recommend the reader to understand and complete this non-trivial proof — it contains many deep and useful ideas. Now it is time for some applications.

Problem 8.5 Let $n > 1$ be an integer. Prove that $2^n - 1$ does not divide $3^n - 1$.

Search. Suppose the opposite. The case when n is even is trivial because $3 = 2^2 - 1 \mid 2^n - 1$. So let us assume that n is odd, $n = 2m + 1$. Each prime p that divides $2^n - 1$ divides $3^n - 1$ as well. Thus

$$3^n \equiv 1 \quad \Rightarrow \quad 3 \equiv 3^{n+1} = \left(3^{m+1}\right)^2 \pmod{p}.$$

Interesting. This means that 3 is a square modulo p.

When is 3 a square modulo an odd prime p? Since $\frac{3-1}{2} = 1$ is odd, the quadratic reciprocity law shows that $\left(\frac{p}{3}\right) = (-1)^{\frac{p-1}{2}}$. Thus if p is a square modulo 3 then $p \equiv 1 \pmod{3}$ and simultaneously $p \equiv 1 \pmod{4}$. Otherwise $p \equiv -1 \pmod{3}$ and simultaneously $p \equiv -1 \pmod{4}$. In other words, $p \equiv \pm 1 \pmod{12}$. This suggests that we need to look carefully at $2^n - 1$ modulo 12. We have $4^2 \equiv 4 \pmod{12}$ and

$$2^{2m+1} - 1 = 2 \cdot 4^m - 1 \equiv 2 \cdot 4 - 1 = 7 \pmod{12}.$$

But a product of primes that are equal to ± 1 modulo 12 cannot be equal to 7 modulo 12 and so we get a contradiction. $\square$

8.3 Continuous fractions and Pell's equation

Each real number α can be approximated by rational numbers. There exists a canonical way to find the best approximations $r_n = \frac{p_n}{q_n}$ using a representation of α as a so-called **continued fraction**

$$\alpha = [a_0; a_1, a_2, a_3, \ldots] = a_0 + \cfrac{1}{a_1 + \cfrac{1}{a_2 + \cfrac{1}{a_3 + \cdots}}}.$$

As a_0 we take the integer part $a_0 = [\alpha]$. If α was an integer we are done, otherwise we consider $\beta = \frac{1}{\alpha - a_0}$ and take $a_1 = [\beta]$. If $\beta = a_1$ we stop here. Otherwise we continue in the same manner, taking $\gamma = \frac{1}{\beta - a_1}$ and $a_2 = [\gamma]$

and so on. We get a finite or infinite sequence (a_n) and it is easy to prove that we get a finite continued fraction if and only if α is rational. If we cut the tail — the sequence after a_n — then the remaining sequence gives a rational number $\frac{p_n}{q_n} = [a_0; a_1, \ldots, a_n]$ and this is the approximation we wanted. Consider as an example what the continued fraction for $\alpha = \sqrt{2}$ looks like. We have $a_0 = [\sqrt{2}] = 1$. After that

$$a_1 = \left[\frac{1}{\sqrt{2}-1}\right] = \left[\sqrt{2}+1\right] = 2; \quad a_2 = \left[\frac{1}{\sqrt{2}+1-2}\right] = \left[\sqrt{2}+1\right] = 2$$

and we see that $a_k = 2$ for all remaining k. Thus $\sqrt{2} = [1; 2, 2, 2, 2, \ldots]$. What are the approximations?

$$\frac{p_0}{q_0} = 1 = \frac{1}{1}; \quad \frac{p_1}{q_1} = [1;2] = 1 + \frac{1}{2} = \frac{3}{2}; \quad \frac{p_2}{q_2} = [1;2,2] = 1 + \cfrac{1}{2 + \cfrac{1}{2}} = \frac{7}{5},$$

and so on giving better and better approximations. If we try π we find that the first digits in the continued fraction are $\pi = [3; 7, 15, 1, 292, 1, 1, 1, 2, 1, 3, \ldots]$ and already $\frac{p_1}{q_1} = \frac{22}{7}$, $\frac{p_3}{q_3} = \frac{355}{113}$ are very good approximations.

If α is a rational number, $\alpha = \frac{a}{b}$, we see that the numbers a_i are the same as those we get when performing the Euclidean algorithm to compute $\gcd(a, b)$. For example, looking at the calculations on page 69, we immediately conclude that $\frac{102}{81} = [1; 3, 1, 6] = \frac{34}{27}$. Thus we can interpret the continued fraction for an irrational number as the result of performing the (infinite) Euclidean algorithm (where we do not know what we are looking for).

The fractions $r_n = \frac{p_n}{q_n}$ have very nice properties. For rational numbers we are not surprised that $[1; 3, 1] = \frac{5}{4}$ gives the numbers 5 and 4 which are exactly the same that we obtained on page 69 (explain why, if you want to improve your understanding of continued fractions). But mostly we are interested in what we get when α is irrational.

First of all there are recursion relations. If we define $p_{-1} = 1$, $q_{-1} = 0$, then for all $n \geq 1$ we have

$$p_n = a_n p_{n-1} + p_{n-2}; \quad q_n = a_n q_{n-1} + q_{n-2}$$

and straight-forward induction shows that this implies

$$q_n p_{n-1} - p_n q_{n-1} = (-1)^n \quad \Leftrightarrow \quad r_{n-1} - r_n = \frac{(-1)^n}{q_n q_{n-1}};$$

$$q_n p_{n-2} - p_n q_{n-2} = (-1)^{n-1} a_n \quad \Leftrightarrow \quad r_{n-2} - r_n = \frac{(-1)^{n-1} a_n}{q_n q_{n-2}}.$$

Thus the subsequence of truncated continued fractions with an even number of terms is increasing: $r_0 < r_2 < r_4 < \cdots$ but are smaller than the corresponding fractions with an odd number of terms, which are decreasing: $r_1 > r_3 > r_5 > \cdots$. Both subsequences are therefore convergent to α. In other words:

$$\alpha = a_0 + \sum_{n=0}^{\infty} \frac{(-1)^n}{q_n q_{n+1}}.$$

As good practice we recommend the reader to prove, using induction, that for any $n \geq 1$,

$$\frac{q_n}{q_{n-1}} = [a_n; a_{n-1}, \ldots, a_1].$$

An error estimate for the approximation is

$$\frac{1}{q_n(q_n + q_{n-1})} < \left|\alpha - \frac{p_n}{q_n}\right| \leq \frac{1}{q_n q_{n+1}}.$$

We say that a rational number $\frac{a}{b}$ is a **best approximation** for α if for any other fraction $\frac{c}{d}$ with $0 < d \leq b$ we have $|d\alpha - c| > |b\alpha - a|$. Note that this implies a weaker inequality: $\left|\alpha - \frac{c}{d}\right| > \left|\alpha - \frac{a}{b}\right|$. The amazing and unexpected result is that every best approximation is equal to some r_n. The converse is true as well (if we exclude r_0 in the case $\alpha = a_0 + \frac{1}{2}$). The proof of this result and many other interesting facts about continued fractions can be found in [57]. Let us consider some corollaries.

Theorem 8.6 *Let $r = \frac{a}{b}$ be a rational number with $(a, b) = 1$. If $|\alpha - r| < \frac{1}{2b^2}$ then $r = r_n$ for some n. On the other hand for every n, either $|\alpha - r_n| < \frac{1}{2q_n^2}$ or $|\alpha - r_{n-1}| < \frac{1}{2q_{n-1}^2}$.*

Proof. It is enough to prove that r is a best approximation. Suppose that $|d\alpha - c| \leq |b\alpha - a| < \frac{1}{2b}$. Then $\left|\alpha - \frac{c}{d}\right| < \frac{1}{2bd}$ and

$$\left|\frac{c}{d} - \frac{a}{b}\right| \leq \left|\alpha - \frac{c}{d}\right| + \left|\alpha - \frac{a}{b}\right| < \frac{1}{2bd} + \frac{1}{2b^2} = \frac{b+d}{2b^2 d}.$$

If $\frac{b}{d} \neq r$ we get

$$\frac{1}{bd} \leq \left|\frac{c}{d} - \frac{a}{b}\right| < \frac{b+d}{2b^2d} \quad \Rightarrow \quad d > b.$$

As to the second statement, note that α is between r_n and r_{n-1} and therefore

$$|\alpha - r_n| + |\alpha - r_{n-1}| = |r_n - r_{n-1}| = \frac{1}{q_n q_{n-1}} < \frac{1}{2q_n^2} + \frac{1}{2q_{n-1}^2},$$

where we have used AM–GM inequality at the end. □

It is not so difficult to prove that the sequence (a_n) is (eventually) periodic if and only if α is an irrational root of a quadratic equation with rational coefficients. If the sequence is finite, we know that α is rational. What is interesting is that there are no other algebraic numbers (e.g. $\alpha = \sqrt[3]{2}$) for which a general formula for a_n is known. Perhaps the reader of this book can find one?

Another important application is **Pell's equation**, $x^2 - my^2 = 1$, where m is a given non-square natural number and we look for integer solutions x and y. Obviously $x = 1$, $y = 0$ is a trivial solution and for other solutions we may restrict ourselves to positive integers. All positive non-trivial solutions can be described as follows. Solutions always exist and they can be obtained from one particular solution, called the **fundamental solution** (x_1, y_1) using recursion. More precisely, if we rewrite $(x_1 + y_1\sqrt{m})^n$ as $x_n + \sqrt{m}y_n$ then (x_n, y_n) is a solution and conversely, all the solutions can be obtained in this way. Thus

$$(x_{n+1}, y_{n+1}) = (x_1 x_n + m y_1 y_n, x_1 y_n + y_1 x_n).$$

For example, the equation $x^2 - 2y^2 = 1$ has a non-trivial solution $(x_1, y_1) = (3, 2)$ and is easy to check that it is the minimal non-trivial solution, and thus the fundamental solution. Therefore the recursion

$$(x_{n+1}, y_{n+1}) = (3x_n + 4y_n, 2x_n + 3y_n)$$

generates all non-trivial solutions:

$$(3, 2), (17, 12), (99, 70), (577, 408), \ldots$$

An unexpected connection with Chebyshev's polynomials $T_n(x)$ is that $x_n = T_n(x_1)$. Try to prove this claim!

The main difficulty is to find a fundamental solution. Because it is the minimal non-trivial solution, it can often be found by trial-and-error, but a better idea is to use the representation of $\sqrt{m}$ as an infinite continued fraction $\sqrt{m} = [a_0; a_1, a_2, \ldots]$. If $r_n = \frac{p_n}{q_n}$ are the rational approximations that we get by truncating the terms after a_n, then it is known that (x_1, y_1) can be found as one of (p_n, q_n), and thus we need to test only very special numbers.

For example, if we need a fundamental solution for $x^2 - 10y^2 = 1$ we begin by computing the continued fraction for $\sqrt{10}$,

$$\sqrt{10} = [3; 6, 6, 6, \ldots].$$

Indeed, $[\sqrt{10}] = 3$ and $\left[\frac{1}{\sqrt{10}-3}\right] = [\sqrt{10}+3] = 6$ and after that we get a period. The first approximation $[3;] = \frac{3}{1}$ does not give a solution to $x^2 - 10y^2 = 1$, but the second one $[3; 6] = 3 + \frac{1}{6} = \frac{19}{6}$ is already good enough: $x_1 = 19$, $y_1 = 6$.

Consider now a difficult problem.

Problem 8.7 Show that the equation

$$x(x+1)(x+2)(x+3) = y(y+1)^2(y+2)^3(y+3)^4$$

has no solutions in natural numbers.

Search. We will skip a natural attempt to use the primes in the factorizations. Here we have another play. Can we find some reasonable variable substitution? At least on the left-hand side we see that $x(x+3) = (x+1)(x+2) - 2$, so if we put $u = x^2+3x+1$ we can rewrite the left hand side as $(u-1)(u+1) = u^2-1$. Can we do something similar in the right-hand side? We have $y(y+2) = (y+1)^2-1$ so if we put $z = y+1$, the right-hand side becomes $(z^2-1)(z(z+1)(z+2)^2)^2$. But now our equation looks like

$$u^2 - 1 = (z^2 - 1)v^2$$

where $v = z(z+1)(z+2)^2$, and this can be interpreted as Pell's equation $u^2 - dv^2 = 1$ with $d = z^2 - 1$. Good. Even better: we immediately identify the fundamental solution, $u_1 = z$, $v_1 = 1$, and therefore we get all solutions (u_n, v_n) using the recursion relations:

$$u_{n+1} = zu_n + dv_n, \quad v_{n+1} = zv_n + u_n.$$

This directly gives us:

$$u_2 = z^2 + d = 2z^2 - 1, \qquad v_2 = z + z = 2z;$$
$$u_3 = 4z^3 - 3z, \qquad v_3 = 4z^2 - 1;$$
$$u_4 = 8z^4 - 8z^2 + 1, \qquad v_4 = 8z^3 - 4z;$$
$$v_5 = 16z^4 - 12z^2 + 1.$$

Now all we need to do is to check that for any $z \geq 2$, the value of $v = z^4 + 5z^3 + 8z^2 + 4z$ is greater than v_4 and less than v_5 and to use that the sequence v_n is increasing. This is routine. □

CHAPTER 9

Elements of linear algebra

Most students at the IMO level have already had some taste of linear algebra and we highly recommend you to look through any introductory textbook in this subject. For reference, we will collect some of the most important facts and definitions in this chapter, but this will probably not be too useful if the reader has not seen them before.

Basically, linear algebra is the art of interpreting and/or solving systems of linear equations in several variables. The most important notions are vectors, matrices and linear maps. The main tools for the beginner are Gaussian elimination, linear dependency, inner and vector product, dimension, determinants and matrix multiplication. We will concentrate on matrices first and consider abstract vectors in the next section.

9.1 Matrices

A **matrix** $A = (a_{ij})$ of **size** $m \times n$ is a table that contains mn elements arranged in m rows and n columns. Here a_{ij} is the element at the intersection of the ith row and the jth column. The **transposed matrix** $B = A^T$ has size $n \times m$ and $b_{ij} = a_{ji}$.

An **upper triangular** matrix is a square matrix A such that $a_{ij} = 0$ for $i > j$. An upper triangular matrix looks like:

$$A = \begin{pmatrix} a_{11} & a_{12} & a_{13} & \cdots & a_{1n} \\ 0 & a_{22} & a_{23} & \cdots & a_{2n} \\ 0 & 0 & a_{33} & \cdots & a_{3n} \\ \dots & \dots & \dots & \dots & \dots \\ 0 & 0 & 0 & \cdots & a_{nn} \end{pmatrix}$$

i.e. has zeros under the main diagonal. A **diagonal matrix,** $\operatorname{diag}(\lambda_1, \ldots, \lambda_n)$, is a square matrix that has non-zero elements on the main diagonal only:

$a_{ii} = \lambda_i$. The most important diagonal matrix is the **identity matrix** $I = \text{diag}(1, 1, \dots, 1)$.

Matrices X of size $n \times 1$ are identified with **column vectors** of size n. For such matrices it is natural to use a shorter, single index, notation:

$$X = \begin{pmatrix} x_1 \\ x_2 \\ \vdots \\ x_n \end{pmatrix}.$$

We also use **row vectors**, i.e. matrices of size $1 \times n$, which we can interpret as transposed column vectors $Y = X^T$.

The parentheses in a matrix is for typographic reasons only and ordinary numbers are nothing else than matrices of size 1×1. Vice versa: matrices can be thought of as "generalized numbers". What we are trying to do in matrix theory is to understand which properties of numbers can be extended to matrices so comparing with numbers helps our intuition. This point of view is useful: instead of remembering a lot of properties of matrices we can sometimes simply say: "this is the same as for numbers" and everything is immediately clear. So it will be necessary to remember only the properties that are different, or rather, more complicated, for matrices than for numbers.

For example, we can add and subtract matrices of the same size:

$$C = A + B, \quad D = A - B,$$

meaning $C = (c_{ij})$, $D = (d_{ij})$, where

$$c_{ij} = a_{ij} + b_{ij}, \quad d_{ij} = a_{ij} - b_{ij}.$$

Note that this is a *definition* of + and − for matrices, but these operations have the same properties as for numbers, because we work with every element independently. So we can, for example, be sure that $A + B = B + A$ and $A - (A - B) = B$, because the addition and the subtraction is "the same as for numbers". We even have a zero-matrix O, where $O_{ij} = 0$, with the usual properties of zero. Note, however, that matrices of *different* sizes cannot be added or subtracted.

Multiplication is more complicated to generalize. In fact there are several different ways to do this.

In the first definition, we multiply *different* kinds of objects: we multiply a number c by a matrix $A = (a_{ij})$. The result is a matrix of the same size as A:

$$cA = Ac = (ca_{ij}).$$

This operation is simple to understand. For example,

$$2\begin{pmatrix} 1 & 2 & 3 \\ 0 & 0 & 4 \end{pmatrix} = \begin{pmatrix} 2 & 4 & 6 \\ 0 & 0 & 8 \end{pmatrix},$$

and natural properties such as

$$(c_1 + c_2)A = c_1A + c_2A; \quad c(A + B) = cA + cB;$$
$$1A = A; \quad 0A = O; \quad (-1)A = -A; \quad cO = O$$

are very easy to prove.

The second kind of multiplication looks like this: If $X = (x_1, \ldots, x_n)$, $Y = (y_1, \ldots, y_n)$ are two row vectors then their **inner product** is the *number*

$$X \cdot Y = x_1y_1 + \cdots + x_ny_n.$$

If X and Y are real vectors the inner product has an important geometrical interpretation:

$$X \cdot Y = |X| \cdot |Y| \cos\alpha,$$

where $|X| = \sqrt{x_1^2 + \cdots + x_n^2}$ is the length and α is the angle between the vectors. (The reader can restrict to the case $n \leq 3$ to see the picture. We can visualise the row vector X as an arrow from the origin to the point with the coordinates X).

As special cases it follows that $X \cdot X = |X|^2$ and that X and Y are **orthogonal** (or perpendicular) iff $X \cdot Y = 0$. The inner product allows us to compute the angle between vectors when we know their coordinates. For example, the vector $X = (a, b, c)$ is orthogonal to the plane $ax + by + cz = d$, because for any two points (x_1, y_1, z_1), (x_2, y_2, z_2) on the plane we see that the vector $Y = (x_1 - x_2, y_1 - y_2, z_1 - z_2)$ is orthogonal to X:

$$X \cdot Y = a(x_1 - x_2) + b(y_1 - y_2) + c(z_1 - z_2) = d - d = 0.$$

Of course, we can define the inner product between column vectors in an analogous fashion, and more importantly, using the same formula we can multiply any row vector X by any column vector Y of the same length. Using this operation we can extend our definition of multiplication so that we may multiply any matrix A of size $m \times n$ by any matrix B of size $n \times k$ (notice the

n occurring in both sizes). The result is a matrix C of size $m \times k$ where c_{ij} is a product of the ith row of A by the jth column of B. In other words,

$$C = (c_{ij}) = \left(\sum_{k=1}^{n} a_{ik} b_{kj} \right).$$

Let us investigate some properties of matrix multiplication. Since the definition is somewhat complicated, we cannot expect that everything works the same as for numbers, but actually it is almost so. We have the usual distributive laws:

$$A(B+C) = AB + AC, \quad (A+B)C = AC + BC,$$

We have the identity $IA = AI = A$, and even the zero matrix works as expected: $OA = O$, $AO = O$, although the zero matrices may have different sizes.

There are two main differences though. First of all, in general, $AB \neq BA$. The two products may have different sizes and it may even happen that only one of the products exists.

Another difference is that the product of two non-zero matrices can be equal to the zero matrix. An unpleasant consequence of this is that we cannot cancel non-zero matrices:

$$AB = AC \quad \not\Rightarrow \quad B = C.$$

One important formula that survives is the associative law $(AB)C = A(BC)$. One of the most important consequences of this is that any *square* matrix A has powers A^k for any integer $k \geq 0$ and these powers have the properties we would expect, such as $A^k A^l = A^l A^k = A^{k+l}$, $A^0 = I$. Consequently, if p is a polynomial, we can form the matrix $p(A)$.

What about negative powers? We say that a matrix B is **inverse** to A if $AB = BA = I$. Inverses, if they exist, are uniquely determined and the inverse of A is denoted by A^{-1}. We say that the matrix is **invertible** if it has an inverse. It follows that invertible matrices can be raised to negative powers. Only square matrices can be invertible, and later on we will see how to recognize them.

It can be checked directly (try!) that a triangular matrix is invertible iff its diagonal elements are non-zero and in this case the inverse is a triangular matrix as well. For example, if $D = \operatorname{diag}(\lambda_1, \lambda_2, \ldots, \lambda_n)$ then $D^{-1} = \operatorname{diag}(\lambda_1^{-1}, \lambda_2^{-1}, \ldots, \lambda_n^{-1})$.

Consider a couple of practical interpretations. The multiplication $X \mapsto AX$ of a column vector X by the matrix

$$A = \begin{pmatrix} \cos\alpha & \sin\alpha \\ -\sin\alpha & \cos\alpha \end{pmatrix}$$

gives the same effect rotating X by the angle α (counter-clockwise). Compare this with complex numbers! Similarly $\operatorname{diag}(k, k)$ produces a homothety with centre at the origin. If V is a row vector, the map $X \mapsto \frac{X \cdot V}{|V|^2} V$ is the orthogonal projection on the line containing V, and thus the map $X \mapsto X - 2\frac{X \cdot V}{|V|^2} V$ describes reflection in the same line.

We see that linear algebra has many applications in geometry, but it is useful in other areas of mathematics as well.

Let G be a finite graph (see Section 6.2) on n vertices. Consider the $n \times n$ matrix A, where $a_{ij} = 1$ if there is an edge from the vertex i to the vertex j and $a_{ij} = 0$ otherwise. If $B = A^k$ with $k \geq 0$, then b_{ij} is the number of paths in G of length k from the vertex i to the vertex j. This works for oriented graphs as well!

Another application is difference equations. For example, the conditions

$$\begin{cases} x_n = x_{n-1} + 2y_{n-1} \\ y_n = 4x_{n-1} + 3y_{n-1} \end{cases}$$

can be rewritten as

$$\begin{pmatrix} x_n \\ y_n \end{pmatrix} = \begin{pmatrix} 1 & 2 \\ 4 & 3 \end{pmatrix} \begin{pmatrix} x_{n-1} \\ y_{n-1} \end{pmatrix} \quad \Rightarrow \quad \begin{pmatrix} x_n \\ y_n \end{pmatrix} = \begin{pmatrix} 1 & 2 \\ 4 & 3 \end{pmatrix}^n \begin{pmatrix} x_0 \\ y_0 \end{pmatrix}.$$

There is a standard method to find powers, which may be found in any linear algebra book. In our case it helps to rewrite our matrix A as QDQ^{-1}, where

$$D = \begin{pmatrix} -1 & 0 \\ 0 & 5 \end{pmatrix}, \quad Q = \begin{pmatrix} 1 & 1 \\ -1 & 2 \end{pmatrix}, \quad Q^{-1} = \frac{1}{3}\begin{pmatrix} 2 & -1 \\ 1 & 1 \end{pmatrix}.$$

The (systematic) process of finding Q and D is called *diagonalization*. Once we have expressed A as QDQ^{-1}, it follows that

$$A^n = \left(QDQ^{-1}\right)^n = QDQ^{-1}QDQ^{-1}QDQ^{-1}\cdots QDQ^{-1} = QD^nQ^{-1}$$
$$= \begin{pmatrix} 1 & 1 \\ -1 & 2 \end{pmatrix}\begin{pmatrix} (-1)^n & 0 \\ 0 & 5^n \end{pmatrix}\cdot\frac{1}{3}\begin{pmatrix} 2 & -1 \\ 1 & 1 \end{pmatrix}$$
$$= \frac{1}{3}\begin{pmatrix} 2\cdot(-1)^n + 5^n & 5^n - (-1)^n \\ 2\cdot 5^n - 2\cdot(-1)^n & (-1)^n + 2\cdot 5^n \end{pmatrix}.$$

and we get the solutions to the difference equations, e.g.

$$x_n = \frac{1}{3}\big((2x_0 - y_0)(-1)^n + (x_0 + y_0)5^n\big).$$

Any system of linear equations can be written as a matrix equation

$$AX = B.$$

For example, if we are looking for x_1, x_2, x_3, x_4, x_5, satisfying the system

$$\begin{cases} x_3 + 2x_4 + 3x_5 = 6 \\ 2x_2 + 4x_3 + 6x_4 + 8x_5 = 20 \\ 4x_2 + 8x_3 + 12x_4 + 16x_5 = c \\ x_2 + 2x_3 + 3x_4 + 5x_5 = 11, \end{cases}$$

we take

$$A = \begin{pmatrix} 0 & 0 & 1 & 2 & 3 \\ 0 & 2 & 4 & 6 & 8 \\ 0 & 4 & 8 & 12 & 16 \\ 0 & 1 & 2 & 3 & 5 \end{pmatrix}, \quad X = \begin{pmatrix} x_1 \\ x_2 \\ x_3 \\ x_4 \\ x_5 \end{pmatrix}, \quad B = \begin{pmatrix} 6 \\ 20 \\ c \\ 11 \end{pmatrix}.$$

If A has an inverse A^{-1}, we can directly find the unique solution:

$$AX = B \quad\Rightarrow\quad A^{-1}AX = A^{-1}B \quad\Rightarrow\quad X = A^{-1}B.$$

Unfortunately, this idea only works when A is invertible, so we need another method. The most common one is **Gaussian elimination**. The idea is straight-forward: simplify the matrix without losing solutions. The permitted operation is to subtract a multiple of one row from another row.

The algorithm works as follows: If a row is non-zero, call its first non-zero element its **pivot element**. Zero rows have no pivot element. If some *column* has more than one pivot element we can, using the permitted operation, annihilate all of them except one. To simplify the work, we can allow other operations at this stage: multiplication/division of one row by a non-zero constant, and rearrangement of the rows in another order.

Once we have at most one pivot element in each column, the matrix is sufficiently simple to find all the solutions. First we will introduce some definitions. The columns $j_1, j_2, \ldots, j_r$ which contain pivot elements we call **pivot columns**. The corresponding variables $x_{j_1}, x_{j_2}, \ldots, x_{j_r}$ in the system are called **basic variables** and the remaining variables are called **free variables.**

To solve the system $AX = B$ when there is at most one pivot element in each column of A is easy. First, if there are equations of the form $0 = b$ for some non-zero b, there is obviously no solutions. Otherwise we may choose the $n-r$ free variables arbitrarily as parameters $t_1, \ldots, t_{n-r}$. The basic variables we express (uniquely) in terms of the parameters starting from the last one.

Let us look at our example, and do all the transformations. First, we create the extended matrix $(A \mid B)$

$$\left(\begin{array}{ccccc|c} 0 & 0 & \underline{1} & 2 & 3 & 6 \\ 0 & \underline{2} & 4 & 6 & 8 & 20 \\ 0 & \underline{4} & 8 & 12 & 16 & c \\ 0 & \underline{1} & 2 & 3 & 5 & 11 \end{array}\right),$$

to have everything under control. We underline the pivot elements and begin transforming the matrix. The first non-zero column is the second one and it contains three pivot elements in the last three rows. Subtracting twice the second row from the third row we get:

$$\left(\begin{array}{ccccc|c} 0 & 0 & \underline{1} & 2 & 3 & 6 \\ 0 & \underline{2} & 4 & 6 & 8 & 20 \\ 0 & 0 & 0 & 0 & 0 & c-40 \\ 0 & \underline{1} & 2 & 3 & 5 & 11 \end{array}\right).$$

Here we can already say that the system has no solutions if $c \neq 40$. So suppose $c = 40$. Now we can “kill” the 1 in the last row subtracting half of the second row:

$$\left(\begin{array}{ccccc|c} 0 & 0 & \underline{1} & 2 & 3 & 6 \\ 0 & \underline{2} & 4 & 6 & 8 & 20 \\ 0 & 0 & 0 & 0 & 0 & 0 \\ 0 & 0 & 0 & 0 & \underline{1} & 1 \end{array}\right)$$

Now we have at most one pivot element in each column. For cosmetic reasons, let us make a small permutation of rows to obtain the result in echelon form:

$$\left(\begin{array}{ccccc|c} 0 & \underline{2} & 4 & 6 & 8 & 20 \\ 0 & 0 & \underline{1} & 2 & 3 & 6 \\ 0 & 0 & 0 & 0 & \underline{1} & 1 \\ 0 & 0 & 0 & 0 & 0 & 0 \end{array}\right).$$

We can solve the system: $x_1 = s$ and $x_4 = t$ are free variables. From the third equation, we immediately get $x_5 = 1$. From the second equation we get $x_3 + 2t + 3 = 6 \Rightarrow x_3 = 3 - 2t$. Finally, from the first equation, we get $x_2 = 10 - 2x_3 - 3x_4 - 4x_5 = 10 - 2(3 - 2t) - 3t - 4 = t$.

Of course, all this could be done much smarter. For example, we could from the very beginning divide the second row by 2 and could delete the zero row when we obtained it, but we wanted to show some of the difficulties that may show up in the process of the elimination. Note as well that we could get our solutions from the echelon form, dividing each non-zero row by its pivot element and performing "complete cleaning" (or Gauss–Jordan elimination) of the columns containing pivot elements using the obtained ones (pivot elements) and starting from the last one (thus in the opposite order):

$$\left(\begin{array}{ccccc|c} 0 & \underline{1} & 2 & 3 & 4 & 10 \\ 0 & 0 & \underline{1} & 2 & 3 & 6 \\ 0 & 0 & 0 & 0 & \underline{1} & 1 \\ 0 & 0 & 0 & 0 & 0 & 0 \end{array}\right) \sim \left(\begin{array}{ccccc|c} 0 & \underline{1} & 2 & 3 & 0 & 6 \\ 0 & 0 & \underline{1} & 2 & 0 & 3 \\ 0 & 0 & 0 & 0 & \underline{1} & 1 \\ 0 & 0 & 0 & 0 & 0 & 0 \end{array}\right)$$

$$\sim \left(\begin{array}{ccccc|c} 0 & \underline{1} & 0 & -1 & 0 & 0 \\ 0 & 0 & \underline{1} & 2 & 0 & 3 \\ 0 & 0 & 0 & 0 & \underline{1} & 1 \\ 0 & 0 & 0 & 0 & 0 & 0 \end{array}\right)$$

and we (of course) get the same solutions.

Gauss–Jordan elimination is also used when we want to find the inverse to a square matrix A. Put the identity matrix I of the same size into the extended matrix $(A \mid I)$ and perform Gaussian elimination (using all allowed operations) to obtain I instead of A, resulting in $(I \mid C)$. The matrix C is the inverse A^{-1}. If, during the elimination, we get a zero row in A we stop and conclude that A is not invertible. Consider a tiny example where we find Q^{-1} for the matrix Q on page 253:

$$\left(\begin{array}{cc|cc} 1 & 1 & 1 & 0 \\ -1 & 2 & 0 & 1 \end{array}\right) \sim \left(\begin{array}{cc|cc} 1 & 1 & 1 & 0 \\ 0 & 3 & 1 & 1 \end{array}\right)$$
$$\sim \left(\begin{array}{cc|cc} 1 & 1 & 1 & 0 \\ 0 & 1 & 1/3 & 1/3 \end{array}\right) \sim \left(\begin{array}{cc|cc} 1 & 0 & 2/3 & -1/3 \\ 0 & 1 & 1/3 & 1/3 \end{array}\right).$$

Each square matrix A has a very important number called the **determinant**, denoted $\det(A)$ or simply $|A|$. We are familiar with determinants for 2×2 matrices from Theorem 1.17. The general definition is less trivial and uses induction.

Suppose that we already know how to calculate the determinant for matrices of size $n - 1$ and want to define the determinant $|A|$ for a matrix of size $n > 1$. First of all, let us introduce some notation. For any indices i, j consider the submatrix obtained from A by deleting the ith row and the jth column. This submatrix has size $n - 1$, and we can calculate its determinant which we denote by D_{ij}. Now we can write our recursive definition:

$$|A| = a_{11}D_{11} - a_{21}D_{21} + a_{31}D_{31} - \cdots \pm a_{n1}D_{n1},$$

where we alternate signs, so the last one will be + if n is odd and − if n is even.

Let us try the case $n = 3$, which is sufficient for most of the cases we are interested in. We already know how to calculate the determinant for 2×2 matrices. Let

$$A = \begin{pmatrix} a_{11} & a_{12} & a_{13} \\ a_{21} & a_{22} & a_{23} \\ a_{31} & a_{32} & a_{33} \end{pmatrix}$$

By definition, $|A| = a_{11}D_{11} - a_{21}D_{21} + a_{31}D_{31}$, i.e.

$$|A| = a_{11}\begin{vmatrix} a_{22} & a_{23} \\ a_{32} & a_{33} \end{vmatrix} - a_{21}\begin{vmatrix} a_{12} & a_{13} \\ a_{32} & a_{33} \end{vmatrix} + a_{31}\begin{vmatrix} a_{12} & a_{13} \\ a_{22} & a_{23} \end{vmatrix}.$$

In this form we see how we can calculate the determinant of a 3×3 matrix A as long as we are able to calculate arbitrary 2×2 determinants.

Of course, we can continue and get a complete expansion for the 3×3 determinant, namely

$$
\begin{aligned}
|A| &= a_{11}\left(a_{22}a_{33} - a_{23}a_{32}\right) - a_{21}\left(a_{12}a_{33} - a_{13}a_{32}\right) + a_{31}\left(a_{12}a_{23} - a_{13}a_{22}\right) \\
&= a_{11}a_{22}a_{33} - a_{11}a_{23}a_{32} - a_{12}a_{21}a_{33} + a_{13}a_{21}a_{32} + a_{12}a_{23}a_{31} - a_{13}a_{22}a_{31},
\end{aligned}
$$

but we have no interest in this formula for the moment. The important thing is that, in principle, we have a way to calculate the determinant of any square matrix of arbitrary size $n \times n$.

A practical way to calculate the determinant is to do Gaussian elimination and use the following rules:

- If we get a zero row, the determinant is zero;
- The determinant of a triangular matrix is the product of the elements on the main diagonal;
- The determinant does not change if we add a multiple of a row to another row. It changes sign if we interchange two rows. It is multiplied by c if we multiply one row by c.

The determinant respects matrix multiplication, i.e. if A, B are square matrices of the same size then $|AB| = |A| \cdot |B|$. The importance of the determinant is seen in the following theorem.

Theorem 9.1 *For a square matrix A the following are equivalent:*

- *A has an inverse,*
- *For each column vector B, the system $AX = B$ has a unique solution,*
- *For each column vector B, the system $AX = B$ has at least one solution,*
- *For each column vector B, the system $AX = B$ has at most one solution,*
- *$AX = O$ implies $X = O$,*
- *The map $X \mapsto AX$ is injective,*
- *The map $X \mapsto AX$ is surjective,*
- *The map $X \mapsto AX$ is bijective,*
- *The column vectors in A are linearly independent (see the next section for the definition).*
- *The row vectors in A are linearly independent.*

- $|A| \neq 0$.

The geometrical interpretation of the det A is (up to sign) the area (for $n = 2$) and volume (for $n = 3$) of the parallelogram (parallelepiped) obtained from the column vectors of A. The sign has to do with the *orientation* of these vectors, but we will not explore this interpretation further. Here is a nice application.

Problem 9.2 The different points A, B, C in the plane have integer coordinates. Show that the triangle ABC is not equilateral.

Search. The vectors $v = \overline{AB}$ and $u = \overline{AC}$ have integer coordinates. How can we use this? One possible way is to look at the area of the triangle. The area is equal to half of the area of the corresponding parallelogram and therefore (if we look at the determinant), the area is an integer or a half of an integer. Good. On the other hand, if the triangle is equilateral, then the area is equal to

$$\frac{|v| \cdot |u| \sin 60°}{2} = |v|^2 \frac{\sqrt{3}}{4},$$

which is irrational because $|v|^2 = v \cdot v$ is an integer. Done! Of course, we could use Pick's formula instead of the determinant (see page 216). □

9.2 Vector algebra

Next, we would like to discuss abstract vectors. We have already mentioned column and row vectors, but even these we would like to consider from another point of view. But first we need to discuss the numbers. Which numbers can we allow in our matrices and vectors?

In fact this is a very important question. We want to have a **field** $\mathbb{K}$, i.e. a set of numbers for which it is possible to perform all four arithmetic operations, including division (the result still belonging to $\mathbb{K}$). Two good examples of fields are the complex numbers $\mathbb{C}$ and $\mathbb{Z}_p$ if p is prime. All other examples that we need will be some subsets of complex numbers, and thus to save a formal definition we note that a subset of complex numbers $\mathbb{K} \subseteq \mathbb{C}$ is a field if it contains 1 and for any two elements $x, y \in \mathbb{K}$ we have $x - y \in \mathbb{K}$ and if $y \neq 0$ then $\frac{x}{y} \in \mathbb{K}$ as well. Thus the real numbers $\mathbb{R}$ and the rational numbers $\mathbb{Q}$ are fields, but the set of integers is not, because we have no division here. For the same reason $\mathbb{Z}_n$ is not a field when n is not a prime.

Informally, a vector space V over a field $\mathbb{K}$ is a set whose elements $v \in V$ we call vectors. These vectors can be added and subtracted. We can also multiply vectors by any number $c \in \mathbb{K}$, and we want all "expected rules" to be valid.

For example, every vector space has a unique zero vector also denoted as 0 for which $v - v = 0 \cdot v = 0$ (note that here we use different zeros with the same notation: the first zero in this formula is $0 \in \mathbb{K}$, the second is the zero vector $0 \in V$). Geometrically, we imagine a vector as an arrow, but it may be any object. The reader can find a formal axiomatic definition in any linear algebra textbook. For us it is enough to recognize the following vector spaces over $\mathbb{K}$.

- $\mathbb{K}$ is a vector space over itself,
- All column vectors $\mathbb{K}^n$ of a given size n,
- All row vectors of a given size n (this space is often denoted $\mathbb{K}^n$ as well).
- All matrices $M_{m,n}(\mathbb{K})$ of a given size,
- All polynomials $f(x)$ with coefficients in $\mathbb{K}$.
- All functions $f : \mathbb{R} \to \mathbb{R}$ form a vector space over all real numbers $\mathbb{R}$.

In all the examples above, we know how to add, to subtract and to multiply by a constant $c \in \mathbb{K}$.

Let V be a vector space over $\mathbb{K}$. A subset $U \subseteq V$ is called a **subspace** (of V) if for every $u, v \in U$ and every $c \in \mathbb{K}$ we have

$$u + v \in U, \quad \text{and} \quad cv \in U.$$

It immediately follows that a subspace is a vector space by itself (using the same operations as in V). This significantly increases our list of vector spaces.

For example we easily can check that the following are vector spaces without checking all axioms of vector spaces:

- all upper triangular matrices of a given size,
- all diagonal matrices of a given size,
- all polynomials of a degree less than n.
- all continuous functions $f : \mathbb{R} \to \mathbb{R}$.

Another useful remark is that if $\mathbb{F} \subseteq \mathbb{K}$ is a smaller field then V can be considered as a vector space over $\mathbb{F}$ as well. For example the complex matrices can be considered as a vector space over $\mathbb{R}$ or $\mathbb{Q}$.

Let V be a vector space over $\mathbb{K}$. Consider a set E of vectors in V. We say that they **span** V if any vector in V can be written as a (finite) sum

$$v = x_1 v_1 + x_2 v_2 + \cdots + x_n v_n$$

where $x_k \in \mathbb{K}$ and $v_j \in E$.

A minimal spanning set (i.e. a set E which spans V but such that no proper subset of E spans V) is called a **basis** of V. It turns out that every basis has the same number of elements. This number is called the **dimension** of V and is denoted by $\dim V$.

For example, the set $\mathbb{C}$ of complex numbers considered as a vector space over $\mathbb{R}$ has a basis $\{1, i\}$ and $\dim_{\mathbb{R}} \mathbb{C} = 2$. On the other hand, if we view $\mathbb{C}$ as a vector space over itself, it is one-dimensional (with 1 as a basis). The set of column vectors $\mathbb{K}^n$ has dimension n over $\mathbb{K}$ — we can choose as a basis columns E_i, which have 1 in the ith row and zeros elsewhere.

The dimension of a vector space may be infinite. For example, the set of all polynomials $\mathbb{K}[x]$ is an infinite dimensional vector space with the basis $1, x, x^2, \ldots$. We get a more interesting example if we consider $\mathbb{R}$ as a vector space over $\mathbb{Q}$. It is relatively easy to show that we cannot have a finite basis (otherwise $\mathbb{R}$ would be countable, which is not the case).

Mostly we are interested in finite dimensional vector spaces — those that have a finite basis $e_1, \ldots, e_n$. Thus we will assume that any vector v can be written as a **linear combination** of the vectors e_i:

$$v = x_1 e_1 + x_2 e_2 + \cdots + x_n e_n.$$

This representation is unique for every v. Indeed, if $\sum y_i e_i$ is another representation then we have $\sum (x_i - y_i) e_i = 0$ and therefore one of e_i can be expressed in terms of the remaining e_j, which contradicts the assumption of minimality.

The numbers x_i are called the **coordinates** of v in this basis.

Coordinates allow us a very important translation: *whenever we have a question about some abstract vectors we can translate this question to the corresponding column vectors formed by their coordinates and perform the necessary calculations in terms of the coordinates.* From this point of view, $\mathbb{K}^n$ are the only vector spaces required to understand any finite dimensional vector space V.

For example, if we want to know whether the given vectors $u_1, \ldots, u_n$ form a basis in V we do the following. First, we create the corresponding column vectors of the coordinates A_i (in any fixed basis) and form the matrix A consisting of these columns. If B is any column vector then $B = \sum x_i A_i \Leftrightarrow B = AX$. Every vector can we written in this way exactly when the given vectors form a basis, so Theorem 9.1 shows that a necessary and sufficient condition is $\det(A) \neq 0$.

Next, we would like to discuss the following important definition.

The vectors $u_1, u_2, \ldots, u_m$ are **linearly independent** if there is only one (trivial) way to write the zero vector as a linear combination. More precisely:

$$x_1 u_1 + x_2 u_2 + \cdots + x_m u_m = 0 \quad \Rightarrow \quad x_1 = x_2 = \cdots = x_m = 0.$$

Note that all vectors in a basis are linearly independent, because there is only one (trivial) way to express the zero vector using them. It is important (and non-trivial) that any set of linearly independent vectors can be completed to a basis. This gives us the following result, which can be interpreted as a vector space analogue of the pigeonhole principle:

Theorem 9.3 *Any $n+1$ vectors in a vector space of dimension n are always linearly dependent. Therefore one of them can be obtained as a linear combination of others.*

Another useful analogue is the following statement.

Theorem 9.4 *Let U, W be subspaces in a finite dimensional vector space V then $U \cap W$ is a subspace as well and* $\dim(U \cap W) \geq \dim U + \dim W - \dim V$.

Note that if A is any matrix then the solutions of the system $AX = O$ form a vector space and its dimension is exactly the number of free variables we get in Gaussian elimination. Another vector space is the set of all possible vectors of form AX. Its dimension coincides with the number of the basic variables. This is quite obvious for an echelon matrix and it is not difficult to see that these dimensions do not change during Gaussian elimination.

9.3 Real numbers as a rational vector space

We want to discuss more advanced subjects that are very deep and beautiful, but may be difficult in the first reading. In any case the reader will need a good understanding of linear algebra otherwise, s|he can skip this section.

We would like to consider $\mathbb{R}$ as a vector space over the rational numbers $\mathbb{Q}$. Because this is a vector space there is a basis which is, as we mentioned, infinite. Only God (and maybe the Devil) can construct such a basis explicitly, but it is sufficient to know that an infinite basis E exists.

Thus every non-zero real number x can be uniquely written as a (finite!) sum $x = r_1 e_1 + \dots + r_k e_k$ for some $k > 0, 0 \neq r_i \in \mathbb{Q}$ and $e_i \in E$.

Consider now an arbitrary map $f : E \to \mathbb{R}$ and extend it to $f : \mathbb{R} \to \mathbb{R}$ by the rule $f(x) = r_1 f(e_1) + \dots + r_k f(e_k)$. If we additionally put $f(0) = 0$, we obtain an amazing function with the properties $f(x + y) = f(x) + f(y)$ and $f(rx) = rf(x)$ for every $x, y \in \mathbb{R}, r \in \mathbb{Q}$. Note that linear functions $f(x) = cx$ are only a very special case, and we can obtain many other functions! For example, if α is an irrational number then 1 and α are linearly independent over $\mathbb{Q}$ and we can complete them to a basis E. Now if we set $f(1) = 0$ but $f(\alpha) = 1$ we end up with a function $f(x)$ which is non-linear.

Note that we still have no idea what this function looks like in general, but we do have a guarantee that it exists. We may only be sure that this function is not continuous (see page 230 for a proof that every continuous function with the property $f(x + y) = f(x) + f(y)$ is a linear function $f(x) = cx$).

The non-linear solutions are *very badly behaved*. For example, every non-linear solution to $f(x + y) = f(x) + f(y)$ has a graph which is dense in $\mathbb{R}^2$.

Next we would like to consider a solution of Hilbert's third problem. We say that two polyhedra are **scissors-congruent** if the first can be cut into finitely many polyhedral pieces that can be reassembled into the other polyhedron. Obviously, any two scissors-congruent polyhedra have the same volume. The Hilbert question was if the converse is true.

The answer is NO, and we would like to show that a cube and a regular tetrahedron of the same volume are not scissors-congruent. The idea of the proof, suggested by Dehn, consists of three steps. First, to define for every polyhedron P a value (now called the **Dehn invariant**) $D(P)$ as follows. If P has n edges then $D(P)$ is a collection of pairs $\{(l_1, \varphi_1), \dots, (l_n, \varphi_n)\}$, where l_i is the length of ith edge and φ_i is the dihedral angle between the two faces meeting at this edge, measured in radians.

The second step is to introduce the notion of similarity between different collections. We say that one collection is *similar* to another if they can be transformed to each other by finitely many of the following steps:

- Replace some angle φ by $\varphi + 2\pi k$, for some $k \in \mathbb{Z}$.
- Replace any pair (l, φ) by two pairs $(l', \varphi), (l'', \varphi)$ with $l = l' + l''$.
- Vice versa, replace two pairs $(l', \varphi), (l'', \varphi)$ by $(l' + l'', \varphi)$.
- Replace any pair (l, φ) by two pairs $(l, \varphi'), (l, \varphi)$ with $\varphi = \varphi' + \varphi''$.
- Vice versa, replace two pairs $(l, \varphi'), (l, \varphi'')$ by $(l, \varphi' + \varphi'')$.
- If $l = 0$ or $\varphi = 0$ then the pair (l, φ) can be deleted from the collection (or, conversely, added).

Note that during these operations we may get collections that can not be obtained from any polyhedron, but we do not care about that. The important fact is that we consider the Dehn invariant up to this similarity and that two scissors-congruent polyhedra have a similar (thus the same!) Dehn invariant. The last statement needs some pictures to check different cases of cutting and gluing polyhedra and we leave this as a reasonably easy exercise.

The third and final step is to show that a cube and a regular tetrahedron have different, non-similar Dehn invariants. For this we use the function $f(x)$ constructed above where we take $\alpha = \frac{1}{2\pi} \arccos \frac{1}{3}$. Note that because $\cos 2\pi\alpha = \frac{1}{3}$ it follows, according to the solution of problem 2.14 on page 78, that α indeed is irrational.

Now, for every collection $\{(l_1, \varphi_1), \dots (l_n, \varphi_n)\}$ we consider the number $l_1 f\left(\frac{\varphi_1}{2\pi}\right) + \dots + l_n f\left(\frac{\varphi_n}{2\pi}\right)$ and note that because $f(x + y) = f(x) + f(y)$ and $f(1) = 0$ this number does not change while we perform our transformations. Thus similar invariants have the same value. But what are the values for a cube and a regular tetrahedron?

The Dehn invariant of a cube consists of 12 pairs $(a, \frac{\pi}{2})$ that are similar to 3 pairs $(a, 2\pi)$ and we get $3af\left(\frac{2\pi}{2\pi}\right) = 3f(1) = 0$.

Now consider a regular tetrahedron. Its Dehn invariant consists of 6 pairs $(b, \arccos \frac{1}{3})$ and we get $6bf\left(\frac{\arccos \frac{1}{3}}{2\pi}\right) = 6bf(\alpha) = 6b \neq 0$ which finishes this nice proof. It was not an easy solution, but you hardly expected one for a problem that was difficult for Hilbert!

9.4 Finite-dimensional vector spaces inside of $\mathbb{R}$

Next we would like to play with finite dimensional vector spaces V that are subsets of $\mathbb{R}$ (see Section 9.2 if you need some help with this notion). Let

$\mathbb{K} \subseteq \mathbb{R}$ be a field. Thus we can do all arithmetic operations with elements in $\mathbb{K}$, and we may consider V as a vector space over $\mathbb{K}$ as well. The main example for a non-experienced reader is $\mathbb{Q}$.

We want to extend $\mathbb{K}$ with an element $\alpha \in \mathbb{R}$ and consider the set $\mathbb{K}[\alpha]$ consisting of all polynomials $p(\alpha)$ with coefficients in $\mathbb{K}$. Naturally, some of these polynomials can be equal, e.g. in $\mathbb{Q}[\sqrt{2}]$ the polynomials x^3 and $2x$ give the same element evaluated in $\alpha = \sqrt{2}$, because $(\sqrt{2})^3 = 2\sqrt{2}$. Thus we want to distinguish two cases.

We say that α is **transcendental** over $\mathbb{K}$ if all $p(\alpha)$ are different, and **algebraic** over $\mathbb{K}$ if this is not the case. When we say just **algebraic** (without specifying the field), we always mean algebraic over $\mathbb{Q}$ and similarly for **transcendental** numbers. Let us present the main theorem that connects these abstract notions with some concrete ones.

Theorem 9.5 *The following statements are equivalent.*

(i) *α is algebraic over $\mathbb{K}$.*

(ii) *There exists a polynomial $f(x) \in \mathbb{K}[x]$ such that $f(\alpha) = 0$.*

(iii) *Either $\alpha = 0$ or α is invertible in $\mathbb{K}[\alpha]$ (thus $\frac{1}{\alpha} \in \mathbb{K}[\alpha]$).*

(iv) *$\mathbb{K}[\alpha]$ is a finite-dimensional vector space over $\mathbb{K}$.*

(v) *Every element in $\mathbb{K}[\alpha]$ is algebraic over $\mathbb{K}$.*

(vi) *$\mathbb{K}[\alpha]$ is a field.*

Proof. If we have $p_1(\alpha) = p_2(\alpha)$ for two different polynomials $p_1(x), p_2(x)$ we can take $f(x) = p_1(x) - p_2(x)$ which proves the implication $(i) \Rightarrow (ii)$. If

$$f(\alpha) = a_0\alpha^n + a_1\alpha^{n-1} + \cdots + a_{n-1}\alpha + a_n = 0$$

we can (cancelling if necessary) suppose that $a_0, a_n \neq 0$. Then

$$\frac{1}{\alpha} = -\frac{a_0}{a_n}\alpha^{n-1} - \frac{a_1}{a_n}\alpha^{n-2} - \cdots - \frac{a_{n-1}}{a_n} \in \mathbb{K}[\alpha]$$

which shows that $(ii) \Rightarrow (iii)$.

If $\frac{1}{\alpha} = p(\alpha)$ then 1 and $\alpha p(\alpha)$ are equal which shows the implication $(iii) \Rightarrow (i)$ and therefore the first three statements are equivalent.

If $\mathbb{K}[\alpha]$ has dimension n then the $n+1$ vectors $\alpha^n, \alpha^{n-1}, \ldots, \alpha, 1$ are linearly dependent: $a_0\alpha^n + a_1\alpha^{n-1} + \cdots + a_{n-1}\alpha + a_n \cdot 1 = 0$ and we get the implication $(iv) \Rightarrow (ii)$.

Vice versa, if (ii) is valid then we can show by induction that every α^k belongs to a finite-dimensional subspace spanned by $\alpha^{n-1}, \dots, \alpha, 1$ where n is the degree of the polynomial $f(x)$. Indeed, it is obvious for $k < n$ and (with the help of $f(\alpha) = 0$) for $k = n$. Then we note that $\alpha^{k+1} = \alpha \cdot \alpha^k$ can by induction be rewritten as a linear combination of $\alpha^n, \dots, \alpha$ and therefore belongs to our subspace as well. Thus the four first statements are equivalent.

To show $(iv) \Rightarrow (v)$, note that if $\beta \in \mathbb{K}[\alpha]$ then $\mathbb{K}[\beta] \subseteq \mathbb{K}[\alpha]$ and therefore is finite-dimensional over $\mathbb{K}$. According to already proven equivalences β is algebraic over $\mathbb{K}$.

The implication $(v) \Rightarrow (i)$ is evident, so the first five statements are equivalent.

This also implies that all non-zero elements in $\mathbb{K}[\alpha]$ are invertible which is essentially the last statement, because other arithmetic operations in $\mathbb{K}[\alpha]$ do not create any problem. The converse, e.g. $(vi) \Rightarrow (ii)$ is trivial. □

Let $\mathbb{L}$ be another field, containing $\mathbb{K}$. Because we can add elements of $\mathbb{L}$ and multiply them by elements from $\mathbb{K}$ we can think of $\mathbb{L}$ as a vector space over $\mathbb{K}$ and therefore has (a finite or infinite) dimension denoted by $(\mathbb{L} : \mathbb{K})$.

A very useful (but difficult) exercise is to show that if $(\mathbb{L} : \mathbb{K})$ is finite then there exists α such that $\mathbb{L} = \mathbb{K}[\alpha]$.

What if we have three fields? The following theorem is fairly simple to prove but has many non-trivial consequences.

Theorem 9.6 *If $\mathbb{K} \subset \mathbb{L} \subset \mathbb{F}$ is a chain of fields then*

$$(\mathbb{F} : \mathbb{K}) = (\mathbb{F} : \mathbb{L})(\mathbb{L} : \mathbb{K}).$$

Proof. We restrict ourselves to the finite dimensional case only. Let $e_1, \dots e_n$ be a basis for $\mathbb{L}$ over $\mathbb{K}$ and $f_1, \dots, f_m$ be a basis for $\mathbb{F}$ over $\mathbb{L}$.

It is enough to prove that the set of all products $e_j f_i$ is a basis for $\mathbb{F}$ over $\mathbb{K}$. If $f \in \mathbb{F}$ then $f = \sum_i l_i f_i$ with $l_i \in \mathbb{L}$. But $l_i = \sum a_{ij} e_j$ with $a_{ij} \in \mathbb{K}$ and therefore $f = \sum_{i,j} a_{ij} e_j f_i$ and $\{e_j f_i\}$ span $\mathbb{F}$ over $\mathbb{K}$. It only remains to prove that they are linearly independent over $\mathbb{K}$.

Suppose that

$$\sum_{i,j} x_{ij} e_j f_i = \sum_i \Big(\sum_j x_{ij} e_j \Big) f_i = 0.$$

Because f_i is a basis it follows that $\sum_j x_{ij} e_j = 0$. But now $x_{ij} = 0$ because e_j is a basis as well. □

Here is a nice application of the theorem. We want to show that it is impossible to obtain $\alpha = \sqrt[3]{2}$ from the field of rational numbers by repeatedly extending this field by adding of roots of quadratic equations. Indeed, adding roots β of a quadratic equation to a field $\mathbb{K}$ is nothing else than considering the extensions of a field with β and $(\mathbb{K}[\beta] : \mathbb{K}) = 2$ or $\beta \in \mathbb{K}$. According to the theorem, after several such operations we get a field $\mathbb{F}$ such that $(\mathbb{F} : \mathbb{Q}) = 2^m$. But on the other hand $(\mathbb{Q}[\alpha] : \mathbb{Q}) = 3$ thus $\mathbb{Q}[\alpha]$ cannot be inside $\mathbb{F}$ by the same theorem again. Thus α does not belong to $\mathbb{F}$ and this is exactly what we wanted.

This sounds simple, but after some consideration we find that this is a way — probably the easiest way — to show that the ancient *Doubling the cube* (with ruler and compass, construct a cube with exactly twice the volume of a given cube) problem is unsolvable: all that we can do using ruler and compass is to create new segments with lengths that are solutions of linear and quadratic equations with the coefficients in rational combinations of lengths of given segments. For the same reason, *the angle trisection problem* is unsolvable for most angles. Of course, we have skipped some details of the proof, but not the main idea.

Let us restrict ourselves to the case $\mathbb{K} = \mathbb{Q}$, thus algebraic and transcendental numbers. First we note that if α and β are algebraic then the same is true for $\alpha \pm \beta$, $\alpha \cdot \beta$ and (if $\beta \neq 0$) for $\frac{\alpha}{\beta}$. Indeed, consider $\mathbb{K} = \mathbb{Q}[\alpha]$ and $\mathbb{L} = \mathbb{K}[\beta]$. Obviously, β is algebraic over $\mathbb{K}$, because the last contains $\mathbb{Q}$. Thus $(\mathbb{L} : \mathbb{K})$ is finite and according to the last theorem $(\mathbb{L} : \mathbb{Q})$ is finite as well and therefore contains algebraic numbers only. Because $\mathbb{L}$ is a field, it contains all the new elements. To estimate how elegant this proof is try to prove directly that $x = \sqrt{2} + \sqrt{3} + \sqrt{5}$ is algebraic (by constructing a polynomial with rational coefficients having x as a zero).

We would like to be more precise about the dimension $n = (\mathbb{Q}[\alpha] : \mathbb{Q})$. If we choose $f(x)$ to be a minimal monic rational polynomial with the property that $f(\alpha) = 0$, then it easy to check that $f(x)$ is irreducible over $\mathbb{K}$ (otherwise its factor has a smaller degree) and that it has degree exactly n (otherwise there is a smaller basis). That is why we say that α is **algebraic of degree** n. Rational numbers, and only those, have degree one. The numbers of degree 2 are exactly those that have (eventually) periodic continued fractions.

The numbers $\phi_k = [k; k, \ldots]$ (e.g. the golden ratio) have another interesting property. We denote by $\{\{x\}\}$ the distance from x to the closest integer.

Theorem 9.7 *Let $\alpha > 1$ be an algebraic real number. The following are equivalent.*

- *α is a root of a polynomial $p(x) = x^m + a_1x^{m-1} + \cdots + a_m = 0$ with integer coefficients where all the other roots (real or complex) are non-zero and have absolute value less than 1 (such polynomials are called* ***Pisot polynomials****. They are always irreducible. The number α is called* ***Pisot number****).*
- $\lim_{n\to\infty} \{\{\alpha^n\}\} = 0.$

Whether the same is true for transcendental numbers is still an open question.

The only rational Pisot numbers are the integers (larger than 1). Each ϕ_k is a Pisot number of degree 2. Other examples are $2+\sqrt{2}$ and the **plastic number** $\rho = \sqrt[3]{1+\sqrt[3]{1+\sqrt[3]{1+\cdots}}}$ — the minimal Pisot number. It is the only real root of $x^3 = x+1$. The Pisot polynomials $x^m - (k+1)x^{m-1} - 1$ produce Pisot numbers in the interval $(k, k+1)$ for $k \geq 2$. The set of all Pisot numbers is closed.

A common property for all non-rational algebraic numbers is that they are badly approximated by rational numbers. More exactly, Liouville has shown that for any algebraic number α of degree $n > 1$ there exists a positive constant C such that for any rational $\frac{p}{q}$ one has

$$\left|\alpha - \frac{p}{q}\right| > \frac{C}{q^n}.$$

This allows us to create transcendental numbers $[a_0; a_1, \ldots]$ as continued fractions. All we need is the property $a_{k+1} > q_k^{k-1}$ (fill in the details!). Historically these were the first concrete examples of transcendental numbers and the first possible way to prove that some known numbers (such as e) are transcendental. But today we know other powerful (and very difficult) results.

Theorem 9.8 *Let α be a non-zero algebraic number. Then*

- ***(Lindemann)*** *The number e^α is transcendental.*
- ***(Gelfond, Schneider)****. If $\alpha \neq 1$ and β is a non-rational algebraic number then α^β is a transcendental number.*

This works for complex algebraic numbers (such as i) as well, and thus π is transcendental because otherwise $i\pi$ is algebraic, which contradicts the fact that $e^{i\pi} = -1$.

CHAPTER 10

Real functions

We would like to mention some interesting results that are normally considered only in university courses. These theorems are not necessary to solve problems at the IMO level, but the approaches and ideas discussed below can still be helpful.

10.1 Lagrange's Mean Value Theorem

If $f(x)$ is a continuous function on the closed interval $[a, b]$ we know that it attains its maximum and minimum value. If $f(a) = f(b)$, we can be sure that at least one of the extreme values is attained somewhere strictly inside the interval. Let's say this happens at the point $\theta \in (a, b)$. If $f(x)$ is differentiable on (a, b), then we can conclude that according to Rolle's theorem $f'(\theta) = 0$ as already discussed in Chapter 7.

An arbitrary function $f(x)$ can be replaced by $g(x) = f(x) - kx$ with $g(a) = g(b)$ and we obtain famous **Mean Value Theorem**.

Theorem 10.1 (Lagrange) *Let $f(x)$ be a continuous function on $[a, b]$, differentiable on (a, b). Then there exist $a < \theta < b$ such that*

$$f(b) - f(a) = f'(\theta)(b - a).$$

Proof. The condition $g(a) = g(b)$ implies $f(b) - f(a) = k(b - a)$. By Rolle's theorem, $g'(\theta) = 0$ for some $\theta \in (a, b)$ and this equality can be rewritten as $f'(\theta) = k$. After multiplication with $b - a$ the statement follows directly. □

We can interpret this theorem as follows: during any car trip, at some point the velocity of the car is the same as the average velocity. The first application is the possibility to estimate differences, e.g. $|\cos x - \cos y| \le |x - y|$, or perhaps more interestingly:

$$|\sin x| = |\sin x - \sin 0| = |(x - 0)\cos\theta| \le |x|.$$

In fact we can do much more.

If we take $a = 0$, $b = x$ we can reformulate the theorem as $f(x) = f(0) + f'(\theta)x$ for some θ between zero and x. This may be generalized as follows (we leave the pleasure of the proof to the reader).

Theorem 10.2 (Taylor's formula with Lagrange remainder) *If $f(x)$ is at least $n+1$ times differentiable, then for each x there exists θ between zero and x such that*

$$f(x) = f(0) + \frac{f'(0)x}{1!} + \frac{f''(0)x^2}{2!} + \cdots + \frac{f^{(n)}(0)x^n}{n!} + \frac{f^{(n+1)}(\theta)x^{n+1}}{(n+1)!}.$$

Because only the last term depends on θ, Taylor's formula is an incredible instrument for estimation. For example, taking $n = 4$, we get $\sin x = x - \frac{x^3}{6} + \frac{\cos(\theta)x^5}{120}$ which immediately produces non-trivial estimates for $0 < x < \frac{\pi}{2}$:

$$x - \frac{x^3}{6} < \sin x < x - \frac{x^3}{6} + \frac{x^5}{120}.$$

Try to obtain these inequalities geometrically!

Some other simple but important applications of the Mean Value Theorem are:

- A function whose derivative is everywhere zero must be constant.
- A function whose derivative is constant must be linear.
- A function whose derivative is positive must be strictly increasing.

10.2 Some elements of topology

In order to talk about functions of several real variables, it is helpful to know some terminology from point-set topology. To keep the discussion at a reasonable level, we restrict ourselves to subsets of $\mathbb{R}^n$.

If $x = (x_1, \ldots, x_n)$ and $y = (y_1, \ldots, y_n)$ are (coordinates of) points in $\mathbb{R}^n$, we define the *distance between x and y* (compare with the Pythagorean theorem for $n = 2$ or $n = 3$) as:

$$d(x, y) = |x - y| = \sqrt{\sum_{j=1}^{n}(x_j - y_j)^2}.$$

Note that $d(x, y) = d(y, x)$ and $d(x, y) = 0$ iff $x = y$. We also have a version of the **triangle inequality**: $d(x, y) + d(y, z) \geq d(x, z)$.

Remark One way to generalize the ideas in this section is to consider so called *metric spaces.* A **metric space** is a set X equipped with a function $d : X \times X \to \mathbb{R}$ satisfying the above three conditions. For example if we define

$$d\big((x_1, y_1), (x_2, y_2)\big) = |x_1 - x_2| + |y_1 - y_2|$$

and draw a "circle" $\{(x, y) \mid d((x, y), O) = 1\}$ we in fact get a square. A more exotic metric space is a chessboard, where the distance between two cells is the minimal number of knight moves. More general (and useful) is the distance between vertices in a connected graph, defined as a minimal number of edges in the connecting path between them. Most analysis can be done in metric spaces just as well as in $\mathbb{R}^n$.

Now we can define some fundamental notions from topology.

Open ball An *open ball* in $\mathbb{R}^n$, *centered* at x_0 with *radius* r is the set:

$$B_r(x_0) = \{x \in \mathbb{R}^n : d(x, x_0) < r\}.$$

Closed ball A *closed ball* in $\mathbb{R}^n$, *centered* at x_0 with *radius* r is the set:

$$\overline{B}_r(x_0) = \{x \in \mathbb{R}^n : d(x, x_0) \leq r\}.$$

Note that balls in $\mathbb{R}^2$ are discs, and in $\mathbb{R}^1$ just intervals.

Open set A subset $X \subset \mathbb{R}^n$ is called *open* if, for every $x \in X$, there is *some* open ball, centered at x (possibly with a very small radius) contained in X.

Closed set A subset $X \subset \mathbb{R}^n$ is called *closed* if its complement is open.

Boundary point A point p is called a *boundary point* of the set X if every open ball centered at p contains points in X **and** points not in X. The set of all boundary points of C is called the *boundary of* X and is usually denoted ∂X (bX is another fairly common notation).

Bounded set A subset $X \subset \mathbb{R}^n$ is called *bounded* if it is contained in some open ball.

Compact set A subset $X \subset \mathbb{R}^n$ is called *compact* if it is closed and bounded.

Connected set A subset $X \subset \mathbb{R}^n$ is called *disconnected* if there exist two disjoint open sets E_1, E_2, such that $X \cap E_j$ is non-empty and $X = (X \cap E_1) \cup (X \cap E_2)$. If X is not disconnected, we way that X is *connected.* If X is open, then X is connected iff for every pair of points $x_1, x_2 \in X$, there is a curve in X from x_1 to x_2.

Note that the intersection of any family of closed sets is closed. Similarly, the union of open sets is open. Also, the union of *finitely many* closed sets is closed and the intersection of finitely many open sets is again open. Give examples showing that the last two statements does not necessarily hold for infinitely many sets!

It is helpful to know some additional connections between continuous functions and the topology of sets. If $f : \mathbb{R}^n \to \mathbb{R}$ is continuous, then the solution set to the non-strict inequality $f(x_1, \ldots, x_n) \le c$ is *closed*. Since the intersection of closed sets is closed, and one equality is equivalent to two non-strict inequalities, it follows that the same is true for equalities. Moreover the simultaneous solution to a system of arbitrarily many non-strict inequalities $\{f_j(x_1, \ldots, x_n) \le c_j\}$ is also closed (assuming the f_j are continuous). For example, the cone

$$\left\{(x_1, x_2, x_3) \in \mathbb{R}^3 \mid x_1^2 + x_2^2 = x_3^2\right\} = \left\{(x_1, x_2, x_3) \in \mathbb{R}^3 \mid x_1^2 + x_2^2 - x_3^2 = 0\right\}$$

and the rectangle $\left\{(x_1, x_2) \in \mathbb{R}^2 \mid -3 \le x_1 \le 2, 5 \le x_2 \le 8\right\}$ are closed, and the same is true if we put the conditions together:

$$X = \left\{(x_1, x_2, x_3) \in \mathbb{R}^2 \mid x_1^2 + x_2^2 = x_3^2, -3 \le x_1 \le 2, 5 \le x_2 \le 8\right\}.$$

The set X and the rectangle are compact, but the cone is not.

Similarly, the solution set to an a strict inequality $f(x_1, \ldots, x_n) > c$ is *open*, and as long as the number of inequalities is *finite*, the same holds for systems of inequalities, again assuming that the functions are continuous. A useful exercise is to show that a set is closed iff it contains all its boundary points and that it is open iff it doesn't contain any of its boundary points. Do there exist sets that are neither open nor closed? Do there exist sets that are both open and closed?

Compact sets play the same role in $\mathbb{R}^n$ as closed intervals do on the real line. For example, if $f : X \to \mathbb{R}$ is a continuous function defined on the *compact* set X, then (cf. Theorem 7.2)

- f is bounded and attains its maximum and minimum values. In other words, there exist $x_{\text{min}}, x_{\text{max}} \in X$ such that for any $x \in X$ we have:

$$m = f(x_m) \le f(x) \le f(x_M) = M.$$

- Assuming that X is connected, f attains all values between m and M, thus for any $m \le v \le M$ there exists $x \in X$ such that $f(x) = v$.

Compact sets can be characterized by the following theorem. In order to formulate the theorem, let us introduce some additional terminology. An **open cover** of a set X is a collection of open sets E_j such that $X \subset \bigcup_j E_j$. A **finite subcover** of an open cover is a *finite* number of E_j's that still cover X.

Theorem 10.3 (Heine–Borel) *A set $X \subset \mathbb{R}^n$ is compact if and only if every open cover of X has a finite subcover.*

For a proof, we refer to any standard textbook on real analysis. We illustrate the result with two examples.

Problem 10.4 Consider an infinite sequence of nested squares: $S_1 \supset S_2 \supset S_3 \supset \cdots$. Show that all of them have a common point.

Search. This sounds obvious, but is it? If we take the intersection of all the squares, we get a closed set, but the problem is that the intersection might be empty. For example, the intersection of all subsets $M_i = \{x \in \mathbb{R} \mid x \geq i\}$ in $\mathbb{R}$ is empty, even though the M_i are nested closed sets. But we can use that S_1 is a compact set. If our intersection is empty then the open subsets U_i — the complements of S_i cover S_1. But then Theorem 10.3 shows that there exists a finite subcover. Returning to S_i it follows that finitely many of them have an empty intersection, which obviously is a contradiction. □

Of course we can replace squares by arbitrary closed non-empty subsets, but at least one of the sets in the sequence should be compact. In the second example we show that the compactness is the essential condition.

Example 10.5 Let $X = (0, 1) \subset \mathbb{R}$, and let $E_j = (\frac{1}{j}, \frac{j-1}{j})$. Then $\{E_j\}$ is an open cover of X, but any finite subset of E_j:s fail to cover X. (If $j_1 \leq \cdots \leq j_n$, then $E_{j_1} \cup \cdots \cup E_{j_n} = E_{j_n}$.)

Finally, we need a few more topological notions. If X is a set, then the **closure** of X, denoted $\overline{X}$, is the smallest closed set containing X. Equivalently, $\overline{X} = X \cup \partial X$. If $\overline{X} \supset A$, we say that X is **dense** in A. If $\overline{X}$ contains no non-empty open ball, we say that X is **nowhere dense**.

Example 10.6 The set of rational numbers $\mathbb{Q}$ is dense in $\mathbb{R}$. (Note that $\partial\mathbb{Q} = \mathbb{R}$.) The set of integers $\mathbb{Z}$ is nowhere dense.

10.3 Completeness and Baire's theorem

A metric space (X, d) is called *complete*, if there are no "holes" or "missing points" in X. Formally, let us consider sequences of points in X. Such a sequence $x_1, x_2, \dots$ is called a **Cauchy sequence** if the points get closer to each other as the index j grows. More precisely, for every $\varepsilon > 0$ there should be some integer N such that $d(x_j, x_k) < \varepsilon$ whenever $j, k \geq N$.

For example: if $r \in \mathbb{R}$ is a real number, then the corresponding sequence of decimal approximations is a Cauchy sequence. On the other hand, the sequence $x_j = (-1)^j$ is *not* Cauchy.

The space X is called **complete** if *every* Cauchy sequence in X has a limit, i.e. a point $x \in X$ such that $\lim_{j\to\infty} d(x_j, x) = 0$. The fact that $\mathbb{R}$ is a complete metric space is crucial for the usual methods of calculus to work: for example the properties of continuous functions (Theorem 7.2) rely on the fact that $\mathbb{R}$ is complete. Similarly $\mathbb{R}^n$ is complete for every natural number n.

For an example of a metric space that is *not* complete, consider $\mathbb{Q}$ with the usual distance. Then the sequence $(3, 3.1, 3.14, 3.141, \dots)$ is a Cauchy sequence in $\mathbb{Q}$ (why?) but fails to have a limit in $\mathbb{Q}$ — the set of rational numbers has a "hole" where π ought to be.

In fact, each subset $X \subset \mathbb{R}^n$ can be viewed as a metric space, with the same distance function as in $\mathbb{R}^n$. As as exercise, show that $X \subset \mathbb{R}^n$ is complete iff X is closed.

An extremely powerful theorem about complete metric spaces is **Baire's theorem**. The theorem is also very abstract, and at first glance it is difficult to appreciate it, but we will discuss some applications that hopefully convinces the reader of its usefulness.

Theorem 10.7 (Baire's category theorem) *If X is a complete metric space and $A_1, A_2, \dots$ is a sequence of dense open sets in X, then the intersection $\bigcap_{n=1}^{\infty} A_n$ is non-empty.*

Proof. Pick some open ball $B_{r_1}(x_1)$ whose closure is contained in A_1. Since A_2 is open and dense, $B_{r_1}(x_1) \cap A_2$ contains the closure of some ball $B_{r_2}(x_2)$. Since A_3 is open and dense, $B_{r_2}(x_2) \cap A_3$ contains the closure of some ball $B_{r_3}(x_3)$.

Continue in the same fashion, producing a nested sequence of open balls $B_{r_k}(x_k)$ such that $B_{r_k}(x_k) \cap A_{k+1}$ contains $\overline{B}_{r_{k+1}}(x_{k+1})$ for each $k = 1, 2, \dots$

We may assume that the radii are chosen so that $r_k \to 0$ as $k \to \infty$. Then $x_1, x_2, \ldots$ form a Cauchy sequence, and since X is complete, there is some $x \in X$ with $x = \lim_{k\to\infty} x_k$.

Finally, for every k, by construction, $x_k \in B_{r_n}(x_n)$ for $n \geq k$. Hence $x \in \overline{B}_{r_n}(x_n) \subset A_n$ for every n, and consequently $x \in \bigcap_{n=1}^{\infty} A_n$, showing that the intersection is non-empty. □

In fact, the conclusion of Baire's theorem can be made considerably stronger: the intersection of a sequence of dense open sets in a compete metric space is itself dense. (But not necessarily open.)

Let us start with something simple (which surely can be proven in a more elementary way).

Problem 10.8 Show that $[0,1]$ is *uncountable*, i.e. that there is no bijection $f : \mathbb{N} \to [0,1]$.

Search. Let us try to use Baire's theorem, and construct an interesting sequence of open, dense sets using the function f.

One way to do this is the following: Take any function $f : \mathbb{N} \to [0,1]$, and let $A_j = [0,1] \setminus \{f(j)\}$. Then each A_j is open and dense, so by Baire's theorem, the intersection $\bigcap_{j=1}^{\infty} A_j$ contains some point y, which is *not* in the image of f. Hence f cannot be a bijection. □

As a more complicated, but perhaps also more typical application, let us turn to differentiable functions. Most functions that we come in contact with are "nice", for example functions in elementary textbooks tend to be differentiable, and the derivative of a typical textbook function is itself differentiable.

For a long time, it was believed that continuous functions are differentiable — not at every point, as shown by the simple example $f(x) = |x|$ — but at least at "most" points. This belief turned out to be completely wrong. In 1872, Weierstrass constructed an example of a continuous function that fails to be differentiable at every point! (And in a certain sense — reminding of Baire's theorem — most continuous functions are like this.)

If we turn to derivatives, it is not entirely obvious that there exists differentiable functions whose derivatives fail to be continuous. One of the simplest examples is

$$f(x) = \begin{cases} x^2 \sin \frac{1}{x}, & x \neq 0 \\ 0, & x = 0 \end{cases}$$

whose derivative

$$f'(x) = \begin{cases} 2x \sin \frac{1}{x} - \cos \frac{1}{x}, & x \neq 0 \\ 0, & x = 0 \end{cases}$$

fails to be continuous at $x = 0$. Does there perhaps exist a differentiable function whose derivative is nowhere continuous? Baire's theorem implies that this cannot happen. In fact, let us prove a slightly stronger result.

Theorem 10.9 *Assume that $f(x) = \lim_{n\to\infty} f_n(x)$ is the pointwise limit of a sequence of continuous functions f_n. Then f must be continuous at some point.*

Proof. Fix $\varepsilon > 0$, and define (for each $n, k \in \mathbb{N}$) the set

$$E_{n,k} = \{x \in \mathbb{R} : |f_n(x) - f_{n+k}(x)| > 2\varepsilon\},$$

and let $E_n = \bigcup_{k=1}^{\infty} E_{n,k}$. Since every f_j is continuous, each $E_{n,k}$ is open, and consequently so is E_n.

Now, let $x \in \mathbb{R}$. Then there is some N (depending on x) such that $|f_j(x) - f(x)| < \varepsilon$ for $j \geq N$. Hence

$$|f_N(x) - f_{N+k}(x)| \leq |f_N(x) - f(x)| + |f_{N+k}(x) - f(x)| < \varepsilon + \varepsilon = 2\varepsilon$$

which shows that $x \notin E_N$. The same holds for every x, so $\bigcap_{n=1}^{\infty} E_n = \varnothing$. Baire's theorem implies that some E_n is *not* dense in $\mathbb{R}$. Let $U = \mathbb{R} \setminus \overline{E_n}$. Note that U is open and non-empty.

Next, take any point $x \in U$. Since f_n is continuous, there is some $r > 0$ such that $|f_n(y) - f_n(x)| < \varepsilon$ for all $y \in B_r(x)$. We can choose this r so small that $\overline{B}_r(x) \subset U$. So, for every $y \in B_r(x)$ we have

$$|f(y)-f(x)| \leq |f(y)-f_n(y)|+|f_n(y)-f_n(x)|+|f_n(x)-f(x)| < 4\varepsilon. \quad (10.1)$$

Thus, for every $\varepsilon > 0$, there is an open ball B_1 on which (10.1) holds.

Now, start over, replacing ε by $\varepsilon/2$ and focus just on the function values on B_1. This will give us a new ball B_2 such that $\overline{B_2} \subset B_1$ on which

$$|f(y) - f(x)| \leq |f(y) - f_n(y)| + |f_n(y) - f_n(x)| + |f_n(x) - f(x)| < 2\varepsilon.$$

Repeating the construction, we end up with a nested sequence of open balls B_j where $\overline{B_{j+1}} \subset B_j$ and

$$|f(y)-f(x)| \le |f(y)-f_n(y)|+|f_n(y)-f_n(x)|+|f_n(x)-f(x)| < 4\varepsilon \cdot 2^{-j+1}.$$

for $x, y \in B_j$. Since the intersection $F = \bigcap_{j=1} \overline{B_j}$ is non-empty (see comments after problem 10.4), and clearly f is continuous at every point $x \in F$ we have finished the proof. □

A small variation (find it!) of the above proof shows that the set of points where f is continuous is in fact dense.

This was certainly not an easy proof, but the result is also very deep.

It remains to show that every derivative can be written as a pointwise limit of continuous functions, but this is easy:

$$f'(x) = \lim_{n\to\infty} \frac{f(x+\frac{1}{n}) - f(x)}{\frac{1}{n}}.$$

As an exercise, try solving the following problem by yourself:

Problem 10.10 Assume that f is smooth function on $\mathbb{R}$ (i.e. that all the derivatives $f', f'', \ldots$ exist) and that for every $x \in \mathbb{R}$ there is some natural number n such that $f^{(n)}(x) = 0$. (Maybe different n for different x.)

Show that f must be a polynomial.

10.4 Fixpoints

A **fixed point** (or shorter, **fixpoint**) of a function f is a point x for which $f(x) = x$. If $f : \mathbb{R} \to \mathbb{R}$ is a continuous function it is sufficient to find x_1, x_2 such that $f(x_1) \le x_1$ and $f(x_2) \ge x_2$ to ensure that a fixpoint exists (somewhere between x_1 and x_2.) On the other hand, the function $f(x) = x+1$ definitely has no fixpoints.

Theorem 7.2 shows that any continuous $f : [a,b] \to [a,b]$ always has a fixed point. The example $f(x) = \frac{x}{2}$ shows that this is not the case for an open interval $(a,b) = (0,1)$. On the other hand each periodic continuous function $f : \mathbb{R} \to \mathbb{R}$ must have a fixpoint, because we can restrict the function to any period on which f is automatically bounded.

We would like to mention a few nice theorems, not because they are important to know, but mostly just because they are nice. The first one, **Brouwer's**

fixed-point theorem is a generalization from a segment to an n-dimensional cube (restrict yourself to $n \leq 3$ if you are not sure what this means).

Theorem 10.11 *Each continuous map from an n-dimensional cube to itself has a fixed point.*

For example, the map $(x, y) \mapsto \left(2^{-xy}, \cos^2(x^2 + y)\right)$ maps the unit square to itself and therefore has a fixpoint, which is far from being obvious. Brouwer's fixed-point theorem also holds for any set that can be continuously transformed to a cube, e.g. for n-dimensional balls. On the other hand, a rotation of a circle has no fixed points. Thus there is no bijective continuous transformation of a circle to a closed interval.

Note that Brouwer's theorem can be interpreted in the following way: If we make a cake and pour all our ingredients in a bowl and stir everything around in such a way that the cake batter occupies the same (convex) region of space when we are done, we can be sure that some point (atom?) of the ingredients is still at the exactly same place as when we started.

Brouwer's theorem has many other serious and useful applications: existence of Nash equilibria in game theory (the theorem giving John Nash the Nobel prize in economics) just to mention one example.

Another theorem on existence of fixpoints is **Banach's theorem.**

Theorem 10.12 (Banach's fixed-point theorem) *Assume that X is compact and that $f : X \to X$ is a contraction, i.e. that $d(f(x), f(y)) < d(x, y)$ for all $x \neq y \in X$. Then f has a unique fixed point in X.*

Proof. Let us start with uniqueness. Assume that $f(x) = x$ and $f(y) = y$ with $x \neq y$. Then

$$d(x, y) = d(f(x), f(y)) < d(x, y)$$

which is a contradiction.

For the existence, consider the function $g : X \to \mathbb{R}$ defined by

$$g(x) = d(f(x), x).$$

It's not hard to show that g is continuous, and since X is compact, g attains its minimum m at some point $x_{\min} \in X$. If $x_{\min}$ is not a fixed point, then

$$g(f(x_{\min})) = d(f(f(x_{\min})), f(x_{\min})) < d(f(x_{\min}), x_{\min}) = g(x_{\min}),$$

but this contradicts $x_{\min}$ being the minimum. Hence $x_{\min}$ is a fixpoint. □

Banach's fixed point theorem has many interesting applications. For example the most common result on existence and uniqueness of solutions of differential equations is proved using (a slightly more general version of) Banach's fixed point theorem. It can also be used to prove the implicit and inverse function theorems.

An easy application can be found in the following problem.

Problem 10.13 Two maps of the same region have different sizes. The smaller map is placed on top of the larger one (inside it). Prove that one can pin the same place on two maps.

Note that Banach's theorem is still true for non-compact sets X if we assume that there is some constant $c < 1$ such that $d(f(x), f(y)) \leq cd(x, y)$.

Our final result in this section concerns periodic points. A **periodic point of order** k is a point x such that $f^k(x) = x$ (and k is the least such natural number). Thus periodic points of order 1 are exactly fixpoints and periodic point of order 2 are solutions to $f(f(x)) = x$ that are not fixpoints.

To formulate the theorem, we need to introduce a very strange order on the set of natural numbers. Every natural number can be uniquely written as $n = q2^k$, where q is odd. First we restrict ourselves to those n for which $q > 1$. If $n_i = q_i 2^{k_i}$ are two such numbers we say that $n_1 \prec n_2$ if $q_1 < q_2$ or $q_1 = q_2$ and $k_1 < k_2$. This is not so strange. But for the remaining 2^l we do something unexpected. First we take $q2^k \prec 2^l$ if $q > 1$. Furthermore, we demand $2^k \prec 2^l$ if $k > l$. Thus the full picture looks like

$$\begin{array}{ccccccccc}
3 & \prec & 6 & \prec & 12 & \prec & 24 & \prec & \cdots \\
5 & \prec & 10 & \prec & 20 & \prec & 40 & \prec & \cdots \\
7 & \prec & 14 & \prec & 28 & \prec & 56 & \prec & \cdots \\
9 & \prec & 18 & \prec & 36 & \prec & 72 & \prec & \cdots \\
\cdots & \prec & & & & & & & \\
\cdots & \prec & 2^n & \prec 2^{n-1} \prec & \cdots & \prec & 1. & &
\end{array}$$

Now we can formulate Sharkovsky's theorem.

Theorem 10.14 *If a continuous function $f : \mathbb{R} \to \mathbb{R}$ has a period of order n and $n \prec m$ then it has a period of order m as well.*

In particular, if we have a point of period 3 we have all possible periods! Also, the existence of any period implies the existence of a fixed point.

In other words, if f^k has a fixpoint for some k then f also has one. A very beautiful and far from trivial result! For a proof of Sharkovsky's theorem, see [59].

10.5 Trigonometric polynomials

Sometimes trigonometric polynomials are useful. Consider some of their properties.

Theorem 10.15 *For every trigonometric polynomial*

$$f(x) = c_0 + c_1 \cos x + c_2 \cos 2x + \cdots + c_n \cos nx$$

there exists a polynomial $T(x)$ with highest coefficient $2^{n-1}c_n$ such that $f(x) = T(\cos x)$. Vice versa, for any polynomial $T(x)$ the expression $T(\cos x)$, can be rewritten as a trigonometric polynomial.

Proof. We use induction, where the base for $n = 1, 2$ is evident and the formula $\cos nx + \cos(n-2)x = 2\cos(n-1)x \cos x$ is sufficient for the induction step. Complete the details. □

Polynomials $T_n(x)$ for which $T_n(\cos x) = \cos nx$ are called **Chebyshev polynomials.** They have one nice property, which we will come back to. But first we will consider a very difficult problem.

Problem 10.16 Let $p(x) = x^n + a_1 x^{n-1} + \cdots + a_n$ be a polynomial. Show that

$$|p(x_0)| \geq \frac{1}{2^{n-1}}$$

for some x_0 from the interval $[-1, 1]$.

Search. We have no idea where to start. Perhaps a change of variables, as usual? The condition $|x| \leq 1$ suggests the most natural substitution: $x = \cos t$. Then $p(\cos t)$ can be expressed as a trigonometric polynomial

$$p(\cos t) = f(t) = c_0 + c_1 \cos t + c_2 \cos 2t + \cdots + c_n \cos nt.$$

Moreover, from Theorem 10.15, it follows that the coefficient $c_n = \frac{1}{2^{n-1}}$ and this is a good sign that we are on right track.

How do we proceed? It would be nice to have all $c_k = 0$ for $k < n$. In this case $t = \frac{2\pi}{n}$ would be perfect. But how do we kill those coefficients? Perhaps we can kill $\cos kt$ instead by choosing the right t? Probably not — we need different t for different k.

But this approach looks promising. What if we take several such t and sum them up? If we succeed with $\sum_j c_k \cos kt_j = 0$ then we can hope for a proof.

We need to use that we have trigonometrical functions with some of their nice properties. For example ,we know how to add

$$\cos t + \cos 2t + \cdots + \cos(n-1)t,$$

all we need is a little help from their close relatives e^{int} — and take a real part of the geometric sum $1 + e^{it} + \cdots + e^{i(n-1)t} = \frac{1-e^{int}}{1-e^{it}}$ to get the sum. Now any choice $t_j = \frac{2\pi j}{n}$ gives a zero sum.

This is all good, but how does it help up? After all, we have arbitrary coefficients c_k, not just ones. It looks hopeless. But if we think a little and combine all the ideas we already have and improve them slightly, taking the alternating sums (to kill the constant term as well), we find a very short but not trivial solution (we leave to the reader to finish the search to check the details): Note that

$$\frac{1}{2n}\sum_{k=0}^{2n-1}(-1)^k f\left(\frac{k\pi}{n}\right) = c_n.$$

Because this is a mean value, the mean value of the absolute values is greater or equal than $|c_n| = \frac{1}{2^{n-1}}$. We conclude that $\left|f\left(\frac{k\pi}{n}\right)\right| \geq \frac{1}{2^{n-1}}$ for some k and take $x_0 = \cos\frac{k\pi}{n}$. □

Of course, to find such a solution we need a lot of experience, for example in working with trigonometric sums and using mean values. But with experience the path we followed is quite natural.

Now what about the property of the Chebyshev polynomials? If $f(x)$ is monic and $|f(x)| \leq \frac{1}{2^{n-1}}$ for all x from the interval $[-1,1]$ then $f(x) = \frac{1}{2^{n-1}}T_n(x)$, which more or less follows from the proof of the theorem (we need all the terms in our mean value to be equal and these values determine the polynomial uniquely — try to understand how).

10.6 Infinite sums

Summing a finite number of terms is simple, at least in theory, but if we have an infinite number of terms, things become more problematic. First of all, we need to understand what an infinite sum (often called *series*) actually means, and even when we have decided this, we should be aware that some "natural" rules of summation may fail. To see what can happen if we are careless, consider the infinite sum

$$S = \sum_{i=0}^{\infty}(-1)^i = 1 - 1 + 1 - 1 + 1 - 1 + \cdots$$

What is S equal to? Probably

$$0 = (1 - 1) + (1 - 1) + (1 - 1) + \cdots.$$

Perhaps

$$1 = 1 - (1 - 1) - (1 - 1) - (1 - 1) - \cdots?$$

There is also a good argument for $\frac{1}{2}$:

$$S = 1 - (1 - 1 + 1 - 1 + 1 - 1 + \cdots) = 1 - S \quad \Rightarrow \quad 2S = 1 \quad \Rightarrow \quad S = \frac{1}{2}.$$

The correct answer is *nothing*: S does not exist, because we *define* the infinite sum $S = \sum_{i=0}^{\infty} a_i$ as $S = \lim S_n$, where $S_n = \sum_{i=0}^{n} a_i$ and it does not exist if the limit does not exist (as in our case — check!) We say that the series is **convergent** in the first case and **divergent** in the second. Moreover, even for convergent series we cannot carelessly rearrange the brackets or change the order of the summands, because the value of the sum may change. In view of this, it is not surprising that we also cannot change the order in a doubly infinite sum. Fortunately we *can do all this if all the terms are non-negative.* To discuss all of this in full detail is beyond the scope of this book, but we will mention a few well-known series.

The series

$$\sum_{n=1}^{\infty} \frac{1}{n^k}$$

is convergent iff $k > 1$. For $k = 2$ the sum is $\frac{\pi^2}{6}$, for $k = 4$ it is $\frac{\pi^4}{90}$ and a great challenge is to find the exact value of this sum for at least some odd $k > 1$ — this is still an open problem in mathematics. In particular, for $k = 1$ the sum is divergent. Moreover, the series

$$\sum_p \frac{1}{p},$$

where the sum is taken over all *primes* p is divergent.

There are two important cases when we can show convergence. First is the **Leibniz criterium** that the alternating series

$$a_0 - a_1 + a_2 - a_3 + a_4 - a_5 + \cdots,$$

converges if a_i is a decreasing sequence of positive numbers that tends to zero. The partial sum S_n differs from the full sum by an amount no greater than a_{n+1}. For example the sum

$$1 - \frac{1}{2} + \frac{1}{3} - \frac{1}{4} + \frac{1}{5} + \cdots$$

exists and differs from $\frac{7}{12} = 1 - \frac{1}{2} + \frac{1}{3} - \frac{1}{4}$ by at most $\frac{1}{5}$ (the exact value of the sum can be shown to be $\ln 2$).

The second case deals with power series

$$\sum_{n=0}^{\infty} a_n x^n = a_0 + a_1 x + a_2 x^2 + \cdots.$$

If the limit

$$R = \lim_{n \to \infty} \left| \frac{a_n}{a_{n+1}} \right|$$

exists or is equal to ∞ then the series converges for all x for which $|x| < R$. Moreover, for such x the sum converges to a differentiable function $f(x)$ and $a_n = \frac{f^{(n)}(0)}{n!}$. In other words,

$$f(x) = f(0) + \frac{f'(0)x}{1!} + \frac{f''(0)x^2}{2!} + \cdots = \sum_{n=0}^{\infty} \frac{f^{(n)}(0)x^n}{n!}.$$

We often use this in the opposite way: take a known function, calculate the derivatives and get (using Theorem 10.2) the series (**Maclaurin series**). This gives us classical identities such as

$$\begin{aligned} e^x &= 1 + x + \frac{x^2}{2!} + \frac{x^3}{3!} + \frac{x^4}{4!} + \frac{x^5}{5!} + \cdots \\ \cos x &= 1 - \frac{x^2}{2!} + \frac{x^4}{4!} - \frac{x^6}{6!} + \frac{x^8}{8!} - \cdots \\ \sin x &= x - \frac{x^3}{3!} + \frac{x^5}{5!} - \frac{x^7}{7!} + \frac{x^9}{9!} - \cdots \end{aligned}$$

which are valid for all x and

$$\begin{aligned} -\ln(1-x) = \ln\left(\frac{1}{1-x}\right) &= x + \frac{x^2}{2} + \frac{x^3}{3} + \frac{x^4}{4!} + \cdots \\ (1+x)^\alpha &= 1 + \alpha x + \binom{\alpha}{2}x^2 + \binom{\alpha}{3}x^3 + \binom{\alpha}{4}x^4 + \cdots \\ \frac{1}{(1-x)^k} &= 1 + kx + \binom{k+1}{2}x^2 + \binom{k+2}{3}x^3 + \binom{k+3}{4}x^4 + \cdots \end{aligned}$$

for $|x| < 1$. Two other useful infinite series are

$$\begin{aligned} \frac{1}{\sqrt{1-4x}} &= 1 + \binom{2}{1}x + \binom{4}{2}x^2 + \binom{6}{3}x^3 + \binom{8}{4}x^4 + \cdots, \\ \frac{1-\sqrt{1-4x}}{2x} &= 1 + \frac{1}{2}\binom{2}{1}x + \frac{1}{3}\binom{4}{2}x^2 + \frac{1}{4}\binom{6}{3}x^3 + \frac{1}{5}\binom{8}{4}x^4 + \cdots, \end{aligned}$$

both of which are convergent for $|x| < \frac{1}{4}$.

All of the above remains true even if x is a complex number with $|x| < R$.

CHAPTER 11

Algebraic constructions

11.1 Formal series

The convergence troubles we considered in Section 10.6 are rather unpleasant. On the other hand power series with the possibility to add, subtract, differentiate and multiply them are so attractive that we would like to be able to do this without worrying about convergence. This is in fact possible, if we consider *formal power series*. The idea is to have a "formal" variable (say t) and to consider a series as some kind of "infinite polynomial" in t. Thus a **formal series** is a formal expression

$$a(t) = \sum_{n=0}^{\infty} a_n t^n,$$

where the word *formal* means that we are not interested in convergence, because t does not represent a number but a formal variable. What we *are* interested in, is how to add, multiply and differentiate such formal series. The rules are the following: if

$$b(t) = b_0 + b_1 t + b_2 t^2 + \cdots = \sum_{n=0}^{\infty} b_n t^n$$

is another formal series then the sum $(a+b)(t)$ is *defined* as

$$\sum_{n=0}^{\infty} (a_n + b_n) t^n$$

and the product $(a \cdot b)(t)$ as

$$\begin{aligned} &a_0 b_0 + (a_0 b_1 + a_1 b_0)t + (a_0 b_2 + a_1 b_1 + a_2 b_0)t^2 + \cdots \\ &= \sum_{n=0}^{\infty} \left(\sum_{i=0}^{n} a_i b_{n-i} \right) t^n. \end{aligned}$$

Differentiation is also quite natural:

$$a'(t) = a_1 + 2a_2t + 3a_3t^2 + 4a_4t^3 + \cdots = \sum_{n=0}^{\infty}(n+1)a_{n+1}t^n.$$

Because all these definitions are formal and do not use any limits we can forget all convergence troubles. But we do need to check that normal rules such as

$$c(t)(a(t) + b(t)) = c(t)a(t) + c(t)b(t);$$
$$(a(t)b(t))' = a'(t)b(t) + a(t)b'(t)$$

are still valid. We will leave this to the interested reader. What is important is that there is also a formal division. Namely, if $a_0 = 1$ then there exists $b(t)$ such that $a(t)b(t) = 1$. Indeed, from the multiplication rule we get $b_0 = 1$. Looking at the coefficient with t^n in the product we see that $b_n = -\sum_{i=1}^{n} a_i b_{n-i}$ can be found using induction. For example,

$$(1 + t + t^2 + t^3 + \cdots)^{-1} = (1 - t) \quad \Leftrightarrow \quad (1 - t)^{-1} = 1 + t + t^2 + t^3 + \cdots,$$

and thus we have the familiar geometric series also as a formal series.

The same arguments work when $a_0 \neq 0$, but if $a_0 = 0$, there is no formal inverse. Similarly, if $a_0 = 0$, the composition $b(a(t))$ is well defined as a formal series, but this is not the case when $a_0 \neq 0$. For example, if $a(t) = 3t - 2t^2$ and $b(t) = 1 + t + t^2 + \cdots$ then

$$b(a(t)) = 1 + (3t - 2t^2) + (3t - 2t^2)^2 + \cdots \tag{11.1}$$

and we see that for any n, we only need finitely many terms to determine the coefficient of t^n.

The nice thing is that natural identities for convergent power series are still valid for formal series, though some of them should be taken as definitions (for example the exponential e^t). We only need to be a little careful with the value of a_0 when we perform division and composition, as explained above. For example, we can get the exact expression for the series in (11.1) using routine transformations:

$$\begin{aligned}\frac{1}{1-(3t-2t^2)} = \frac{1}{1-3t+2t^2} = \frac{1}{(1-t)(1-2t)} &= \frac{2}{1-2t} - \frac{1}{1-t}\\ &= 2(1 + 2t + 4t^2 + 8t^3 + \cdots) - (1 + t + t^2 + \cdots)\\ &= 1 + 3t + 7t^2 + \cdots + \left(2^{n+1} - 1\right)t^n + \cdots\end{aligned}$$

and this is a typical illustration of how to work with formal power series.

If $a_0, a_1, a_2, \ldots$ is an arbitrary sequence, we call the formal series

$$A(t) = \sum_{n=0}^{\infty} a_n t^n$$

its **generating function**. Thus $\frac{1}{1-t}$ is the generating function for the constant sequence $1, 1, 1, \ldots$ and more generally, $\frac{1}{(1-t)^k}$ is the generating function for the sequence $a_n = \binom{n+k-1}{k-1} = \binom{n+k-1}{n}$.

The advantage of generating functions is that they contain, in compact form, simultaneous information about all the terms in the sequence and sometimes operations on the sequences have very nice interpretations as operations on the corresponding generating functions.

For example, the sequences $b_n = a_{n-1}$ (i.e. $0, a_0, a_1, a_2, \ldots$) and $c_n = a_{n+1}$ (i.e. $a_1, a_2, \ldots$) have the generating functions

$$B(t) = tA(t), \quad C(t) = \frac{A(t) - a_0}{t},$$

respectively. Because $A'(t)$ is, obviously, the generating function for $b_n = (n+1)a_{n+1}$ the sequence na_n has the generating function $tA'(t)$.

For example, the sequence $0, 1, 2, 3, \ldots$ has the generating function

$$t\left(\frac{1}{1-t}\right)' = \frac{t}{(1-t)^2}$$

and the sequence $0, 1^2, 2^2, 3^2, \ldots$ has the generating function

$$t\left(\frac{t}{(1-t)^2}\right)' = \frac{t+t^2}{(1-t)^3}.$$

Last but not least, the sequence of partial sums $a_0, a_0 + a_1, a_0 + a_1 + a_2, \ldots$ has the generating function $A(t) \cdot \frac{1}{1-t}$ which follows from the rule for the multiplication of formal series.

To show how to use these observations we will consider some examples.

First we deduce the formula for the **Fibonacci numbers** F_n, that are defined as

$$F_0 = 0, \quad F_1 = 1, \quad F_{n+2} = F_{n+1} + F_n.$$

Recognizing Fibonacci numbers is often helpful, so we write down several of the first numbers:

$$0, 1, 1, 2, 3, 5, 8, 13, 21, 34, 55, 89, 144, 233, 377, 610, 987, \dots$$

If $F(t)$ is the generating function for F_n, we can multiply the defining equation with t^{n+2}, sum over n and rewrite as:

$$\begin{aligned}
&\sum_{n=0}^{\infty} F_{n+2}t^{n+2} = t\sum_{n=0}^{\infty} F_{n+1}t^{n+1} + t^2\sum_{n=0}^{\infty} F_n t^n && \Leftrightarrow\\
&F(t) - 0 - t = t(F(t) - 0) + t^2F(t) && \Leftrightarrow\\
&F(t) = \frac{t}{1 - t - t^2}.
\end{aligned}$$

Factorizing $1 - t - t^2 = (1 - \varphi t)(1 - \psi t)$, where $\varphi = \frac{1+\sqrt{5}}{2}$ is the well-known **golden ratio** and $\psi = -\varphi^{-1} = \frac{1-\sqrt{5}}{2}$ we get

$$\frac{t}{1 - t - t^2} = \frac{1}{\sqrt{5}}\left(\frac{1}{1 - \varphi t} - \frac{1}{1 - \psi t}\right) \quad \Rightarrow \quad F_n = \frac{1}{\sqrt{5}}(\varphi^n - \psi^n).$$

Since $|\psi| < 1$ we see that $\varphi = \lim \frac{F_{n+1}}{F_n}$, but the generating function is more interesting because it contains all the information about the Fibonacci numbers.

In the second example, we find the sum of the first n squares. The generating function for the sequence $s_n = 0 + 1^2 + \cdots + n^2$ is, according to the remarks above, equal to

$$\frac{t + t^2}{(1 - t)^3} \cdot \frac{1}{1 - t} = \frac{t + t^2}{(1 - t)^4}.$$

We need only to rewrite this as a sum of known series. We do this by clever guesswork (also known as partial fractions decomposition):

$$\begin{aligned}
\frac{t + t^2}{(1 - t)^4} = \frac{A}{1 - t} + \frac{B}{(1 - t)^2} + \frac{C}{(1 - t)^3} + \frac{D}{(1 - t)^4} \qquad \Leftrightarrow\\
t + t^2 = A(1 - t)^3 + B(1 - t)^2 + C(1 - t) + D.
\end{aligned}$$

Putting $t = 1$ gives $D = 2$. Comparing degrees, we get $A = 0$. Looking at the coefficient for t^2 we find $B = 1$. Thus $t + t^2 = (1 - t)^2 + C(1 - t) + 2 \Rightarrow C = -3$. Hence, the generating function for s_n is

$$\frac{1}{(1 - t)^2} - \frac{3}{(1 - t)^3} + \frac{2}{(1 - t)^4}.$$

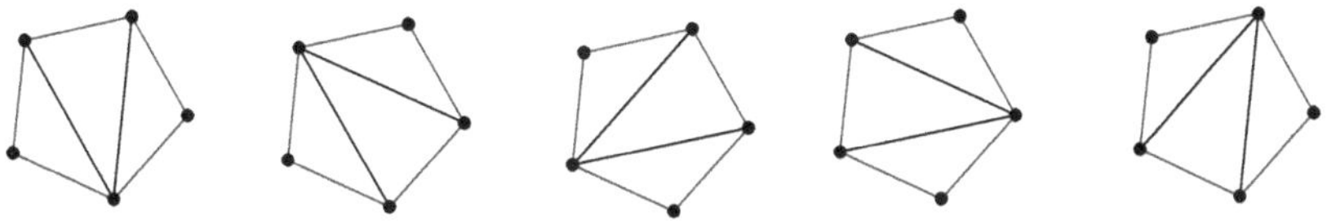

Figure 11.1 Dividing a polygon into triangles

Consequently,

$$\begin{aligned} s_n &= \binom{n+1}{n} - 3\binom{n+2}{n} + 2\binom{n+3}{n} \\ &= (n+1) - \frac{3(n+2)(n+1)}{2} + \frac{(n+3)(n+2)(n+1)}{3} \\ &= (n+1)\frac{6 - 9(n+2) + 2(n+3)(n+2)}{6} \\ &= \frac{(n+1)(2n^2+n)}{6} = \frac{n(n+1)(2n+1)}{6}. \end{aligned}$$

Not an exciting proof, but the calculations were quite straightforward!

Next we will consider a more advanced example.

Problem 11.1 (Euler's polygon division problem) What is the number c_n of ways to cut a convex polygon with $n + 2$ vertices into triangles using the diagonals?

Search. If $n = 1$ we have a triangle, thus $c_1 = 1$. A quadrilateral has two diagonals and we can choose either one, so $c_2 = 2$. For $n = 3$ there are five possibilities — see Figure 11.1. We can try to find a recursive formula for c_{n+1} in a polygon with $n + 3$ vertices.

Number the vertices (in order) from 1 to $n + 3$ and consider the triangle containing the edge from 1 to $n + 3$. If the third vertex of the triangle is at $k + 2$, the shaded triangle cuts the polygon into two smaller polygons with $k + 2$ and $n - k + 2$ vertices respectively. Hence for a fixed k, there are $c_k c_{n-k}$ ways to cut the two smaller polygons into triangles, and summing over k, we obtain (assuming $c_0 = 1$)

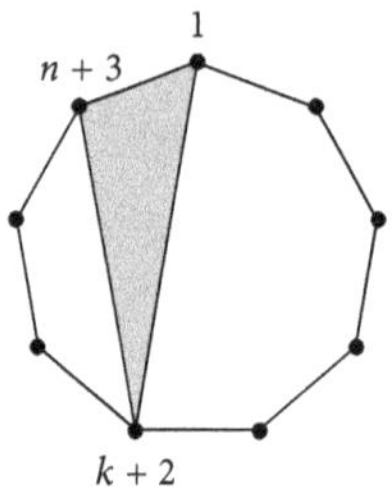

Figure 11.2 The recursion for c_n in problem 11.1

$$c_{n+1} = \sum_{k=0}^{n} c_k c_{n-k}.$$

Multiplying by t^{n+1} and summing over n we get

$$\sum_{n=0}^{\infty} c_{n+1} t^{n+1} = t \sum_{n=0}^{\infty} \left(\sum_{k=0}^{n} c_k c_{n-k} \right) t^n.$$

Thus for the generating function $C(t) = \sum c_n t^n$ we found the equation

$$C(t) - 1 = tC(t)^2 \quad \Rightarrow \quad C(t) = \frac{1 \pm \sqrt{1-4t}}{2t}.$$

Comparing with the series on page 284 we find that we should choose the minus sign so that

$$c_n = \frac{1}{n+1}\binom{2n}{n}.$$

The numbers c_n are called the **Catalan numbers** and appear quite often in different enumeration problems (you can find more than 60 different interpretations in [40]). □

11.2 Combinatorics on words

An important part of algebra is devoted to formal manipulations with words, written in some alphabet. In fact there are many very difficult problems in this area and some of them are simply unsolvable (and this has been proven!). In this text, we are unable to look that deeply, and will restrict ourselves to a couple of interesting facts that are useful to be familiar with.

So, let X be an arbitrary finite alphabet and f, g be two words in this alphabet. When we write f and g together getting the word fg we can interpret this as a multiplication.

We start from an easy but useful fact. Let $|f|$ denote the length of f.

Theorem 11.2 *Let $f = gh$ be a word such that the word h is at the same time the beginning of f. If for some natural number n we have $|f| \geq n|g|$, then $f = g^n s$ for some word s.*

Proof. Easy induction on n. For $n = 1$, the conclusion follows directly from the assumption. For larger n we note that $|h| = |f| - |g| \geq (n-1)|g|$. Thus g is not longer than h and therefore must be the beginning of h, because h is the beginning of $f = gh$. It follows that $h = gw$ and gw is the beginning of $gh = ggw$ which means that w is the beginning of gw. Now we can apply induction to $h = gw$ and conclude that $h = g^{n-1}s \Rightarrow f = g^n s$. □

The next interesting question is to determine when two words f and g commute, that is when $fg = gf$. For example, the words MA and $MAMA$ commute but the words MA and MU do not.

Theorem 11.3 *The following are equivalent for words f and g.*

- *The words f and g commute, $fg = gf$.*
- *There exists a word w such that $f = w^k, g = w^l$.*
- *There exist natural numbers m and n such that $f^m = g^n$.*

In particular, if the words f and g commute and the words g and h commute, then the words f and h commute as well.

Proof. WLOG $|f| \leq |g|$. For the first implication we use induction.

The condition $fg = gf$ implies that f is the beginning of gf and therefore the beginning of g. If f and g have the same length then they are equal and we are done. Otherwise $g = fv$ and we get

$$ffv = fvf \quad \Rightarrow \quad fv = vf.$$

By the induction hypothesis f and v are powers of the same w and therefore the same is true for their product $g = fv$, which finishes the proof. It remains to check that the base for our induction, when f and g are letters, is valid, but this is obvious.

For the second implication, we simply take $m = l$, $n = k$. Finally, suppose that $f^m = g^n$. If $n = 1$, then f clearly commutes with $g = f^m$. Otherwise, if $n > 1$ we also have $m > 1$ because $|f| \leq |g|$. Then g is not longer than f^{m-1} and must be its beginning. Furthermore, f (being no longer than g) is the beginning of g^{n-1}. Consider now the beginning of $ff^{m-1} = f^m = g^n = gg^{m-1}$ of length $|f| + |g|$. It can be written both as fg and and gf, so those products are equal.

Now we can return to the last statement. We can rewrite f and g as powers of some w and simultaneously we can represent g and h as powers of some u. Because $g = w^l = u^m$ we have that w and u commute and the same is true for their powers f and h. □

Suppose now that our alphabet consists of two letters only, say x and y. Note first that any word f of length 4 contains a *square* — i.e. a subword of the form gg. Indeed, if the word does not contain squares and starts with x, then the next letter must be y (to avoid xx), after that we have x (to avoid the subword yy) and again y. Hence, $f = xyxy$, but this is the square of xy and we have reached our contradiction. Similarly, f cannot start with y.

What about cubes? How long can words that do not contain a subword of the form ggg be? Surprisingly enough, cube-free words can be arbitrary long! Here is a famous construction, the **Prouhet–Thue–Morse–Arshon sequence** (or shortly **Thue sequence**).

Consider the following words.

$$f_0 = x, \quad f_{n+1} = f_n\overline{f_n},$$

where $\overline{f_n}$ is obtained from f_n by replacing $x \mapsto y$, $y \mapsto x$. Hence,

$$\begin{aligned}
f_0 &= x,\\
f_1 &= xy,\\
f_2 &= xyyx,\\
f_3 &= xyyxyxxy,\\
f_4 &= xyyxyxxyyxxyxyyx,\\
&\vdots
\end{aligned}$$

Let f_∞ be the obtained infinite word:

$$f_\infty = xyyxyxxyyxxyxyyxyxxyx\ldots$$

Note that this is one-sided infinite word and we cannot concatenate such words. We do not plan to go deeper inside the attractive area of infinite words and restrict ourselves by the following statement.

Theorem 11.4 *The infinite word f_∞ does not contain cubes.*

We recommend the reader to prove the theorem as a useful exercise in combinatorics on words. As a hint, it is easier to prove the stronger statement that f_∞ has no subwords of the form g^2z, where z is the first letter of the word g.

The Thue sequence has many other nice properties. Here are some of them.

- The substitution $x \mapsto xy, y \mapsto yx$ transforms f_n to f_{n+1}.
- The word f_∞ can be uniquely decomposed with the help of the three following words $a = xyy$, $b = xy$, $c = x$ as:

$$f_\infty = abcacbabcb\ldots$$

  Viewed as a word in the letters a, b, c, the word f_∞ is square-free.
- Write n in base 2 and put $t_n = x$ if the sum of the binary digits of n is even and $t_n = y$ if it is odd. The sequence $t_0t_1t_2\cdots$ is the Thue sequence.
- Consider the word f_{n+1}. Let S_x consist of the indices i for which $t_i = x$ and S_y be the remaining index set. Then

$$\sum_{i \in S_x} i^k = \sum_{j \in S_y} j^k$$

  for $k = 1, 2, \ldots, n$.

Let us verify this final statement for f_4. A trivial equality is

$$0 + 3 + 5 + 6 + 9 + 10 + 12 + 15 = 1 + 2 + 4 + 7 + 8 + 11 + 13 + 14 = 60.$$

Non-trivial are different decompositions of 620 and 7200 :

$$0^2 + 3^2 + 5^2 + 6^2 + 9^2 + 10^2 + 12^2 + 15^2 = 1^2 + 2^2 + 4^2 + 7^2 + 8^2 + 11^2 + 13^2 + 14^2;$$
$$0^3 + 3^3 + 5^3 + 6^3 + 9^3 + 10^3 + 12^3 + 15^3 = 1^3 + 2^3 + 4^3 + 7^3 + 8^3 + 11^3 + 13^3 + 14^3.$$

11.3 More about sequences

In this section, we would like to mention some other useful theorems.

First we would like to describe a rather wide, and commonly occurring, class of sequences. Let $x_0, x_1, \dots, x_n, \dots$ be a sequence described by a *linear recursive equation of order k*:

$$x_n + a_1 x_{n-1} + \cdots + a_k x_{n-k} = 0,$$

where the a_i are constants. The **Fibonacci sequence** $F_n = F_{n-1} + F_{n-2}$, $n > 1$, is a classical example, but we can change F_0 and F_1 to get other sequences satisfying the same equation. The **Lucas numbers**

$$2, 1, 3, 4, 7, 11, 18, 29, 47, 76, 123, 199, \dots$$

are especially important.

The **characteristic polynomial** of a linear recursive equation is defined by

$$p(\lambda) = \lambda^k + a_1\lambda^{k-1} + \cdots + a_k.$$

This polynomial can be used to find a closed formula for x_n.

Theorem 11.5 *If all (complex) roots λ_i of the characteristic polynomial are different, then*

$$x_n = c_1\lambda_1^n + c_2\lambda_2^n + \cdots + c_k\lambda_k^n$$

for some constants c_i (which can be determined from the first k values $x_0, \dots, x_{k-1}$). More generally, if there are multiple roots (e.g. λ_i has the multiplicity m_i) then x_n is a sum of terms $p_i(n)\lambda_i^n$ where p_i are (fixed) polynomials with $\deg p_i < m_i$.

Proof. Idea of the proof is quite elementary. First we check that the set of all solutions of a linear recursive equation is a vector space, because the sum of two solutions is a solution and the same is true if we multiply a solution by the constant. The dimension of this vector space is k because its first k values, which can be used as independent parameters, determine the sequence. Thus all we need is to find a basis. Essentially the statement is that the sequences $n^j\lambda_i^n$ with $j < m_i$ form such a basis. To check that this is the case (at least when all λ_i are different) is a good exercise in linear algebra. □

For example, the equation $x_n = 2x_{n-1} + 3x_{n-2}$ has the characteristic polynomial $p(\lambda) = \lambda^2 - 2\lambda - 3$ so x_n can be written as $x_n = c_1 3^n + c_2(-1)^n$. If we know two different terms in the sequence, we can find the constants c_i from the corresponding equations.

Similarly, the equation $x_{n+2} = 4x_{n+1} - 4x_n$ has the solutions $x_n = (an + b)2^n$.

Note that if $p_n = \lambda_1^n + \lambda_2^n + \cdots + \lambda_k^n$ are arbitrary power sums we can obtain a recursive formula connecting p_n with the elementary symmetric polynomials $s_i = s_i(\lambda_1, \ldots, \lambda_k)$ in k variables:

$$p_n - s_1 p_{n-1} + s_2 p_{n-2} + \cdots + (-1)^k s_k p_{n-k} = 0$$

for $n > k$ which follows directly from Vieta's theorem. It is more convenient to use the more general **Newton's identities**

$$p_n = (-1)^{n-1} n s_n + \sum_{i=1}^{n-1} (-1)^{n-1+i} s_{n-i} p_i$$

for arbitrary n where we put $s_j = 0$ for $j > k$. For example

$$\begin{aligned}
p_1 &= s_1, \\
p_2 &= s_1 p_1 - 2s_2, \\
p_3 &= s_1 p_2 - s_2 p_1 + 3s_3, \\
p_4 &= s_1 p_3 - s_2 p_2 + s_3 p_1 - 4s_4.
\end{aligned}$$

To find all solutions of the *inhomogeneous* recursive equation

$$x_n + a_1 x_{n-1} + \cdots + a_k x_{n-k} = f(n)$$

we need to guess one *particular solution* and add to it the general solution of the corresponding *homogeneous* equation where $f = 0$.

It is natural to look for particular solutions that "look like $f(n)$". For example, to solve $x_{n+1} - 2x_n = 3^n$ we try the *ansatz* $x_n = A3^n$, and get

$$A3^{n+1} - 2 \cdot 3^n = 3^n \quad \Rightarrow \quad 3A - 2 = 1 \quad \Rightarrow \quad A = 1.$$

The characteristic polynomial is equal to $\lambda - 2$. Thus all the solutions are

$$x_n = 3^n + C2^n.$$

Another approach is to use generating functions (try this on your own!).

Next we would like to formulate two theorems about arithmetic progressions. The theorems are extremely difficult to prove but very useful to know.

Theorem 11.6 (Van der Waerden's theorem) *For any given positive integers r and k, there is some number N such that if the integers $\{1, 2, \ldots, N\}$ are coloured, each with one of r different colours, then there are at least k integers in arithmetic progression all of the same colour.*

Note that there is no guarantee that we can get an infinite progression (already for two colours – construct an example!). For infinite progressions there is another useful theorem.

Theorem 11.7 (Dirichlet's theorem) *If a and d are relatively prime positive integers, then there are infinitely many primes in the arithmetic progression $a + nd$. In other words, there exist infinitely many primes p for which $p = a \pmod{d}$.*

Here is one possible application.

Problem 11.8 Show that for any natural n there exists a prime p such that its distance to any other prime is greater than n.

Search. It is not too difficult to construct n consecutive composite numbers: one classical construction is

$$(n + 1)! + 2, (n + 1)! + 3, \ldots, (n + 1)! + (n + 1).$$

This is not quite enough to solve our problem, but from this sequence we get a hint. We need $p \pm k$ to be composite for all $k \leq n$. We can simply choose different primes P_k, Q_k and demand $p = k \pmod{P_k}$, and $p = -k \pmod{Q_k}$ for all $k < n$. By the Chinese remainder theorem such a system has a solution and all the solutions form an infinite arithmetic progression. Dirichlet's theorem finishes the proof. □

CHAPTER 12

Advanced geometry

12.1 Identities and inequalities in triangles

Here we would like to consider a powerful method that allows us to prove a lot of identities and inequalities between different objects in an arbitrary triangle. The only demand is a symmetry.

Suppose that we have some symmetric (in sides) expression in a triangle. Because R, r and p are such expressions and we need exactly three parameters to describe a triangle completely, a natural question is if we can express any symmetric expression in these variables. The answer is YES and we will show how to do this efficiently.

We will start from symmetric expressions in a, b, c. The idea is to deduce a cubic equation that they satisfy. The way to do this is not trivial. First look at a. We know that $a = 2R\sin\alpha = 4R\sin\frac{\alpha}{2}\cos\frac{\alpha}{2}$. On the other hand $p-a = r\cot\frac{\alpha}{2}$, (see Theorem 4.1). Multiplying and dividing these equalities we get

$$\left.\begin{aligned} \cos^2\frac{\alpha}{2} &= \frac{a(p-a)}{4Rr} \\ \sin^2\frac{\alpha}{2} &= \frac{ar}{4R(p-a)} \end{aligned}\right\} \quad\Rightarrow\quad 1 = \frac{a(p-a)}{4Rr} + \frac{ar}{4R(p-a)}$$

which gives $a^3 - 2pa^2 + (p^2 + r^2 + 4Rr)a - 4pRr = 0$. By symmetry, a, b and c satisfy the cubic equation

$$x^3 - 2px^2 + (p^2 + r^2 + 4Rr)x - 4pRr = 0.$$

Now we can use Vieta's theorem to get

$$a + b + c = 2p; \quad ab + bc + ac = p^2 + r^2 + 4Rr; \quad abc = 4pRr.$$

The first of these is no surprise, but already the second equation is non-trivial. Using these equations and the theory of symmetric polynomials (see

Table 12.1 Cubic equations satisfied by triples of values in a triangle

Parameters	Cubic equation
a, b, c	$x^3 - 2px^2 + (p^2 + r^2 + 4Rr)x - 4pRr = 0$
$p - a, p - b, p - c$	$x^3 - px^2 + r(4R + r)x - pr^2 = 0$
h_a, h_b, h_c	$2Rx^3 - (p^2 + r^2 + 4Rr)x^2 + 4p^2rx - 4p^2r^2 = 0$
r_a, r_b, r_c	$x^3 - (4R + r)x^2 + p^2x - p^2r = 0$
$\sin\alpha, \sin\beta, \sin\gamma$	$4R^2x^3 - 4Rpx^2 + (p^2 + r^2 + 4Rr)x - 2pr = 0$
$\cos\alpha, \cos\beta, \cos\gamma$	$4R^2x^3 - 4R(R + r)x^2 + (p^2 + r^2 - 4R^2)x + (2R + r)^2 - p^2 = 0$
$\tan\alpha, \tan\beta, \tan\gamma$	$(p^2 - (2R + r)^2)x^3 - 2prx^2 + (p^2 - 4Rr - r^2)x - 2pr = 0$
$\tan\frac{\alpha}{2}, \tan\frac{\beta}{2}, \tan\frac{\gamma}{2}$	$px^3 - (4R + r)x^2 + px - r = 0$

Section 3.3), we can get as many new relations as we want, for example

$$(a + b)(b + c)(a + c) = 2p(p^2 + r^2 + 2Rr);$$
$$\frac{1}{ab} + \frac{1}{ac} + \frac{1}{bc} = \frac{1}{2Rr}.$$

Check this! We can play this game with other expressions. For example, replacing a by $p - (p - a)$ in the equation above it follows that $p - a$, $p - b$ and $p - c$ satisfy the equation

$$x^3 - px^2 + r(4R + r)x - pr^2 = 0.$$

Now Vieta's theorem directly implies that $(p - a)(p - b)(p - c) = pr^2$ which gives us, free of charge, a proof of Heron's formula. Playing with the symmetric polynomials, we can easily get relations such as

$$\frac{a}{p - a} + \frac{b}{p - b} + \frac{c}{p - c} = \frac{4R - 2r}{r}. \tag{12.1}$$

Furthermore, we can use inequalities such as $R \geq 2r$ to see that the left-hand side of (12.1) is not less than 6, which is not at all obvious. We recommend the reader to play the same game with the altitudes, trigonometric functions and external circles. Some of the associated cubic equations are collected in Table 12.1.

Without any effort we can now prove the triangle identities considered in

Problem 4.7. More than this, we can create our own, for example:

$$\frac{1}{\sin\alpha} + \frac{1}{\sin\beta} + \frac{1}{\sin\gamma} = \frac{p^2 + r^2 + 4Rr}{2pr};$$

$$\frac{1}{r_a} + \frac{1}{r_b} + \frac{1}{r_c} = \frac{1}{h_a} + \frac{1}{h_b} + \frac{1}{h_c};$$

$$\tan\frac{\alpha}{2}\tan\frac{\beta}{2} + \tan\frac{\beta}{2}\tan\frac{\gamma}{2} + \tan\frac{\gamma}{2}\tan\frac{\alpha}{2} = 1.$$

Practice this art! The cosine line in Table 12.1 gives us a useful result.

Theorem 12.1 *The value $p - 2R - r$ serves to discriminate between triangles:*

- *The triangle is right-angled iff $p = 2R + r$.*
- *The triangle is acute iff $p > 2R + r$.*
- *The triangle is obtuse iff $p < 2R + r$.*

Proof. From Table 12.1,

$$4R^2\cos\alpha\cos\beta\cos\gamma = p^2 - (2R + r)^2 = (p + 2R + r)(p - 2R - r)$$

which implies the conclusion (note that exactly one of cosines is negative in an obtuse triangle). □

To get some other useful inequalities from the table, we can use $R \geq 2r$ as well as the following estimates:

$$27r^2 \leq 16Rr - 5r^2 \leq p^2 \leq 4R^2 + 4Rr + 3r^2 \leq \frac{27}{4}R^2.$$

Moreover, for acute triangles we can use the additional inequality $p > 2R + r$. Here are some examples of possible inequalities:

$$\frac{3r}{R} \leq \frac{R + r}{R} = \cos\alpha + \cos\beta + \cos\gamma \leq \frac{3}{2};$$

$$\frac{1}{R^2} \leq \frac{1}{ab} + \frac{1}{bc} + \frac{1}{ac} \leq \frac{1}{4r^2};$$

$$27r^2 \leq h_a^2 + h_b^2 + h_c^2 \leq p^2 \leq \frac{27}{4}R^2.$$

We refer to the book [54] for more details of this nice approach.

12.2 Directed angles

Sometimes we can read that the angles considered in the solution are directed. What does this mean and how can it be used in our own solutions? We use directed angles when we are forced to consider many different cases, where ordinary angles change and formally we need separate proofs, though they look quite similar. Honestly, in most cases it means that we are too lazy to consider all the cases, and consider only one of them and hope that magic words "directed angles" allow us to skip the rest. This works amazingly well if the notations for the angles are correctly chosen, so we need only to learn this art.

First we need a notation. For ordinary angles we use the ordinary notation $\angle$, for the directed angles we will use the sign $\measuredangle$. If A, B, C are three different points we define the **directed angle** $\measuredangle ABC$ as the angle through which we needs to rotate the *line* AB in positive direction (i.e. counterclockwise) around B in order to coincide with the line BC. Note that according to the definition, $\measuredangle CBA = \pi - \measuredangle ABC$, so the order is important. To understand the difference look at Figure 4.12 on page 161. We have, obviously, $\alpha = \angle ABC = \angle AB'C'$. Though $\measuredangle ABC = \alpha$ as well, we have $\measuredangle AB'C' = \pi - \alpha$, because the rotation would be in the opposite direction.

Directed angles are defined modulo π, which allows for quite natural arithmetic, e.g. $\measuredangle ABC + \measuredangle CBD = \measuredangle ABD, \quad -\measuredangle ABC = \pi - \measuredangle ABC = \measuredangle CBA$. The sum of the directed angles in a triangle is not necessary π, but because we work modulo π we can rewrite this as

$$\measuredangle ABC + \measuredangle BCA + \measuredangle CAB = 0 \quad \Leftrightarrow \quad \measuredangle ABC = \measuredangle ACB + \measuredangle BAC.$$

The advantage of using directed angles is that we do not need to care about the order of the points and can freely choose points A, C in $\measuredangle ABC$ on the corresponding lines. Here is a typical statement (and one of the main tools).

Theorem 12.2 *The (distinct) points A, B, C are collinear iff $\measuredangle ABD = \measuredangle CBD$ for any other point D.*

The reader can check that it does not matter in which order the points lie or where we choose the point D. We could reformulate the theorem for ordinary angles, but we would need to consider many different cases and the statement would become much longer. The advantage is even greater when A, B, C are not collinear.

Theorem 12.3 *Four different points A, B, C, D are concyclic (in other words belong to the same circle) iff* $\measuredangle ABD = \measuredangle ACD$.

Again it does not matter how the points are arranged. For example, if they form a convex quadrilateral $ABCD$, then the condition is the same as the saying $\angle ABD = \angle ACD$, But if the line AD is between the points B and C then the condition for ordinary angles is $\angle ABD + \angle ACD = \pi$, whereas the equation for the directed angles does not change. If we look again at Figure 4.12 we can see how the equalities $\measuredangle ABC = \measuredangle C'BC = \measuredangle C'B'C = \pi - \measuredangle CB'C' = \pi - AB'C'$ work independently of the picture we choose.

We can define the directed angle between any two lines: it is zero if they are parallel, otherwise it is $\measuredangle ABC$ where B is the intersection point, A is another point on the first line and C is another point on the second line.

As a little practice with measured angles, we will prove a couple of statements. The first one is easy.

Theorem 12.4 *If O is the circumcentre of the triangle ABC then* $\measuredangle OAB = \measuredangle ACB + \frac{\pi}{2}$.

Proof. Let AD be a diameter. Then O belongs to the line AD and

$$\measuredangle OAB = \measuredangle DAB = \measuredangle DCB = \measuredangle DCA + \measuredangle ACB = \frac{\pi}{2} + \measuredangle ACB. \qquad \square$$

The next result is known as **Miquel's theorem.**

Theorem 12.5 *Let ABC be a triangle and A_1, B_1, C_1 be arbitrary points on the lines BC, AC and AB respectively. Then the circumcircles of the triangles AB_1C_1, BA_1C_1 and CA_1B_1 have a common point (called the* ***Miquel point****).*

Proof. We draw an arbitrary picture, for example Figure 12.1 where the points are chosen on the segments. We see here already that there are too many cases to handle, and we try to find a proof using this picture but interpreting everything in the language of directed angles. One solution is quite natural: we find the intersection point P of the circumcircles AB_1C_1 and BA_1C_1 and try to prove that it belongs to the third circumcircle. Angle chasing is evident, but we need to rewrite it using directed angles.

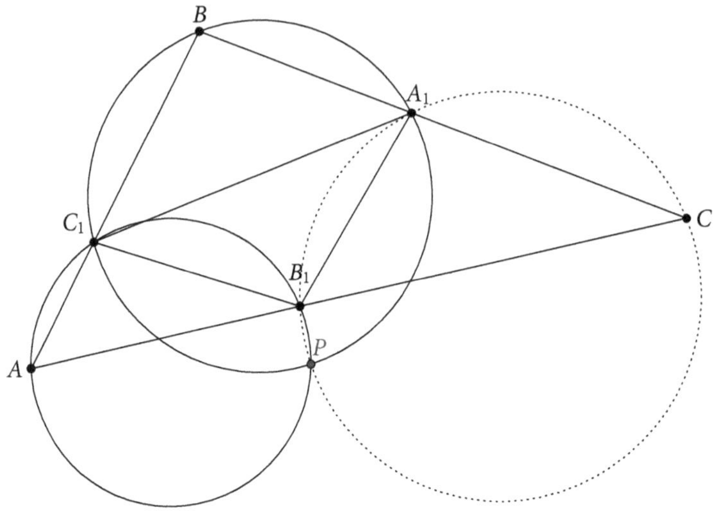

Figure 12.1 Miquel point.

First we have $\measuredangle C_1AB_1 = \measuredangle C_1PB_1$ and $\measuredangle A_1BC_1 = \measuredangle A_1PC_1$. We need a similar equality for $\measuredangle A_1CB_1$. We have

$$\begin{aligned}\measuredangle A_1CB_1 &= \measuredangle BCA = \measuredangle BAC + \measuredangle CBA = \measuredangle C_1AB_1 + \measuredangle A_1BC_1\\ &= \measuredangle C_1PB_1 + \measuredangle A_1PC_1 = \measuredangle A_1PC_1 + \measuredangle C_1PB_1 = \measuredangle A_1PB_1\end{aligned}$$

and that is all! □

12.3 Inversion

The easiest way to understand how **inversion** works is to interpret the plane as the set of complex numbers $z = x + iy$ and consider the map

$$z \mapsto \frac{1}{\bar{z}} = \frac{z}{|z|^2}.$$

It will be even more clear in trigonometric form:

$$Re^{i\varphi} \mapsto \frac{1}{R}e^{i\varphi}.$$

How does inversion map lines and circles?

Note that there is a common equation for lines and circles in $z, \overline{z}$-language:

$$az\overline{z} - w\overline{z} - \overline{w}z + c = 0$$

for some real a, c and some complex number w. Indeed, $a = 0$ gives the equation for a line (rewrite the equation in x, y-language if you do not see this). If $a \neq 0$ we can divide and rewrite the equation as

$$\left(z - \frac{w}{a}\right)\overline{\left(z - \frac{w}{a}\right)} = \frac{w\overline{w}}{a^2} - \frac{c}{a} \quad \Leftrightarrow \quad \left|\left(z - \frac{w}{a}\right)\right|^2 = \frac{|w|^2 - ac}{a^2}.$$

If $|w|^2 > ac$ we can write the right-hand side as R^2 (for some real $R > 0$). Note that for $c = 0$ we get a circle (or a line) that passes through the origin. We recommend the reader to check that, vice versa, each line or circle can be written in this way.

Because inversion is its own inverse map, we can easily see how to map our objects putting $\frac{1}{\overline{z}}$ instead for z in the equation (see page 61).

We get after multiplication by $z\overline{z}$:

$$cz\overline{z} - w\overline{z} - \overline{w}z + a = 0$$

and draw some easy conclusions:

- Lines passing through the origin are mapped to themselves.
- If $OA : OB = m : n$ and A, B and the origin O lie on the same line then A' (the image of A) and B' (the image of B) will be on the same line and $OA' : OB' = n : m$ (in the reverse order!)
- A line that does not contain the origin maps to a circle that contains the origin.
- Vice versa, a circle that contains the origin maps to a line that does not contain the origin.
- Every circle that does not pass through the origin maps to another circle that does not pass through the origin. If R, R' are their radii and P, P' are their centres then $\overrightarrow{OP'} = \frac{R'}{R}\overrightarrow{OP}$. Note (this is important) that the centres should not map to each other!

Check all of this on a few examples. An important property of inversion is that *it preserves angles* between lines and circles. However, this does not mean that the triangle ABC is similar to $A'B'C'$ (because the image of line AB may be a circle through A' and B'). Naturally, the angle between two intersecting

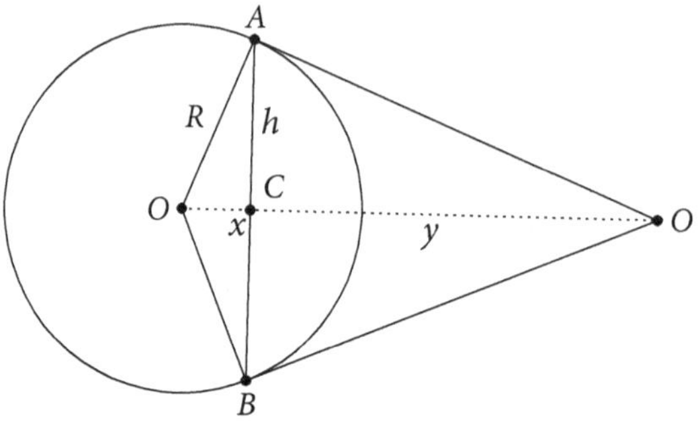

Figure 12.2 Inversion

circles is the angle between their tangent lines at the intersection point. For example, a circle that passes through the origin and a line that contains the origin and the circle's centre map to two orthogonal lines.

Normally we do not involve complex numbers and understand the **inversion** relative to a circle with centre O and radius R as the following: The inversion sends each point $A \neq O$ to the point A' on the line OA such that A, A' lie on the same side of O and $|AO| \cdot |A'O| = R^2$. Thus points on the circle itself do not move, and all points outside the circle map inside (and vice versa). What about O? In some sense it maps to "the point at infinity", but normally we do not care about O.

To have some geometrical visualization of the inversion relative to a circle with the centre O consider Figure 12.2. If C is the midpoint of the chord AB, then for its image C' we have that $C'A$ and $C'B$ are the tangents to the circle. Vice versa, the image of C' is C. To see this, note that $xy = h^2$ (the altitude of a right-angled triangle is the geometric mean of the two parts in which it divides the hypotenuse). Thus

$$|OC| \cdot |OC'| = x(x + y) = x^2 + h^2 = R^2.$$

The importance of inversion is that it can dramatically change the problem. As an example, we will look at problem 4.15 again. How does Figure 4.17 change after inversion with centre at A? The lines AB, AC will stay the same, the circles will become two parallel lines and the line BC will become a circle tangent to one of these lines at point D', intersecting another line at the

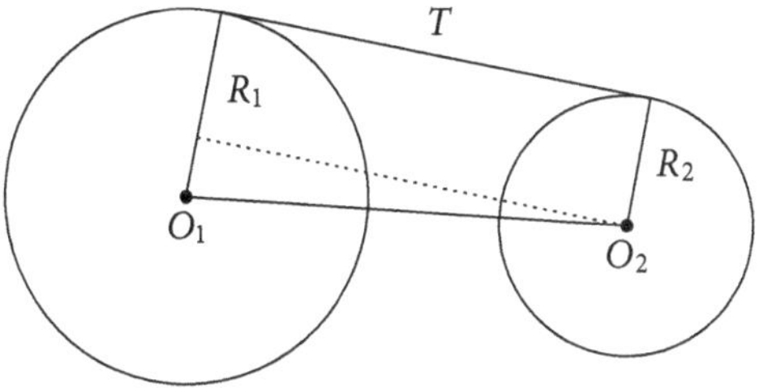

Figure 12.3 The length of the exterior tangent

points B', C'. It is obvious now that AD' is a bisector of $B'AC'$ and because the angles do not change, the same is true for AD.

Even if the new problem is not necessarily easier, it will be quite different and this itself can be helpful. The advantage is, as we have seen, that we can transform some circles to lines. The disadvantage, of course, is that some lines are transformed to circles, and we have rather bad control over the lengths.

For example, if we have two circles with centres at O_1 and O_2 and radii R_1 and R_2, it is not difficult to calculate the length T of their common exterior tangent (see Figure 12.3): according to the Pythagorean theorem we have $T^2 = |O_1O_2|^2 - (R_1 - R_2)^2$. The reader should check that for the interior tangent we need only to change one sign: $T^2 = |O_1O_2|^2 - (R_1 + R_2)^2$.

Suppose that after inversion, the circles map to new circles with radii R_1', R_2'. What is the length T' of their common exterior or interior tangent? Note that the tangent line probably maps to a circle that tangents our circles, but has nothing in common with their common tangent. Fortunately, there is a nice answer obtained by Casey:

Theorem 12.6 $\dfrac{T^2}{R_1R_2} = \dfrac{T'^2}{R_1'R_2'}$.

Proof. The proof is rather direct, but boring. We can calculate all the values involved in this formula using complex numbers. Suppose that our circles are defined as

$$a_i z\bar{z} - w_i\bar{z} - \overline{w_i}z + c_i = 0, \quad i = 1, 2.$$

Their centres are $O_i = \dfrac{w_i}{a_i}$, their radii $R_i = \dfrac{|w_i|^2 - a_i c_i}{a_i^2}$, and thus we have

$$\begin{aligned}
T^2 = |O_1 - O_2|^2 - (R_1 \pm R_2)^2 &= \left|\frac{w_1}{a_1} - \frac{w_2}{a_2}\right|^2 - R_1^2 - 2R_2^2 \mp 2R_1R_2 \\
&= \frac{|w_1a_2 - w_2a_1|^2}{a_1^2a_2^2} - \frac{|w_1|^2 - a_1c_1}{a_1^2} - \frac{|w_2|^2 - a_2c_2}{a_2^2} \mp 2R_1R_2 \\
&= \frac{-w_1\overline{w_2}a_1a_2 - \overline{w_1}w_2a_1a_2 + a_2^2a_1c_1 + a_1^2a_2c_2}{a_1^2a_2^2} \mp 2R_1R_2 \\
&= \frac{-w_1\overline{w_2} - \overline{w_1}w_2 + a_2c_1 + a_1c_2}{a_1a_2} \mp 2R_1R_2.
\end{aligned}$$

Because after the inversion we only change the places of a_i and c_i, we get

$$T'^2 = \frac{-w_1\overline{w_2} - \overline{w_1}w_2 + a_2c_1 + a_1c_2}{c_1c_2} \mp 2R_1'R_2'$$

and hence

$$\begin{aligned}
\frac{T'^2}{R_1'R_2'} &= \frac{-w_1\overline{w_2} - \overline{w_1}w_2 + a_2c_1 + a_1c_2}{R_1'R_2'c_1c_2} \mp 2 \\
&= \frac{-w_1\overline{w_2} - \overline{w_1}w_2 + a_2c_1 + a_1c_2}{R_1R_2a_1a_2} \mp 2 = \frac{T^2}{R_1R_2}.
\end{aligned}$$

Really boring! □

Consider now the following nice, useful but not so well-known result.

Theorem 12.7 (Casey's theorem) *Let k_1, k_2, k_3, k_4 be four circles that lie inside another circle K and tangent to it (in this order). Let T_{ij} be the length of the common exterior tangent between k_i and k_j. Then*

$$T_{12}T_{34} + T_{14}T_{23} = T_{13}T_{24}.$$

Proof. The problem looks computationally very unattractive and needs some cosmetics. Note that according to the previous theorem, if the formula is valid it will still be valid after inversion (we only multiply each term by $\sqrt{\frac{R_1'R_2'R_3'R_4'}{R_2R_2R_3R_4}}$), so we can try to improve our circles. The best circle is a line — so let us consider an inversion that sends K to a line and do the calculations here (see Figure 12.5). But this is trivial:

$$ac + (a + b + c)b = (a + b)(b + c).$$ □

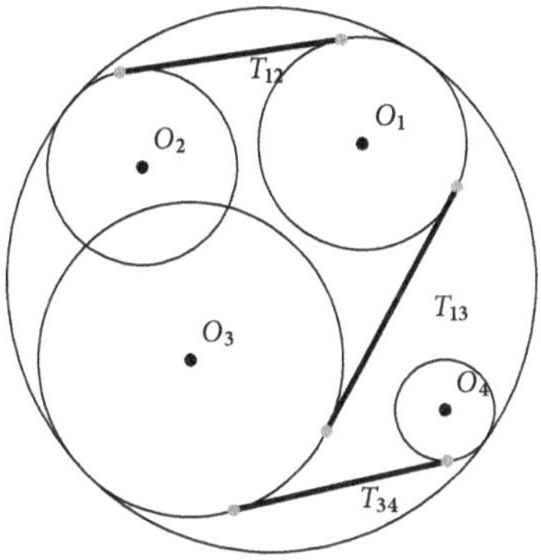

Figure 12.4 Casey's theorem

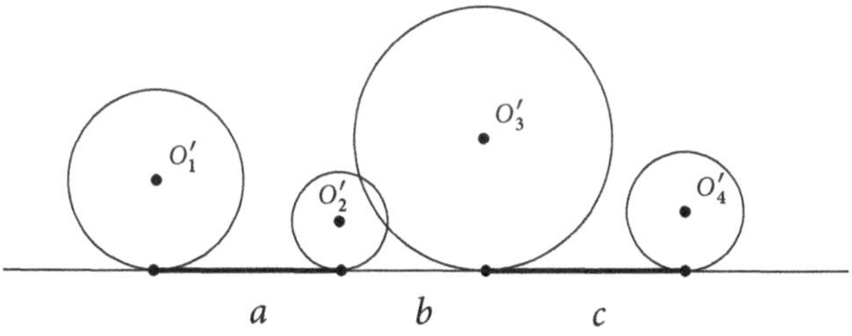

Figure 12.5 Casey's theorem after inversion

The converse to Casey's theorem is valid as well, namely if the equality holds, then the circles are tangent to a circle or to a line. This is a very useful instrument. Moreover we **can allow some circles to be outside** K, but in this case we use the **interior** common tangent if the circles are on the opposite sides of K. Note that our proof does not change! The proof of the converse can also be reduced to the case where three circles are tangent to a line but is more complicated. See [36] for the details. But mostly this theorem is used when some circles are ... points. For example, if all of them are points then we get Ptolemy's theorem.

To see how this works, consider the classic **Feuerbach's theorem**.

Theorem 12.8 *Consider an arbitrary triangle ABC. Its nine-point circle is externally tangent to the triangle's three excircles and internally tangent to its incircle.*

Proof. We do not even need a picture! Consider the excircle that is tangent to the triangle at the side BC. As we have seen on page 144, the tangents to this circle from A are equal to p, and thus the tangents from the midpoints of AB and AC are equal to $p - \frac{c}{2} = \frac{a+b}{2}$ and $p - \frac{b}{2} = \frac{a+c}{2}$ respectively. WLOG $b \le c$ and, according the computations on the same page it follows that the length of the tangent from the midpoint of BC is equal to $(p - b) - \frac{a}{2} = \frac{c-b}{2}$.

Now consider three midpoints as circles with zero radius. All three belong to the nine-point circle and thus touch it. To show that our excircle is tangent to this circle as well, we need only to combine the tangents found and the lengths between the midpoints (which are obviously equal to half the triangle sides and can be interpreted as tangents between the midpoints):

$$\frac{a}{2} \cdot \frac{c-b}{2} + \frac{b}{2} \cdot \frac{a+c}{2} = \frac{c}{2} \cdot \frac{a+b}{2}.$$

It remains for us to apply the converse of Casey's theorem. We leave it to the reader as good practice to find a similar proof for the incircle. □

Let us go back to inversion and study how the distance between two points A and B changes. Of course, this depends on their position. More precisely, if we have an inversion with centre at P and radius R, then for the images A', B' we have

$$|PA| \cdot |PA'| = R^2 = |PB| \cdot |PB'| \quad \Rightarrow \quad \frac{|PA|}{|PB|} = \frac{|PB'|}{|PA'|}$$

which shows that the triangles PAB and $PB'A'$ are similar (AB is antiparallel to $A'B'$). Thus

$$|A'B'| = |AB| \cdot \frac{|PB'|}{|PA|} = \frac{R^2|AB|}{|PA| \cdot |PB|}.$$

The following problem (taken from the IMO-2003 short list) looks very attractive from this point of view.

Problem 12.9 Let $\Gamma_1, \Gamma_2, \Gamma_3, \Gamma_4$ be distinct circles such that Γ_1, Γ_3 are externally tangent at P and Γ_2, Γ_4 are externally tangent at the same point P. Suppose that Γ_1 and Γ_2; Γ_2 and Γ_3; Γ_3 and Γ_4; Γ_4 and Γ_1 meet at A, B, C, D, respectively, and that all these points are different from P. Prove that

$$\frac{|AB| \cdot |BC|}{|AD| \cdot |DC|} = \frac{|PB|^2}{|PD|^2}.$$

Search. There are too many circles — inversion looks like an appropriate approach. Perform an inversion with centre P and radius R. Then the quadrilateral $ABCD$ turns into a parallelogram $A'B'C'D'$. We have

$$\begin{aligned} 1 &= \frac{|A'B'| \cdot |B'C'|}{|A'D'| \cdot |D'C'|} \\ &= \frac{R^2|AB|}{|PA| \cdot |PB|} \cdot \frac{R^2|BC|}{|PB| \cdot |PC|} \cdot \frac{|PA| \cdot |PD|}{R^2|AD|} \cdot \frac{|PC| \cdot |PD|}{R^2|CD|} \\ &= \frac{|AB| \cdot |BC| \cdot |PD|^2}{|AD| \cdot |DC| \cdot |PB|^2}. \end{aligned}$$

Hence

$$\frac{|AB| \cdot |BC|}{|AD| \cdot |DC|} = \frac{|PB|^2}{|PD|^2}.$$ □

Remark It is not necessary to assume that Γ_1, Γ_3 and Γ_2, Γ_4 are *externally* tangent. We may change the first sentence in the problem to the following:

Let Γ_1, Γ_2, Γ_3, Γ_4 be distinct circles such that Γ_1, Γ_3 are tangent at P, and Γ_2, Γ_4 are tangent at the same point P.

To estimate how efficient this was, let us consider a purely geometrical approach.

Solution (Geometric approach). Let O_1, O_2, O_3, O_4 be the centres of Γ_1, Γ_2, Γ_3, Γ_4, respectively, and let A', B', C', D' be the midpoints of PA, PB, PC, PD, respectively. Since Γ_1, Γ_3 are externally tangent at P, it follows that O_1, O_3, P are collinear. Similarly we see that O_2, O_4, P are collinear.

Put $\alpha_1 = \angle O_4O_1O_2$, $\alpha_2 = \angle O_1O_2O_3$, $\alpha_3 = \angle O_2O_3O_4$, $\alpha_4 = \angle O_3O_4O_1$ and $\varphi_1 = \angle PO_1O_4$, $\varphi_2 = \angle PO_2O_3$, $\varphi_3 = \angle PO_3O_2$, $\varphi_4 = \angle PO_4O_1$. By the law of sines, we have

$$|O_1O_2| : |O_1O_3| = \sin\varphi_3 : \sin\alpha_2, \qquad |O_3O_4| : |O_2O_4| = \sin\varphi_2 : \sin\alpha_3,$$
$$|O_3O_4| : |O_1O_3| = \sin\varphi_1 : \sin\alpha_4, \qquad |O_1O_2| : |O_2O_4| = \sin\varphi_4 : \sin\alpha_1.$$

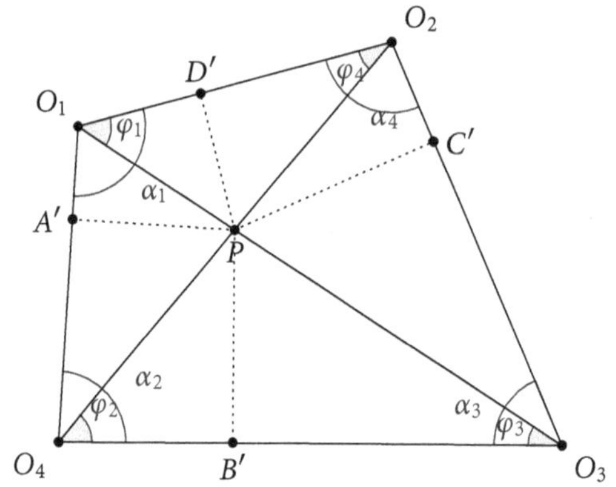

Figure 12.6 Solution 2

Since the segment PA is the common chord of Γ_1 and Γ_2, the segment PA' is the altitude from P to O_1O_2. Similarly PB', PC', PD' are the altitudes from P to O_2O_3, O_3O_4, O_4O_1, respectively. Then O_1, A', P, D' are concyclic. So, again by the law of sines, we have

$$|D'A'| : |PD'| = \sin\alpha_1 : \sin\varphi_1.$$

Likewise

$$\frac{|A'B'|}{|PB'|} = \frac{\sin\alpha_2}{\sin\varphi_2}, \qquad \frac{|B'C'|}{|PB'|} = \frac{\sin\alpha_3}{\sin\varphi_3}, \qquad \frac{|C'D'|}{|PD'|} = \frac{\sin\alpha_4}{\sin\varphi_4}.$$

Since $|A'B'| = \frac{1}{2}|AB|$, $|B'C'| = \frac{1}{2}|BC|$, $|C'D'| = \frac{1}{2}|CD|$, $|D'A'| = \frac{1}{2}|DA|$, $|PB'| = \frac{1}{2}|PB|$ and $|PD'| = \frac{1}{2}|PD|$, it follows that

$$\begin{aligned}\frac{|AB|\cdot|BC|}{|AD|\cdot|DC|}\cdot\frac{|PD|^2}{|PB|^2} &= \frac{|A'B'|\cdot|B'C'|}{|A'D'|\cdot|D'C'|}\cdot\frac{|PD'|^2}{|PB'|^2}\\ &= \frac{\sin\alpha_2\sin\alpha_3\sin\varphi_4\sin\varphi_1}{\sin\varphi_2\sin\varphi_3\sin\alpha_4\sin\alpha_1}\\ &= \frac{|O_1O_3|}{|O_1O_2|}\cdot\frac{|O_2O_4|}{|O_3O_4|}\cdot\frac{|O_1O_2|}{|O_2O_4|}\cdot\frac{|O_3O_4|}{|O_1O_3|} = 1\end{aligned}$$

and the conclusion follows. □

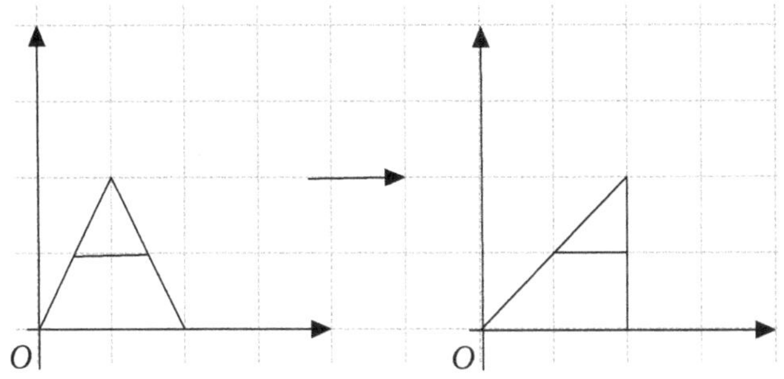

Figure 12.7 An affine map changes the angles but preserves midpoints.

12.4 Affine and projective maps

An **affine map** Φ is a map that preserves barycentric coordinates. In other words, if $\sum \lambda_i = 1$ then for any points P_i we have

$$\Phi(\sum \lambda_i P_i) = \sum \lambda_i \Phi(P_i).$$

In particular, Φ maps lines to lines, parallel lines to parallel lines and midpoints of segments to midpoints. In coordinates, Φ can be written as $\Phi : X \mapsto AX + B$ where A is a constant matrix and B is a constant column vector. In Figure 12.7, where we used the affine map $(x, y) \mapsto (x + \frac{y}{2}, y)$ (write down the matrix A! Here, B is obviously the zero vector), we see that the angles can change.

Most of the transformations we have considered earlier (homothety, rotation, translation, symmetry, but not inversion) are affine, so this is a very general notion. But being too general also makes it less useful. Because a circle is generally transformed to an ellipse, affine maps are not so often used in mathematics competitions. The main application is that any two triangles can be transformed to each other. So if the problem is not about circles or angles we can WLOG choose one triangle in our problem as equilateral. But this is often too restrictive a condition.

Nevertheless affine transformations are very important in image processing, where affine transformations are used to change or move the images so they can be used on the computer screens with different resolutions.

If we want to work with lines and points only, strangely enough it is more convenient to go one step farther and work with *projective maps*. But first we need to introduce the **projective plane** $\mathbb{RP}^2$. The idea is quite simple: we want to add some extra, **ideal points**, to the usual plane $\mathbb{R}^2$ in such a way that parallel lines intersect in these new points. In fact, every set of parallel lines have a common ideal point, and parallel lines in different directions correspond to different ideal points. The set of all ideal points form a new line: the **ideal line**.

To realize this idea formally, we will consider three-dimensional non-zero vectors (p_1, p_2, p_3) and identify those that have proportional coordinates. We define a point P in our projective plane with homogeneous coordinates $(p_1 : p_2 : p_3)$ as a class of such identified vectors. For example, the point $P = (4 : 4 : 2)$ is the same as $(-2 : -2 : -1)$, whereas $Q = (2 : 1 : 0)$ is a different point. Our ordinary two-dimensional points (x, y) can be identified with points with homogeneous coordinates $(x : y : 1)$. Thus the point P above corresponds to the point $(2, 2)$ in ordinary coordinates and generally a **point** $(p_1 : p_2 : p_3)$ with $p_3 \neq 0$ corresponds to point (x, y), where $x = \frac{p_1}{p_3}$, $y = \frac{p_2}{p_3}$. Points for which $p_3 = 0$, are interpreted as the ideal point corresponding to lines that are parallel to the vector (p_1, p_2). For example, the point Q above is an ideal point common to all the lines $x - 2y = c$ for different constants c.

A **line** l in the projective plane also has projective coordinates $(l_1 : l_2 : l_3)$ and consists of all points satisfying the equation

$$l_1 p_1 + l_2 p_2 + l_3 p_3 = 0.$$

Dividing by p_3 we get the equation of an ordinary line in (x, y)-coordinates.

For example, the line $p_1 - 2p_2 + 2p_3 = 0$ contains both P and Q and ordinary points satisfy the equation $x - 2y + 2 = 0$.

This slightly abstract definition is very convenient for calculations. Here are some examples.

Theorem 12.10 *Let A, B, C be three points in $\mathbb{RP}^2$.*

- *If $A = (a_1 : a_2 : a_3)$ and $B = (b_1 : b_2 : b_3)$ are different then there exists exactly one line $l = (l_1 : l_2 : l_3)$ containing them, which can be found from* *the determinant equation:*

$$\begin{vmatrix} p_1 & p_2 & p_3 \\ a_1 & a_2 & a_3 \\ b_1 & b_2 & b_3 \end{vmatrix} = 0 \quad \Leftrightarrow \quad \begin{cases} l_1 = \begin{vmatrix} a_2 & a_3 \\ b_2 & b_3 \end{vmatrix}, \\ l_2 = -\begin{vmatrix} a_1 & a_3 \\ b_1 & b_3 \end{vmatrix}, \\ l_3 = \begin{vmatrix} a_1 & a_2 \\ b_1 & b_2 \end{vmatrix}. \end{cases}$$

- *A, B, C are collinear iff the determinant of the matrix containing their homogeneous coordinates is equal to zero.*
- *if A, B, C are different collinear points and a vector $\mathbf{c} = (c_1, c_2, c_3)$ is one of the representatives of the point C then one can choose the row vectors $\mathbf{a} = (a_1, a_2, a_3)$, $\mathbf{b} = (b_1, b_2, b_3)$ representing the points in such a way that $\mathbf{c} = \mathbf{a} + \mathbf{b}$.*

Proof. Replacing p_i by a_i in the determinant, we get two equal rows which explains why the determinant is equal to zero. For similar reasons, the point B satisfies the determinant equation. Note that $l_1 p_1 + l_2 p_2 + l_3 p_3$ is simply the expansion of the determinant, which explains the equivalence sign (see Section 9.2 if you need to review the definitions). Note that not all $l_i = 0$ because A and B are different points.

If we put the coordinates of C instead of p_i we get the collinearity conditions. They remains valid even when $A = B$.

As for the last statement, note that according to Theorem 9.1 any vectors representing points A, B and C are linearly dependent. Because v and λv represent the same point we can choose the coefficients in the dependence as we like. The only possible trouble could be if some coefficient in the dependence is zero, but this would imply that two points are the same. □

Now let us see how all this machinery creates nice geometrical results.

Theorem 12.11 (Desargues' theorem) *Consider six different points A, B, C and A', B', C' in $\mathbb{RP}^2$. Suppose that three lines AA', BB' and CC' have a common point P (thus they are concurrent). Then the intersection points $D = AB \cap A'B'$, $E = AC \cap A'C'$ and $F = BC \cap B'C'$ are collinear (and thus belong to the same line). The converse statement (collinearity implies concurrency) is valid as well.*

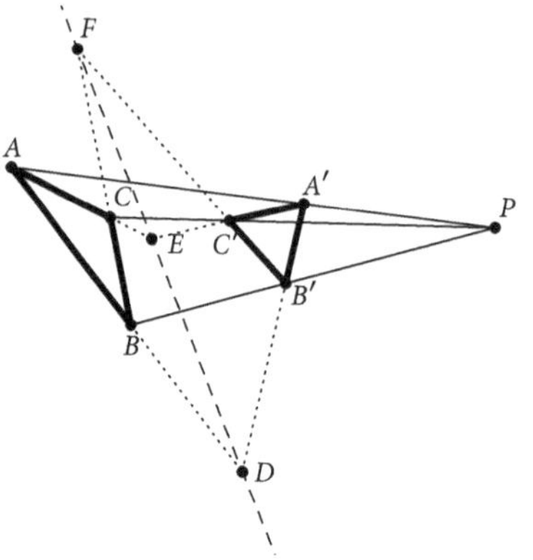

Figure 12.8 Desargues' theorem.

Proof. For reasons that will soon be clear, we will present only the direct proof. Using the previous theorem we can find representative vectors for the six given points and the point P such that

$$\mathbf{p} = \mathbf{a} + \mathbf{a}' = \mathbf{b} + \mathbf{b}' = \mathbf{c} + \mathbf{c}' \quad \Rightarrow \quad \mathbf{b} - \mathbf{a} = \mathbf{a}' - \mathbf{b}'.$$

Consider the non-zero vector $\mathbf{d} = \mathbf{b} - \mathbf{a}$. It is a representative for some point D'. We want to show that $D = D'$.

Indeed, because $\mathbf{a}, \mathbf{b}, \mathbf{d}$ are linearly dependent, the determinant of their coordinates is equal to zero. Thus, by the same theorem A, B, D' are collinear. But simultaneously $\mathbf{d} = \mathbf{a}' - \mathbf{b}'$ and for the same reasons A', B', D' are collinear as well which shows that $D' = D$.

Similarly $\mathbf{e} = \mathbf{a} - \mathbf{c}$, $\mathbf{f} = \mathbf{c} - \mathbf{b}$. But $\mathbf{d} + \mathbf{e} + \mathbf{f} = \mathbf{0}$ and the same theorem now shows that D, E, F are collinear. □

One very nice thing in projective geometry is **duality**. Note that we denoted lines and points in the same way — using homogeneous coordinates: if we look at $(1 : 2 : 3)$ we cannot say whether this is a line or a point. Moreover, in any equation for a line, the coordinates of the line and the coordinates of the point are used symmetrically. This leads to a very interesting observation: Doing the opposite, calling lines points and points lines, we will not see the difference in any of our proofs! Of course, at first sight this contradicts our geometrical interpretation, but if we think a little we find that we are simply proving two theorems instead of one. The only thing we need to do is to translate phrases such as "line l through two points A, B" into "point L — the intersection of two lines a and b" and vice versa. For example, we translate Theorem 12.10 to a statement about three lines and find conditions for concurrency and the coordinates for the common point of two different lines. Check this!

Let us try to translate Desargues' theorem. We have six lines a, b, c, a', b', c'. We need to consider points $a \cap a'$, $b \cap b'$ and $c \cap c'$ and suppose that they belong to the same line p, and thus are collinear. Now we will consider three new lines: line d through the points $a \cap b$ and $a' \cap b'$. We define the lines e and f similarly. The conclusion is that d, e, f are concurrent. This statement we get free of charge because it is dual to the proven statement. But the interesting fact is that this is exactly the converse of Desargues' theorem! Check this! For example, the line a we can interpret as the line BC in our picture, the line a' as $B'C'$ and so on. The line p is the line through D, E, F. So in fact we have proven Desargues' theorem in both directions simultaneously. Isn't this amazing?

Because we interpreted the ordinary plane $\mathbb{R}^2$ as a subset of the projective plane we can apply Desargues's theorem here as well, but the statement needs to be adjusted to deal with special cases, such as when a pair of sides are parallel.

The idea of duality is much deeper than it looks and can sometimes be applied in ordinary geometry as well. For example, Ceva's theorem and Menelaus' theorem are in fact dual to each other, though they cannot be considered as statements in projective geometry (we use lengths), and thus they do not immediately imply each other. Try to be on the lookout for ways to interpret problems in a dual way. Here is another pair of useful (and dual) statements.

Theorem 12.12 (Brianchon) *Let $ABCDEF$ be a hexagon formed by six tangent lines of a circle. Then the lines AD, BE and CF are concurrent.*

Theorem 12.13 (Pascal) *Three pairs of opposite sides of a cyclic hexagon (extended if necessary) meet in three points which are collinear.*

Again those theorems cannot be formulated in this form in projective geometry, because there is no notion of circle in projective geometry. In fact this

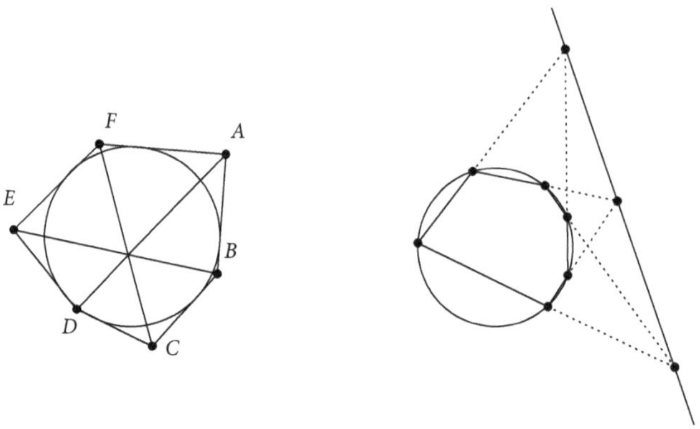

Figure 12.9 Brianchon's and Pascal's theorems.

means that the theorems are valid in the more general situations where we use a conic (e.g. a parabola) instead of a circle, but we will skip this interesting subject and note only that the hexagons in Pascal's theorem do not need to be convex because this is in a statement about six arbitrary points on a conic.

We will finish this section with some comments about projective maps. A formal definition is easy: each such map is determined by a 3×3 matrix A and consists of a multiplication of the column vector containing the homogeneous coordinates of a given point by A. Every affine map $\Phi : X \mapsto AX + B$ can be extended to a projective map using the matrix

$$\begin{pmatrix} A & B \\ O & 1 \end{pmatrix},$$

so projective maps are more general. Projective maps still send lines to lines, but (restricted to $\mathbb{R}^2$) it can send parallel lines to non-parallel lines and does not preserve the ratio AB/BC between three points A, B, C on a line. Interestingly enough, projective maps preserve the **cross-ratio**

$$\frac{AC/BC}{AD/BD}$$

between four points on a line and in fact any map that sends lines to lines and preserves cross-ratios is projective. A classical projective map is **central perspectivity** . If there are two lines in a plane and a point O outside them then the perspectivity sends each point P on the first line to the intersection of the line OP with the second one. One point is sent to an ideal point at infinity so it is more natural to consider this map from the projective point of view. To check that the cross-ratio is invariant is an exercise which we recommend to reader.

We finish this section with **Pappus theorem.**

Problem 12.14 Given two set of collinear points A_i, B_i, C_i for $i = 1, 2$, show that the intersection points of line pairs A_1B_2 and A_2B_1, A_1C_2 and A_2C_1, B_1C_2 and B_2C_1 are collinear (the corresponding line is called the **Pappus line**).

We leave the proof as a good exercise in projective coordinates.

12.5 Geometry in Figures

The title of this section is the same as the title of an amazing book written by Aseniy Akopyan [58]. The book is almost completely without text, it contains only pictures. You need to guess for yourself what the corresponding problem is and how to find the proofs.

Working with this book is one of the best way to learn geometry. We copy a couple of figures to convince the reader that this book is very useful. The existence of Miquel point in Figure 12.10 is easy to prove. But Verrièr's lemma in Figure 12.12 is really challenging. Try!

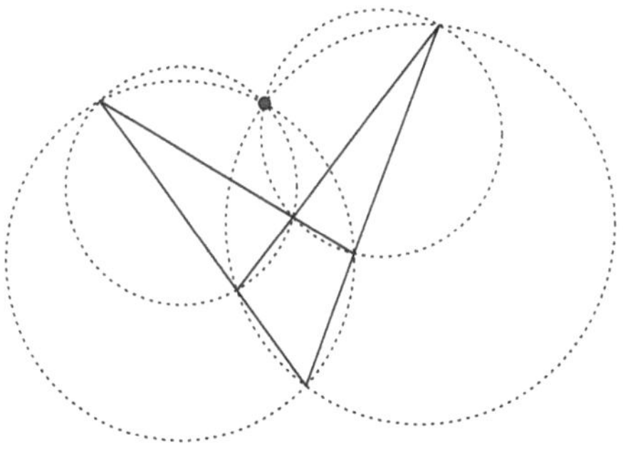

Figure 12.10 Miquel point

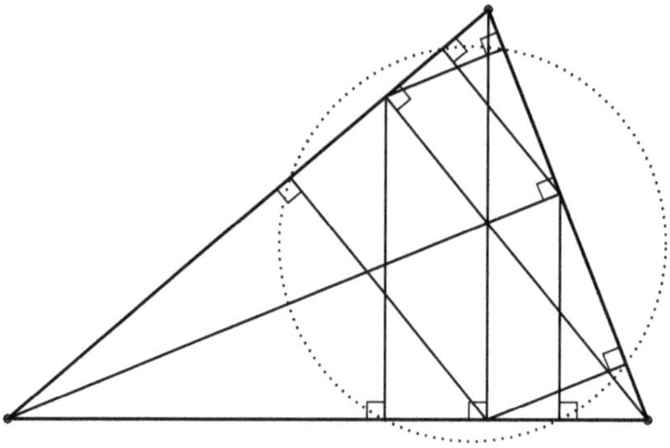

Figure 12.11 Altitudes

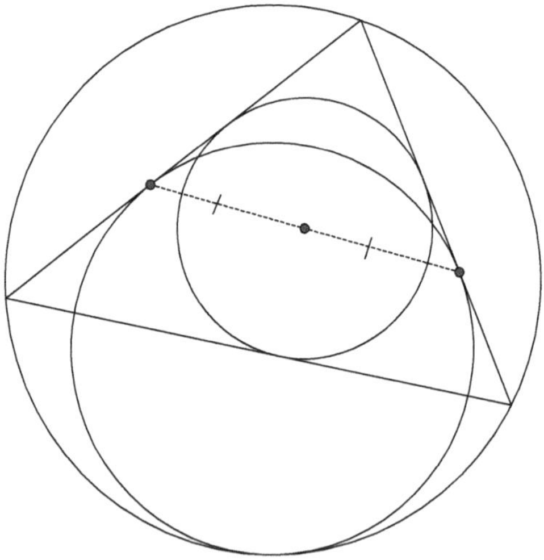

Figure 12.12 Verrièr's lemma

CHAPTER 13

Combinatorial arguments

13.1 König's lemma

We would like to formulate the lemma we mentioned on page 46. It is the proof that is important to remember.

Theorem 13.1 (König's lemma) *Let G be a connected graph with infinitely many vertices such that each vertex has finite degree. Then G contains an infinitely long path with no repeated vertices (a simple path).*

Most often König's lemma is applied to infinite trees, where we can conclude that either we have a vertex with an infinite degree or have an infinite simple path. We will prove this theorem for trees only, leaving the reader to handle the general case. See Section 6.2 if you need some definitions. We remind you only that a **tree** T is a connected (non-oriented) graph without cycles. and that the **degree of a vertex** is the number of vertices adjacent to it.

Proof. Suppose that all vertices in T have finite degree. We will construct a path by induction.

Consider any vertex x_1. This is the beginning of the path. Consider all the edges $x_1 \to y_j$, $j = 1, \ldots, k$ starting from x_1. If we throw away those edges we get k different trees $T_1, \ldots, T_k$. At least one of them, say T_j, is infinite as well. Call the corresponding vertex y_j by x_2 and repeat the same argument to the tree T_j where $y_j = x_2$ plays the role of x_1. We find the vertex x_3 adjacent to x_2 and new infinite tree, containing x_3 and repeat this again and again. As a result we construct an infinite path

$$x_1 \to x_2 \to x_3 \to \cdots$$

□

13.2 Ramsey's theorem

Now we would like, following [32], to extend the pigeonhole principle to the next, much deeper level, called Ramsey's theory. We start from its infinite variant, because it is slightly simpler. Consider an infinite set X and suppose that we are interested in its k-element subsets (shortly k-sets). For simplicity we suppose that X is just the set of non-negative integers, but the theory works for arbitrary infinite sets.

Suppose now that we have r colours and that each k-set has one of these colours. Note: we do not paint the elements, but the subsets! For example, if $k = 2$ we paint all the pairs $\{i, j\}$ and this we can visualize as a colouring of the edges in the infinite graph G where X is the set of the vertices and we draw the edges $i \leftrightarrow j$ between any different integers.

For $k = 3$ we (in some sense) paint the triangles $\{i, j, k\}$ but not the edges or vertices. This is already difficult to visualize (suppose that the triangle $\{1, 2, 3\}$ has one colour and the triangle $\{1, 2, 4\}$ has another. Which colour does the common edge $\{1, 2\}$ have? Answer — none, we do not paint the edges! For $k = 4$ the visualization is hardly possible.

Now we will formulate **Ramsey's theorem.**

Theorem 13.2 *For any colouring of all k–sets in an infinite set X by r colours there exists an infinite subset $S \subseteq X$ such that all its k-subsets have the same colour.*

Proof. We will start from small k. If $k = 1$ then we are painting the elements and claim that some colour was used infinitely many times which is an obvious infinite interpretation of the pigeonhole principle. The case $k = 2$ is already non-trivial and we want to study it carefully.

Consider $x_0 = 0$ and all the edges $0 \leftrightarrow i$. From all the colours they have at least one of them must appear infinitely many times (in fact, this follows from the statement for $k = 1$). Call this colour c_1 and the corresponding infinite set Y. For any $y \in Y$ we have now $x_0 \overset{c_1}{\leftrightarrow} y$.

Consider now the minimal element $y_0 \in Y$ and all the edges $y_0 \leftrightarrow y_j$. From all the colours they have, at least one of them (say c_2) must appear infinitely many times and we consider the corresponding infinite subsets $Z \subset Y$ such that $y_0 \overset{c_2}{\leftrightarrow} z$ for any $z \in Z$. We can continue this to infinity and get the sequence $x_0 < y_0 < z_0 < \cdots$. The infinite set $T = \{x_0 < y_0 < z_0 < \cdots\}$

has an important property: if $t < t' \in T$ then the colour c for the edge $t \overset{c}{\leftrightarrow} t'$ depends on t only. Now we paint each element $t \in T$ in this colour c. Some infinite subset $S \subseteq T$ is painted in the same colour c_0 and this is exactly the set we need, because for any $s < s' \in S$ we have $s \overset{c_0}{\leftrightarrow} s'$. The key idea here was the reduction to the previous case $k = 1$.

To understand the idea better consider now the case $k = 3$. Now all the triangles $\{i, j, k\}$ are coloured. Consider again $x_0 = 0$ and all the triangles $\{0, i, j\}$. Consider now the infinite set N of positive integers and create an induced colouring on its *edges* $i \leftrightarrow j$. That is, we choose the same colour for this edge as for the triangle $\{0, i, j\}$. Applying the theorem with $k = 2$ we find an infinite subset $Y \subseteq N$ for which all the edges have the same colour c_1. This means that the all the triangles $\{0, i, j\}$ with $i, j \in Y$ are painted in c_1.

Consider now the minimal element $y_0 \in Y$ and the triangles $\{y_0, y_i, y_j\}$ with $y_0 < y_i < y_j \in Y$. Again we can create an induced colouring on the edges $y_i \leftrightarrow y_j$, choosing the same colour as for the triangle and again by the induction hypothesis ($k = 2$) find an infinite subset Z such that for any $z < z' \in Z$ all the triangles $\{y_0, z, z'\}$ are painted in the same colour c_2. We can continue this to infinity and get a sequence $x_0 < y_0 < z_0 < \cdots$. The infinite set $T = \{x_0 < y_0 < z_0 < \cdots\}$ has an important property: if $t < t' < t'' \in T$ then the colour c for the triangle $\{t, t', t''\}$ depends on t only. Now we paint each element $t \in T$ in the corresponding colour c. Some infinite subset $S \subseteq T$ is painted in the same colour c_0 and this is exactly the set we need, because for any $s < s' < s'' \in S$, the triangle $\{s, s', s''\}$ is painted in c_0.

Hopefully the reader can understand the main idea and have no trouble in making a similar inductive proof for an arbitrary k. □

In many cases a statement for an infinite set can be modified to a statement for a sufficiently large set (but with a longer and less readable formulation). Let us formulate the corresponding finite variant of Ramsey's theorem. You probably need to read it more than once to understand it.

Theorem 13.3 *Let k, n, r be some natural numbers. There exists a number $R(k, n, r)$ such that if a set X has at least $R(k, n, r)$ elements then for any colouring of all its k-subsets in r colours there exists a subset $S \subseteq X$ of size n such that all its k-subsets have the same colour.*

Proof. We could of course imitate the previous proof, replacing the infinite sets by finite ones. For example, to start with $k = 1$ we simply take $R(1, n, r) = (n-1)r + 1$ and use the pigeonhole principle. It is a good exercise to do this, but we prefer another way, because it shows another important way of reasoning in similar situations.

Suppose the opposite. For any natural m there exist only finitely many ways to paint a set with m elements and we supposed that for any m there exists a bad painting without any subset S of the size n with all its k-subsets of the same colour. Consider a graph with all the bad paintings (for different m) as a vertices. Naturally, if we throw away an element from a badly painted set we again get a bad painting of a set with $m-1$ elements. We draw an edge between these two paintings. What we get is obviously a tree with a root in the empty set.

Now König's lemma (see Theorem 13.1) gives us an infinite set of paintings, each of which is included in the next. If we take their union we get an infinite coloured set X without any subset S of the size n with all its k-subsets of the same colour (because such S should be part of some of our finite paintings). But this contradicts the infinite variant of Ramsey's theorem. □

This was not easy, but nothing is easy in Ramsey's theory which we in fact are discussing now. We do not know too much how to calculate the numbers $R(k, n, r)$. Here are some known optimal values: $R(2, 3, 2) = 6$, $R(2, 4, 2) = 18$, $R(2, 3, 3) = 17$. One easy estimate is the following.

Theorem 13.4 $R(2, 3, r) \le x_r = [r! \cdot e] + 1$. *In other words if each edge in complete graph G on x_r vertices is coloured in one of r colours then G contains a monochromatic triangle.*

Proof. Note first that the numbers

$$x_m = [m! \cdot e] + 1 = \left[m! \sum_{n=0}^{\infty} \frac{1}{n!} \right] + 1 = m! \sum_{n=0}^{m} \frac{1}{n!} + 1$$

satisfy the recursive equation $x_m - 1 = m(x_{m-1} - 1) + 1$. Indeed:

$$m\left((m-1)! \sum_{n=0}^{m-1} \frac{1}{n!}\right) + 1 = m! \sum_{n=0}^{m-1} \frac{1}{n!} + \frac{m!}{m!} = m! \sum_{n=0}^{m} \frac{1}{n!}$$

Now if we consider an arbitrary vertex in G then according to pigeonhole principle at least x_{m-1} of its edges have the same colour. If the corresponding vertices are connected with the edge of the same colour we get a monochromatic triangle. Otherwise we can use induction and get a monochromatic triangle in any case. □

Now it is time for some applications. The first one is simple.

Problem 13.5 Let m, n be some natural numbers. Show that there exists a number $R(m,n)$ such that for any graph G with at least $R(m,n)$ vertices either G contains a complete subgraph K_m or its complement $\overline{G}$ contains a complete subgraph K_n. The minimal such number is usually called **Ramsey's number.**

Search. We start with a little make-up. Obviously it is sufficient to prove the statement for $m = n$ by simply increasing the smaller number. The number $R(2,n,2)$ from the previous theorem looks very appropriate. Indeed, if we consider the edges in G as painted black and the absent edges as painted in an invisible white colour, then we find that the statement is simply a reformulation of the previous theorem for this case. Good! □

It is relatively easy to check (try!) that $R(3,3) = 6$, and thus from any six persons there exist either three familiar or three unfamiliar ones with each other. It is interesting that in fact we can find at least two such triples (not necessarily of the same kind) — try to prove this as well.

Our next example is more complicated. Recall that a **tournament** is a complete graph where each edge has an orientation. If vertices correspond to teams then we can interpret an oriented edge $x \to y$ as team x beats team y, which explains the name. A tournament is **transitive** if for any three teams x, y, z such that x beats y and y beats z we automatically have that x beats z. Clearly, not all tournaments are transitive.

Problem 13.6 Let n be some natural numbers. Show that there exists a number N such that any tournament with at least N teams contains a transitive subtournament with n teams.

Search. The smell of Ramsey's theory is strong. Let us try to apply it directly. We paint, obviously, triangles in our tournament in nice colour if the triangle $\{x, y, z\}$ corresponds to a transitive triple and paint in an ugly colour if not. We try $N = R(3,n,2)$ and find a subset with n vertices with all triangles painted in the same colour. If this colour is nice we are happy. What if all the triangles are ugly? Consider a small example. Case $n = 3$ is too trivial; what if $n = 4$? So we have four vertices, say x, y, z, t and all the triangles are ugly. WLOG we can suppose that $x \to y \to z$ then $z \to x$. If $t \to x$ then we should have that $y \to t$, because the triangle $\{x, y, t\}$ is ugly. But in this case the triangle $\{y, z, t\}$ cannot be ugly! Good. Thus $x \to t$. But then $\{x, y, t\}$ cannot be ugly. So, we cannot have all the triangles ugly if $n > 3$. Is $n = 3$ a problem? No, we find a transitive tournament for $n = 4$ and throw away one team. □

Here is another example.

Problem 13.7 Let n be some natural numbers. Show that there exists a number N such that from any N points on the plane in a *general position* (which means that no three of them are collinear) one can choose n which form a convex polygon.

Search. How we do use Ramsey's theorem here? There are (at least) two ways. In one of them we paint triangles depending on their orientation. In another we paint 4-subsets depending on their convexity. To start we need to understand the case $n = 5$ first. It helps to understand that if any four points in a polygon form a convex polygon then the polygon itself is convex. These hints should be sufficient for the experienced reader. □

If we look at IMO problems then we find that the problem 4 in IMO-1964 is a special case of Theorem 13.4 for $r = 3$. More challenging is the last problem from IMO-1978.

Problem 13.8 An international society has its members in 6 different countries. The list of members contains 1978 names, numbered $1, 2, \ldots, 1978$. Prove that there ia at least one member whose number is the sum of the numbers of two, not necessarily distinct, of his compatriots.

Search. We need a sum so let us look at the differences! An obvious identity $i - j = (i - k) + (k - j)$ suggests that it is a good idea to consider a complete graph with 1979 vertices numbered from 1 to 1979 where the edge between i and j is coloured in one of six colours depending on which country the member with the number $|i - j|$ belongs to. Because $6!e = 720e < 1978$ we can according to Theorem 13.4 find a monochromatic triangle and the problem is solved. □

13.3 Playing games

Game problems are popular in competitions. Here we would like to discuss some ideas that can be used in such problems. We will restrict ourselves by finite games between two persons, Alex and Bob. Alex makes a first move and the person who makes the last move wins the game. As a simple example we consider the game G_n where in the beginning there are n coins and in one move it is possible to take one, two or three coins. The question is naturally who is going to win if both are playing optimally.

Such a game is quite easy to analyze. We consider small numbers first. For $n = 1, 2, 3$, Alex obviously wins taking all the coins. For $n = 4$ he loses the game: after any move Bob gets a winning position. Easy induction shows that Alex loses for all n that are divisible by 4 and wins in the other cases.

Consider now a more complicated game $G_{m,n}$, where there are two piles with m and n coins in each. The rules are the same — in one move it is allowed to take one, two or three coins but only from one of the piles. Who is going to win now? Can we find a winning strategy when there are more piles? What if it is allowed to take an arbitrary number of coins from one of the piles? (This game is called Nim). These questions are more difficult and to answer them we will introduce some useful notations.

First of all, we will introduce the notation $G \oplus H$ for a game consisting of two parallel games G, H where each move can be done only in one of them. For example, $G_{m,n} = G_m \oplus G_n$.

Next we want to assign a number $p(G)$ for each game. We do it recursively. For a game G without possible moves we put $p(G) = 0$. When we have made some move $G \to H$ we transform the game G to a new game H (where Alex and Bob exchange their roles). We define $p(G)$ as a minimal non-negative integer that is not equal to any $p(H)$ for all possible moves $G \to H$.

For example $p(G_0) = 0$, $p(G_1) = 1$, because the only possible move is to G_0 and $1 \neq p(G_0) = 0$. From G_2 there are two possible moves to G_0 and G_1. Thus $p(G_2)$ is not 0, 1 and, therefore $p(G_2) = 2$. Similarly $p(G_3) = 3$. But $p(G_4) = 0$ because we have no moves to a position H with $p(H) = 0$. Easy induction shows that $p(G_n) = n \pmod 4$.

Obviously, Alex loses the game G if $p(G) = 0$ and wins if $p(G) > 0$, but we can say more than that. To understand what, we will introduce a new operation on non-negative integers which we also denote $\oplus$. We put

$$0 \oplus 0 = 0; \quad 0 \oplus 1 = 1 \oplus 0 = 1; \quad 1 \oplus 1 = 0.$$

To find $c = a \oplus b$ for larger integers a, b we need to rewrite both numbers in the binary system, say $a = \sum_k a_k 2^k, n = \sum b_k 2^k$. Then $c = \sum_k c_k 2^k$ where $c_k = a_k \oplus b_k$. For example,

$$\left.\begin{array}{l} 5 = 1 \cdot 4 + 0 \cdot 2 + 1 \cdot 2 \\ 6 = 1 \cdot 4 + 1 \cdot 2 + 0 \cdot 2 \end{array}\right\} \quad \Rightarrow \quad 6 \oplus 5 = 0 \cdot 4 + 1 \cdot 2 + 1 \cdot 2 = 3.$$

Now we can formulate the main result which is a good exercise for the reader.

Theorem 13.9 $p(G \oplus H) = p(G) \oplus p(H)$.

Using this theorem, it is easy to prove that Alex wins in $G_{m,n}$ iff $m \neq n \pmod 4$. If he is playing Nim where k piles have n_i coins then he wins iff $n_1 \oplus \cdots \oplus n_k \neq 0$.

A more complicated game (call it H_n) is the following. A rectangle $n \times 1$ consists of n squares. In one move one can delete two squares that have a common edge. Obviously after one move we get the game $H_i \oplus H_j$, where $i + j = n - 2$. The theorem allows us to find $p(H_n)$ for small values:

n	0	1	2	3	4	5	6	7	8	9	10
$p(H_n)$	0	0	1	1	2	0	3	1	1	0	3

For example, to get $p(H_6)$ we calculate

$$p(H_4 \oplus H_0) = 2 \oplus 0 = 2; \quad p(H_3 \oplus H_1) = 1 \oplus 0 = 1; \quad p(H_2 \oplus H_2) = 1 \oplus 1 = 0,$$

and find that 3 is the first non-negative number that is not here.

But we need a computer to suspect that after $n = 53$ the sequence $p(H_n)$ has a period of length 34. It is sufficient to check on the computer that $p(H_{n+34}) = p(H_n)$ for $53 < n \leq 108$, because after that an easy induction shows that the same is true for larger n: if $i + j = n - 2$ then at least one of the numbers is greater than 53 and we can use the induction hypothesis. After that it is not difficult to check the table and conclude that Alex loses the game H_n iff $n = 5, 9, 21, 25, 29 \pmod{34}$ or $n = 0, 1, 15, 35$. Far from being trivial!

Next we would like to consider another game, "Queen's Move". A chess queen is placed on a chessboard. In one move it is allowed to travel south, west, or southwest. The player who gets the queen to the lower left corner is the winner. If we introduce the coordinates such that this corner has coordinates $(0,0)$ and the queen has the coordinates (m,n) then we can denote the game as $W_{m,n}$.

Who wins in this game? Our first observation is that this game is mathematically the same as Wythoff's Game (it has many other different names; the oldest one is *tsyan-shidzi* from China). In this game we have two piles with m and n coins. In one move a player either takes an arbitrary number of coins from one of the piles or an *equal* number from both piles. Can we apply our binary instruments here? Unfortunately not; we cannot split the game into two parallel games and need quite different tools.

First we need some experimental information. Naturally we need only to know for which m, n Alex is in a losing position. WLOG we suppose that $n \le m$. After several games we find small critical values and put them in a table.

n	0	1	3	4	6	8	9	11
m	0	2	5	7	10	13	15	18

Looking at these two nice sequences in the rows we can immediately make a lot of nice conjectures (and most of them will be correct). We leave this pleasure to the reader and try to use the modern approach. The most efficient way to analyze the result is to use the On-line Encyclopedia of Integer Sequences [62] — every mathematician should know this web page. It is sufficient to input 1, 3, 4, 6, 8, 9 to get the answer: these are the first terms of the well-known Beatty sequence, $a_k = [k\varphi]$. We will find a lot of useful references and recognize the second sequence as well. In one of the references we will probably find a solution to our problem as well, but to have some pleasure out of being a mathematician we will restrict ourselves by suggested hypothesis: the pairs $n_k = [k\varphi]$, $m_k = [k\varphi^2]$ for $k = 0, 1, 2, \dots$ describe all losing positions.

To prove this we need to check two facts: a) from any other position the queen can achieve one of these in one move; b) if the queen is already in one of these positions it cannot move to another. Both statements do not look trivial, because it is not clear how to use our golden ratio φ.

But note that $\varphi^2 = \varphi + 1$ which gives us $m_k - n_k = k$. This shows that using diagonal movements the queen leaves the losing positions. Besides that $\frac{1}{\varphi} + \frac{1}{\varphi^2} = 1$ and we can use our old knowledge — namely problem 1.8. The problem states that all numbers n_k, m_k obtained above for $k > 0$ are different and each natural number is equal either to some n or to some m. The reader should not have difficulties to finish the proof now.

We could not use the binary system here, but we could use another one — the **Fibonacci number system**, in which every natural number is written as a sum of different Fibonacci numbers. To have uniqueness we skip $F_0 = 0, F_1 = 1$ and do not allow ourselves to have two consequent Fibonacci numbers (we can replace them by their sum). The reader should prove that each number now has a unique representation. For example, $50 = 34 + 13 + 3 = F_9 + F_7 + F_4$ or 10100100 if we prefer a binary code. If s|he likes this system then it will be a pleasure to prove that the losing positions are exactly those where n has an even number of last zeros in this code, but m is obtained from n by an additional zero at the end. For example, $n = 6$ is written as 1001 and the corresponding m is 10010, or 10, exactly as in the table. So many beautiful approaches to the same problem!

13.4 Combinatorial geometry

Combinatorial geometry is a subject that is difficult to describe. More or less it involves problems where the object has a geometrical nature (e.g. points, lines), but the questions — and often the methods — are mostly combinatorial. One of the difficulties is that we need to combine the continuous geometrical arguments with the discrete arguments that are typical for combinatorics. In some sense, graph theory can be considered as a very close relative, because it uses our geometrical intuition to visualise purely combinatorial objects. The best we can do to present the subject is to solve a few nice problems.

We start with a classic example, the Sylvester–Gallai problem.

Problem 13.10 In any configuration of n points in the plane, not all on a line, there is a line which contains exactly two of the points.

It sounds like an easy exercise, but if you have never seen this problem before, we urge you to spend at least an hour to find the solution before you continue reading. This hour will be sufficient to understand the subject of combinatorial geometry.

The main difficulty is that we have to use Euclidean geometry, because the same statement is wrong in projective geometry. Because projective geometry is a very exciting area, let us take a small detour before we return to our problem. We begin with the definition.

A **projective plane** is a set whose elements we call *points*. Some of its subsets we call *lines*. We assume that there exist at least three different lines and points and have only two demands:

- For any two distinct points, there exists exactly one line containing both of them.
- For any two distinct lines, there exists exactly one point belonging them (their intersection).

A perfect example is $\mathbb{RP}^2$, but more interesting, especially from a combinatorial point of view, are **finite projective planes**, which contain only finitely many points (and lines). If we take any two lines ℓ_1 and ℓ_2 and a point $P \notin \ell_1$ (why does P exist?), then we can create an obvious bijection between points on those lines. For each point $Q \in \ell_1$, draw the unique line containing P and Q and find its unique intersection with ℓ_2. This argument shows that every line contains the same number of points which we denote $n + 1$. The number n is called the **order of the projective plane**. We leave to the reader as an easy exercise to show that the total number of points is exactly $n^2 + n + 1$ and that every point belongs to exactly $n + 1$ different lines.

The projective plane of order 1 can be identified with a triangle ABC, where A, B, C are the points and the subsets $\{A, B\}, \{B, C\}, \{A, C\}$ are the lines. The projective plane of order 2 has 7 points and is known as the **Fano plane**. The Fano plane is impossible to realize as a collection of usual lines and points. To visualize it, we can consider a triangle ABC and and its three medians AA_1, BB_1, CC_1. This is already six points and six lines. The last point is the triangle's centre of gravity G, but the last line should go through A_1, B_1 and C_1 and cannot be an ordinary real line. Instead we identify the final line with the nine-points circle. So already for the projective plane of order 2, the construction is non-trivial.

For which values of n does there exist a projective plane of order n? This is still an open question and not much is known.

First of all, if $n = p^k$ is a power of a prime p, there exists a (often several non-isomorphic) projective plane of order n. One possible construction is to take $\mathbb{FP}^2$, where $\mathbb{F}$ is a finite field with n elements. We simply repeat the arguments in Section 12.4, replacing $\mathbb{R}$ by $\mathbb{F}$. In other words, we consider non-zero triples $(p_1 : p_2 : p_3)$, where $p_i \in \mathbb{F}$ and we identify $(p_1 : p_2 : p_3)$ and $(cp_1 : cp_2 : cp_3)$. These are our points and we see that there are exactly $n^3 - 1$ triples and $\frac{n^3-1}{n-1} = n^2 + n + 1$ points. The lines are also parameterized by triples $(a_1 : b_1 : c_1)$ and consist of the solutions to the equation $a_1p_1 + a_2p_2 + a_3p_3 = 0$. Again there are $n^2 + n + 1$ such lines and repeating the arguments in Section 12.4, we not only prove that the projective plane axioms are satisfies, but we also see that Desargues' theorem remains valid. This is a very nice mixture of algebra, geometry and combinatorics! Note that there exist other constructions of finite projective planes for which Desargues' theorem does not hold. For example, for $n = 9$, there exist exactly four different (non-isomorphic) projective planes, three of which are non-Desarguesian.

Every known finite projective plane has prime power order, and it has been conjectured that there does not exist any finite projective plane whose order is not a prime power. The main result on non-existence of projective planes, obtained by Bruck and Ryser in 1949, is that if $n = 1, 2 \pmod 4$ is the order of a finite projective plane, then n must be a sum of two squares (see [33] for details). So for $n = 6, 14, 21, 22$ there are no projective planes of order n. A massive computer search concluded in 1989 that there is no projective plane of order 10. It is still a wide-open problem whether there is one of order 12.

After this detour, let us return to our problem.

Search. We understand now that the solution cannot use the points and lines alone, because such a solution would also work in any projective plane, where the conclusion does not hold.

Hence, we need to use some Euclidean tools such as angles or distances to find a solution. Let us try to get a contradiction supposing that each line in our configuration contains at least three points. As usual, we should try to find something extremal and this something should involve angles or distances. A natural candidate is a pair consisting of a point P and a line ℓ where the distance between P and ℓ is minimal.

The line must contain at least three other points. If P' is the orthogonal projection of P on ℓ, then at least two of those points, say X and Y must be on the same side relative P'. Suppose that X is closer to P'. What is the distance d between X and the line PY? If we manage to prove that d is less than PP'

we will obtain the desired contradiction. The reader should not have any difficulty doing this. □

One of the most beautiful problems in combinatorial geometry appeared in IMO-2011.

Problem 13.11 Let S be a finite set containing at least two points in the plane. Assume that no three points of S are collinear. A *windmill* is a process that starts with a line l going through a single point $P \in S$. The line rotates clockwise about the *pivot* P until the first time that the line meets some other point in S. This point, Q, takes over as the new pivot, and the line now rotates clockwise about Q, until it next meets a point in S. This process continues indefinitely. Show that we can choose a point $P \in S$ and a line l going through P such that the resulting windmill uses each point of S as a pivot infinitely many times.

Search. The first idea that comes to mind is to consider the **convex hull**. This is the minimal convex figure containing our set. We remind that a set is **convex** if for any of its points, the connecting segment belongs to the set as well. Obviously the intersection of two convex sets is convex, so we can obtain the convex hull of a set S as the intersection of all convex subsets containing S.

In our case, S is finite and the hull is a convex polygon. If we choose the starting point P on the boundary of the convex hull and the line outside the convex hull (except P), we can never reach the points in the interior of the convex hull. Such a situation must be avoided and we should choose a starting line that intersect the convex hull.

If we erase the points on the boundary of the convex hull, the remaining set has a convex hull of its own, and continuing in this way we can write our set as a union of convex polygons included in each other, reminding of nested Russian dolls, matreshka. This already gives a possible approach using induction on the number of such dolls. Nevertheless this attempt to prove the statement does not lead to immediate success. We need something more.

We have a process, and it is reasonable to look for some kind of invariants. One evident observation is that as long as the windmill contains only one of our points, the number of points in each half-plane is the same. Interesting. What happens when we change the point? In fact nothing! Those numbers do not change: we simply replace one point by another but they are in the same half-plane (of course we need to formulate and prove this rigorously, for example, by introducing an orientation of the line to calculate the numbers in a similar way. But we obviously have found the invariant we are looking for).

So we have our invariant. If the windmill passes through all the points, in each step the invariant must stay the same. What are the possible choices? A natural hypothesis (suggested by previous arguments and examples) is that the difference between the numbers of points in the half-planes should be as small as possible (zero or one depending of parity). And such a line obviously exists for each point. Indeed, take any line through the point. If the difference is larger we can rotate this line: after 180°, the half-planes have replaced each other, so somewhere in between there must be a line that we are looking for.

Very good! So, suppose we start at some point with such a nice line. Why do we reach another point? It has a line, with the same invariant, but why do we get there? Suppose not. Sooner or later our windmill will be parallel to this line and we obtain two parallel lines with the same invariant. Is this possible? Definitely not. Done! □

Problem 13.12 Two friends, let us call them Amy and Bob, are about to share a pizza. To make it somewhat fair, they decide to do it like this: Amy cuts the pizza into a number of slices (circular sectors), perhaps of different size. Bob chooses the first slice freely, and then they take turns, each one picking one of the two "available" slices, i.e. one of the slices adjacent to a slice that's already been chosen.

a) Show that if Amy cuts the pizza into an *even* number of slices, Bob can always guarantee to get at least half of the pizza.

b) Show that it is possible for Amy to get more than half the pizza!

c) In fact, show that Amy can cut the pizza in such a way that she is guaranteed at least $5/9$ of the whole pie. Conversely, show that Bob can get at least $4/9$ for himself, regardless of how the pizza is cut. This is a *very* difficult problem. If you get stuck, see [37] for the full solution.

PART III

Problems

CHAPTER 14

Papers

14.1 Paper 1

1. At an international conference there are four official languages. Any two participants can discuss in one of these languages. Show that at least 60 % of the participants can speak the same language.

2. Solve the following system in real numbers

$$\begin{cases} x^2 + y^2 - z(x + y) = 2 \\ y^2 + z^2 - x(y + z) = 4 \\ z^2 + x^2 - y(z + x) = 8 \end{cases}$$

3. Let $ABCD$ be a parallelogram of centre O. The points M and N are the midpoints of BO and CD, respectively. Prove that if the triangles ABC and AMN are similar, then $ABCD$ is a square.

4. Prove that for any real numbers a, b, c such that $0 < a, b, c < 1$, the following inequality holds

$$\sqrt{abc} + \sqrt{(1 - a)(1 - b)(1 - c)} < 1.$$

5. Five points are given in the plane such that each of the ten triangles they define has area greater than 2. Prove that there at least one of the triangles has area greater than 3.

6. Do there exist four polynomials such that the sum of any three of them has a real zero, but the sum of any two of them has no real zeroes?

14.2 Paper 2

7. A mouse eats its way through a $3 \times 3 \times 3$ cube of cheese by tunneling through all of the twenty-seven $1 \times 1 \times 1$ sub-cubes. If it starts at one of the corner sub-cubes and always moves onto an uneaten adjacent sub-cube can it finish at the centre of the cube? (Assume that the mouse can tunnel through walls but not edges or corners.)

8. $12k$ persons were invited to a party. Each person shakes hands with $3k+6$ persons. Also, we know that the number of the persons who shake hands with any two persons is constant. Find the number of persons invited.

9. Suppose that A, B and C are the angles of a triangle. Prove that

$$\sin A + \sin B + \sin C \leq \sqrt{\frac{15}{4} + \cos(A-B) + \cos(B-C) + \cos(C-A)}.$$

10. A *Pythagorean triangle* is a right-angled triangle whose sides are of integral length. Prove that

(i) The inscribed circle of a Pythagorean triangle has integer radius.

(ii) For any arbitrary positive integer N there exists a Pythagorean triangle with an incircle of radius N.

11. Let $a_1, a_2, \ldots, a_n$ be positive real numbers in arithmetic progression. Prove that

$$\sum_{k=1}^{n} \frac{1}{a_k a_{n-k+1}} \geq \frac{4n}{(a_1+a_n)^2}.$$

12. Let ABC be a triangle and let P be a point in its interior. Denote by D, E, F the feet of the perpendiculars from P to the lines BC, CA, AB respectively. Suppose that

$$AP^2 + PD^2 = BP^2 + PE^2 = CP^2 + PF^2.$$

Denote by I_A, I_B, I_C the excentres of the triangle ABC. Prove that P is the circumcentre of the triangle $I_A I_B I_C$.

14.3 Paper 3

13. Prove that if an equilateral triangle is divided into a finite number of triangles, then at least one of the triangles has all angles less than or equal to 120°.

14. Prove that

a) There exists a triangle ABC (not equilateral) such that the angle bisector AL, the median BM and the altitude CH intersect in the same point.

b) For every triangle ABC with the above property, the triangles ABC and HBL are similar.

15. Find all positive real solutions of the system

$$\begin{cases} x_1 + x_2 + \cdots + x_n = 9, \\ \dfrac{1}{x_1} + \dfrac{1}{x_2} + \cdots + \dfrac{1}{x_n} = 1. \end{cases}$$

16. A bijective function $f : M \to M$ is called a *reflection* if $f(f(x)) = x$ for all $x \in M$. Prove that for any bijective function $F : M \longrightarrow M$ there exist two reflections f_1, f_2 such that $F = f_1 \circ f_2$, i.e. $F(x) = f_1(f_2(x))$ for all $x \in M$.

17. In the triangle ABC, the midpoint of BC is D. Given that $\angle ADB = 45^\circ$ and $\angle ACB = 30^\circ$, determine $\angle BAD$.

18. A triangle has sides of length a, b, c and its circumcircle has radius R. Prove that the triangle is right-angled if and only if $a^2 + b^2 + c^2 = 8R^2$.

14.4 Paper 4

19. Let $n \geq 4$ be an integer and $a_1, a_2, \ldots, a_n$ be positive real numbers such that $a_1^2 + a_2^2 + \cdots + a_n^2 = 1$. Show that the following inequality holds:

$$\frac{a_1}{a_2^2+1} + \frac{a_2}{a_3^2+1} + \cdots + \frac{a_n}{a_1^2+1} \geq \frac{4}{5}(a_1\sqrt{a_1} + a_2\sqrt{a_2} + \cdots + a_n\sqrt{a_n})^2.$$

20. Given the triangle $A_0B_0C_0$, a sequence of triangles $\{\triangle A_nB_nC_n\}_{n\in\mathbb{N}}$ is formed by each time choosing the points $A_{k+1}, B_{k+1}, C_{k+1}$ on the sides B_kC_k, C_kA_k, A_kB_k, respectively, of the triangle $A_kB_kC_k$ in such a way that

$$\frac{A_{k+1}B_k}{A_{k+1}C_k} = \frac{B_{k+1}C_k}{B_{k+1}A_k} = \frac{C_{k+1}A_k}{C_{k+1}B_k} = 2.$$

Prove that the triangles $A_kB_kC_k$ and $A_{k+2}B_{k+2}C_{k+2}$ are similar. Show that the vertices of $\triangle A_{k+3}B_{k+3}C_{k+3}$ lie on the segments A_kA_{k+1}, B_kB_{k+1}, C_kC_{k+1} respectively and divide them in the ratio $2:1$, counted from the point with lower index.

21. Find the least positive real number r with the property: Whatever four disks are considered with centres in the four vertices of a unit square and with sum of their radii equal to r, there exists an equilateral triangle which has its vertices in three of the four disks.

22. After elections, every member of parliament (PM) has his own absolute rating. When the parliament is set up, he enters a group and gets a relative rating. The relative rating is the ratio of his own absolute rating to the sum of the absolute ratings of the PMs in the group. A PM can move from one group to another only if his relative rating in the new group is higher. In a given day, only one PM can change his group. Show that only a finite number of group moves is possible. (Remark: a rating is a positive real number.)

23. Let $\triangle ABC$ be a triangle such that $AC \neq BC$, $AB < AC$ and let K be its circumcircle. The tangent line to K at the point A intersects the line BC in the point D. Let K_1 be the circle tangent to K and to the segments AD, BD. We denote by M the point where K_1 touches BD. Show that $AC = MC$ if and only if AM is the bisector line of the angle $\angle DAB$.

24. Find all functions f, defined and continuous on the whole real line, and such that

$$f\left(\frac{x+y}{2}\right) = \frac{f(x)+f(y)}{2} \qquad \forall x, y \in \mathbb{R}.$$

14.5 Paper 5

25. A list of complex numbers $(x_1, \ldots, x_n)$ is said to be a *cycle of length n* for a polynomial f if all the x_i's are distinct and if

$$x_2 = f(x_1), \quad x_3 = f(x_2), \quad \ldots, \quad x_n = f(x_{n-1}), \quad x_1 = f(x_n).$$

Let $c \in \mathbb{C}$ be given, and let $f(z) = z^2 + c$. Let $n \geq 2$ be an integer, and let $(x_1, \ldots, x_n)$ be a cycle of length n for f. Prove that

$$\prod_{i=1}^{n} (f(x_i) + x_i) = 1.$$

26. Let $ABCD$ be a cyclic quadrilateral such that AD is not parallel to BC. Given points E and F on the line CD, let G and H be the circumcentres of $\triangle BCE$ and $\triangle ADF$. Prove that the lines AB, CD and GH are concurrent or parallel if and only if there is a circle through A, B, E and F.

27. Find the smallest constant $k > 0$ such that

$$\frac{ab}{a+b+2c} + \frac{bc}{b+c+2a} + \frac{ca}{c+a+2b} \leq k(a+b+c)$$

for all $a, b, c > 0$.

28. Let G be a graph with n vertices. For each vertex v, let $f(v)$ be the maximum cardinality of an independent set of neighbours of v. Show that $\sum f(v) \leq n^2/2$, where the sum is taken over all vertices of G. (The *cardinality of a finite set* is the number of its elements. An *independent* set of vertices is a subset V' of G:s vertices such that no two vertices in V' are joined by an edge in G.)

29. Consider a convex quadrilateral with no parallel sides. On each side AB, select a point T as follows: Draw lines from A and B parallel to the opposite side. Let A' and B' be the new points where these lines intersect the sides neighbouring AB. Let T be the point where AB intersects $A'B'$. Prove that the four points selected in this way are the corners of a parallelogram.

30. What is the largest cubical present that can be completely wrapped (without cutting) by a unit square of wrapping paper?

14.6 Paper 6

31. Consider an isosceles triangle KL_1L_2 with $|KL_1| = |KL_2|$, and let KA, L_1B_1, L_2B_2 be its angle bisectors. Prove that $\cos \angle B_1AB_2 < \frac{3}{5}$.

32. Let $0 < \alpha < \frac{\pi}{2}$ and $x_1, x_2, \ldots, x_n$ be real numbers such that

$$\sin x_1 + \sin x_2 + \cdots + \sin x_n \geq n \sin \alpha.$$

Prove that

$$\sin(x_1 - \alpha) + \sin(x_2 - \alpha) + \cdots + \sin(x_n - \alpha) \geq 0.$$

33. Place a pebble at each *non-positive* integer point on the real line, and let n be a fixed positive integer. At each step, we choose some n consecutive integer points, remove one of the pebbles located at these points and rearrange all others arbitrarily within these points (placing at most one pebble at each point). Determine whether there exists a positive integer n such that for any given $N > 0$, we can place a pebble at a point with coordinate greater than N in a finite number of steps described above.

34. A conference hall has n chairs placed in a circle. There are n delegates at the conference. The first delegate chooses his or her seat arbitrarily. Thereafter the $(k+1)^{\text{th}}$ delegate sits k places to the right of the kth delegate, for $1 \leq k \leq n-1$. No chair can be occupied by more than one delegate.

Find the set of values for which this is possible.

35. Does there exist a solid in 3D space satisfying the following condition: for any natural $n \geq 3$ there is a parallel projection such that the image of the solid is a convex n-gon?

36. Let $\mathbb{R}^+$ denote the set of all positive real numbers. Find all functions $f : \mathbb{R}^+ \to \mathbb{R}^+$ satisfying for all $x, y \in \mathbb{R}^+$ the equality

$$f(xf(y)) = f(xy) + x.$$

14.7 Paper 7

37. Find all positive integers n such that the equation

$$x + y + u + v = n\sqrt{xyuv}$$

has a positive integer solution x, y, u, v.

38. Consider the equation

$$\frac{1}{x-1} + \frac{1}{4x-1} + \cdots + \frac{1}{k^2x-1} + \cdots + \frac{1}{n^2x-1} = \frac{1}{2},$$

where n is a positive integer parameter.

a) Prove that for every $n \in \mathbb{N}$, the equation has a unique solution greater than 1; denote this solution x_n.

b) Prove that $\lim_{n\to\infty} x_n = 4$.

39. Let $\triangle ABC$ and $\triangle PQR$ be two triangles such that

(i) P is the midpoint of BC and A is the midpoint of QR;

(ii) QR bisects $\angle BAC$ and BC bisects $\angle QPR$.

Prove that $AB + AC = PQ + PR$.

40. Let $P(x)$ be a polynomial of degree n and let $x_1, \ldots, x_n$ be its n different zeroes. Prove that

$$\sum_{k=1}^{n} \frac{P''(x_k)}{P'(x_k)} = 0.$$

41. Prove that given any $2n - 1$ natural numbers it is possible to choose n of them such that their sum will be divisible by n.

42. Let AA_1, BB_1 and CC_1 intersect each other in the point O inside the triangle ABC (A_1 belongs to BC, B_1 to AC and C_1 to AB.) Prove that if $OB_1 + AC_1 = AB_1 + OC_1$ and $OC_1 + BA_1 = OA_1 + BC_1$, then $OB_1 + CA_1 = OA_1 + CB_1$.

14.8 Paper 8

43. Can you cover a 10×10 "chessboard" with

a) L-shaped pieces, consisting of 4 squares each;

b) T-shaped pieces, consisting of 4 squares each;

c) both L- and T-shaped pieces, consisting of 4 squares each?

44. Given a triangle such that the sines of all three angles are rational numbers, prove that the cosines of all three angles are rational too.

45. Prove the inequality

$$\frac{a}{b^2+1} + \frac{b}{c^2+1} + \frac{c}{a^2+1} \geq \frac{3}{2}$$

for all non-negative numbers a, b, c such that $a + b + c = 3$.

46. Let $f : [0,1] \to \mathbb{R}$ be a function such that

(a) $f(x) \geq 0$ for all $x \in [0,1]$;

(b) $f(1) = 1$;

(c) f satisfies the inequality $f(x_1) + f(x_2) \leq f(x_1 + x_2)$ for all $x_1, x_2 \in [0,1]$ and such that $x_1 + x_2 \leq 1$.

Show that $f(x) \leq 2x$ for all $x \in [0,1]$.

47. Let C be a point on the semicircle k with endpoints A and B. Let the point D on the diameter AB be such that $CD \perp AB$. Consider the circles inscribed in the curvilinear "triangles" ADC and DBC, where ADC is formed by the segments AD and DC and the k-arc CA, and DBC is formed by the segments DB and CD and the k-arc BC. Denote by E the tangent point of the circle inscribed in ADC and the segment AD, and denote by F the tangent point of the circle inscribed in DBC and the segment DB. Prove that $\angle ECF = \frac{\pi}{4}$.

48. A herd consists of 101 cows. Any 100 of them can be split into two groups of 50 cows each such that the weights of the two groups are equal. Prove that all the cows have the same weight.

14.9 Paper 9

49. The polynomial $p(x) = a_n x^n + a_{n-1}x^{n-1} + \cdots + a_1 x + a_0$, has integer coefficients and satisfies $p(-1) = 0$, $p(\sqrt{2}) \in \mathbb{Z}$. Prove that there exists an integer k, $1 \leq k \leq n$, such that the number $p(k) + a_k$ is even.

50. The four points A, B, C, D are chosen inside a square of side 6 in such a way that any two of them are at a distance of at least 5. Prove that A, B, C, D form a convex quadrilateral and that its area is greater than 21.

51. The sequence $\{a_n\}_{n\in\mathbb{N}}$ is formed by taking

$$a_1 = \frac{1}{2}, \qquad a_{n+1} = \frac{a_n^2}{a_n^2 - a_n + 1}.$$

Prove that

$$\sum_{k=1}^{n} a_k < 1 \qquad \forall n \in \mathbb{N}.$$

52. The diagonals AC and BD of the convex quadrilateral $ABCD$ are perpendicular and intersect at the point O. The circles S_1, S_2, S_3, S_4 with centres O_1, O_2, O_3, O_4 are inscribed in the triangles AOB, BOC, COD, DOA, respectively. Prove that a) the sum of the diameters of S_1, S_2, S_3, S_4 is less than or equal to

$$(2 - \sqrt{2})(AC + BD);$$

b) $O_1O_2 + O_2O_3 + O_3O_4 + O_4O_1 < 2(\sqrt{2} - 1)(AC + BD)$.

53. Consider all words consisting of n letters from the alphabet $\{a, b, c, d\}$. We shall call a word complicated if and only if it has two consecutive identical groups of letters (i.e. *caab* or *cababdc* are complicated words, whereas *abcab* is not a complicated word). A word which is not complicated will be called simple.

Prove that the number of simple words consisting of n letters from the above alphabet is greater than 2^n.

54. The real numbers a, b, c are positive and such that

$$a^2 + b^2 + c^2 + abc = 4.$$

Prove that

$$a + b + c \leq 3.$$

14.10 Paper 10

55. For each positive t, find the number of positive solutions to the system of equations

$$\begin{cases} a(1-b^2) = t \\ b(1-c^2) = t \\ c(1-d^2) = t \\ d(1-a^2) = t \end{cases}$$

56. A young man walks into a "7-eleven" store and asks for four items. The shop assistant tells him that his bill is € 7.11, since the product of the four prices is exactly € 7.11. The young man explains indignantly that one is supposed to add together the four prices, not to multiply them. "Oh dear!", exclaims the shop assistant and sums the four numbers. But, can you imagine, the right sum turns out to be € 7.11 too! How much did each item cost?

57. A polygon inscribed in a circle is triangulated by drawing non-intersecting diagonals. Prove that the sum of the radii of the circles inscribed in all resulting triangles does not depend on the choice of triangulating diagonals.

58. The set F consists of all functions $f : \mathbb{R}^+ \to \mathbb{R}^+$, such that

$$f(3x) \geq f(f(2x)) + x \qquad \forall x \in \mathbb{R}^+ \quad (\text{i.e. } x > 0).$$

Find the largest real constant α such that

$$f(x) \geq \alpha x \qquad \forall f \in F, \quad \forall x \in \mathbb{R}^+.$$

59. Prove the inequality

$$(ad + bc)^5 + (ac + bd)^5 \leq (a + b)^5(c^5 + d^5)$$

for all positive numbers a, b, c, d. When does equality occur?

60. Every point with integer coordinates in the plane is the centre of a disc with radius $\frac{1}{1000}$.

a) Prove that there exists an equilateral triangle whose vertices lie in different discs.

b) Prove that every equilateral triangle with vertices in different discs has side-length greater than 96.

14.11 Paper 11

61. How many points with integer coordinates can be at the distance $\sqrt{p}$ from the point $(2014, 2015)$ if p is a prime number?

62. Given that $x_0 = 5$ and that

$$x_{n+1} = x_n + \frac{1}{x_n},$$

prove the estimate

$$45 < x_{1000} < 45.1\,.$$

63. Let $p(x) = x^n + a_1x^{n-1} + \cdots + a_{n-1}x + 1$ be a polynomial with non-negative real coefficients and n real roots. Prove that

$$p(x) \geq (x+1)^n$$

for all $x \geq 0$.

64. Let D be the midpoint of the hypotenuse AB of the right triangle ABC. Denote by O_1 and O_2 the circumcentres of the triangles ADC and DBC, respectively. Prove that AB is tangent to the circle with diameter O_1O_2.

65. Let n be a natural number and let S be a subset of $\{1, 2, \ldots, n\}$ such that no pair of elements of S are relatively prime and no element of S is divisible by another element of S. Find the maximal number of elements in S (as a function of n).

66. Suppose that the inscribed circle of the triangle ABC is tangent to the sides AB, BC, CA at the points D, E, F respectively. Prove the inequality

$$\frac{BC}{DE} + \frac{CA}{EF} + \frac{AB}{FD} \geq 6.$$

14.12 Paper 12

67. For the positive real numbers $a_1, a_2, \ldots, a_n$, $n \geq 2$, let $s = a_1+a_2+\cdots+a_n$. Prove the inequality

$$\frac{a_1}{s-a_1}+\frac{a_2}{s-a_2}+\cdots+\frac{a_n}{s-a_n} \geq \frac{n}{n-1}.$$

68. One hundred ants are placed randomly on a pole of length 1 m. One additional ant (Alice) is placed in the middle of the pole (and is alone there). The ants start to crawl immediately with the speed 1 m/min, each ant choosing its direction arbitrarily. When two ants meet, both turn and start crawling in the opposite direction with the same speed as before. When an ant reaches an end of the pole, it also turns around and starts back with its speed unchanged. What is the probability that after one minute, Alice will be in the middle of the pole again?

69. Given a triangle, find a line through the centre of its inscribed circle such that the product of the areas of the two parts into which it divides the triangle is minimal.

70. Let m be a positive integer. Prove that if $p = 2^{m+1}+1$ divides $3^{2^m}+1$ then p is a prime. Is the converse true?

71. Does there exist a function $f : \mathbb{R} \to \mathbb{R}$ such that $f(f(x)) = x^2 - 2$ for all real x?

72. A new chess piece called "prince" is allowed to move horizontally and vertically, one step at the time. Starting on an arbitrary cell of the (8×8) chessboard, the prince covers all cells and returns to the initial one after 64 moves. Show that the number of horizontal moves is not equal to the number of vertical moves.

14.13 Paper 13

73. Each of the players in a tennis tournament played one match against each of the others. If every player won at least one match, show that there is a group A, B, C of three players for which A beat B, B beat C and C beat A.

74. Let $a, b \geq 0$. Prove that

$$\sqrt{2}\left(\sqrt{a(a+b)^3} + b\sqrt{a^2+b^2}\right) \leq 3(a^2+b^2),$$

with equality if and only if $a = b$.

75. A and B are distinct points on a circle T. C is a point distinct from B such that $AB = AC$, and such that BC is tangent to T at B. Suppose that the bisector of $\angle ABC$ meets AC at a point D inside T. Show that $\angle ABC > 72°$.

76. Suppose p, q are distinct primes and S is a subset of $\{1, 2, \ldots, p-1\}$. Let $N(S)$ denote the number of solutions of the equation

$$\sum_{i=1}^{q} x_i \equiv 0 \pmod{p},$$

where $x_i \in S$, $i = 1, 2, \ldots, q$. Prove that $N(S)$ is a multiple of q.

77. For a natural number n, consider all possible "words" of length n composed of the two letters A and B. Divide these words into two groups, E_n and O_n, according to whether the given word contains an even or an odd number of "syllables" BA (the number 0 is considered to be even). For instance, the words $BABBBBA$ and $AAAAAAB$ both belong to the group E_7, and the words $AABBABB$ and $BABAABA$ both belong to O_7. Determine all n such that the groups E_n and O_n have the same number of elements.

78. The points A, B, C, D, E, F on a circle of radius R are such that $AB = CD = EF = R$. Show that the middle points of BC, DE, FA are vertices of an equilateral triangle.

14.14 Paper 14

79. Find all functions $f : \mathbb{N} \cup \{0\} \to \mathbb{R}$ such that

$$f(n+m) + f(n-m) = f(kn), \qquad \forall n, m \in \mathbb{N} \cup \{0\}, \quad m \le n,$$

where k is a fixed nonnegative integer.

80. For positive real numbers a, b, c, such that $abc \le 1$, prove that

$$\frac{a}{b} + \frac{b}{c} + \frac{c}{a} \ge a + b + c.$$

81. Let $\triangle ABC$ be an isosceles triangle with $AB = AC$ and $\angle A = 20°$. The point D on AC is such that $AD = BC$. Determine the angle $\angle BDC$.

82. Let $ABCD$ be a cyclic quadrilateral whose interior angle at the vertex B is $60°$.

a) Show that if $BC = CD$, then $CD + DA = AB$.

b) Does the converse implication also hold?

83. Let a_1, a_2, a_3, a_4 be the lengths of the sides of a quadrilateral and let s be its semi-perimeter. Prove that

$$\sum_{i=1}^{4} \frac{1}{s + a_i} \le \frac{2}{9} \sum_{1 \le i < j \le 4} \frac{1}{\sqrt{(s - a_i)(s - a_j)}}.$$

When does equality hold?

84. Prove that there does not exist an integer $n > 1$ such that n divides $3^n - 2^n$.

14.15 Paper 15

85. The numbers $a_1, a_2, \dots, a_n$ are positive and such that $a_1 + a_2 + \cdots + a_n = 1$. Show that

$$\frac{a_1}{1 + a_1\sqrt{2}} + \frac{a_2}{1 + a_2\sqrt{2}} + \cdots + \frac{a_n}{1 + a_n\sqrt{2}} \le \frac{n}{n + \sqrt{2}}.$$

When does equality occur?

86. In the real domain, solve the system of equations

$$x^2 = \frac{1}{y} + \frac{1}{z}, \quad y^2 = \frac{1}{z} + \frac{1}{x}, \quad z^2 = \frac{1}{x} + \frac{1}{y}.$$

87. Find all functions $f : \mathbb{R} \to \mathbb{R}$ such that

$$f(xf(x) + f(y)) = (f(x))^2 + y,$$

for all real numbers x and y.

88. In a triangle ABC with orthocenter H it is known that $AH^2 + BH^2 + CH^2 = 7$ and $AH \cdot BH \cdot CH = 3$. Find:

a) the circumradius of $\triangle ABC$;

b) the sides of the triangle ABC with maximal possible area.

89. Find all real numbers α with the property that all members of the sequence

$$\cos\alpha, \ \cos 2\alpha, \ \cos 2^2\alpha, \dots, \ \cos 2^n\alpha, \dots$$

are negative.

90. Let $p_1 < p_2 < \cdots < p_n < \cdots$, be the sequence of all primes. Show that for all $n \in \mathbb{N}$ the interval $(p_1 + p_2 + \cdots + p_n, \ p_1 + p_2 + \cdots + p_n + p_{n+1})$ contains the square of a natural number.

14.16 Paper 16

91. Denote by $\mathbb{R}^+$ the set of all positive real numbers. Find all functions $f : \mathbb{R}^+ \to \mathbb{R}^+$ such that the equality

$$x^2(f(x) + f(y)) = (x + y)f(f(x)y)$$

holds for any pair of positive numbers x, y.

92. For positive real numbers a, b, c, such that $a + b + c = 1$, show that

$$\frac{1+a}{1-a} + \frac{1+b}{1-b} + \frac{1+c}{1-c} \le 2\left(\frac{b}{a} + \frac{c}{b} + \frac{a}{c}\right).$$

(You do not have to state when equality holds.)

93. Prove that for all positive integers n there exists an arithmetic progression consisting of n composite numbers and such that any two of its terms are relatively prime.

94. The points A, B, C are given in the plane. Let D, E, F, G, H and I be points in the same plane such that the triangles ABD, BAE, CAF, DFG, ECH, GHI are positively oriented equilateral triangles. Prove that the point E is the midpoint of the segment AI.

95. Given a triangle, denote by R and r the radii of the circumscribed and of the inscribed circles respectively, and by l_a, l_b, l_c and h_a, h_b, h_c the lengths of the angle bisectors and of the altitudes from A, B, C respectively. Prove that

$$\frac{l_a^2}{h_a} + \frac{l_b^2}{h_b} + \frac{l_c^2}{h_c} \le 4R + r.$$

96. Eight problems were proposed to each of 30 contestants of a Mathematical Olympiad. To take into account the difficulty level of the problems the Jury, when the contest was over, assigned each problem its "worth" according to the following rule: a problem is worth n points if it is not solved by exactly n contestants. (For example, a problem solved by all contestants is worth 0 points.)

a) Is it possible that the highest-scoring contestant at the same time has solved fewer problems than any other contestant?

b) Is it possible that the lowest-scoring contestant at the same time has solved more problems than any other contestant?

14.17 Paper 17

97. Let a, b, c be positive real numbers such that $a^2 + b^2 + c^2 = 3$. Prove that

$$\frac{1}{1+2ab} + \frac{1}{1+2bc} + \frac{1}{1+2ca} \geq 1.$$

98. The alphabet of the language BAU consists of the letters B, A and U. All BAU words of length n can be constructed starting from any BAU word of length n, using iteratively the following rules:

(1) invert the order of the letters in the word;

(2) replace two consecutive letters: BA → UU, AU → BB, UB → AA, BB → AU, UU → BA or AA → UB.

Given that BBAUABAUUABAUUUABAUUUUABB is a BAU word, does BAU have

a) the word BUABUABUABUABAUBAUBAUBAUB?

b) the word ABUABUABUABUAUBAUBAUBAUBA?

99. Let L be an arbitrary interior point of the shorter of the arcs CD of the circumcircle of the square $ABCD$. Denote by K the point of intersection of the lines AL and CD, by M the intersection of the lines AD and CL, and by N the intersection of the lines MK and BC. Show that the points B, L, M and N are concyclic.

100. An ordered set a consisting of k digits is said to be stable if the decimal representation of the product of any two positive integers such that their decimal representations end with a, also ends with a. (For example, the sets 0 and 25 are stable.)

Prove that for any positive integer k there exist exactly four stable sets consisting of k digits.

101. A point O is situated in the plane of the triangle ABC. A circle $\mathcal{C}$ passing through O meets the second time OA, OB, OC in P, Q, R, respectively, and $\mathcal{C}$ intersects the second time the circles $(B, O, C), (A, O, C), (A, O, B)$ in K, L, M respectively. Prove that PK, QL, RM are concurrent.

102. Let $f = f(x)$ and $g = g(x)$ be polynomials of degree 2 and 3 respectively. Show that x cannot be represented as a polynomial of f and g, i.e. show that there exists no polynomial $P = P(s, t)$ such that $x = P(f(x), g(x))$. What if f and g are polynomials of degree 2 and 4 respectively?

14.18 Paper 18

103. Find the smallest real number p such that the inequality

$$\sqrt{1^2+1}+\sqrt{2^2+1}+\sqrt{3^2+1}+\cdots+\sqrt{n^2+1}\le\frac{n(n+p)}{2},$$

holds for any natural number n.

104. In a country, there are 2004 cities connected by roads so that there are ways to go from any city to any other city. It is also known that after closing any one of the roads, it is still possible to get to any city from any other city. The Minister of Transportation and the Minister of Internal Affairs declare in turn one-way traffic on the roads (one road at the time). The minister, whose move makes it impossible to reach any city from any other city, resigns immediately. The Minister of Transportation moves first. Can either minister attain the resignation of the other, regardless of how the other plays?

105. The diagonals AD, BE, CF of a convex hexagon $ABCDEF$ meet at the point O. Determine the smallest possible value of the area of this hexagon if the areas of the triangles AOB, COD, EOF equal 4, 6, 9, respectively.

106. Find all non-negative integers which can be written as

$$\frac{a^2+ab+b^2}{ab-1},$$

where a, b are non-negative integers and $ab \neq 1$.

107. Let P be a point outside the circle k. The lines l_1, l_2 and l_3 intersect the circle in A, B; C, D; and E, F respectively so that the segments PA, PC, PE lie outside the circle. Denote by M the intersection of the lines DE and PB. Prove that the length of the tangent to k through M is equal to PM if and only if CF is parallel to PB.

108. Let $f_0(x) \equiv 1, f_1(x) = x$ and $f_{n+1}(x) = \frac{f_n^2-x^n}{f_{n-1}}$, $n \geq 1$. Show that for every natural number n, the function f_n is a polynomial with integer coefficients.

14.19 Paper 19

109. Let $x_1, x_2, \ldots, x_n$, $n > 3$, be positive numbers such that $x_1 x_2 \cdots x_n = 1$. Prove that

$$\frac{1}{1 + x_1 + x_1 x_2} + \frac{1}{1 + x_2 + x_2 x_3} + \cdots + \frac{1}{1 + x_n + x_n x_1} > 1.$$

110. Find all injective functions $f : \mathbb{N} \to \mathbb{N}$, such that for each $n \in \mathbb{N}$

$$f(f(n)) \leq \frac{n + f(n)}{2}.$$

111. Consider a triangle ABC and let O be a point in its interior. The straight lines OA, OB, OC meet the sides of the triangle in A_1, B_1, C_1, respectively. Let R_1, R_2, R_3 be the radii of the circumcircles of $\triangle OBC$, $\triangle OCA$, $\triangle OAB$ respectively and let R be the circumradius of $\triangle ABC$. Prove that

$$\frac{OA_1}{AA_1} R_1 + \frac{OB_1}{BB_1} R_2 + \frac{OC_1}{CC_1} R_3 \geq R.$$

112. Let $ABCD$ be a convex quadrilateral with perpendicular diagonals AC, BD. Denote the point of intersection of AC and BD by O. Let O_1, O_2, O_3, O_4 be the orthogonal projections of O on the sides AB, BC, CD, DA, respectively. Prove that the lines O_1O_2, O_3O_4 and AC are concurrent, as are the lines O_2O_3, O_1O_4 and BD.

113. Suppose that G is a graph on n vertices without triangles. Prove that G has at most $n^2/4$ edges.

114. Let P_i be some numbers and define by induction

$$P_{i,0} = P_i, \quad P_{i,n+1} = (\cos^2 t) P_{i,n} + (\sin^2 t) P_{i+1,n}.$$

Determine P_i for $i = 0, \ldots, 2005$, if $P_{0,2005} = \cos 2t$.

14.20 Paper 20

115. Prove the inequality

$$\sqrt[3]{\frac{a}{b}}+\sqrt[3]{\frac{b}{a}} \le \sqrt[3]{2(a+b)\left(\frac{1}{a}+\frac{1}{b}\right)},$$

for all positive numbers a and b. When does equality hold?

116. The circles C_1 and C_2 intersect at A and B. The tangent line to C_2 at A meets C_1 at the point C and the tangent line to C_1 at A meets C_2 at the point D. A ray from A, interior to the angle $\angle CAD$, intersects C_1 at M, C_2 at N and the circumcircle of the triangle $\triangle ACD$ at P. Prove that $AM = NP$.

117. Given a triangle, find the line through the centre of its inscribed circle that divides the triangle into two parts such that the product of their areas is maximal.

118. Let A be a set of k natural numbers. Denote by $A + A$ the set

$$A + A = \{n : n = a + b,\ a, b \in A\}.$$

Show that $A + A$ has exactly $2k - 1$ (distinct) elements if and only if A contains an arithmetic progression of length k.

119. The sequence $\{a_n\}_{n\ge 0}$ is constructed in the following way

$$a_0 = a_1 = a_2 = 1, \qquad a_{n+3} = a_{n+2}a_{n+1} + a_n.$$

Show that for any $k \in \mathbb{N}$ there exists $n_k \in \mathbb{N}$ such that a_{n_k} is divisible by k.

120. The total area of a set of finitely many squares is equal to 4. Prove that the squares can cover a square with area 1. (The squares may overlap.) Is this still true if we replace 4 by 3?

PART IV

Solutions

CHAPTER 15

Solutions to all papers

15.1 Solutions to Paper 1

1. *Solution 1.* Denote, for simplicity, by 1, 2, 3, 4 the spoken languages and let n be the number of the participants. We distinguish the following three cases.

A. There is a participant who speaks only one language. It is then clear that all participants have to speak it.

B. Any participant speaks at least two languages and there is a language that is common to those speaking exactly two languages. WLOG this is the first language. Thus this case is described by the following groups of languages (12), (13), (14), (123), (124), (134), (234), (1234). Assume the required condition on each of the languages 1, 2, 3, 4 is contradicted, and denoting by $x_{ij\ldots}$ the number of those who speak simultaneously the languages $i, j, \ldots$, we obtain:

$$x_{12} + x_{13} + x_{14} + x_{123} + x_{124} + x_{134} + x_{1234} < \frac{3}{5}n$$

$$x_{12} + x_{123} + x_{124} + x_{234} + x_{1234} < \frac{3}{5}n$$

$$x_{13} + x_{123} + x_{124} + x_{234} + x_{1234} < \frac{3}{5}n$$

$$x_{14} + x_{124} + x_{134} + x_{234} + x_{1234} < \frac{3}{5}n.$$

Multiplying the first inequality by 2 and summing up, we get:

$$3(x_{12} + x_{13} + x_{14}) + 4(x_{123} + x_{124} + x_{134}) + 3x_{234} + 5x_{1234} < 3n.$$

This is a contradiction. Note that we covered simultaneously the case when everybody speak at least three languages.

C. Any participant is speaking at least two of the given languages but there is no language spoken by all those who speak exactly two. WLOG $x_{12} \neq$

$0 \Rightarrow x_{34} = 0$. There should be something else who speaks two languages. WLOG $x_{23} \neq 0 \Rightarrow x_{14} = 0$. If $x_{24} \neq 0$ then $x_{13} = 0$ and everybody who speaks two languages speaks the second language, which is impossible. Thus $x_{24} = 0$ and we should have

$$x_{12} + x_{13} + x_{123} + x_{124} + x_{134} + x_{1234} < \frac{3}{5}n$$

$$x_{12} + x_{23} + x_{123} + x_{124} + x_{234} + x_{1234} < \frac{3}{5}n$$

$$x_{13} + x_{23} + x_{123} + x_{134} + x_{234} + x_{1234} < \frac{3}{5}n.$$

Summing up, we get $2n + x_{123} + x_{1234} < \frac{9}{5}n$, which is a contradiction.

Solution 2. We use the same notation. Consider a minimal counterexample, thus suppose nobody can forget a language without breaking the conditions. If M_i is the set of languages known by the participant i this mean that no M_i is a proper subset of another M_j. In particularly, nobody speaks four languages and everybody speaks at least two languages.

Again we will consider different cases. Suppose first, that somebody speaks at least 3 languages. WLOG $x_{234} \neq 0$. Then $x_{23} = x_{24} = x_{34} = 0$ and all other speak the first language thus $x_{234} > \frac{2n}{5}$. Put $a_2 = x_{123} + x_{12}$, $a_3 = x_{124} + x_{13}$, $a_4 = x_{134} + x_{14}$.

If $a_i \geq \frac{n}{5}$ then we have at least $\frac{3n}{5}$ persons speaking the language i. But on the other hand if $a_3 < \frac{n}{5}, a_4 < \frac{n}{5}$ then we have $x_{234} + a_2 = n - a_3 - a_4 > \frac{2n}{5}$ who speaks the second language. In any case we get a contradiction.

Now suppose that everybody speaks two languages. WLOG $x_{12} > 0 \Rightarrow x_{34} = 0$. Because not everybody speaks the first language WLOG $x_{23} > 0 \Rightarrow x_{14} = 0$. If $x_{24} \neq 0$ then $x_{13} = 0$ and everybody speaks the second language. Thus $x_{24} = 0$ and nobody speaks the fourth language. Now we have

$$x_{12} + x_{13} < \frac{3n}{5};\ x_{12} + x_{23} < \frac{3n}{5};\ x_{13} + x_{23} < \frac{3n}{5} \quad \Rightarrow \quad 2n < \frac{9n}{5}.$$

The contradiction.

2. Subtracting pairs of equations of the system we get the system

$$(z - x)(x + y + z) = 2$$

$$(y - z)(x + y + z) = -6$$

$$(x - y)(x + y + z) = 4$$

or, equivalently (since $x + y + x \neq 0$ for all solutions),

$$\frac{1}{x+y+z} = \frac{x-y}{4} = \frac{y-z}{-6} = \frac{z-x}{2} = t.$$

By adding up the equations of the given system we find

$$(x-y)^2 + (y-z)^2 + (z-x)^2 = 14$$

The two last equalities give $t = \frac{1}{2}$ or $t = -\frac{1}{2}$. Finally, for $t = \frac{1}{2}$, we find $(x, y, z) = (1, -1, 2)$ and for $t = -\frac{1}{2}$ we find $(x, y, z) = (-1, 1, -2)$.

3. From the similarity of the triangles AMN and ABC, we obtain

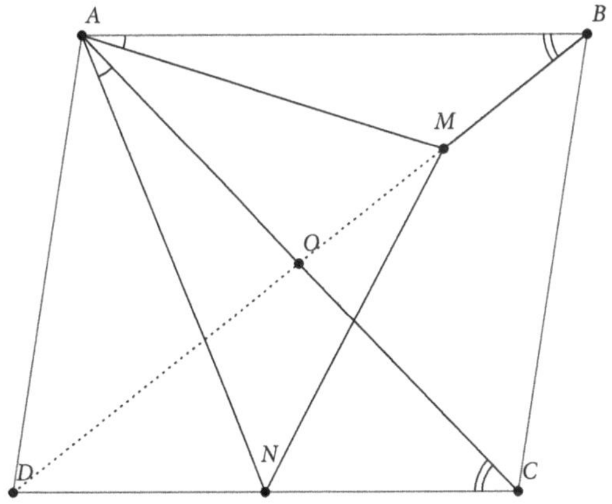

Figure 15.1 Problem 3

$$\frac{AM}{AB} = \frac{AN}{AC} \quad \text{and} \quad \angle MAN = \angle BAC \text{ or } \angle BAM = \angle CAN.$$

Those relations imply the similarity of the triangles BAM and CAN. Hence, we obtain the proportions:

$$\frac{AM}{AN} = \frac{AB}{AC} = \frac{BM}{CN} \tag{15.1}$$

and $\angle ABM = \angle ACN$. The last equality implies that $ABCD$ is a rectangle, because it is an inscribed parallelogram.

To conclude the proof, notice that $BM = \frac{1}{4}BD = \frac{1}{4}AC$ and $CN = \frac{1}{2}AB$. Hence the last equality in 15.1 becomes $\frac{AB}{AC} = \frac{AC}{2AB}$, that is $2AB^2 = AC^2 = AB^2 + BC^2$, which proves that $ABCD$ is a square.

In the **alternative solution** we use complex numbers. We can choose $A = 0$, $D = 1$. Let $B = z$, which means that $C = z + 1$. We then have

$$N = \frac{((z+1)+1)}{2}, \quad O = \frac{z+1}{2}, \quad M = \frac{z + \frac{z+1}{2}}{2} = \frac{3z+1}{4}.$$

The similarity and the same orientation of $\triangle ABC$ and $\triangle AMN$ gives us (see page 90) the equation

$$\frac{C-A}{B-A} = \frac{N-A}{M-A} \Leftrightarrow \frac{z+1}{z} = \frac{\frac{z+2}{2}}{\frac{3z+1}{4}}$$

which is equivalent to the equation $z^2 + 1 = 0$ with solutions $z = i$ and $z = -i$. In both cases $ABCD$ is a square.

Another alternative solution would be to show that the circumscribed circles of $\triangle ABC$ and $\triangle AMN$ intersect in (A and) D, which means that $ABCD$ is a rectangle.

4. Observe first that $\sqrt{x} < \sqrt[3]{x}$ for $x \in [0,1]$. Thus. we have $\sqrt{abc} < \sqrt[3]{abc}$ and

$$\sqrt{(1-a)(1-b)(1-c)} < \sqrt[3]{(1-a)(1-b)(1-c)}.$$

By the AM–GM inequality, we get

$$\sqrt{abc} < \sqrt[3]{abc} \leqslant \frac{a+b+c}{3},$$

and

$$\sqrt{(1-a)(1-b)(1-c)} \leqslant \sqrt[3]{(1-a)(1-b)(1-c)} \leqslant \frac{(1-a)+(1-b)+(1-c)}{3}.$$

Summing up, we obtain

$$\sqrt{abc} + \sqrt{(1-a)(1-b)(1-c)} < \frac{a+1-a+b+1-b+c+1-c}{3} = 1,$$

as desired.

Solution 2. We can substitute $a = \sin^2 x, b = \sin^2 y, c = \sin^2 z$, with $0 < x, y, z < \frac{\pi}{2}$, which gives

$$\sin x \sin y \sin z + \cos x \cos y \cos z < \sin x \sin y + \cos x \cos y \leq 1.$$

5. Denote by A, B, C, D, E the five given points. If the pentagon $ABCDE$ is concave, one of the points (say D) is situated inside the triangle consisting of other points (say ABC). Then

$$\text{area}(ABC) = \text{area}(ABD) + \text{area}(DBC) + \text{area}(DAC) > 6 > 3.$$

Consider now the case when $ABCD$ is a convex pentagon (see figure 15.2). Let M and N be the intersection points of BE with AC and AD respectively.

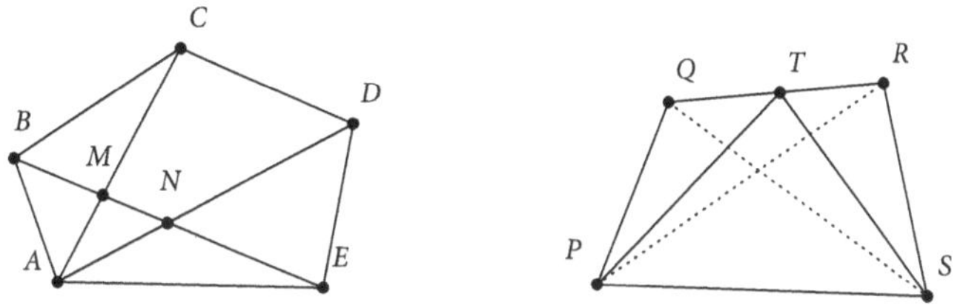

Figure 15.2 Problem 5

The following result will be useful.

Lemma 15.1 *Let $PQRS$ be a quadrilateral and T a point on the side QR. Then*

$$\text{area}(TPS) \geqslant \min(\text{area}(QPS), \text{area}(QRS)).$$

Proof. Following from the simple observation that the distance from T to PS is bounded up and below by the distances from Q and R to PS and those distances are the altitudes in the triangles.

In our case, suppose first that $BM \geqslant \frac{1}{3}BE$, which yields $BM \geqslant \frac{1}{2}ME$. Then

$$\begin{aligned}\text{area}(BDE) &= \text{area}(BDM) + \text{area}(MDE)\\ &\geqslant \frac{1}{2}\,\text{area}(MDE) + \text{area}(MDE) = \frac{3}{2}\,\text{area}(MDE)\\ &\geqslant \frac{3}{2}\min(\text{area}(CDE), \text{area}(ADE)) > \frac{3}{2}\cdot 2 = 3.\end{aligned}$$

The case when $NE \geqslant \frac{1}{3}BE$ is similar. It remains to consider the case when

$MN \geqslant \frac{1}{3}BE$. We then have:

$$\begin{aligned}
\text{area}(AMN) &\geqslant \frac{1}{3}\,\text{area}(ABE) > \frac{2}{3},\\
\text{area}(MND) &\geqslant \frac{1}{3}\,\text{area}(BED) > \frac{2}{3}\\
\text{area}(MCD) &\geqslant \min(\text{area}(BCD), \text{area}(ECD)) > 2.
\end{aligned}$$

Summing up, we conclude area$(ACD) > 2 + \frac{2}{3} + \frac{2}{3} > 3$ and the proof is complete.

6. Suppose that there do exist such polynomials. If a polynomial has no real zeroes, it is either positive for all real x, or negative for all real x. Consider the complete graph with the given polynomials as vertices. Colour the edge P_iP_j in white if $P_i + P_j > 0$ and in black if $P_i + P_j < 0$. We cannot have a triangle of the same colour (otherwise the sum of the three corresponding polynomials would always have constant sign and thus would not have a real zero). At least three of the edges have to be of the same colour. If they have a common vertex, then the other three vertices would form a triangle of constant (the opposite) colour — a contradiction. So, without loss of generality, we can consider the case when P_1P_2, P_2P_3, P_3P_4 have the same colour. Then P_1P_3, P_2P_4 must be of the opposite colour and the sum $(P_1 + P_3) + (P_2 + P_4)$ must have the opposite sign to the sum $(P_1 + P_2) + (P_3 + P_4)$, which is impossible since both sums are the same polynomial. The contradiction means that the answer is NO.

15.2 Solutions to Paper 2

7. Colour the 3×3 cube in this manner: colour each corner sub-cube black. Every other sub-cube is coloured black or white so that each sub-cube is a different colour than all the other sub-cubes that it shares a face with.

   Now, notice that the corner sub-cube is black, and the centre sub-cube is white. But as the mouse goes through from sub-cube to sub-cube, the destination sub-cube is a different colour from his original cube. Since there are 13 white cubes and 14 black cubes, the mouse's path must go $BWBW\ldots BWB$. The last cube must be black. Thus, it cannot finish in the centre sub-cube.

8. Let the fixed quantity referred to be n. Now consider a fixed person a. Let B be the set of people who have shaken hands with a, and C the set of those who have not. Then $|B| = 3k + 6$ and $|C| = 9k - 7$.

For any $b \in B$, people who have shaken hands with a and b must be in B. Hence, b has shaken hands with n people in B, and thus with $3k + 5 - n$ people in C.

For any $c \in C$, people who have shaken hands with a and c must be in B. Hence, c has shaken hands with n people in B.

The total number of handshakes between B and C is given by $(3k+6)(3k+5-n) = (9k-7)n$, which simplifies to $9k^2 - 12kn + 33k + n + 30 = 0$. It follows that $n = 3m$ for some positive integer m, and $4m = k + 3 + \frac{9k+43}{12k-1}$. Since 3 is not a divisor of 44, we have $\frac{9k+43}{12k-1} \neq 1$ for any choice of k. Furthermore, if $k > 3$ then $2(12k-1) > 9k + 43$ and thus, $\frac{9k+43}{12k-1} < 2$. Therefore, we need only consider $k = 1, 2, 3$. Only $k = 3$ yields an integer value for $\frac{9k+43}{12k-1}$, and there are 36 people at the party.

It remains the most difficult part — to show that $k = 3$, $n = 6$ really provides a solution to the problem. Consider 36 persons P_{ij} sitting in 6 rows and 6 columns. Suppose that P_{ij} and P_{kl} exchanged greetings iff $i = k$ or $j = l$ or $i - j = k - l \pmod 6$ thus if they are in the same row, column or a line $x - y = c \pmod 6$. Now it easy to check directly that P_{00} and P_{ij} have 6 common persons and it is sufficient to note that the construction is invariant under addition modulo 6, thus we can replace P_{00} by arbitrary P_{st}.

An **alternative approach** to find k is to use divisibility in the following way. We have $(3k+6)(3k+5) = n(12k-1)$. On the other hand

$$
\begin{aligned}
(3k+5, 12k-1) &= (12k+20, 12k-1) = (21, 12k-1) = (7, 12k-1),\\
(3k+6, 12k-1) &= (12k+24, 12k-1) = (25, 12k-1),
\end{aligned}
$$

since $(a-b, b) = (a, b)$ and $(ka, b) = (a, b)$, provided $(k, b) = 1$. We get that $12k - 1 \equiv -1 \pmod{12}$ and $12k-1$ is the product of two numbers, one of them a divisor of 7, the other a divisor of 25. The only possibility is $12k - 1 = 5 \cdot 7$. The problem was on the IMO short list in 1995.

9. Since $\sin A + \sin B + \sin C > 0$ the desired is equivalent to

$$
(\sin A + \sin B + \sin C)^2 \leq \frac{15}{4} + \cos(A-B) + \cos(B-C) + \cos(C-A).
$$

We have

$$\Bigl(\sum_{\text{cyclic}} \sin A\Bigr)^2 \le \frac{15}{4} + \sum_{\text{cyclic}} \cos(A-B) \quad \Leftrightarrow$$

$$\sum_{\text{cyclic}} \sin^2 A + 2\sum_{\text{cyclic}} \sin A \sin B \le \frac{15}{4} + \sum_{\text{cyclic}} (\cos A \cos B + \sin A \sin B) \quad \Leftrightarrow$$

$$3 - \sum_{\text{cyclic}} \cos^2 A \le \frac{15}{4} + \sum_{\text{cyclic}} (\cos A \cos B - \sin A \sin B) \quad \Leftrightarrow$$

$$\sum_{\text{cyclic}} \cos(A+B) + \sum_{\text{cyclic}} \cos^2 A + \frac{3}{4} \ge 0 \quad \Leftrightarrow$$

$$\sum_{\text{cyclic}} \cos(\pi - A) + \sum_{\text{cyclic}} \cos^2 A + \frac{3}{4} \ge 0 \quad \Leftrightarrow$$

$$-\sum_{\text{cyclic}} \cos A + \sum_{\text{cyclic}} \cos^2 A + \frac{3}{4} \ge 0 \quad \Leftrightarrow \quad \sum_{\text{cyclic}} \Bigl(\cos A - \frac{1}{2}\Bigr)^2 \ge 0.$$

Clearly, the last inequality is true, so that the initial inequality is also true. Equality holds when $\cos A = \cos B = \cos C = \frac{1}{2}$; that is, when $A = B = C = 60°$.

10. (i) Let C be a right angle. By Theorem 4.3

$$r = p - c = \frac{1}{2}(a + b - c).$$

This will be an integer if $(a+b-c)$ is even. This follows, of course, from the description of the Pythagorean triples in Section 2.3, but can be shown directly:

$$\begin{aligned}(a+b-c)^2 &= a^2 + b^2 + c^2 + 2ab - 2bc - 2ca \\ &= 2c^2 + 2ab - 2bc - 2ca = 2(c^2 + ab - bc - ca)\end{aligned}$$

(which is an even number). Hence $a + b - c$ is even and r is an integer.

(ii) Let $a = N^2 + 2N, b = 2N + 2, c = N^2 + 2N + 2$. Then $a^2 + b^2 = c^2$ and $r = \frac{1}{2}(a + b - c) = N$.

As an alternative you can choose $a = 3N$, $b = 4N$, $c = 5N$, but the previous solution is more sophisticated.

11. Let d denote the common difference of the arithmetic progression. Then for all $k = 1, 2, \ldots, n$, we have $a_k = a_1 + (k-1)d$ and $a_{n-k+1} = a_1 + (n-k)d$.

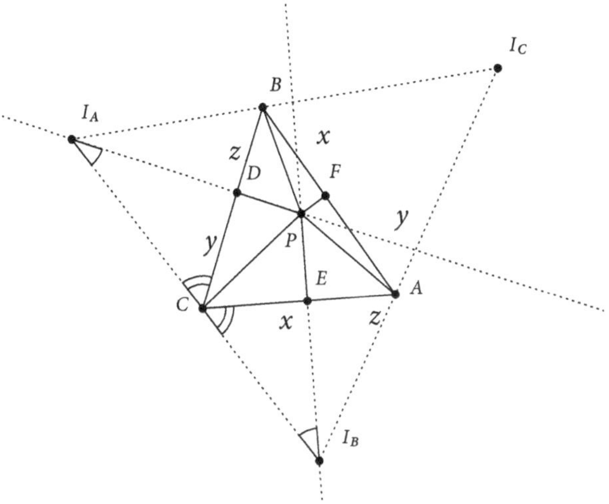

Figure 15.3 Problem 12

By the AM–GM inequality we have

$$\sqrt{a_k a_{n-k+1}} \le \frac{a_k + a_{n-k+1}}{2} = \frac{2a_1 + (n-1)d}{2} = \frac{a_1 + a_n}{2},$$

whence it follows that

$$\sum_{k=1}^{n} \frac{1}{a_k a_{n-k+1}} \ge \sum_{k=1}^{n} \frac{4}{(a_1 + a_n)^2} = \frac{4n}{(a_1 + a_n)^2}.$$

We get equality only when $d = 0$.

12. Since the given condition implies

$$\begin{aligned} 0 &= (BP^2 + PE^2) - (CP^2 + PF^2) \\ &= (BP^2 - PF^2) - (CP^2 - PE^2) = BF^2 - CE^2, \end{aligned}$$

we may put $x = BF = CE$. Similarly we may put $y = CD = AF$ and $z = AE = BD$.

If one of the three points D, E, F were to lie not on a side of the triangle ABC, then this would contradict the triangle inequality. Indeed, if, for example, B, C, D lie in this order, then E is between A and C and we have $AB + BC = (z - y) + (x + y) = x + z = AC$, a contradiction. Thus all three points lie on the sides of the triangle ABC.

Putting $a = BC$, $b = CA$, $c = AB$ and $p = \frac{a+b+c}{2}$, we have $x = p - a$, $y = p - b$, $z = p - c$. Since $BD = p - c$ and $CD = p - b$, we see that D is the point at which the excircle of the triangle ABC opposite to A meets BC (see page 144). Similarly E and F are the points at which the excircles opposite to B and C meet CA and AB, respectively. Since both PD and I_AD are perpendicular to BC, the three points P, D, I_A are collinear. Analogously, P, E, I_B are collinear and P, F, I_C are collinear.

The three points I_A, C, I_B are collinear because they belong to the bisector of the external angle to C. By the same reason $\angle PI_AC = \angle PI_BC = \frac{1}{2}\angle C$ and the triangle PI_AI_B is isosceles. Likewise we have $PI_A = PI_C$ and hence $PI_A = PI_B = PI_C$. Thus P is the circumcentre of the triangle $I_AI_BI_C$.

Remark The conclusion is true even if the point P lies outside the triangle ABC.

Remark In fact, the common value of $AP^2 + PD^2$, $BP^2 + PE^2$, $CP^2 + PF^2$ is equal to $8R^2 - p^2$, where R is the circumradius of the triangle ABC. We can prove this as follows:

Observe that the circumradius of the triangle $I_AI_BI_C$ is equal to $2R$ since its orthic triangle is ABC. It follows that $PD = PI_A - DI_A = 2R - r_a$, where r_a is the radius of the excircle of the triangle ABC opposite to A. In a similar manner, we have $PE = 2R - r_b$. Now we have

$$AP^2 + PD^2 = AE^2 + PE^2 + PD^2 = (p-c)^2 + (2R - r_b)^2 + (2R - r_a)^2.$$

Since

$$\begin{aligned}
(2R - r_a)^2 &= 4R^2 - 4Rr_a + r_a^2 \\
&= 4R^2 - 4\cdot\frac{abc}{4\,\text{area}\,(\triangle ABC)}\cdot\frac{\text{area}\,(\triangle ABC)}{p-a} + \left(\frac{\text{area}\,(\triangle ABC)}{p-a}\right)^2 \\
&= 4R^2 + \frac{p(p-b)(p-c) - abc}{p-a} \\
&= 4R^2 + \frac{(p-a)bc + p(p-b)(p-c) - pbc}{p-a} \\
&= 4R^2 + bc + \frac{p^2(p-b-c)}{p-a} = 4R^2 + bc - p^2,
\end{aligned}$$

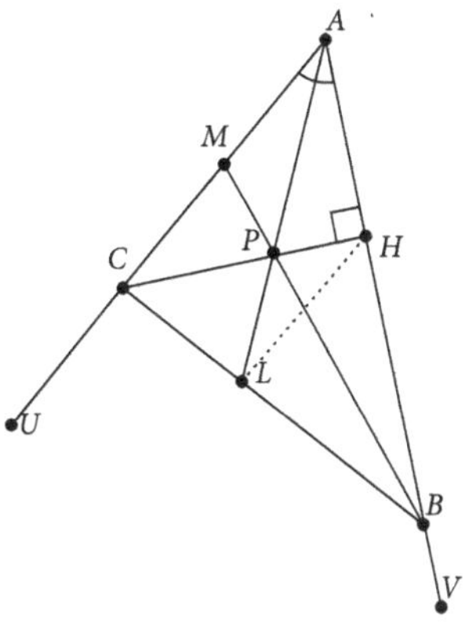

Figure 15.4 Problem 14

and we can obtain $(2R - r_b)^2 = 4R^2 + ca - p^2$ in a similar way, it follows that

$$AP^2 + PD^2 = (p - c)^2 + (4R^2 + ca - p^2) + (4R^2 + bc - p^2) = 8R^2 - p^2.$$

15.3 Solutions to Paper 3

13. There are three different types of vertices: the three original vertices, n vertices inside the triangle and m on its boundary that are not vertices of the original triangle. If N is the total number of triangles, then the total sum of angles in all the triangles is equal to $180°\cdot N$ and can be calculated as $60°\cdot 3 + 360°n + 180°m$, and we get $N = 1 + 2n + m$. Let us count the maximal number of angles, larger than 120°. Two such angles (and no more than two) can only meet at an inner vertex. At each of the vertices on the sides we can have at most one such angle, none of them in the original vertices. This means that we cannot have more than $2n + m$ angles larger than 120°. Since this number is less than $N = 1 + 2n + m$, at least one triangle must have angles not larger than 120°.

14. a) Let $\angle UAV$ be an acute angle, $\angle UAV \neq 60°$. Draw its angle bisector. Let C, $(C \neq A)$ be a point on the side AU and $CH \perp AV$ (where H is on the side AV). Denote by P the intersection of CH with the angle bisector and by M the midpoint of AC. Draw the straight line MP. It intersects AV in B, that is

to the right of point H, because in the right-angled triangle AHC we have:

$$\frac{CP}{PH} = \frac{CA}{AH} > 1,$$

which gives $PH < \frac{1}{2}CH$. In that manner we construct a non-equilateral triangle ABC, that has the required property.

b) From what is given we have the identities: $CM = MA$, $\frac{AH}{HB} \cdot \frac{BL}{LC} \cdot \frac{CM}{MA} = 1$ (Ceva's Theorem). From these relations we obtain:

$$\frac{AH}{HB} = \frac{CL}{LB},$$

which shows that $AC \parallel HL$ and, consequently, $\triangle ABC \sim \triangle HBL$.

15.

$$\begin{aligned} 9 &= (x_1 + x_2 + \cdots + x_n)\left(\frac{1}{x_1} + \frac{1}{x_2} + \cdots + \frac{1}{x_n}\right) \\ &= \underbrace{1 + 1 + \cdots + 1}_{n} + \underbrace{\left(\frac{x_1}{x_2} + \frac{x_2}{x_1}\right)}_{\geq 2} + \cdots + \underbrace{\left(\frac{x_n}{x_{n-1}} + \frac{x_{n-1}}{x_n}\right)}_{\geq 2} \\ &\geq n + 2 \cdot \frac{n(n-1)}{2} = n^2. \end{aligned}$$

Hence $n \in \{1, 2, 3\}$.

Note that the same conclusion follows from the AM–HM inequality.

Let $n = 1$. It is obvious that the system of equations $x_1 = 9$, $\frac{1}{x_1} = 1$ has no solution.

Let $n = 2$. We get $x_1 + x_2 = 9$, $\frac{1}{x_1} + \frac{1}{x_2} \frac{x_1+x_2}{x_1 x_2} = \frac{9}{x_1 x_2} = 1$, which leads to the quadratic equation $x_1^2 - 9x_1 + 9 = 0$. The solutions are $x_{1,2} = \frac{1}{2}(9 \pm 3\sqrt{5})$.

Let $n = 3$. All inequalities should be equalities. Thus $x_1 = x_2 = x_3 (= 3)$.

16. For any $x \in M$, consider the orbit of x: $M_x = \{F^k(x) \mid k \in \mathbb{Z}\}$ (where $F^2 = F \circ F$, etc.).

It is easy to see that for any $x, y \in M$ either $M_x = M_y$, or $M_x \cap M_y = \emptyset$. Thus M is a union of pairwise disjoint orbits and it is sufficient to solve the problem for the case $M = M_x$. In this case the two maps $f_1 : F^k(x) \to F^{1-k}(x)$ and $f_2 : F^k(x) \to F^{-k}(x)$ are (well-defined) reflections and

$$f_1(f_2(F^k(x))) = f_1(F^{-k}(x)) = F^{1+k}(x) = F(F^k(x)).$$

Note that this construction works both when M_x is finite and when it is infinite.

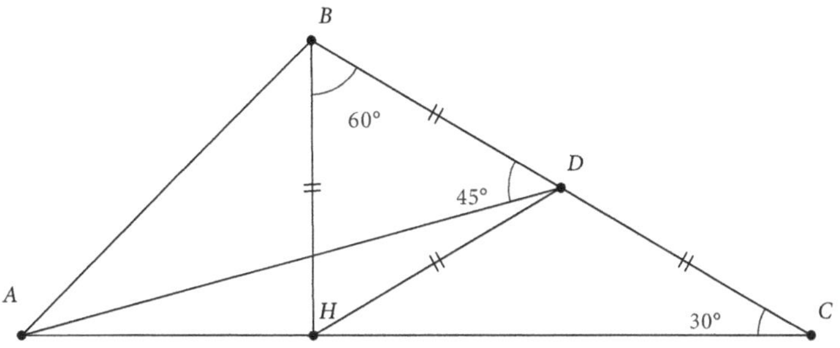

Figure 15.5 Problem 17, solution 1.

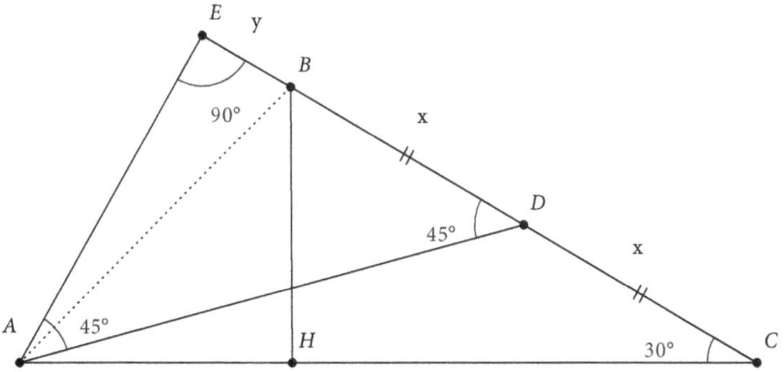

Figure 15.6 Problem 17, solution 2.

17. *Solution 1.* Look at figure 15.5. Let BH be the altitude from the point B, H belongs to AC. The triangle BCH has a right angle at H and an angle of 30° at the vertex C. It follows that $BH = \frac{1}{2}BC = BD = DC$. We also have that $HD = \frac{1}{2}BC$ as the median to the hypotenuse. Hence the triangle BDH is equilateral. Then $\angle HDA = 60° - 45° = 15°$, $\angle HAD = \angle BDA - \angle BCA = 45° - 30° = 15°$ and it follows that $AH = HD = HB$ and $\angle BAH = 45°$. We then have $\angle BAD = \angle BAH - \angle DAH = 45° - 15° = 30°$.

Solution 2. Look at figure 15.6. Let E be on CB and such that AE is perpendicular to CB. Write $CD = DB = x$, $EB = y$. Then, since $\triangle ADE$ is right-angled and isosceles, $AE = x + y$.

In $\triangle ACE$ we have

$$\tan 30^\circ = \frac{1}{\sqrt{3}} = \frac{x+y}{2x+y} \quad \Rightarrow \quad 2x+y = x\sqrt{3} + y\sqrt{3}$$

$$\Rightarrow \quad y(\sqrt{3}-1) = x(2-\sqrt{3})$$

$$\Rightarrow \quad \frac{x}{y} = \frac{\sqrt{3}-1}{2-\sqrt{3}}.$$

If $\angle BAD = \theta$, then $\angle ABE = 45^\circ + \theta$ (exterior angle of $\triangle ABD$) and

$$\tan(45^\circ + \theta) = \frac{x+y}{y} = \frac{x}{y} + 1 = \frac{\sqrt{3}-1}{2-\sqrt{3}} + 1$$
$$= \frac{1}{2-\sqrt{3}} = 2 + \sqrt{3} = \tan 75^\circ$$

(see page 155). It then follows that $\theta = 30^\circ$.

Solution 3. Having equal bases and a common height, area $\triangle ABD$ = area $\triangle ADC$ so that $\frac{1}{2}AD \cdot BD \sin 45^\circ = \frac{1}{2}AC \cdot CD \sin 30^\circ \Rightarrow AD\sqrt{2} = AC$ (since $BD = CD$). Then, using the cosine rule in $\triangle ADC$,

$$AC^2 = AD^2 + CD^2 - 2AD \cdot CD \cos 135^\circ \quad \Rightarrow$$
$$2AD^2 = AD^2 + CD^2 + AD \cdot CD\sqrt{2} \quad \Rightarrow$$
$$AD^2 = CD^2 + \sqrt{2}AD \cdot BD = BD^2 + \sqrt{2}AD \cdot BD.$$

In $\triangle ADB$,

$$AB^2 = AD^2 + BD^2 - 2AD \cdot BD \cos 45^\circ$$
$$= (BD^2 + \sqrt{2}AD \cdot BD) + BD^2 - AD \cdot BD\sqrt{2}$$
$$= 2BD^2 = BC \cdot BD \quad (\text{since } BC = 2BD).$$

It follows that $\frac{AB}{BC} = \frac{BD}{AB}$ and the triangles ABD and CBA are similar (with $\angle B$, the included angle, common; the larger triangle is a scale factor $\sqrt{2}$ times larger than the smaller), and $\angle BAD = \angle BCA = 30^\circ$.

Solution 4. Solution 2 employed a rather obvious "creation" — the point E — but here is a more imaginative one.

Construct the point F on AC such that $\angle FDC = 45^\circ$ (draw the picture!). Then, by the sine rule in $\triangle ADC$ we have

$$\frac{AD}{\sin 30^\circ} = \frac{DC}{\sin 15^\circ} \quad \Rightarrow \quad AD = \frac{DC}{2\sin 15^\circ}$$

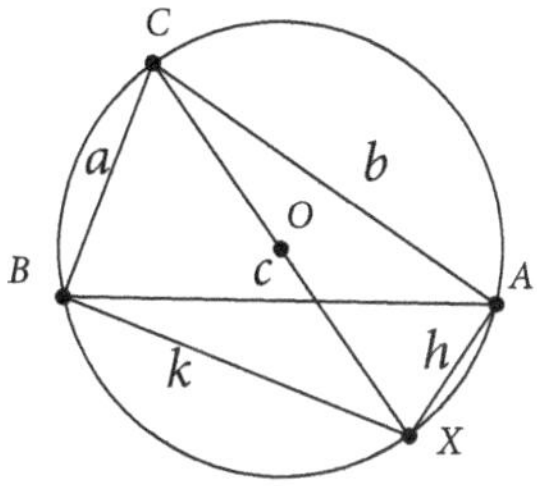

Figure 15.7 Problem 18

and, in $\triangle FDC$,

$$\frac{FD}{\sin 30°} = \frac{DC}{\sin 105°} \quad \Rightarrow \quad FD = \frac{DC}{2\cos 15°}.$$

Together, these give

$$AD \cdot FD = \frac{DC^2}{4\sin 15° \cos 15°} = \frac{DC^2}{2\sin 30°} = DC^2.$$

Hence

$$\frac{AD}{DC} = \frac{CD}{FD}, \quad \text{i.e.} \quad \frac{AD}{BD} = \frac{CD}{FD}$$

and, since $\angle ADB = 45° = \angle CDF$, it follows that triangles ABD and CFD are similar, and so $\angle BAD = \angle FCD = 30°$.

18. If the triangle is right-angled, then the longest side, say c, is the diameter of the circle. In this case $c = 2R$ and, by the Pythagorean theorem, $a^2 + b^2 = c^2$. Hence $a^2 + b^2 + c^2 = 2c^2 = 2 \cdot (2R)^2 = 8R^2$ as required.

It remains to prove the converse, namely if $a^2 + b^2 + c^2 = 8R^2$ then the triangle is right-angled.

Proof (Geometric). Draw the diameter passing form C through the centre of the circle and label its other end X (this is called the *antipodal* of C). Now add the segments AX and BX. We label their lengths $|AX| = h$ and $|BX| = k$. Note that if $h = 0$ then points A and X are coincident, so that CA is a diameter and $\angle ABC = 90°$. Similarly if $k = 0$ then points B and X are coincident, so that BC is a diameter and $\angle CAB = 90°$. We need only consider the case when $k \neq 0$ and $h \neq 0$.

The triangles BCX and ACX are right-angled since CX is a diameter. Hence, by Pythagorean Theorem, $a^2 + k^2 = (2R)^2$ and $b^2 + h^2 = (2R)^2$. Adding these two equalities gives $a^2 + b^2 + h^2 + k^2 = 8R^2$.

If the condition of the question is met, i.e. $a^2 + b^2 + c^2 = 8R^2$, then we have $h^2 + k^2 = c^2$. But h, k and c are the sides of the triangle ABX. It follows, by the converse of Pythagorean Theorem, that ABX is a right-angled triangle with the right angle at X, i.e. the chord AB subtends a right angle at the circumference of the circle. Hence AB is a diameter and $\angle ACB = 90°$ as required. □

Proof (Trigonometric). We will refer to the angles of the triangle themselves simply as A, B and C.

We have $a = 2R\sin A$, $b = 2R\sin B$, $c = 2R\sin C = 2R\sin(A + B)$ and get

$$\begin{aligned}
&\sin^2 A + \sin^2 B + \sin^2 C = 2 \quad \Leftrightarrow \\
&-\cos^2 A - \cos^2 B + \sin^2 A\cos^2 B \\
&\qquad + 2\sin A\sin B\cos A\cos B + \cos^2 A\sin^2 B = 0 \quad \Leftrightarrow \\
&-2\cos^2 A\cos^2 B + 2\sin A\sin B\cos A\cos B = 0 \quad \Leftrightarrow \\
&2\cos A\cos B\cos(A + B) = 0.
\end{aligned}$$

Therefore the triangle contains a right angle as required.

Alternatively we could rewrite this as

$$\cos 2A + \cos 2B + \cos 2C + 1 = 0.$$

and use the factor formula $\cos x + \cos y = 2\cos\frac{x+y}{2}\cos\frac{x-y}{2}$ (try!). □

Proof (Algebraic). We use Heron's formula:

$$a^2+b^2+c^2=\frac{8(abc)^2}{16S^2} \quad \Leftrightarrow$$
$$16(a^2+b^2+c^2)\cdot\frac{a+b+c}{2}\cdot\frac{a+b-c}{2}\cdot\frac{a-b+c}{2}\cdot\frac{-a+b+c}{2}=8(abc)^2 \quad \Leftrightarrow$$
$$(a^2+b^2+c^2)((a+b)^2-c^2)(c^2-(a-b)^2)=8(abc)^2 \quad \Leftrightarrow$$
$$(a^2+b^2+c^2)(a^2+b^2-c^2+2ab)(c^2-a^2-b^2+2ab)=8(abc)^2 \quad \Leftrightarrow$$
$$(a^2+b^2+c^2)((2ab)^2-(a^2+b^2-c^2)^2)=8(abc)^2 \quad \Leftrightarrow$$
$$(a^2+b^2+c^2-2c^2)(2ab)^2-(a^2+b^2+c^2)(a^2+b^2-c^2)^2=0 \quad \Leftrightarrow$$
$$(a^2+b^2-c^2)(4(ab)^2-(a^2+b^2+c^2)(a^2+b^2-c^2))=0 \quad \Leftrightarrow$$
$$(a^2+b^2-c^2)(4(ab)^2-((a^2+b^2)^2-c^4))=0 \quad \Leftrightarrow$$
$$(a^2+b^2-c^2)(c^4-(a^2-b^2)^2)=0 \quad \Leftrightarrow$$
$$(a^2+b^2-c^2)(c^2-a^2+b^2)(c^2+a^2-b^2)=0. \qquad \square$$

Proof (Advanced algebraic). We use the approach described on page 297.

$$\begin{aligned} a^2+b^2+c^2 &= (a+b+c)^2-2(ab+bc+ac) \\ &= (2p)^2-2(p^2+r^2+4Rr) \\ &= 2(2p^2-r^2-4Rr) \end{aligned}$$

Thus the condition $a^2+b^2+c^2=8R^2$ is equivalent to

$$2p^2-r^2-4Rr=4R^2 \quad \Leftrightarrow \quad 2p^2=2(R+2R)^2 \quad \Leftrightarrow \quad p=r+2R$$

and by Theorem 12.1 the last equality is valid in the right triangle only.

Remark The last solution can be obviously modified to show that the value of $a^2+b^2+c^2$ in comparison with $8R^2$ serves to discriminate between acute, right-angled and obtuse triangles, e.g.

$$a^2+b^2+c^2>8R^2 \quad \Leftrightarrow \quad p>r+2R \quad \Leftrightarrow \quad \text{triangle is acute.}$$

15.4 Solutions to Paper 4

19. By the Cauchy-Schwarz inequality, we get

$$\left((\sqrt{x_1})^2+\cdots+(\sqrt{x_n})^2\right)\left(\left(\frac{a_1}{\sqrt{x_1}}\right)^2+\cdots+\left(\frac{a_n}{\sqrt{x_n}}\right)^2\right)$$
$$\geqslant (a_1+a_2+\cdots+a_n)^2,$$

for any positive $x_1, x_2, \cdots, x_n$, that is

$$\frac{a_1^2}{x_1}+\frac{a_2^2}{x_2}+\cdots+\frac{a_n^2}{x_n}\geqslant\frac{(a_1+a_2+\cdots+a_n)^2}{x_1+x_2+\cdots+x_n}.$$

When applying this inequality to the left member of the given inequality, one gets

$$\frac{a_1}{a_2^2+1}+\frac{a_2}{a_3^2+1}+\cdots+\frac{a_n}{a_1^2+1}$$
$$=\frac{a_1^3}{a_1^2a_2^2+a_1^2}+\frac{a_2^3}{a_2^2a_3^2+a_2^2}+\cdots+\frac{a_n^3}{a_n^2a_1^2+a_n^2}$$
$$\geqslant\frac{(a_1\sqrt{a_1}+a_2\sqrt{a_2}+\cdots+a_n\sqrt{a_n})^2}{a_1^2a_2^2+a_2^2a_3^2+\cdots+a_n^2a_1^2+1}.$$

Therefore, it will suffice to prove the following inequality

$$a_1^2a_2^2+a_2^2a_3^2+\cdots+a_n^2a_1^2\leqslant\frac{1}{4},$$

where $n\geqslant 4$ and $a_1^2+a_2^2+\cdots+a_n^2=1$.

This inequality is general: if $n\geqslant 4$ and given the positive numbers $x_1, x_2, \cdots x_n$ such that $x_1+x_2+\cdots+x_n=1$, then

$$x_1x_2+x_2x_3+\cdots+x_nx_1\leqslant\frac{1}{4}.$$

If n is even, the proof is immediate, as

$$x_1x_2+x_2x_3+\cdots+x_nx_1\leq(x_1+x_3+\cdots+x_{n-1})(x_2+x_4+\cdots x_n)\leqslant\frac{1}{4},$$

since the product of two positive numbers of sum 1 has $\frac{1}{4}$ as its maximum.

If n is odd, $n\geqslant 5$, we may assume that $x_1\geqslant x_2$, and since

$$x_1x_2+x_2x_3+x_3x_4\leqslant x_1(x_2+x_3)+(x_2+x_3)x_4,$$

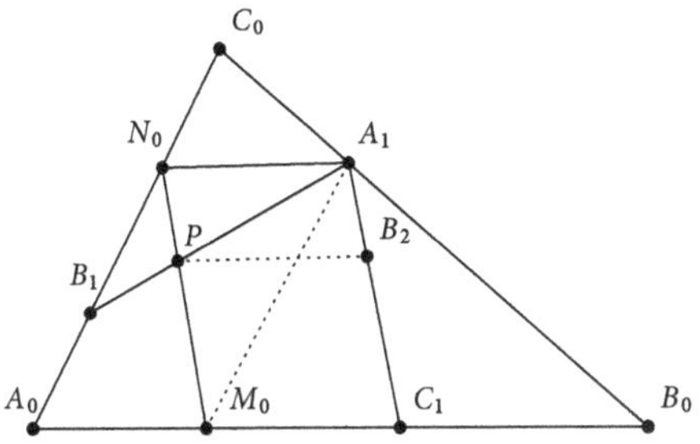

Figure 15.8 Problem 20

we may replace the numbers $x_1, x_2, \ldots, x_n$ by the $n-1$ numbers $x_1, x_2+x_3, x_4, \ldots, x_n$. The left member of the given inequality increases, the sum remains one, and we have an even set of numbers, so the preceding argument can be applied.

20. *Solution 1.* It is enough to consider the case $k=0$. Let the points M_0, N_0 lie on the segments A_0B_0, C_0A_0 and divide them in the ratio $1:2$ counted from A_0 and C_0 respectively. It is obvious that $N_0A_1 \parallel A_0B_0$ and that $N_0A_1 = \frac{1}{3}A_0B_0 = M_0C_1$. It follows that the quadrilateral $M_0C_1A_1N_0$ is a parallelogram. On the other hand, $M_0A_1 \parallel A_0C_0$ and $M_0A_1 = \frac{2}{3}A_0C_0 = 2B_1N_0$. Denote by P the intersection point between M_0N_0 and A_1B_1. By the similarity of the triangles M_0A_1P and B_1PN_0 we have

$$\frac{A_1P}{PB_1} = \frac{A_1M_0}{B_1N_0} = \frac{2}{1},$$

which shows that P coincides with C_2. The equalities $\frac{C_2M_0}{C_2N_0} = \frac{2}{1}$ and $\frac{C_1B_2}{A_1B_2} = \frac{2}{1}$ then imply that $B_2C_2 \parallel A_0B_0$ and that $B_2C_2 = M_0C_1$. In the same way we can prove that $A_2C_2 \parallel B_0C_0$ and that $A_2B_2 \parallel A_0C_0$. It follows that $\triangle A_0B_0C_0 \sim \triangle B_2C_2A_2$.

Let us now denote by S the intersection point between A_0A_1 and B_2C_2. We then have

$$\frac{A_0S}{SA_1} = \frac{C_1B_2}{B_2A_1} = \frac{2}{1}.$$

On the other hand, the equalities

$$\frac{1}{3} = \frac{A_1B_2}{A_1C_1} = \frac{SB_2}{A_0C_1} = \frac{SB_2}{2M_0C_1} = \frac{SB_2}{2B_2C_2}$$

imply

$$\frac{SB_2}{B_2C_2} = \frac{2}{3},$$

which means that S coincides with A_3. Thus we get that A_3 lies on the segment A_0A_1 and divides it in the ratio $2:1$ counted from A_0.

Solution 2. We interpret everything as complex numbers. Again, the case $k = 0$ is sufficient. We have:

$$A_1 = \frac{2}{3}C_0 + \frac{1}{3}B_0; \quad B_1 = \frac{2}{3}A_0 + \frac{1}{3}C_0; \quad C_1 = \frac{2}{3}B_0 + \frac{1}{3}A_0;$$

Consequently,

$$\begin{aligned} A_2 &= \frac{2}{3}C_1 + \frac{1}{3}B_1 = \frac{2}{3}\left(\frac{2}{3}B_0 + \frac{1}{3}A_0\right) + \frac{1}{3}\left(\frac{2}{3}A_0 + \frac{1}{3}C_0\right) \\ &= \frac{4}{9}A_0 + \frac{4}{9}B_0 + \frac{1}{9}C_0 \end{aligned}$$

and similarly

$$B_2 = \frac{4}{9}B_0 + \frac{4}{9}C_0 + \frac{1}{9}A_0; \quad C_2 = \frac{4}{9}C_0 + \frac{4}{9}A_0 + \frac{1}{9}B_0.$$

Now we have

$$\frac{A_2 - B_2}{C_2 - B_2} = \frac{\frac{1}{3}A_0 - \frac{1}{3}C_0}{\frac{1}{3}A_0 - \frac{1}{3}B_0} = \frac{C_0 - A_0}{B_0 - A_0}$$

which shows the similarity (where B_2 maps to A_0!) It remains to check that

$$\begin{aligned} A_3 = \frac{1}{3}A_0 + \frac{2}{3}A_1 \quad &\Leftrightarrow \quad \frac{4}{9}A_1 + \frac{4}{9}B_1 + \frac{1}{9}C_1 = \frac{1}{3}A_0 + \frac{2}{3}A_1 \\ &\Leftrightarrow \quad 4B_1 + C_1 = 3A_0 + 2A_1 \\ &\Leftrightarrow \quad 4\left(\frac{2}{3}A_0 + \frac{1}{3}C_0\right) + \frac{2}{3}B_0 + \frac{1}{3}A_0 = 3A_0 + 2\left(\frac{2}{3}C_0 + \frac{1}{3}B_0\right) \end{aligned}$$

which is evidently true. The rest follows from the symmetry.

21. Let us denote by $D(A, x)$ the disk of centre A and radius x, the border included. We shall use the following lemma.

Lemma 15.2 *Let $D_1 = D(O_1, r_1)$ and $D_2 = D(O_2, r_2)$ be two disks and O_3 be the third vertex of the equilateral triangle $O_1O_2O_3$. Let ABC be an equilateral triangle with variable vertices B, C, in the disks D_1, D_2 respectively. Then, the locus of the vertex A is the disk $D_3(O_3, r_1 + r_2)$, and also its reflection with respect to the line O_1O_2.*

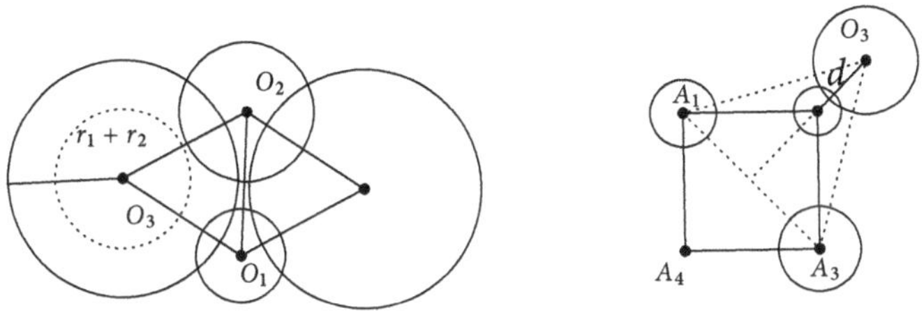

Figure 15.9 Problem 21.

Proof. The vertices of the equilateral triangles O_1CA when $C \in D_2$ can be obtained by rotating the disk D_2 around O_2 with an angle of rotation $\pm 60°$. When B_1 is arbitrary, one may add a translation by a vector of length at most r_1. Since the triangles are equilateral, it is clear that any point D_3 is obtained in this way. □

For the solution of the problem, let $A_1A_2A_3A_4$ be the unit square, let $D_i = D(A_i, r_i)$, $i = 1, 2, 3, 4$, be the disks and denote $r = \sum r_i$. We may assume without loss of generality that $r_1 + r_2 + r_3 \geqslant \frac{3}{4}r$, this being the best evaluation for a minimal sum of three radii.

Consider the disk $D(O_3, r_1 + r_3)$, by using the disks D_1, D_3 in lemma. It intersects D_2 if and only if

$$r_1 + r_3 + r_2 \geqslant d = \sqrt{2} \cdot \frac{\sqrt{3}}{2} - \frac{\sqrt{2}}{2}.$$

We conclude that the minimal required value is $r = \frac{2(\sqrt{6}-\sqrt{2})}{3}$.

It can be seen that the same value is obtained when two consecutive vertices are considered.

22. Suppose that an arbitrary MP has rating R. When he enters a group, say group G_i, he carries the relative rating $r_i = \frac{R}{S_i}$, where $S_i = \sum_{x_k \in G_i} R_k$. When he moves from group G_i to group G_j, the given condition translates into

the inequality

$$\frac{R}{S_i} < \frac{R}{S_j + R}.$$

It is obviously equivalent to $R + S_j - S_i < 0$.

After the move from G_i to G_j, the new sums of absolute ratings are:

$$S_i' = S_i - R, \quad S_j' = S_j + R.$$

The idea is to associate to the Parliament a daily invariant:

$$L = \sum_i S_i^2.$$

When an MP moves from one group to another, the invariant L changes value to a new one

$$L' = \sum_{k \neq i,j} S_k^2 + (S_i - R)^2 + (S_j + R)^2.$$

It is easy to see that

$$L' - L = (S_i - R)^2 + (S_j + R)^2 - S_i^2 - S_j^2 = 2R(R + S_j - S_i) < 0.$$

Therefore $L' < L$. Since there are only finitely many possibilities to arrange the MPs in groups, the invariant L can take only a finite number of values. Thus it cannot decrease infinitely.

Solution 2. Here is another argument. If S is the least possible sum of ranks in a group, no PM can move from a group with the sum S. Assume that there are counter examples to the statement of the problem. Of all such counter examples, let us choose one of possibly several which have the minimal number of groups, the maximal value of S and the minimal number of groups with sum S. In this example all groups must participate in the process an infinite number of times (else we can disregard the groups which participate a finite number of times and get a smaller counter example). Then a group of sum S has to participate. The only possibility is that a PM moves to this group. But then either the minimal sum will get larger, or the number of groups of sum S will decrease.

23. This is an advanced angle chasing involving many non-trivial ideas (but all of them you can find in the book). The main difficulty is to how use the

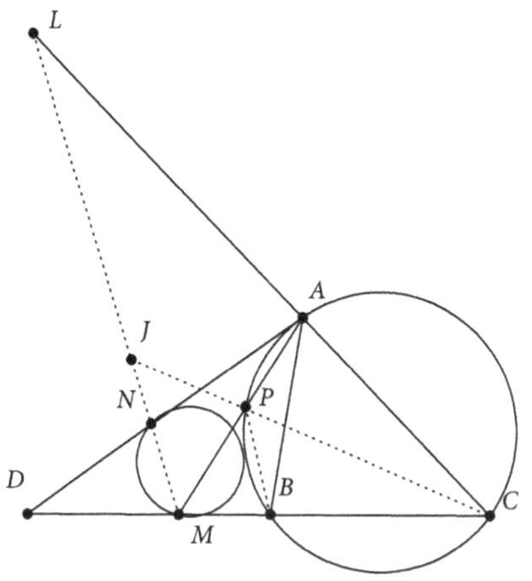

Figure 15.10 Problem 23.

conditions that K and K_1 touch each other. Here it is done with the help of Casey's theorem. Such solutions are almost impossible to find, but very useful to study.

Let N be the tangent point of K_1 and AD and J be the centre of the excribed circle which is tangent to the side AB. For convenience, we will denote by A, B, C the measures of the angles of the triangle ABC, respectively. Then $\angle JAC = 90° + \frac{1}{2}A$, $\angle BAD = C$ and $\angle ADC = B - C$. It follows that $\angle DMN = \frac{1}{2}(180° - B + C)$.

We shall use the following preliminary result.

Lemma 15.3 *The points J, M, N are collinear.*

Proof. Let L be the intersection point of the lines MN and AC. By Casey's theorem, applied when A, B, C are considered as circles of radius zero, we have $aNA + bMB = cMC$, and also

$$a\frac{NA}{MC} + b\frac{MB}{MC} = c. \tag{15.2}$$

By Menelaos' theorem applied to the line ML and triangle ADC, we have

$$\frac{LA}{LC} \cdot \frac{MC}{MD} \cdot \frac{ND}{NA} = 1$$

and then $\dfrac{LA}{LC} = \dfrac{NA}{MC}$. From (15.2) and the last relation, we obtain

$$a\frac{LA}{LC} + b\frac{MB}{MC} = c.$$

Let $LC = x$, $MC = y$. Then we have

$$a \cdot \frac{x-b}{x} + b \cdot \frac{y-a}{y} = c \quad \Leftrightarrow \quad a - \frac{ab}{x} + b - \frac{ab}{y} = c$$

$$\Leftrightarrow \quad \frac{1}{x} + \frac{1}{y} = \frac{a+b-c}{ab}$$

It follows from theorem 4.24 that MN passes through J. □

From the lemma and the formula for $\angle DMN$ we get:

$$\angle JMC = B + \frac{1}{2}A. \tag{15.3}$$

Let us assume $AC = MC$; thus we have $\angle AMC = \angle MAC$ and CJ is the perpendicular bisector of the segment AM. The triangles JAC and JMC are congruent and thus $\angle JAC = \angle JMC$. Since $\angle JAC = 90° + \frac{1}{2}A$, taking into account (15.3), we get $90° + \frac{1}{2}A = B + \frac{1}{2}A$, that is $B = 90°$. Then AC is a diameter and therefore $\angle CAD = 90°$ too. Since $B = \angle CAD$, we have

$$\angle AMC + \angle MAB = \angle MAC + \angle MAD,$$

and thus $\angle MAD = \angle MAB$. So, AM is the bisector line of $\angle DAB$.

Let us assume that $\angle MAB = \angle MAD$. Let P be the intersection point of AM and K. Since $\angle MAB = \angle MAD$, it follows that the arcs AP and BP on the circle K are equal, CP is the bisector of $\angle ACD$ and CP passes through J. Moreover $PA = PB$.

For the angle $\angle JBP$ we have $\angle JBP = \angle JBA - \angle PBA$ and

$$\angle JBP = \frac{A+C}{2} - \frac{C}{2} = \frac{A}{2}. \tag{15.4}$$

In the triangle JBC, $\angle BJC = 180° - \frac{1}{2}C - B - \frac{A+C}{2} = \frac{1}{2}A$. This together with (15.4) gives $PJ = PB$. Moreover $\angle JMP = \angle JMC - \angle AMC = \angle JMC - B + \angle BAM = \left(B + \frac{1}{2}A\right) - B + \frac{1}{2}C$ (we have used (15.3) and tangent-chord theorem). Hence $\angle JMP = \frac{A+C}{2}$. In the triangle JMC we have $\angle MJC = 180° - \frac{C}{2} - B - \frac{A}{2}$, that is $\angle MJC = \frac{A+C}{2}$. Hence $PJ = PM$. Since we also have $PA = PB = PM$ it follows that in $\triangle AMC$ the segment CP is median and bisector of the angle C. One obtains $AC = MC$.

Solution 2. Here is a simpler approach. We present only one of the implications, assuming that AM is the bisector. Consider figure 15.11. Let us first compute

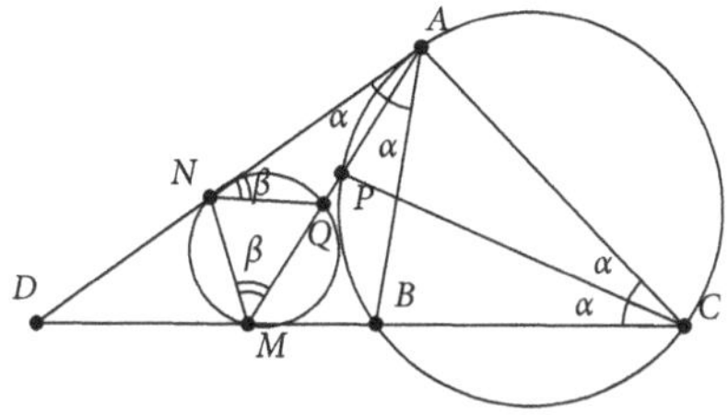

Figure 15.11 Problem 23. Solution 2.

all possible angles. Let $\angle DAM = \angle MAB = \alpha$. We have that

$$\angle BCP = \angle PAB = \alpha; \quad \angle PCA = \angle DAP = \alpha.$$

We see at once that CP is a bisector too and we have to prove that it is also the altitude. This means that AC is a diameter. How much freedom do we actually have? If AC and α are fixed, we can vary only the radius R of the larger circle. This means that we can compute all the rest. Let us introduce a suitable parameter and find an equation it satisfies. Let Q be the point where the line AM interests K_1 and $\angle NMQ = \beta$. Then $\angle ANQ = \beta$ and $\angle DMN = \angle NQM = \alpha + \beta$ (we got the last equality from $\triangle NQA$.). Basically, we can now compute all angles. The only assumption we have not used is the one of the two circles being tangent. For the length of the common tangent we have

$$AN^2 = 2R \cdot 2r,$$

if R and r are their radii. We can now compute all lengths. We shall express them in terms of AM. First we have

$$MN = 2r\sin(\alpha+\beta) \quad\Rightarrow\quad 2r = \frac{MN}{\sin(\alpha+\beta)};$$

$$AB = 2R\sin 2\alpha \quad\Rightarrow\quad 2R = \frac{AB}{\sin 2\alpha}.$$

From $\triangle AMN$ we get

$$\frac{MN}{\sin\alpha} = \frac{AN}{\sin\beta} = \frac{AM}{\sin(\pi-\alpha-\beta)} \quad\Rightarrow\quad AN = \frac{AM\sin\beta}{\sin(\alpha+\beta)}$$

and

$$MN = \frac{AM\sin\alpha}{\sin(\alpha+\beta)} \quad\Rightarrow\quad 2r = \frac{AM\sin\alpha}{\sin^2(\alpha+\beta)}.$$

Finally, from $\triangle AMB$ we get

$$\angle MBA = \angle DMA - \alpha = (\alpha+\beta)+\beta-\alpha = 2\beta$$

and, according to the sine rule,

$$\frac{AM}{\sin 2\beta} = \frac{AB}{\sin(\pi-\alpha-2\beta)} \quad\Rightarrow\quad AB = \frac{AM\sin(\alpha+2\beta)}{\sin 2\beta}$$

$$\Rightarrow\quad 2R = \frac{AM\sin(\alpha+2\beta)}{\sin 2\alpha\sin 2\beta}.$$

We now have all we need to write down the main equation:

$$\left(\frac{AM\sin\beta}{\sin(\alpha+\beta)}\right)^2 = \frac{AM\sin(\alpha+2\beta)}{\sin 2\alpha\sin 2\beta}\cdot\frac{AM\sin\alpha}{\sin^2(\alpha+\beta)}.$$

It is equivalent to

$$\sin^2\beta\sin 2\alpha\sin 2\beta = \sin(\alpha+2\beta)\sin\alpha \quad\Leftrightarrow$$

$$2\sin^2\beta\cos\alpha\sin 2\beta = \sin(\alpha+2\beta) \quad\Leftrightarrow$$

$$(1-\cos 2\beta)\cos\alpha\sin 2\beta = \sin\alpha\cos 2\beta + \cos\alpha\sin 2\beta \quad\Leftrightarrow$$

$$0 = \cos 2\beta\cos\alpha\sin 2\beta + \sin\alpha\cos 2\beta \quad\Leftrightarrow$$

$$\cos 2\beta(\cos\alpha\sin 2\beta + \sin\alpha) = 0.$$

Since $\alpha < \pi/2, 2\beta < \pi$ the only solution is

$$\cos 2\beta = 0 \quad\Leftrightarrow\quad 2\beta = \pi/2.$$

Then $\angle APC = \angle ABC = \pi - \angle MBA = \pi - \pi/2 = \pi/2$ and PC is both the bisector and the altitude in $\triangle ACM$.

The problem can also be solved by using inversion with A as a centre.

24. Denote $f(0) = b$, $g(x) = f(x) - b$. It is then obvious that $g(0) = 0$ and that g satisfies the same equality as f (since constants have the same property). Further, choose $x = 2t$, $y = 0$. This gives $2g(t) = g(2t)$. If we now take $x = 2u$ $y = 2v$, we get

$$g(u+v) = \frac{g(2u)+g(2v)}{2} = \frac{2g(u)+2g(v)}{2} = g(u)+g(v).$$

We see that $g(x)$ satisfies the Cauchy's equation and, being continuous, $g(x) = ax + c$. Because $g(0) = 0$ we have $c = 0$ and finally, $f(x) = ax + b$, where a and b are real constants. It is easy to check that every such function satisfies the conditions.

Remark It is enough to require continuity at one single point, f:s property then yields continuity on the whole line.

15.5 Solutions to Paper 5

25. For any distinct $u, v \in \mathbb{C}$, we have

$$u+v = \frac{u^2-v^2}{u-v} = \frac{(u^2+c)-(v^2+c)}{u-v} = \frac{f(u)-f(v)}{u-v}.$$

By hypothesis, $x_{i+1} \neq x_i$ for every i. In particular, if we denote $f(x_n)$ by x_{n+1}, then

$$\prod_{i=1}^{n}(x_{i+1}+x_i) = \prod_{i=1}^{n}\frac{f(x_{i+1})-f(x_i)}{x_{i+1}-x_i} = \frac{\prod_{i=1}^{n}(f(x_{i+1})-x_{i+1})}{\prod_{i=1}^{n}(f(x_i)-x_i)} = 1,$$

where the last equality is by observing that the product in the numerator has exactly the same factors as the one in the denominator, but in rotated order.

26. Assume first that AB is not parallel to CD. Let K be the point where the lines AB and CD intersect. Suppose there is a circle through A, B, E and F. Consider the configuration of the Figure 15.12. (Other configurations, with E or F between C and D, are handled similarly.) Since $\angle ABC = \angle ADF$ and $\angle KBE = \angle KFA$, we have $\angle DAF = \angle KFA - \angle ADF = \angle KBE - \angle ABC = \angle CBE$. Since G and H are the circumcentres of $\triangle BCE$ and $\triangle ADF$, we have $\angle DHF = 2\angle DAF = 2\angle CBE = \angle CGE$. It follows that the isosceles triangles DHF and CGE are similar.

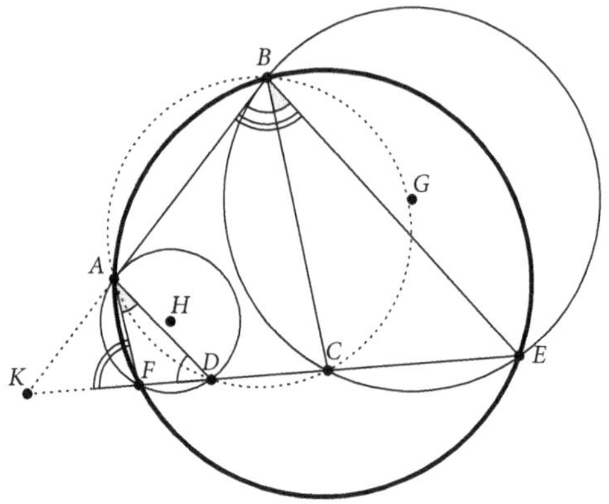

Figure 15.12 Problem 26

By the power-of-the-point theorem, we have $KE \cdot KF = KA \cdot KB$, $KC \cdot KD = KA \cdot KB$, and therefore $KE \cdot KF = KC \cdot KD$. Hence $(KF - KD)/KD = (KC - KE)/KE$, or $FD/KD = CE/KE$. From the similarity of $\triangle DHF$ and $\triangle CGE$ we have $DH/EG = FD/CE$, hence $DH/EG = KD/KE$. The similarity of $\triangle DHF$ and $\triangle CGE$ also implies that $\angle KDH = \angle KEG$. Since $DH/EG = KD/KE$, the triangles KDH and KEG are similar, with $\angle DKH = \angle EKG$. This means that the lines AB, CD, and GH are concurrent.

To prove the converse, suppose that AB, CD and GH are concurrent, in other words that H, G, and K are collinear. Let F' be the point other than E where the circumcircle of $\triangle ABE$ intersects the line CD, and let H' be the circumcentre of $\triangle ADF'$. We know that H', G, and K are collinear, so that both H and H' lie on the line KG and on the perpendicular bisector of AD. Since AD is not parallel to BC, KG is not parallel to the perpendicular bisector of AD. Hence, $H' = H$, which implies that $F' = F$.

Assume now that AB is parallel to CD. Suppose that there is a circle through A, B, E, and F. By symmetry, GH is parallel to AB and CD. To prove the converse, suppose that AB, CD, and GH are parallel. Let F' be the point other than E where the circumcircle of $\triangle ABE$ intersects the line CD, and let H' be the circumcentre of $\triangle ADF'$. Since GH' is parallel to AB, both H and

H' lie on the line parallel to AB through G and on the perpendicular bisector of AD. Hence $H' = H$, which implies that $F' = F$.

27. The smallest such value of k is $1/4$. First note that for $x, y > 0$,

$$\frac{1}{x+y} = \frac{4xy}{x+y} \cdot \frac{1}{4xy} \le \frac{(x+y)^2}{x+y} \cdot \frac{1}{4xy} = \frac{1}{4}\Big(\frac{1}{x} + \frac{1}{y}\Big),$$

with equality if and only if $x = y$. (Actually this is the inequality between the arithmetic and the harmonic means.) We then have

$$\begin{aligned}
&\frac{ab}{a+b+2c} + \frac{bc}{b+c+2a} + \frac{ca}{c+a+2b} \\
&\quad \le \frac{ab}{4}\Big(\frac{1}{a+c} + \frac{1}{b+c}\Big) + \frac{bc}{4}\Big(\frac{1}{c+a} + \frac{1}{b+a}\Big) + \frac{ca}{4}\Big(\frac{1}{c+b} + \frac{1}{a+b}\Big) \\
&\quad = \frac{1}{4}\Big(\frac{ab+bc}{a+c} + \frac{ab+ac}{b+c} + \frac{ac+bc}{a+b}\Big) = \frac{1}{4}(a+b+c),
\end{aligned}$$

with equality if and only if $a+b = b+c = c+a$, that is, if and only if $a = b = c$.

28. Let S be a maximal independent set in G. If $v \notin S$, then $f(v) \le |S|$, since $f(v)$ is the size of an independent set. If $v \in S$, then $f(v) \le n - |S|$, since v has at most $n - |S|$ neighbours. Hence, using AM–GM inequality, we get

$$\sum_{v \in V(G)} f(v) \le (n - |S|)|S| + |S|(n - |S|) \le n^2/2.$$

If G is the complete bipartite graph $K_{n/2,n/2}$, then $f(v) = n/2$ for all v, and hence the inequality is sharp when n is even.

29. Let A' be on the line BC (see Figure 15.13). By analogy with the construction of T, draw lines from B and C parallel to side AD. Let B'' and C'' be the new points where these lines intersect lines CD and AB, respectively. Let S be the point of intersection of lines BC and $C''B''$.

We will prove that ST is parallel to AC.

Extend DA and CB so that they meet at X, and extend AB and DC so that they meet at Y. Similarity of $\triangle A'BA$ and $\triangle CBY$ implies that $A'B/CB = BA/BY$. Also, similarity of $\triangle C''BC$ and $\triangle ABX$ implies that $C''B/AB = BC/BX$. Thus $A'B \cdot BY = BA \cdot CB = C''B \cdot BX$, and therefore $A'B/BX = C''B/BY$. By Thales's Theorem, $A'C''$ is parallel to XY and $XA'/XB = YC''/YB$. Similarity of $\triangle XA'A$ and $\triangle XBB'$ implies that $XA'/XB = A'A/BB'$, and similarity of $\triangle YC''C$ and $\triangle YBB''$ implies that

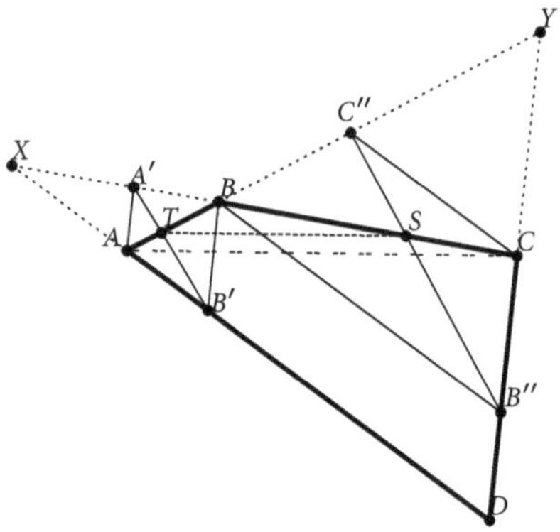

Figure 15.13 Problem 29

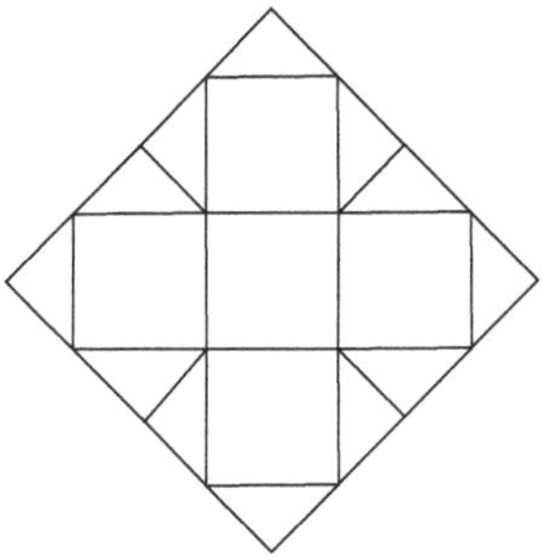

Figure 15.14 Problem 30

$YC''/YB = C''C/BB''$. Similarity of $\triangle A'AT$ and $\triangle B'BT$ implies that $A'A/B'B = AT/BT$, and similarity of $\triangle C''CS$ and $\triangle B''BS$ implies that $C''C/B''B = CS/BS$. Therefore $AT/TB = A'A/BB' = XA'/XB = YC''/YB = C''C/BB'' = CS/SB$, $AT/TB = CS/SB$, and TS is indeed parallel to AC. The rest follows.

30. The side of the largest possible cubical present is $\sqrt{2}/4$.

Consider two points in the square separated by a distance d. After the square is wrapped around the cube, the surface distance between the two

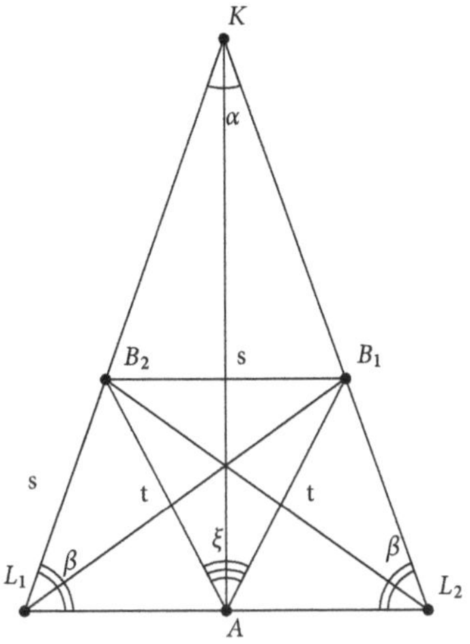

Figure 15.15 Problem 31

points will be less than or equal to d, since the paper cannot be stretched or cut. For any point on the cube, there is another point on the cube that has surface distance at least $2s$ from it, where s is the length of the edge of the cube.

Therefore, for any point on the square there is another point on the square such that the distance between them is at least $2s$. In particular, there is a point at a distance of at least $2s$ from the centre of the square, so the diagonal of the square is at least $4s$. Hence the side length of the square is at least $2\sqrt{2}s$. Since the side length of the square is 1, the edge length of the cube is at most $\sqrt{2}/4$.

Figure 15.14 shows a wrapping with indicated folds which achieves the lower bound.

15.6 Solutions to Paper 6

31. Denote $\alpha = \angle L_1KL_2$, $\beta = \angle KL_1L_2 = \angle KL_2L_1$ and $\xi = \angle B_1AB_2$. Since B_1 and B_2 are symmetric relative to KA, we have $B_1B_2 \perp KA$ and $B_1B_2 \parallel L_1L_2$, i.e. $\triangle AB_1B_2$ is isosceles and AK is its angle bisector. Since $\angle B_2B_1L_1$

$= \angle L_2L_1B_1 = \angle B_2L_1B_1$, we have that $\triangle B_2L_1B_1$ is also isosceles. Denote $s = |B_1B_2| = |B_2L_1|$ and $t = |AB_1| = |AB_2|$.

The sine rule in $\triangle AL_1B_2$ yields

$$\frac{s}{t} = \frac{\sin \angle B_2AL_1}{\sin \beta} = \frac{\cos \xi/2}{\cos \alpha/2}$$

and $(\frac{s}{t})^2 = \frac{1+\cos \xi}{1+\cos \alpha}$. The cosine rule in $\triangle AB_1B_2$ yields

$$s^2 = t^2 + t^2 - 2t^2 \cos \xi = 2t^2(1 - \cos \xi),$$

and $(\frac{s}{t})^2 = 2(1 - \cos \xi)$. Hence $\frac{1+\cos \xi}{1+\cos \alpha} = 2(1 - \cos \xi)$ and

$$\cos \xi = \frac{1 + 2\cos \alpha}{3 + 2\cos \alpha} = 1 - \frac{2}{3 + 2\cos \alpha} < 1 - \frac{2}{5} = \frac{3}{5}.$$

Solution 2. Note that the cosine of an angle between two sides of equal length in a triangle is less than $\frac{3}{5}$ if and only if the altitude h from the corresponding vertex is shorter than the third side. Indeed:

$$\begin{aligned} h < a \quad &\Leftrightarrow \quad \tan\frac{\alpha}{2} > \frac{1}{2} \quad \Leftrightarrow \quad \tan^2\frac{\alpha}{2} > \frac{1}{4} \\ &\Leftrightarrow \quad \frac{1}{\cos^2\frac{\alpha}{2}} = 1 + \tan^2\frac{\alpha}{2} > \frac{5}{4} \quad \Leftrightarrow \quad \cos^2\frac{\alpha}{2} < \frac{4}{5} \\ &\Leftrightarrow \quad \cos\alpha = 2\cos^2\frac{\alpha}{2} - 1 < \frac{3}{5}. \end{aligned}$$

We must now understand why the altitude from A is shorter than the base B_1B_2. Since B_1B_2 is parallel to L_1L_2, we have $\angle B_2L_1B_1 = \angle L_2L_1B_1 = \angle B_2B_1L_1$, the triangle $\triangle L_1B_2B_1$ is isosceles and $B_1B_2 = L_1B_2$. It remains to note that L_1B_2 is greater than the height of the trapezoid.

32. Suppose the claim does not hold, i.e.

$$\sin(x_1 - \alpha) + \sin(x_2 - \alpha) + \cdots + \sin(x_n - \alpha) < 0$$

which gives

$$\cos x_1 + \cdots + \cos x_n > \frac{\cos\alpha}{\sin\alpha} \cdot (\sin x_1 + \cdots + \sin x_n) \geq n \cdot \cos\alpha$$

and hence

$$(\sin x_1 + \ldots + \sin x_n)^2 + (\cos x_1 + \cdots + \cos x_n)^2 > n^2.$$

On the other hand,

$$(\sin x_1 + \cdots + \sin x_n)^2 + (\cos x_1 + \cdots + \cos x_n)^2$$
$$= \sum_{i=1}^{n}\sum_{j=1}^{n} \cos(x_i - x_j) \leq n^2,$$

which is a contradiction.

33. There is no such n.

For $n = 1$ and $n = 2$ we cannot place a pebble in any point with a positive coordinate — hence let $n \geq 3$. Consider the infinite sum

$$S = a^{x_1} + a^{x_2} + a^{x_3} + \cdots,$$

where $x_1, x_2, x_3, \ldots$ are the coordinates of the points with pebbles at some given moment and a is a positive real number to be determined later. We shall show that it is possible to choose a according to n so that the initial sum $S_0 = a^0 + a^{-1} + a^{-2} + a^{-3} + \ldots$ converges (for this it suffices to have $a > 1$) and so that at each step the sum S can only decrease. Hence we always have $S \leq S_0$ and it is impossible to place a pebble at a point with an arbitrarily large positive coordinate N, since for sufficiently large N we have $a^N > S_0$.

To show this consider for $n = 2k - 1$ the equation

$$1 + x + x^2 + \cdots + x^{k-1} = x^k + \cdots + x^{2k-2} \tag{15.5}$$

and for $n = 2k$ the equation

$$1 + x + x^2 + \cdots + x^{k-1} = x^{k+1} + \cdots + x^{2k-1}. \tag{15.6}$$

For $0 \leq x \leq 1$ the left side exceeds the right side but for sufficiently large positive x the right side exceeds the left side. Hence the equation has a root $a > 1$. It remains to show that for any integers m and t such that $1 \leq t \leq \left[\frac{n+1}{2}\right]$ the sum of any $t - 1$ elements of $A = \{a^m, a^{m+1}, \ldots, a^{m+n-1}\}$ does not exceed the sum of any t elements of A (here $m, \ldots, m + n - 1$ are the chosen n consecutive integer points and t is the number of points having a pebble before this step and no pebble after this step — hence $t - 1$ points have no pebble before this step and have a pebble after this step). Note that it suffices to have $m = 0$ and prove that the sum of the t smallest elements of A is not less then the sum of the $t - 1$ largest elements, i.e.

$$1 + a + a^2 + \cdots + a^{t-1} \geq a^{n-t+1} + \cdots + a^{n-1},$$

where $1 \le t \le [\frac{n+1}{2}] = k$. This directly follows from (15.5) or (15.6) for $x = a, t = k$ and from the fact that since $a > 1$, deleting an equal number of terms from each side makes the left side larger than the right side.

34. If we label the chairs $0, 1, 2, 3, \ldots, n-1$, with the chair occupied by the 1st candidate labeled 0, then the $(k+1)^{\text{th}}$ delegate is in the chair labeled $\sum_{r=0}^{k} r \pmod{n}$. Now

$$\sum_{r=1}^{k} r = \frac{k(k+1)}{2} \quad \text{and} \quad \frac{p(p+1)}{2} - \frac{q(q+1)}{2} = \frac{(p+q+1)(p-q)}{2}.$$

Hence two delegates will occupy one chair under this scheme if and only if we can find two integers p and q between 0 and $n-1$ such that $\frac{(p+q+1)(p-q)}{2}$ is a multiple of n. We will assume, without loss of generality, that $p > q$ so that $\frac{(p+q+1)(p-q)}{2}$ is positive.

In table 15.1, we check some small values of n to try to establish a pattern.

Table 15.1 Problem 34. Some small values of n

k	$\sum_{r=0}^{k} r$	$(\sum_{r=0}^{k} r) \bmod n$							
		$n=2$	$n=3$	$n=4$	$n=5$	$n=6$	$n=7$	$n=8$	$n=9$
0	0	0	0	0	0	0	0	0	0
1	1	1	1	1	1	1	1	1	1
2	3		0	3	3	3	3	3	3
3	6			2	1	0	6	6	6
4	10				0	4	3	2	1
5	15					3	1	7	6
6	21						0	5	3
7	28							4	1
8	36								0
Different?		Yes		Yes				Yes	

The values of n in table 15.1 for which no chair is occupied by more than one delegate are powers of two: 1, 2, 4 and 8.

We start by considering integers n which are not a power of 2, in which case we can write $2n = uv$, with $u > v$, where one of u, v is a power of 2, at least 2, and the other is an odd number, at least 3. Let $p = \frac{u+v-1}{2}$ and $q = \frac{u-v-1}{2}$.

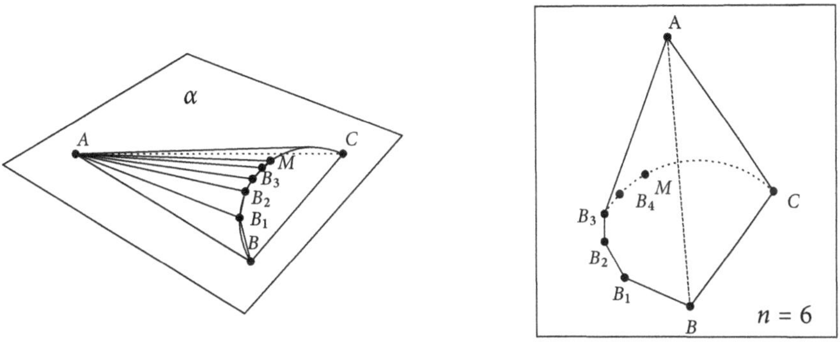

Figure 15.16 Problem 35

Since the numerators are even, p and q are both integers. Since the numerators are less than $2u$, p and q are distinct integers between 0 and $n-1$. We have

$$\frac{(p+q+1)(p-q)}{2} = \frac{uv}{2} = \frac{2n}{2} = n.$$

Therefore for integers n which are not a power of 2 we will have more than one delegate in some seats.

Now consider integers n that are powers of 2. In this case any multiple of n has an even factor of size n. Two delegates sitting in the same place correspond to two integers p and q between 0 and $n-1$ such that $\frac{(p+q+1)(p-q)}{2}$ is a multiple of n. The difference between the two factors in the numerator is $(p+q+1)-(p-q) = 2q+1$, an odd number. Therefore only one of the factors can be even. The value of this even factor is at most $p+q+1$ which is less than $2n$. Hence any even factor of $\frac{(p+q+1)(p-q)}{2}$ is less than n. Thus, when n is a power of 2, $\frac{(p+q+1)(p-q)}{2}$ cannot be a multiple of n and every chair is occupied by a single delegate.

The set of values for which no chair is occupied by more than one delegate is the set of powers of 2, namely $1, 2, 4, 8, 16, \ldots$.

35. The answer is yes. We construct this solid in the following way. To begin with, consider a half of a cone K obtained by intersecting a cone with some axial plane α. Choose on the semicircle BMC (M is the midpoint) a countable set of points $B_0 = B, B_1, B_2, \ldots$ so that $\smile B_iB_{i+1} = \frac{1}{3^{i+1}} \smile BC$, $i \in \mathbb{N} \cup \{0\}$. Then

all points B_i lie on the arc BM. Now consider a solid obtained by cutting K with the planes B_iAB_{i+1}, $i \in \mathbb{N} \cup \{0\}$. This solid has the required property. Indeed, its projection on α is a triangle; now slightly rotate it about the line AB until the projection of B_1 arises on α, thus getting a quadrilateral. Continuing this process further, we can get n-gons for any $n \geq 3$. The case $n = 6$ is shown in the figure 15.16.

36. Substituting $f(x)$ for x in the given equality, we obtain

$$f(f(x)f(y)) = f(f(x)y) + f(x),$$

from which we express $f(f(x)y) = f(f(x)f(y)) - f(x)$. Another expression for the same quantity $f(f(x)y)$ can be obtained by interchanging the variables x and y in the original equality; this gives $f(f(x)y) = f(yx) + y$. Comparing both expressions, we thus arrive at

$$f(f(x)f(y)) = f(yx) + y + f(x),$$

where the left-hand side remains unchanged upon interchanging x and y. The right-hand side must therefore enjoy the same property, that is, we must have

$$f(yx) + y + f(x) = f(xy) + x + f(y), \quad \text{or} \quad y + f(x) = x + f(y).$$

This gives $f(x) - x = f(y) - y$, which is to be satisfied for any $x, y \in \mathbb{R}^+$. This means that the function $x \mapsto f(x) - x$ is constant on $\mathbb{R}^+$, so the sought function f must be of the form $f(x) = x + c$ for a suitable real number c. Substituting this recipe into both sides of the original equation,

$$f(xf(y)) = xf(y) + c = x(y + c) + c = xy + cx + c,$$
$$f(xy) + x = (xy + c) + x = xy + x + c$$

we find that the latter is satisfied if and only if $c = 1$. Thus the sought function f is unique and is given by $f(x) = x + 1$.

Solution 2. Consider the set M of all z such that $f(z) = z + 1$. If we put $x = 1$ in the main equation we find that M contains the set V_f of all values $f(y)$ of the function f. Thus $a \in M \Rightarrow a + 1 = f(a) \in M$ and by induction $a + n \in M$ for all natural n. Next, we show by induction that $a, b \in M \Rightarrow \frac{ab}{a+n} \in M$

for all integer $n \geq 0$. Indeed:

$$\begin{aligned} f\left(\frac{ab}{a+n+1}\right) &= f\left(\frac{ab}{(a+n+1)(a+n)} \cdot (a+n)\right) \\ &= f\left(\frac{ab}{(a+n)(a+n+1)} \cdot f(a+n)\right) - \frac{ab}{(a+n)(a+n+1)} \\ &= f\left(\frac{ab}{(a+n)(a+n+1)} \cdot (a+n+1)\right) - \frac{ab}{(a+n)(a+n+1)} \\ &= \frac{ab}{a+n} + 1 - \frac{ab}{(a+n)(a+n+1)} = \frac{ab}{a+n+1} + 1. \end{aligned}$$

This means that M contains very small positive numbers ε. For any $x > 0$ let $y = \varepsilon/x$. By the formula $f(xf(y)) = f(\epsilon) + x = \epsilon + 1 + x$ we get that all numbers greater than $1+\varepsilon$ belong to M (since the left-hand side is in $V_f \subseteq M$). But ϵ can be chosen arbitrarily small and therefore any $s > 1$ belongs to M. Finally, if $t \leq 1$ we choose z such that $zt > 1$, $zf(t) > 1$ and it follows that

$$zf(t) + 1 = f(zf(t)) = f(zt) + z = zt + 1 + z$$

Hence $f(t) = t + 1$, so $t \in M$.

15.7 Solutions to Paper 7

37. For positive x, y, u, v, we can write the given equation in the form

$$(x + y + u + v)^2 = n^2xyuv$$

or in the form

$$x^2 + 2(y + u + v)x + (y + u + v)^2 = n^2xyuv. \tag{15.7}$$

Let n be a positive integer such that (15.7) has positive integer solutions (x, y, u, v). Let (x_0, y_0, u_0, v_0) be a positive integer solution of (15.7) such that $x_0 + y_0 + u_0 + v_0$ is minimal. We can suppose WLOG that $x_0 \geq y_0 \geq u_0 \geq v_0$. It is easily seen that:

(i) $(y_0 + u_0 + v_0)^2$ is divisible by x_0.

(ii) x_0 is a positive integer zero of the polynomial of degree two:

$$f(x) = x^2 - n^2y_0u_0v_0x + 2(y_0 + u_0 + v_0)x + (y_0 + u_0 + v_0)^2.$$

From *(i)*, *(ii)* and Vieta's theorem we deduce that $x_1 = (y_0 + u_0 + v_0)^2/x_0$ is also a positive integer zero of $f(x)$. Thus (x_1, y_0, u_0, v_0) is a positive integer solution of (15.7). From the definition of the root (x_0, y_0, u_0, v_0), we have

$$x_1 \geq x_0 \geq y_0 \geq u_0 \geq v_0.$$

Note that $f(x) < 0$ only between the roots x_0, x_1. Thus $f(y_0) \geq 0$ and we have

$$\begin{aligned}0 &\leq f(y_0) = y_0^2 - n^2 y_0^2 u_0 v_0 + 2(y_0 + u_0 + v_0)y_0 + (y_0 + u_0 + v_0)^2 \\ &\leq 16y_0^2 - n^2 y_0^2 u_0 v_0.\end{aligned}$$

Therefore $n^2 y_0^2 u_0 v_0 \leq 16 y_0^2$, so $n^2 \leq n^2 u_0 v_0 \leq 16$ and $n \in \{1, 2, 3, 4\}$.

Conversely, $(x, y, u, v) = (4, 4, 4, 4), (2, 2, 2, 2), (1, 1, 2, 2), (1, 1, 1, 1)$ are positive integer solutions to the equation for $n = 1, 2, 3, 4$, respectively.

So the demanded values of n are 1, 2, 3, 4.

38. The given equation can be rewritten in the form:

$$f_n(x) = -\frac{1}{2} + \frac{1}{x-1} + \frac{1}{4x-1} + \cdots + \frac{1}{k^2x-1} + \cdots + \frac{1}{n^2x-1} = 0.$$

a) It is easily seen that for every $n \in \mathbb{N}$, the function $f_n(x)$ is continuous and decreasing on the interval $(1, +\infty)$. Moreover, $f_n(x) \to +\infty$ when $x \to 1^+$ and $f_n(x) \to -1/2$ when $x \to +\infty$. Therefore, for every $n \in \mathbb{N}$, the equation has a unique root $x_n > 1$.

b) For $n \in \mathbb{N}$, we have:

$$\begin{aligned}f_n(4) &= -\frac{1}{2} + \frac{1}{2^2-1} + \frac{1}{4^2-1} + \cdots + \frac{1}{(2k)^2-1} + \cdots + \frac{1}{(2n)^2-1} \\ &= \frac{1}{2}\left(-1 + 1 - \frac{1}{3} + \cdots + \frac{1}{2k-1} - \frac{1}{2k+1} + \cdots + \frac{1}{2n-1} - \frac{1}{2n+1}\right)\end{aligned}$$

so

$$f_n(4) = \frac{-1}{2(2n+1)} < 0 = f_n(x_n).$$

But $f_n(x)$ is decreasing on $(1, +\infty)$, so $x_n < 4$ for all $n \in \mathbb{N}$.

On the other hand, since for all $n \in \mathbb{N}$ the function $f_n(x)$ is differentiable on the interval $(1, +\infty)$ and $[x_n, 4] \subset (1, +\infty)$, the Lagrange mean value theorem (theorem 10.1) shows that there exists $t = t_n \in (x_n, 4)$ such that

$$\frac{f_n(4) - f_n(x_n)}{4 - x_n} = f_n'(t) = \frac{-1}{(t-1)^2} + \frac{-4}{(4t-1)^2} + \cdots + \frac{-n^2}{(n^2t-1)^2} < -\frac{1}{9}$$

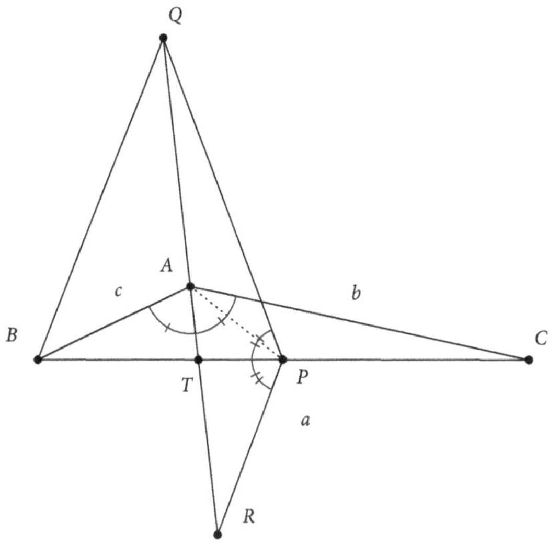

Figure 15.17 Problem 39

so that

$$\frac{-1}{2(2n+1)(4-x_n)} < -\frac{1}{9}, \quad \text{i.e.} \quad x_n > 4 - \frac{9}{2(2n+1)} \qquad \forall n \in \mathbb{N}$$

and it follows that $\lim\limits_{n\to\infty} x_n = 4$.

39. Consider Figure 15.17. Let, as usual, $BC = a$, $CA = b$, $AB = c$. WLOG we can put $AP = 1$. Let BC intersects QR in T and denote $AT = x$ and $TP = y$. The idea is to express $b + c$ as a function $f(x, y)$ and to check that $f(x, y) = f(y, x)$. Because in the triangle PQR, the segments x and y are used in a similar way (but the roles of x and y are reversed) it would be sufficient to conclude that $AB + AC = c + b = r + q = PQ + PR$. But first we do the opposite and express x^2 and y^2 in terms of a, b, c. Because AT divides a in relation $c : b$ it follows that $BT = \frac{ac}{b+c}$ and hence

$$y^2 = (BP - BT)^2 = \left(\frac{a}{2} - \frac{ac}{b+c}\right)^2 = \frac{a^2}{4} \cdot \left(\frac{b-c}{b+c}\right)^2$$

For the bisector $AT = x$ we know (see page 150) that

$$x^2 = bc - BT \cdot CT = bc - \frac{a^2bc}{(b+c)^2} = bc\left(\frac{(b+c)^2 - a^2}{(b+c)^2}\right).$$

At last, for the median $AP = 1$ we know that

$$4 = (2AP)^2 = 2(b^2 + c^2) - a^2.$$

We get a system with three equations and now want to solve it to find $f(x, y)$. For simplicity, let us introduce u and v by $u = (b + c)^2$ and $v = bc$, i.e.

$$b^2 + c^2 = u - 2v; \quad (b - c)^2 = u - 4v.$$

We can rewrite our system as

$$\begin{cases} 4uy^2 = a^2(u - 4v) \\ ux^2 = v(u - a^2) \\ 4 = 2u - 4v - a^2 \end{cases}$$

From the first two equations we have

$$4uy^2 - 4ux^2 = a^2u - 4vu \quad \Rightarrow \quad 4y^2 - 4x^2 = a^2 - 4v.$$

Adding and subtracting this to the third equation we get

$$\begin{aligned} 4 + 4y^2 - 4x^2 = 2u - 8v \quad &\Rightarrow \quad 4v = u - 2(y^2 - x^2) - 2, \\ -4 + 4y^2 - 4x^2 = 2a^2 - 2u \quad &\Rightarrow \quad a^2 = u - 2 + 2(y^2 - x^2) \end{aligned}$$

Now we are able to eliminate both v and a^2 from the first equation:

$$\begin{gathered} 4uy^2 = (u - 2 + 2(y^2 - x^2))(2 + 2(y^2 - x^2)) \quad \Rightarrow \\ u(2y^2 + 2x^2 - 2) = 4(y^2 - x^2)^2 - 4. \end{gathered}$$

Note that $x^2 + y^2 \neq 1$ (otherwise AT is the altitude and the bisector and therefore the median which gives $T = P$). Thus we can divide and express $u = (b + c)^2$ as a symmetric expression in x and y. The same is true for $b + c = \sqrt{u}$, which finishes the proof.

Solution 2. Extend BA to C' such that $AC' = AC$. Extend RP to Q' such that $PQ' = PQ$ (figure 15.18). It follows that $PQ + PR = RQ'$, $AB + AC = BC'$.

We shall now prove that $RQ' = BC'$. The angle $\angle QPR$ is external to isosceles triangle PQQ' and therefore $\angle BPQ = \angle PQQ'$ and $QQ' \parallel BC$. We

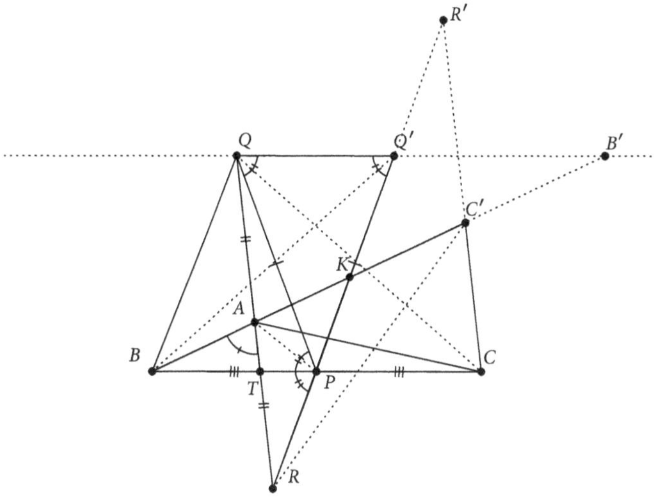

Figure 15.18 Problem 39. Solution 2.

have $\angle BPQ' = \angle RPC = \angle QPC$ and also $PQ' = PQ$, $PB = PC$. This gives us equal triangles $\triangle PBQ'$ and $\triangle CPQ$ and $BQ' = CQ$.

Similarly, $CC' \parallel QR$ and $RC' = QC$.

We now have $BQ' = RC'$. Let BC' and RQ' meet in K. Now to prove $RQ' = BC'$, it is enough to prove that the triangles $BQ'K$ and CRK are similar, because in this case they will be congruent. For this we need only to prove that

$$\frac{Q'K}{KR} = \frac{C'K}{KB}$$

and this is our aim now.

Let the extension of AB meet QQ' in B'. Consider $\triangle RQQ'$ and the transversal AKB'. By Menelaus' theorem

$$\frac{QB'}{B'Q'} \cdot \frac{Q'K}{KR} \cdot \frac{RA}{AQ} = 1,$$

so that

$$\frac{Q'K}{KR} = \frac{B'Q'}{B'Q}.$$

Similarly, considering $\triangle BCC'$ and the transversal PKR' we get

$$\frac{C'K}{KB} = \frac{C'R'}{R'C}.$$

Thus it remains to show that

$$\frac{B'Q'}{B'Q} = \frac{C'R'}{R'C}.$$

Since $PT \parallel QQ'$ and $BT \parallel QB'$ we have

$$\frac{PT}{QQ'} = \frac{RT}{RQ}; \quad \frac{B'Q}{BT} = \frac{AQ}{AT}.$$

From this follows that

$$\frac{B'Q}{Q'Q} = \frac{BT}{PT} \cdot \frac{RT}{RQ} \cdot \frac{AQ}{AT} = \frac{1}{2}\frac{BT}{PT} \cdot \frac{RT}{AT}.$$

The obtained result is now quite symmetric, and similar calculations for $\frac{C'R'}{R'C}$ give the same result, which finishes the proof.

40. Note that $P(x) = C\prod(x - x_i)$ for some constant C and, according to the Leibniz rule,

$$P'(x) = C\sum_{j=1}^{n}\prod_{i\neq j}(x - x_i),$$

$$P''(x) = C\sum_{j=1}^{n}\sum_{l\neq j}\prod_{i\neq j,l}(x - x_i).$$

This means that all the terms containing the factor $(x - x_k)$ vanish when $x = x_k$, so only $j = k$ or $l = k$ give non-zero terms and we get

$$P'(x_k) = C\prod_{i\neq k}(x_k - x_i),$$

$$P''(x_k) = C\sum_{l\neq k}\prod_{i\neq k,l}(x_k - x_i) + C\sum_{j\neq k}\prod_{i\neq j,k}(x_k - x_i) = 2C\sum_{l\neq k}\prod_{i\neq k,l}(x_k - x_i).$$

That is why

$$\frac{P''(x_k)}{P'(x_k)} = 2\sum_{l\neq k}\frac{1}{x_k - x_l} \quad \text{and} \quad \sum_{k=1}^{n}\frac{P''(x_k)}{P'(x_k)} = 2\sum_{k=1}^{n}\sum_{l\neq k}\frac{1}{x_k - x_l} = 0,$$

since $\dfrac{1}{x_k - x_l} + \dfrac{1}{x_l - x_k} = 0.$

Solution 2. Here is a proof by induction. Assume that the statement is established for $P(x)$ and let us prove that it follows for $Q(x) = P(x)(x - x_{n+1})$. We have

$$Q'(x) = P(x) + (x - x_{n+1})P'(x);$$
$$Q''(x) = 2P'(x) + (x - x_{n+1})P''(x);$$

so

$$\frac{Q''(x_{n+1})}{Q'(x_{n+1})} = \frac{2P'(x_{n+1})}{P(x_{n+1})} = \sum_{k=1}^{n} \frac{2}{x_{n+1} - x_k}$$

and for $k \leq n$ we have (since $P(x_k) = 0$)

$$\frac{Q''(x_k)}{Q'(x_k)} = \frac{2}{x_k - x_{n+1}} + \frac{P''(x_k)}{P'(x_k)}.$$

Summing over k and using the induction assumption finishes the argument.

41. We start with a lemma.

Lemma 15.4 *Let p be a prime number and let $a_1, a_2, \ldots, a_k$ be $k < p$ (not necessarily different) non-zero residues modulo p. For every non-empty subset $M \subset \{1, 2, \ldots, k\}$ let $S_M = (\sum_{i \in M} a_i) \pmod{p}$. Then there exist at least k different non-zero residues S_{M_j}.*

Proof. Induction on k. For $k = 1$, the sum of one element a_1 is non-zero. For the induction step, consider $k + 1 < p$ non-zero residues $a_1, a_2, \ldots, a_k, a_{k+1}$ and let $0 < S_{M_1} < S_{M_2} < \ldots S_{M_k} < p$ be k different sums that we can obtain (by the induction assumption) using the first k numbers only. First consider the case when $a_{k+1} = 1$. If for some j we have $S_{M_j} + 1 < S_{M_{j+1}}$, then the sum $S_{M_j} + 1 = S_{M_j \cup \{k+1\}}$ gives a new residue. Otherwise $S_{M_k} - S_{M_1} = k - 1$ and either $S_{M_k} < p - 1$ so that $S_{M_k} + 1 = S_{M_k \cup \{k+1\}}$ is a new residue, or $1 < S_1$ and then a_{k+1} itself is a new residue (here we used that $k + 1 < p$).

Finally, consider the case when $a_{k+1} \neq 1$. Since we can divide modulo a prime number we can consider the $k + 1$ residues $b_i = \frac{a_i}{a_{k+1}} \pmod{p}$ and use the previous case to obtain $k + 1$ different sums (note that $b_{k+1} = 1$). Multiplying those sums by a_{k+1} we get the desired result and the lemma is proved. □

Now we are ready to prove the main result. Consider first the case when $n = p$ is a prime number. Let $x_1 \leq x_2 \leq \cdots \leq x_{2p-1}$ be the ordered residues of the given numbers. We have to choose p of them such that their sum is equal to zero modulo p. If for some i we have $x_i = x_{p+i-1}$, then $x_i = x_{i+1} = \cdots = x_{p+i-1}$ and their sum is equal to zero modulo p. Otherwise the $p-1$ numbers

$$\begin{aligned} a_1 &= x_{p+1} - x_2, \\ a_2 &= x_{p+2} - x_3, \\ a_3 &= x_{p+3} - x_4, \\ &\vdots \\ a_{p-1} &= x_{p+p-1} - x_p \end{aligned}$$

are non-zero and we can, according to the lemma, create all $p-1$ non-zero residues by summing them. If $S = a_1 + \ldots + a_p = 0 \pmod{p}$, we have found the desired p elements. If it is not zero, we can add S_M such that $S + S_M = 0 \pmod{p}$ and this sum is the desired one — it is equal to

$$x_1 + \sum_{i \notin M} x_{i+1} + \sum_{i \in M} x_{p+i}.$$

It remains to consider the case when $n = ab$ is not a prime number. Here we can use induction by the number of prime factors (and what we already proved). From some $2a-1$ of the $2n-1$ numbers ($2a-1 < 2n-1$) we can choose a such that their sum A_1 is divisible by a. We can pick up the next $2a-1$ numbers from the remaining $2n-a-1$ numbers and choose the next a from them with sum A_2 divisible by a. Continuing this process we stop when the number of remaining numbers $2n-am-1$ gets less than $2a-1$; the sum corresponding to the last step is A_m. In other words,

$$\begin{aligned} 2n - am - 1 < 2a - 1 \quad &\Leftrightarrow \quad 2n < a(m+2) \\ &\Leftrightarrow \quad 2b < m+2 \quad \Leftrightarrow \quad 2b - 1 \le m. \end{aligned}$$

So we can get $2b-1$ numbers $A_1 = ax_1, \ldots, A_{2b-1} = ax_{2b-1}$. Using the induction assumption once again we choose b of the $2b-1$ indices such that the sum of the corresponding x_i numbers is divisible by b. This means that we get ab numbers such that their sum is divisible by ab.

Solution 2. Consider the sum

$$\sum (x_{\sigma(1)} + x_{\sigma(2)} + \cdots + x_{\sigma(p)})^{p-1},$$

where the summation is over all ordered choices of p different indices between 1 and $2p-1$. The number of such choices is $\binom{2p-1}{p}$, a number that is not

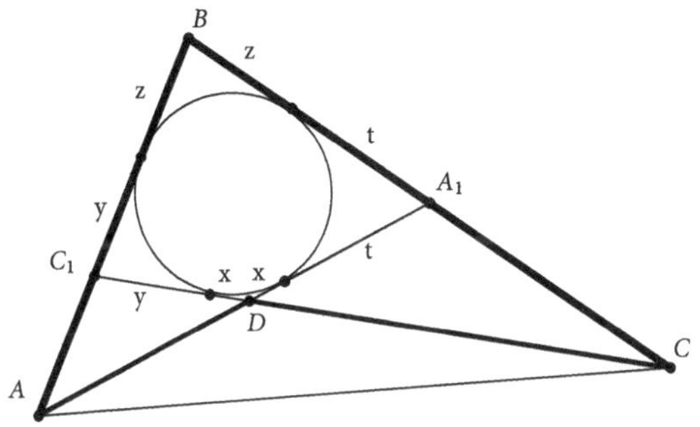

Figure 15.19 Problem 42

divisible by p. If the statement is false, then by Fermat's little theorem the sum will not be divisible by p. On the other hand, each term $Cx_{i_1}^{e_1}x_{i_2}^{e_2}\dots x_{i_k}^{e_k}$ with $e_1+e_2+\ldots+e_k = p-1$ and k nonzero e_i is included $\binom{2p-1-k}{p-k}$ times (the number of ways to choose the remaining $p-k$ indices), which is divisible by p.

42. We start with a useful fact.

Lemma 15.5 *Let D be a point inside the triangle ABC, let C_1 be the intersection point of the lines AB and CD, and let A_1 be the intersection point of the lines AD and BC. It is possible to inscribe a circle in DA_1BC_1 if and only if $AB + CD = AD + BC$. (So, even if $ABCD$ is not convex the condition for the existence of a suitably defined inscribed circle remains the same.)*

Proof. To prove the necessity, assume that it is possible to inscribe a circle in DA_1BC_1 (see figure 15.19). Let x, y, z, t be the lengths of the common tangents from the points D, C_1, B, A_1 correspondingly. Then

$$AD + x = AC_1 + y, \quad CD + x = CA_1 + t$$

as lengths of common tangents. Now

$$\begin{aligned} AB + CD &= AC_1 + y + z + CA_1 + t - x \\ &= (AD + x - y) + y + z + (BC - z - t) + t - x = AD + BC. \end{aligned}$$

To prove the sufficiency of the above condition, assume the contrary, i.e. assume that $AB + CD = AD + BC$, but that it is impossible to inscribe a circle in DA_1BC_1. We can then inscribe a smaller circle that is tangent to only three of the sides, including BC_1 and BA_1. Suppose that it is not tangent to A_1A (the other case can be considered similarly). Let AA_2 be a tangent from the point A, different from AB. Let D_1 be the common point for AA_2 and C_1C. According to the first part of the proof we have

$$AD_1 + BC = CD_1 + AB \quad \Rightarrow \quad AD - AD_1 = CD - CD_1 = -DD_1,$$

so that $AD = AD_1 - DD_1$ – a contradiction. □

Returning to the original problem, we have two inscribed circles and have to prove the existence of a third. Using the lemma we get:

$$\left.\begin{aligned} BO + AC &= CO + AB \\ AO + BC &= CO + AB \end{aligned}\right\} \quad \Rightarrow \quad BO + AC = AO + BC$$

and we can apply the lemma once again.

15.8 Solutions to Paper 8

43. *a)* and *b)* The answer is no. For T-shaped pieces, use the usual colouring of a chessboard. For L-shaped pieces, colour every second row white and every second black (it doesn't matter how you start). In both cases now each piece will cover three cells of one colour and one cell of the other colour. It follows that we need as many pieces covering 3 black and 1 white cells, as we need pieces covering 3 white and 1 black cells. But the total number of pieces is in both cases $\frac{100}{4} = 25$, an odd number, which means both coverings are impossible.

    c) The answer is yes, by construction, see Figure 15.20.

44. Denote the angles of the triangle by α, β, γ and the side lengths by a, b, c (a lies opposite α, etc.). Similarity does not affect the angles, nor does it affect their sines and cosines, which means we are free to choose the length of one of the sides. Choose $a = 1$. By the law of sines we then have

$$\frac{1}{\sin\alpha} = \frac{b}{\sin\beta} = \frac{c}{\sin\gamma},$$

Figure 15.20 Problem 43c

which means all side lengths are rational numbers. Then, by the cosine rule,

$$\begin{aligned} 1 &= b^2 + c^2 - 2bc\cos\alpha, \\ b^2 &= c^2 + 1 - 2c\cos\beta, \\ c^2 &= 1 + b^2 - 2b\cos\gamma, \end{aligned}$$

and hence all three cosines are rational numbers too.

Solution 2.(essentially the same). Put $a = 2R\sin\alpha$, $b = 2R\sin\beta$, $c = 2R\sin\gamma$ in the cosine rule and get

$$\sin^2\alpha + \sin^2\beta - 2\sin\alpha\sin\beta\cos\gamma = \sin^2\gamma,$$

which implies that $\cos\gamma$ is rational.

Solution 3. We get that $\cos^2\alpha = 1 - \sin^2\alpha; \cos^2\beta = 1 - \sin^2\beta$ are rational numbers. The same is true for

$$\begin{aligned} \cos^2\gamma &= \cos^2(\alpha+\beta) \\ &= \cos^2\alpha\cos^2\beta - 2\cos\alpha\cos\beta\sin\alpha\sin\beta + \sin^2\alpha\sin^2\beta. \end{aligned}$$

Thus $\cos\alpha\cos\beta$ and $\cos\gamma = -\cos(\alpha+\beta) = \sin\alpha\sin\beta - \cos\alpha\cos\beta$ are rational as well.

45. We have

$$\frac{a}{b^2+1} = \frac{a+ab^2-ab^2}{b^2+1} = a - \frac{ab^2}{b^2+1}.$$

Since the sum of all three numbers is 3, the inequality is equivalent to

$$\frac{ab^2}{b^2+1} + \frac{bc^2}{c^2+1} + \frac{ca^2}{a^2+1} \le \frac{3}{2}.$$

Consider three cases.

(i) Let all three numbers be positive. We then have

$$\begin{aligned}\frac{ab^2}{b^2+1} + \frac{bc^2}{c^2+1} + \frac{ca^2}{a^2+1} &= \frac{ab}{b+\frac{1}{b}} + \frac{bc}{c+\frac{1}{c}} + \frac{ca}{a+\frac{1}{a}}\\ &\le \frac{1}{2}(ab+bc+ca)\\ &= \frac{1}{2}\cdot\frac{1}{2}\left((a+b+c)^2-(a^2+b^2+c^2)\right)\\ &= \frac{1}{4}\left(9-(a^2+b^2+c^2)\right)\\ &\le \frac{1}{4}\left(9-3\left(\frac{a+b+c}{3}\right)^2\right) = \frac{3}{2}.\end{aligned}$$

(ii) If two of the numbers are equal to zero, a strict inequality obviously holds.

(iii) Suppose only one of the numbers is equal to zero, e.g. $a = 0$, $b, c \neq 0$. We have

$$0 + \frac{bc^2}{c^2+1} + 0 = \frac{bc}{c+\frac{1}{c}} \le \frac{1}{2}bc = \frac{1}{2}b(3-b) \le \frac{9}{8} < \frac{3}{2}.$$

Equality can be obtained if and only if

$$a + \frac{1}{a} = b + \frac{1}{b} = c + \frac{1}{c} = 2 \quad \text{and} \quad a = b = c,$$

which is equivalent to $a = b = c = 1$.

Solution 2. Multiplication by all the denominators gives that the inequality is equivalent to

$$\begin{aligned}&2a(a^2+1)(c^2+1) + 2b(b^2+1)(a^2+1) + 2c(c^2+1)(b^2+1)\\ &\quad\ge 3(a^2+1)(b^2+1)(c^2+1).\end{aligned}$$

Since $x^2+1 \geq 2x$ for all real x (with equality iff $x=1$), it suffices to prove that

$$\begin{aligned}4a^2(c^2+1)+4b^2(a^2+1)+4c^2(b^2+1)\\ \geq 3(a^2b^2c^2+a^2b^2+b^2c^2+c^2a^2+a^2+b^2+c^2+1),\end{aligned}$$

which is equivalent to

$$a^2b^2+b^2c^2+c^2a^2+a^2+b^2+c^2 \geq 3(a^2b^2c^2+1). \tag{15.8}$$

The AM–GM inequality gives that $abc \leq 1$ (with equality iff $a=b=c=1$) and hence it suffices to prove that

$$a^2b^2+b^2c^2+c^2a^2+a^2+b^2+c^2 \geq 6.$$

We have

$$\begin{aligned}&a^2b^2+b^2c^2+c^2a^2+a^2+b^2+c^2\\ &\quad= a^2b^2+b^2c^2+c^2a^2+(a+b+c)^2-2ab-2bc-2ca\\ &\quad= (ab-1)^2+(bc-1)^2+(ca-1)^2-3+9 \geq 6.\end{aligned}$$

Equality holds if and only if $a=b=c=1$.

Remark A shorter way to prove (15.8) is to apply Jensen's inequality to the convex functions x^2 and x^{-2}:

$$\frac{a^2+b^2+c^2}{3} \geq \left(\frac{a+b+c}{3}\right)^2 = 1 \quad \Rightarrow \quad a^2+b^2+c^2 \geq 3;$$

$$\frac{a^{-2}+b^{-2}+c^{-2}}{3} \geq \left(\frac{a+b+c}{3}\right)^{-2} = 1 \quad \Rightarrow \quad a^2b^2+a^2c^2+b^2c^2 \geq 3a^2b^2c^2.$$

Adding this we obtain (15.8).

The difficulty in the problem is that everything was placed "wrongly". For example, we could not use the inequality $x^2+1 \geq 2x$ in the denominators from the very beginning (but have used it in the numerators later). The substitution that we applied in the first solution helped to change the directions of inequalities and this is a standard trick in such cases.

46. For all $x \in [0, \frac{1}{2}]$ we have $x+x \leq 1$ and hence $2f(x) \leq f(2x)$. For all $x \in [0, \frac{1}{4}]$ we also have $2x \in [0, \frac{1}{2}]$, so that $4f(x) \leq 2f(2x) \leq f(4x)$. By induction, for all $x \in [0, \frac{1}{2^k}]$ we have $2^k f(x) \leq f(2^k x)$, $k \in \mathbb{N}$. (From any of these inequalities we immediately get that $f(0)=0$.)

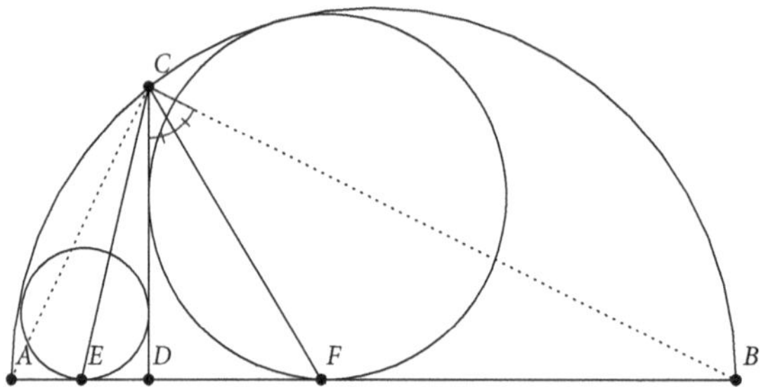

Figure 15.21 Problem 47

Also, for all $x \in [0,1]$ we have $x + (1 - x) = 1$ and hence $f(x) + f(1 - x) \le f(1) = 1$. Since all values of f are non-negative, it follows that $f(x) \le 1 - f(1 - x) \le 1$ for all $x \in [0,1]$.

Assume now that there exists $x_0 \in [0,1]$ and such that $f(x_0) > 2x_0$. We know that $f(0) = 0$, which means that $x_0 > 0$. Since $\frac{1}{2^k} \to 0$ as $k \to \infty$, there must exist $k_0 \in \mathbb{N}$ such that $x_0 > \frac{1}{2^{k_0}}$. The inequality $2x_0 < f(x_0)$ implies that $2x_0 < 1$, which means that $x_0 < \frac{1}{2}$. We then have $2f(x_0) \le f(2x_0) \le 1$, which gives $f(x_0) \le \frac{1}{2}$. The inequality $2x_0 < f(x_0)$ then implies that $x_0 < \frac{1}{4}$, so that $f(x_0) \le \frac{1}{4}$, hence $x_0 < \frac{1}{8}$, etc. By induction we get that $x_0 < \frac{1}{2^n}$ for all $n \in \mathbb{N}$, which contradicts the existence of k_0 as discussed above. The contradiction means that there exists no $x_0 \in [0,1]$ such that $f(x_0) > 2x_0$, hence $f(x) \le 2x$ for all $x \in [0,1]$.

47. We shall prove that $AF = AC$. By this equality and by the similarity of $\triangle ABC$ and $\triangle CBD$, it will follow that

$$\angle BCF = \angle BCD - \angle FCD = \angle A - \angle FCD = \angle A - (\frac{\pi}{2} - \angle AFC)$$
$$= \angle AFC - \angle B = \angle ACF - \angle ACD = \angle DCF.$$

Similarly, $BE = BC$ and hence $\angle ACE = \angle ECD$. It then follows that

$$\angle ECF = \angle ECD + \angle DCF = \frac{1}{2}(\angle ACD + \angle DCB) = \frac{\pi}{4}.$$

So, let us prove that $AF = AC$. Denote the centre of k by O, the centres of the inscribed circles in the curvilinear triangles CAD and BCD by O_1 and O_2, respectively. Denote the corresponding radii by R, r_1 and r_2. Without loss of generality we can assume that $BC \geq AC$, which means that D lies between A and O, or, in the case of an isosceles triangle, D will coincide with O. Also, F and O will lie between D and B (in some order). Since $OO_2 = R - r_2$, $O_2F = r_2$ and $\angle OFO_2 = \frac{\pi}{2}$, we have

$$AF = AO \pm OF = R \pm OF = R \pm \sqrt{(R - r_2)^2 - r_2^2} = R \pm \sqrt{R^2 - 2Rr_2}.$$

Since the circle of radius r_2 and centre O_2 is tangent to both DB and DC and since $\angle D = \frac{\pi}{2}$, we have that $O_2F = DF = r_2$. Hence by Theorem 4.3,

$$\begin{aligned} AC^2 &= AB \cdot AD = 2R(R - OD) \\ &= 2R(R - (DF \mp OF)) = 2R\left(R - r_2 \pm \sqrt{R^2 - 2Rr_2}\right) \\ &= (R^2 - 2Rr_2) \pm 2R\sqrt{R^2 - 2Rr_2} + R^2 \\ &= (R \pm \sqrt{R^2 - 2Rr_2}\,)^2 = AF^2, \end{aligned}$$

which gives the desired equality $AC = AF$.

Solution 2. Again $ED = r_1$. We have that $OA - ED = OO_1$ and hence we have $(OA - ED)^2 = OO_1^2 = OE^2 + r_1^2$, which gives $(R - r_1)^2 = (r_1 + OD)^2 + r_1^2$, i.e. $r_1^2 + OD^2 + 2(R + OD)r_1 - R^2 = 0$. By Theorem 4.3, $AC^2 = AD \cdot AB$ and $DC^2 = AD \cdot DB$. Division yields

$$\frac{AC^2}{DC^2} = \frac{AB}{DB} = \frac{2R}{R + OD}.$$

We also have

$$\frac{EA^2}{ED^2} = \frac{(R - OD - r_1)^2}{r_1^2}.$$

We shall prove that

$$\frac{AC}{DC} = \frac{EA}{ED}.$$

Equivalent transformations of

$$\frac{2R}{R + OD} = \frac{(R - OD - r_1)^2}{r_1^2}$$

give $2Rr_1^2 = (R + OD)(R^2 + OD^2 + r_1^2 - 2R \cdot OD - 2Rr_1 + 2OD \cdot r_1)$. Combining this with the previously proved equality $r_1^2 + OD^2 + 2(R + OD)r_1 - R^2 = 0$, we get

$2Rr_1^2 = (R+OD)(2R^2-2R\cdot OD-4Rr_1)$, so that $r_1^2 = (R+OD)(R-OD-2r_1)$, or, $r_1^2 + OD^2 + 2(R + OD)r_1 - R^2 = 0$, which we already know to be true. Since we only used equivalent transformations, the desired equality

$$\frac{AC}{DC} = \frac{EA}{ED}$$

is proved. By Theorem 1.15 this means that CE is the bisector of $\angle ACD$. Similarly, CF is the bisector of $\angle DCB$ and we are done.

48. Denote the weights of the cows by $x_1, x_2, \ldots, x_{101}$. These numbers then satisfy a linear 101×101 system of equations with all coefficients along the main diagonal equal to zero, 50 coefficients in each equation equal to 1, and 50 coefficients in each equation equal to -1. All right-hand-sides are equal to zero, i.e. the system is homogeneous. For a homogeneous system, the sum of any two solutions is a solution and any multiple of a solution is also a solution.

Obviously $x_1 = x_2 = \cdots = x_{101} = c$ solves the system for any real c. We need to prove that these equalities occur for all (non-trivial) solutions. (Actually, to make any sense, all solutions to the problem need to be strictly positive.)

Assume the contrary, i.e. assume that there exists a solution $(y_1, y_2, \ldots, y_{101})$ such that not all y_i are equal to each other. Since the system is homogeneous, we can add any solution to $(y_1, y_2, \ldots, y_{101})$ and still get a solution; let us add $(c, c, \ldots, c)$ in such a way that in the new solution (denoted again by $(y_1, \ldots, y_{101})$) we have $y_k = 0$ for some k. By our assumption above, this solution has to be non-trivial (i.e. at least one $y_i \neq 0$). The system has integer coefficients. By induction (the details can be found in the Lemma below), we conclude that there is a rational non-trivial solution $(y_1, y_2, \ldots, y_{101})$ with the above property (some $y_k = 0$). Multiplying if necessary all equations by a suitable natural number, we can further assume that all y_i are integers. Also, without loss of generality, we can assume that $y_{101} = 0$ and that the largest common divisor of all y_k is 1. Consider the last equation: since all coefficients are ± 1, there must be an even number m of odd numbers among y_k, $k = 1, \ldots, 100$. We have assumed the solution to be non-trivial, hence $m \geq 2$. But then $(y_1, y_2, \ldots, y_{101})$ couldn't satisfy any of the equations where 0 stands as coefficient in front of an odd y_k. Contradiction!

The contradiction means that $x_1 = x_2 = \cdots = x_{101}$ must hold for all solutions.

Lemma 15.6 *If a homogeneous linear system of equations in n variables with rational coefficients has a non-trivial solution such that $x_n = 0$, then it also has a rational non-trivial solution with the same property.*

Proof. Note that a solution is rational iff all free variables are rational (because we get the basic variables from them by division with rational pivot elements — see Section 9.1 for more details). For the same reasons, the solution of a homogeneous system is trivial iff all free variables are zero.

Thus to have a non-trivial solution, we have to have free variables. Note that at least one of them is different from x_n. Indeed, x_n cannot be the only free variable because otherwise the only solution with $x_n = 0$ would be trivial. So let x_i with $i \neq n$ be a free variable. Let $x_i = 1$ and other free variables be zero. The obtained solution is non-trivial and because x_n was the last variable we still have $x_n = 0$ both in the cases when x_n is free and basic. □

15.9 Solutions to Paper 9

49. We have that $a_n(-1)^n + \cdots + a_2 - a_1 + a_0 = 0$, which means that

$$\sum_{0\le 2k\le n} a_{2k} = \sum_{1\le 2k+1\le n} a_{2k+1}.$$

Also,

$$a_n(\sqrt{2})^n + \cdots + a_2 \cdot 2 + a_1\sqrt{2} + a_0 = \sum_{0\le 2k\le n} a_{2k} \cdot 2^k + \sqrt{2}\sum_{1\le 2k+1\le n} a_{2k+1} \cdot 2^k$$

is an integer and hence

$$\sum_{1\le 2k+1\le n} a_{2k+1} \cdot 2^k = 0.$$

From the last equality we get that

$$a_1 = -\sum_{3\le 2k+1\le n} a_{2k+1} \cdot 2^k,$$

which means that a_1 is an even number. On the other hand,

$$p(1) = \sum_{0\le m\le n} a_m = 2\sum_{0\le 2k\le n} a_{2k},$$

so that $p(1) + a_1$ is an even number as the sum of two even numbers.

50. First of all we observe that no angle formed by three of the four points can be greater than or equal to 120°, because if we assume that $\angle ABC \geq 120°$, then from $AB \geq 5$ and $BC \geq 5$ we deduce $AC \geq 5\sqrt{3} > 6\sqrt{2}$, which is a contradiction. Now, if the quadrilateral $ABCD$ is not convex, then one of the four points must lie inside the triangle formed by the other three. Suppose that D lies in the interior of $\triangle ABC$. Then one of the angles $\angle ADB$, $\angle BDC$ and $\angle CDA$ has to be greater than or equal to 120°, a contradiction. Thus $ABCD$ is a convex quadrilateral. Since each angle of the triangle ABC is smaller than 120° and at least one of its angles, say $\angle ABC$, must be greater than or equal to 60°, it follows that

$$\sin \angle ABC \geq \frac{\sqrt{3}}{2},$$

so that for the area S_{ABC} of the triangle we get

$$S_{ABC} = \frac{1}{2} AB \cdot BC \cdot \sin \angle ABC \geq \frac{\sqrt{3}}{4} \cdot 25 > \frac{21}{2}.$$

Analogously one can prove that the area of $\triangle ACD$ is greater than $\frac{21}{2}$ and thus the area of the quadrilateral $ABCD$ is greater than 21.

51. If $b_n = \frac{1}{a_n}$ then $b_1 = 2$ and $b_{n+1} = b_n^2 - b_n + 1$ for any $n \geq 1$. We deduce $b_{n+1} - 1 = b_n(b_n - 1) = b_n b_{n-1}(b_{n-1} - 1) = \ldots = b_n b_{n-1} \cdots b_1$ and

$$b_{n+1} = b_1 \cdots b_n + 1.$$

Hence

$$\frac{1}{b_1 \cdots b_n} = \frac{1}{b_{n+1}} + \frac{1}{b_1 \cdots b_{n+1}} \quad \Rightarrow \quad \frac{1}{b_k} = \frac{1}{b_1 \cdots b_{k-1}} - \frac{1}{b_1 \cdots b_k}$$

for $k > 1$. Summing up and using $\frac{1}{b_1} = \frac{1}{2}$ we obtain

$$\frac{1}{b_2} + \cdots + \frac{1}{b_n} = \frac{1}{b_1} - \frac{1}{b_1 \cdots b_n} \quad \Rightarrow \quad \frac{1}{b_1} + \cdots + \frac{1}{b_n} + \frac{1}{b_1 \cdots b_n} = 1,$$

whence the conclusion holds.

52. a) By Theorem 4.3, given a right-angled triangle with legs a, b and hypotenuse c, the diameter d of its inscribed circle satisfies $d = a + b - c$.

Using this fact and the inequality between arithmetic and quadratic means,

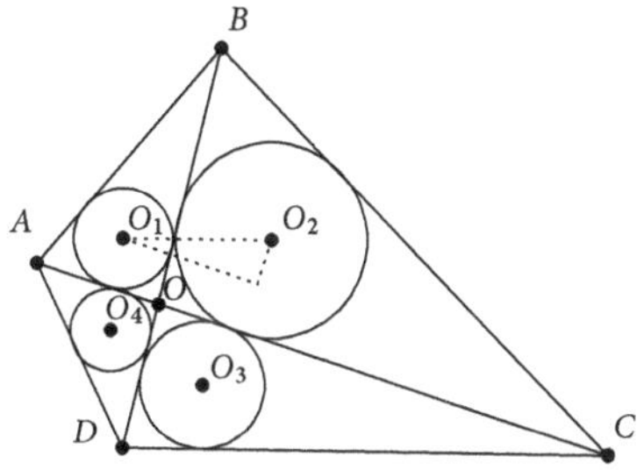

Figure 15.22 Problem 52.

we get for the diameters of S_1, S_2, S_3, S_4

$$\begin{aligned}
&d_1 + d_2 + d_3 + d_4 \\
&\quad = 2OA + 2OB + 2OC + 2OD - AB - BC - CD - DA \\
&\quad = 2(AC + BD) - (AB + BC + CD + DA) \\
&\quad = 2(AC + BD) - \Big(\sqrt{OA^2 + OB^2} + \sqrt{OB^2 + OC^2} \\
&\qquad\qquad + \sqrt{OC^2 + OD^2} + \sqrt{OD^2 + OA^2}\Big) \\
&\quad \le 2(AC + BD) - \frac{1}{\sqrt{2}}\big(OA + OB + OB + OC \\
&\qquad\qquad + OC + OD + OD + OA\big) \\
&\quad = 2(AC + BD) - \sqrt{2}(AC + BD) = (2 - \sqrt{2})(AC + BD).
\end{aligned}$$

b) The Pythagorean theorem implies that

$$(O_1O_2)^2 = (r_1 + r_2)^2 + (r_1 - r_2)^2 = 2(r_1^2 + r_2^2).$$

Hence

$$\begin{aligned}
&O_1O_2 + O_2O_3 + O_3O_4 + O_4O_1 \\
&\quad = \sqrt{2}\left(\sqrt{r_1^2 + r_2^2} + \sqrt{r_2^2 + r_3^2} + \sqrt{r_3^2 + r_4^2} + \sqrt{r_4^2 + r_1^2}\right) \\
&\quad \le \sqrt{2}\,(r_1 + r_2 + r_2 + r_3 + r_3 + r_4 + r_4 + r_1) \\
&\quad = \sqrt{2}\,(d_1 + d_2 + d_3 + d_4) \le 2(\sqrt{2} - 1)(AC + BD).
\end{aligned}$$

53. Let us denote by $S(n)$ the set of simple words with n letters and by $f(n)$ the number of elements in $S(n)$. If we put a letter at the end of each of the simple words from $S(n)$ we obtain a set $T(n+1)$ of t_{n+1} words of length $n+1$, their number being $t_{n+1} = 4f(n)$. Obviously,

$$S(n+1) \subset T(n+1), \ S(n+1) \neq T(n+1).$$

Let $T_1(n+1)$ be the set of those words from $T(n+1)$ that have the last two letters the same, $T_k(n+1)$ the set of words from $T(n+1)$ that end in two consecutive identical groups of k letters, for each $k \in \{1, 2, \ldots, m\}$, where $m = \left[\frac{n+1}{2}\right]$. Obviously

$$f(n+1) \geq t_{n+1} - |T_1(n+1)| - |T_2(n+1)| - \cdots - |T_m(n+1)|.$$

We have $t_{n+1} = 4f(n)$, $|T_1(n+1)| = f(n)$, and furthermore $|T_k(n+1)| \leq f(n+1-k)$, because of the fact that the mapping of $S(n+1-k)$ into $T_k(n+1)$ given by adding to a simple word of $n+1-k$ letters its own last k letters is obviously surjective.

We want to prove by induction that $f(n) > 2f(n-1)$. We have $f(1) = 4$, $f(2) = 12 > 2f(1)$ as the base case. Induction hypothesis $f(n) > 2f(n-1)$ implies $f(n) > 2^i f(n-i)$ and we have

$$\begin{aligned} f(n+1) &\geq 4f(n) - f(n) - f(n-1) - f(n-2) - f(n-3) - \cdots \\ &> 4f(n) - f(n) - \frac{1}{2}f(n) - \frac{1}{4}f(n) - \frac{1}{8}f(n) - \cdots > 2f(n), \end{aligned}$$

from which the conclusion follows.

54. It is obvious that each of the three numbers needs to be less than 2. We can therefore write $a = 2\cos\alpha$, $b = 2\cos\beta$, $c = 2\cos\gamma$, where $\alpha, \beta, \gamma \in (0, \frac{\pi}{2})$. The condition on a, b, c can then be rewritten as

$$\cos^2\alpha + \cos^2\beta + \cos^2\gamma + 2\cos\alpha\cos\beta\cos\gamma = 1.$$

As we have seen on page 156, this is equivalent to the condition $\alpha + \beta + \gamma = \pi$. The cosine function is concave on $(0, \frac{\pi}{2})$, which means that by Jensen's inequality

$$\frac{\cos\alpha + \cos\beta + \cos\gamma}{3} \leq \cos\frac{\alpha+\beta+\gamma}{3} = \cos\frac{\pi}{3} = \frac{1}{2}.$$

Hence $a + b + c = 2(\cos\alpha + \cos\beta + \cos\gamma) \leq 3$.

15.10 Solutions to Paper 10

55. Since a, b, c, d are positive and so is t, we note that $1 - a^2, 1 - b^2, 1 - c^2$ and $1 - d^2$ must also be positive, and it follows that $0 < a, b, c, d < 1$.

If any two variables are not equal, say $a \neq b$, we can assume $a < b$. Since $a(1 - b^2) = b(1 - c^2)$, it follows that $1 - b^2 > 1 - c^2$, and therefore $b < c$. Similarly we obtain $c < d$ and $d < a$, and therefore $a < b < c < d < a$, which is a contradiction. Since an analogous argument holds for $a > b$, we see that $a = b = c = d$ must hold with $a(1 - a^2) = t$.

We note that the value of $f(a) = a(1 - a^2)$ is negative for $a > 1$. Furthermore, we have $f(0) = f(1) = 0$. The equation

$$f(a) = a - a^3 = t$$

therefore has no real solutions if t is larger than the maximum value M of $f(a)$ for $0 \leq a \leq 1$, exactly one solution if t is equal to M, and two solutions if t is greater than 0 but less than M (draw the picture or study the derivative to be sure). In order to determine the value of M, we solve the equation

$$\begin{aligned} f'(x) = 0 \quad &\Leftrightarrow \quad 1 - 3x^2 = 0 \\ &\Rightarrow \quad x = \frac{\sqrt{3}}{3} \quad \Rightarrow \quad M = f(x) = \frac{2\sqrt{3}}{9}. \end{aligned}$$

Note that $f(x)$ must have a maximum being continuous and x is the only possible choice because $f(0) = f(1) < f(x)$.

It follows that the given system of equations has two solutions of the form (a, a, a, a) for $0 < t < \frac{2\sqrt{3}}{9}$, exactly one such solution for $t = \frac{2\sqrt{3}}{9}$ and no real solutions for $t > \frac{2\sqrt{3}}{9}$.

56. Let A, B, C, D be the prices in cents. We have

$$\begin{cases} \dfrac{A}{100} + \dfrac{B}{100} + \dfrac{C}{100} + \dfrac{D}{100} = \dfrac{711}{100} \\ \dfrac{A}{100} \cdot \dfrac{B}{100} \cdot \dfrac{C}{100} \cdot \dfrac{D}{100} = \dfrac{711}{100} \end{cases}$$

so that

$$\begin{cases} A + B + C + D = 711 \\ A \cdot B \cdot C \cdot D = 79 \cdot 2^6 \cdot 3^2 \cdot 5^6. \end{cases}$$

One of the numbers has to be divisible by 79, assume that $A = 79a,\ a < 9$. Not all of them are divisible by 5, so one of them has to be divisible by 25, let $B = 25b$. Evidently $b < 25$, so one of the numbers C, D must be divisible by 5, let $C = 5c$. We get

$$\begin{cases} 79a + 25b + 5c + D = 711 \\ a \cdot b \cdot c \cdot D = 2^6 \cdot 3^2 \cdot 5^3. \end{cases}$$

Consider two cases. Suppose first that $D \equiv 0 \pmod 5$, meaning that $D = 5d$. Then $a \equiv -1 \pmod 5$ and since $a < 9$, we get that $a = 4$. Now we have

$$\begin{cases} 5b + c + d = 79 \\ b \cdot c \cdot d = 2^4 \cdot 3^2 \cdot 5^2. \end{cases}$$

Since $b < 25$ we can choose $c = 5s$, which implies $d \equiv 4 \pmod 5$, so that $d = 4 + 5t$ and

$$\begin{cases} b + s + t = 15 \\ b \cdot s \cdot (4 + 5t) = 2^4 \cdot 3^2 \cdot 5. \end{cases}$$

From the first equation we have $t \leq 13$, so possible values of $d = 4 + 5t$ are

$$4,\ 9,\ 14,\ 19,\ 24,\ 29,\ 34,\ 39,\ 44,\ 49,\ 54,\ 59,\ 64,\ 69,$$

only 4, 9 and 24 being divisors of $2^4 \cdot 3^2 \cdot 5$. If $d \leq 9$, then $b \cdot s \geq 16 \cdot 5$ and $\frac{b+s}{2} \geq \sqrt{16 \cdot 5} > 8$, which contradicts the first equation. If $d = 24$ we get $b + s = 11,\ b \cdot s = 30$, which leads to the solution

$$A = 79 \cdot 4 = 316, \quad B = 25 \cdot 6 = 150,$$
$$C = 25 \cdot 5 = 125, \quad D = 5 \cdot 24 = 120.$$

In the second case $D \not\equiv 0 \pmod 5$. Then c must be divisible by 5, i.e. $c = 5s$. Using symmetry between s and b we can also assume that b is divisible by 5, $b = 5t$ (since $a < 9$ and therefore cannot be divisible by 25). We get

$$\begin{cases} 79a + 125t + 25s + D = 711 \\ a \cdot t \cdot s \cdot D = 2^6 \cdot 3^2 \cdot 5. \end{cases}$$

If a is not divisible by 5 then $s = 5r$ and, using symmetry, we can suppose that $t \geq r$. Since $125(t + r) < 625$, we have only three choices: $r = 1$ and $t = 1, 2, 3$. None of them leads to an integer solution.

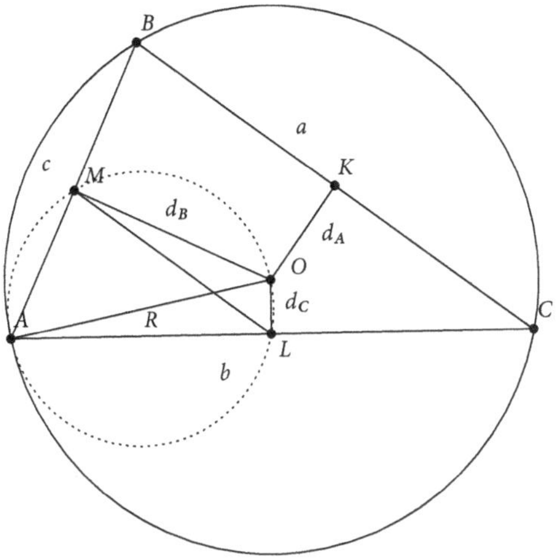

Figure 15.23 Carnot's theorem.

So a should be divisible by 5 and hence $a = 5$. Then $D = 16 \pmod{25}$ and hence D is one of the numbers in the sequence $16, 41, 66, 91, 116, 141, \ldots$, only one of those numbers being a divisor of $2^6 \cdot 3^2$, namely $D = 16$. This possibility does not give any integer solution either.

Hence, the only integer solution to the problem is

$$A = 316, \quad B = 150, \quad C = 125, \quad D = 120.$$

57. We shall first prove the **Carnot's theorem.**

Theorem 15.7 *Let the triangle ABC be inscribed in the circle of centre O and radius R. Define the quantity d_a as $\pm$ the distance from O to the side BC, the plus sign in case O and A lie in the same half-plane with respect to the line through B and C; the minus sign in case O and A lie in different half-planes with respect to this line ($d_a = 0$ in case $O \in BC$). Analogously, define d_b and d_c. Denote by r the radius of the inscribed circle in $\triangle ABC$. Then*

$$d_a + d_b + d_c = R + r.$$

Proof. Denote, as usual, the side lengths of the triangle by a, b, c. Let K, L, M be the orthogonal projections of O on BC, CA, AB respectively. Consider the quadrilateral $ALMO$. The points M, L lie on the circle with diameter AO. Ptolemy's theorem shows that

$$AL \cdot MO + AM \cdot LO = ML \cdot AO.$$

With the introduced notation, this means that

$$\frac{bd_b}{2} + \frac{cd_c}{2} = \frac{aR}{2}$$

and, analogously,

$$\frac{bd_b}{2} + \frac{ad_a}{2} = \frac{cR}{2}, \qquad \frac{ad_b}{2} + \frac{cd_c}{2} = \frac{bR}{2}.$$

Whatever the signs of d_a, d_b, d_c, for the area of $\triangle ABC$ we have

$$S = \frac{ad_a}{2} + \frac{bd_b}{2} + \frac{cd_c}{2} = \frac{(a+b+c)r}{2}.$$

Summing all four obtained equalities and dividing both sides by $\frac{a+b+c}{2}$, we deduce that $d_a + d_b + d_c = R + r$.

An alternative proof. The signed distance d_a to the side BC opposite the angle α is the distance taken with plus if α is acute and minus otherwise. If α is acute then $BOC = 2\alpha$ which gives us $d_a = R\cos\alpha$ as the length altitude in the triangle BOC. The same formula is valid when $\alpha \geq \frac{\pi}{2}$ (check this!) which explains why we want signed distances.

As result we get $d_a + d_b + d_c = R(\cos\alpha + \cos\beta + \cos\gamma)$ and Carnot's theorem follows directly from formula on page 299. □

We can now return to the original problem. Denote by n the number of vertices of the polygon. Observe that any triangulation by non-intersecting diagonals results in triangles inscribed in the same circle as the polygon itself; denote the centre and radius of this circle by O and R respectively. Choose one triangulation; denote by $r_1, r_2, \ldots, r_{n-2}$ the radii of the circles inscribed in the triangles (the number $n-2$ of such triangles follows easily by computing the sum of their angles and comparing that sum with the sum of the angles of the polygon). Sum the equalities from the Carnot's theorem for all such triangles. In this sum the distance from O to any side of the polygon will appear only once with the same sign for each triangulation, whereas the distance from

O to any of the chosen diagonals will appear twice, once with plus sign and once with minus sign, since each such diagonal is a side in two of the chosen triangles and these two triangles lie in different half-planes with respect to the diagonal. Hence we get

$$d_1 + d_2 + \cdots + d_n = (n-2)R + r_1 + r_2 + \cdots + r_{n-2},$$

where $d_1, d_2, \ldots, d_n$ are the distances from O to the n sides of the polygon. It immediately follows that

$$r_1 + r_2 + \cdots + r_{n-2}$$

does not depend on the triangulation.

Solution 2. It actually suffices to prove the statement for an inscribed quadrilateral $ABCD$ and use induction to establish the result for arbitrary n. We shall use the following result: given a triangle of side lengths a, b, c, angles α, β, γ (opposite to a, b, c respectively), area S, inradius r and circumradius R, we have

$$\begin{aligned} r &= \frac{2S}{a+b+c} = \frac{ab\sin\gamma}{a+b+c} = \frac{(2R\sin\alpha)(2R\sin\beta)\sin\gamma}{2R\sin\alpha + 2R\sin\beta + 2R\sin\gamma} \\ &= \frac{2R\sin\alpha\sin\beta\sin\gamma}{\sin\alpha + \sin\beta + \sin\gamma} = 4R\sin\frac{\alpha}{2}\sin\frac{\beta}{2}\sin\frac{\gamma}{2}. \end{aligned}$$

Here we have used the identity

$$\sin\alpha + \sin\beta + \sin\gamma = 4\cos\frac{\alpha}{2}\cos\frac{\beta}{2}\cos\frac{\gamma}{2}$$

from the problem on page 189.

We shall use the index 1 for quantities corresponding to $\triangle ABC$, the index 2 for $\triangle CDA$, the index 3 for $\triangle ABD$ and the index 4 for $\triangle BCD$. What we have to prove is then $r_1 + r_2 = r_3 + r_4$. Denote $\alpha = \angle BAC = \angle BDC$; $\beta = \angle CAD = \angle CBD$; $\gamma = \angle ABD = \angle ACD$; $\delta = \angle ADB = \angle ACB$.

Hence

$$r_1 + r_2 = 4R\left(\sin\frac{\alpha}{2}\sin\frac{\delta}{2}\sin\frac{\beta+\gamma}{2} + \sin\frac{\beta}{2}\sin\frac{\gamma}{2}\sin\frac{\alpha+\delta}{2}\right).$$

Similar computations for $r_3 + r_4$ and the relation $\beta + \gamma = \pi - (\alpha + \delta)$ (since the quadrilateral is inscribed in a circle) yield the result (check if you want to train trigonometrical transformations).

The statement in the problem is called sometimes the **Japanese theorem.**

58. It is obvious that

$$f(x) \geq \frac{1}{3}x \quad \forall x \in \mathbb{R}^+, \quad \forall f \in F.$$

Also, for the function $f_{\mathrm{id}}(x) = x$ we have $3x = 2x + x$, meaning that $f_{\mathrm{id}} \in F$. Hence there exists a positive constant α with the required property and

$$\frac{1}{3} \leq \alpha \leq 1.$$

Furthermore, for all $f \in F$ and all $x \in \mathbb{R}^+$ we have

$$f(3x) \geq f(f(2x)) + x \geq \alpha f(2x) + x \geq \alpha^2 \cdot 2x + x = x(2\alpha^2 + 1).$$

Hence

$$f(x) \geq \frac{2\alpha^2 + 1}{3}x \geq \frac{2\left(\frac{1}{3}\right)^2 + 1}{3}x \qquad \forall x \in \mathbb{R}^+.$$

We can repeat this process and can thus deduce that

$$\alpha \geq \alpha_0 = \lim_{n\to\infty} g^{\circ n}\left(\frac{1}{3}\right) = \lim_{n\to\infty} g \circ g \circ \cdots \circ g\left(\frac{1}{3}\right),$$

where $g(t) = \frac{1}{3}(2t^2 + 1)$ provided the above limit exists. (We have used the standard notation $g \circ g(t) = g(g(t))$.)

We shall now prove that the above limit does in fact exist. Assume that α_0 exists (and is finite). This means that

$$\alpha_0 = \frac{2\alpha_0^2 + 1}{3},$$

so that the two candidates for a limit are $\alpha_0 = 1$ and $\alpha_0 = \frac{1}{2}$.

Denote $\alpha_n = g^{\circ n}\left(\frac{1}{3}\right)$. We have

$$\alpha_1 = g\left(\frac{1}{3}\right) < \frac{1}{2}; \qquad \alpha_n \leq \frac{1}{2} \quad \Rightarrow \quad \alpha_{n+1} \leq \frac{2 \cdot \frac{1}{4} + 1}{3} = \frac{1}{2},$$

hence, $\alpha_n \leq \frac{1}{2}$ for all $n \in \mathbb{N}$. (In particular this means that if the limit exists, it has to be $\frac{1}{2}$.)

Consider next the difference between two consecutive elements of the sequence

$$\alpha_{n+1} - \alpha_n = \frac{2\alpha_n^2 - 3\alpha_n + 1}{3} = \frac{(\alpha_n - 1)(2\alpha_n - 1)}{3} \geq 0,$$

since $\alpha_n \le \frac{1}{2}$ for all $n \in \mathbb{N}$. We have proved that the sequence $\{\alpha_n\}_{n\in\mathbb{N}}$ is increasing and bounded from above by $\frac{1}{2}$. Hence, the sequence has to be convergent and its limit is $\alpha_0 = \frac{1}{2}$.

Let us return to the original problem. We now know that a number α with the required property exists and that $\alpha \ge \alpha_0 = \frac{1}{2}$. We shall prove that $\alpha = \frac{1}{2}$. A simple computation shows that functions of the type $f_c(x) = cx$ belong to F if and only if $\frac{1}{2} \le c \le 1$. In particular, the function $f_{\frac{1}{2}}(x) = \frac{1}{2}x$ belongs to F. This means that $\alpha \le \frac{1}{2}$ and hence $\alpha = \frac{1}{2}$.

59. Divide both sides by $(a+b)^5 > 0$; what we now have to prove is the equivalent inequality

$$\left(\frac{a}{a+b}d + \frac{b}{a+b}c\right)^5 + \left(\frac{a}{a+b}c + \frac{b}{a+b}d\right)^5 \le c^5 + d^5.$$

With $\alpha = \frac{a}{a+b}$, $0 < \alpha < 1$, we can rewrite the above inequality as

$$(\alpha d + (1-\alpha)c)^5 + (\alpha c + (1-\alpha)d)^5 \le c^5 + d^5.$$

The function $f(x) = x^5$ is strictly convex in $\mathbb{R}^+$, so that

$$\begin{aligned}(\alpha d + (1-\alpha)c)^5 &+ (\alpha c + (1-\alpha)d)^5 \\ &\le \alpha d^5 + (1-\alpha)c^5 + \alpha c^5 + (1-\alpha)d^5 = c^5 + d^5,\end{aligned}$$

and according to the strict convexity equality occurs if and only if $c = d$.

Solution 2. The form of the expressions leads the thoughts to the trigonometric identities. We can therefore set $a = r\cos\varphi$, $b = r\sin\varphi$, $c = \rho\cos\theta$, $d = \rho\sin\theta$, where $r = \sqrt{a^2+b^2}$, $\rho = \sqrt{c^2+d^2}$; $\varphi, \theta \in (0, \frac{\pi}{2})$. We now have to prove that

$$\cos^5(\varphi - \theta) + \sin^5(\varphi + \theta) \le (\cos\varphi + \sin\varphi)^5(\cos^5\theta + \sin^5\theta).$$

Now we need only factorization and some standard trigonometry to lead this to an equivalent (and easily proved) inequality. For example, the factor in the left-hand side

$$\begin{aligned}&\cos(\varphi - \theta) + \sin(\varphi + \theta) \\ &\quad = \cos\varphi\cos\theta + \sin\varphi\sin\theta + \sin\varphi\cos\theta + \cos\varphi\sin\theta \\ &\quad = (\cos\varphi + \sin\varphi)(\cos\theta + \sin\theta)\end{aligned}$$

is canceled with two factors in the right-hand side. The remaining expression factors into

$$\cos^4\theta - \cos^3\theta\sin\theta + \cos^2\theta\sin^2\theta - \cos\theta\sin^3\theta + \sin^4\theta$$
$$= (\cos^2\theta + \sin^2\theta)^2 - \cos^2\theta\sin^2\theta - (\cos^2\theta + \sin^2\theta)\cos\theta\sin\theta$$

and can be rewritten as $1 - \frac{y^2}{4} - \frac{y}{2}$ with $y = \sin 2\theta$. We leave it to the reader to finish the argument.

60. We shall use Kronecker's approximation theorem (theorem 6.11).

The existence of a triangle such as required is equivalent to the existence of such a triangle with one vertex in a $\frac{1}{1000}$-neighbourhood of the origin and the two remaining vertices in $\frac{1}{1000}$-neighbourhoods of (m, n) and (p, q), for some $m, n, p, q \in \mathbb{Z}$. We need to prove the existence of $m, n, p, q \in \mathbb{Z}$ such that

$$|p + qi - (m + ni - \rho e^{i\varphi})e^{i\frac{\pi}{3}}| < \frac{1}{500},$$

for some $\rho \in [0, \frac{1}{1000})$, $\varphi \in [0, 2\pi]$. By the triangle inequality it suffices to prove the existence of $m, n, p, q \in \mathbb{Z}$ such that

$$|p + qi - (m + ni)(\frac{\sqrt{3}}{2} + i\frac{1}{2})| < \frac{1}{1000},$$

equivalent to

$$(p + \frac{1}{2}n - m\frac{\sqrt{3}}{2})^2 + (q - \frac{1}{2}m - n\frac{\sqrt{3}}{2})^2 < \frac{1}{10^6},$$

which is an immediate consequence of Kronecker's theorem.

The proof below is a slight variation on the same theme.

a) Define $f : \mathbb{Z} \to [0, 1)$ by $f(x) = x\sqrt{3} - [x\sqrt{3}]$. By the pigeonhole principle, there exist distinct integers x_1 and x_2 such that $|f(x_1) - f(x_2)| < 0.001$. Put $a = |x_1 - x_2|$. Then the distance either between $(a, a\sqrt{3})$ and $(a, [a\sqrt{3}])$ or between $(a, a\sqrt{3})$ and $(a, [a\sqrt{3}] + 1)$ is less than 0.001. Therefore the points $(0, 0), (2a, 0), (a, a\sqrt{3})$ lie in different discs and form an equilateral triangle.

b) Suppose that $P'Q'R'$ is a triangle such that $P'Q' = Q'R' = R'P' = l \le 96$ and P', Q', R' lie in discs with centres P, Q, R, respectively. Then

$$l - 0.002 \le PQ, QR, RP \le l + 0.002.$$

Since PQR is not an equilateral triangle (see page 216), we may assume that $PQ \neq QR$.

Therefore

$$\begin{aligned} |PQ^2 - QR^2| &= (PQ + QR)|PQ - QR| \\ &\le 2(l + 0.002)((l + 0.002) - (l - 0.002)) \\ &\le 2 \cdot 96.002 \cdot 0.004 < 1. \end{aligned}$$

However, $PQ^2 - QR^2 \in \mathbb{Z}$. This is a contradiction.

15.11 Solutions to Paper 11

61. We can put the centre of the circle in $(0, 0)$ without changing the required number of points. Thus we need to calculate the number of integer solutions to the equation $x^2 + y^2 = p$. For $p = 2$ it is 4. If p is an odd prime number then according to theorem 8.3 the equation is solvable if and only if $p \equiv 1 \pmod 4$.

What remains is to prove that there exists essentially only one solution and that all other solutions are obtained by permutation of x and y and by change of signs (so that there are 8 different solutions).

Suppose that

$$a^2 + b^2 = c^2 + d^2 = p,$$

where a, c are odd positive numbers and b, d are even positive numbers. We want to prove that $a = c$ and $b = d$. We use the same approach that was used on the page 66. Since $a^2 - c^2 = d^2 - b^2$ we have $(a - c)(a + c) = (d - b)(d + b)$. If a and c are different numbers we can restrict ourselves to the case $a > c$. We can write $a - c = uv$ where u is the part of $a - c$ which is a factor in $d - b$. Similarly $a + c = zt$, where z is the part of $a + c$ which is a factor in $d - b$. Thus

$$a - c = uv, \quad a + c = zt, \quad d - b = uz, \quad d + b = vt.$$

Solving the above as a system of equations with respect to a, b, c, d, we get:

$$a = \frac{uv + zt}{2}, \quad b = \frac{vt - uz}{2}, \quad c = \frac{-uv + zt}{2}, \quad d = \frac{vt + uz}{2},$$

and hence

$$a^2 + b^2 = \frac{u^2v^2 + z^2t^2 + v^2t^2 + u^2z^2}{4} = \frac{(u^2 + t^2)(v^2 + z^2)}{4}.$$

If $u = t = 1$ or $v = z = 1$ then $a = d$, which is impossible. Thus we get a factorization of the prime p – a contradiction which completes the proof.

62. For all $n \in \mathbb{N}$ we have

$$x_n^2 = 2 + x_{n-1}^2 + \frac{1}{x_{n-1}^2},$$

and hence

$$x_{1000}^2 = 1000 \cdot 2 + x_0^2 + \sum_{k=0}^{999} \frac{1}{x_k^2} > 2025 = 45^2.$$

Since all elements of the sequence are positive, it follows that $x_{1000} > 45$.

Similar reasoning shows that $x_{100} > 15$, $x_{300} > 25$ and $x_{600} > 35$. Also, it is obvious that the sequence is increasing. It follows that

$$\begin{aligned} x_{1000}^2 &= 2025 + \sum_{k=0}^{999} \frac{1}{x_k^2} = \sum_{k=0}^{99} \frac{1}{x_k^2} + \sum_{k=100}^{299} \frac{1}{x_k^2} + \sum_{k=300}^{599} \frac{1}{x_k^2} + \sum_{k=600}^{999} \frac{1}{x_k^2} \\ &< 2025 + 100 \cdot \frac{1}{25} + 200 \cdot \frac{1}{225} + 300 \cdot \frac{1}{625} + 400 \cdot \frac{1}{1225} \\ &< 2025 + 4 + 1 + \frac{1}{2} + \frac{1}{3} = 2030 + \frac{5}{6} < 2034.01 = (45.1)^2. \end{aligned}$$

Thus, both estimates are proved.

63. As the polynomial p has real roots, non-negative real coefficients and as $p(0) = 1$, all its roots have to be negative. Thus we can write:

$$p(x) = (x + x_1)\cdots(x + x_n),$$

where $x_l, \ldots x_n$ are some positive real numbers. Thus a_i coincide with the elementary symmetrical polynomials s_i (see page 105). Because $s_n = 1$ here we get, using Maclaurin's inequality (though AM–GM inequality would be sufficient) that $a_k = s_k \geq \binom{n}{k}$.

Let us denote $q(x) = (x + 1)^n = \sum_{k=0}^{n} \binom{n}{k} x^k$. As the corresponding coefficients of the polynomial p are greater than those of the polynomial q, we have indeed $p(x) > q(x)$ for all $x > 0$.

64. Both O_1 and O_2 lie on the perpendicular bisector of the segment CD. Apart from that O_1 lies on the perpendicular bisector of AC and O_2 on the perpendicular bisector of BC. Consider the perpendicular bisector of AC. It is parallel to BC and passes through the midpoint of AC, hence it must pass through the midpoint of AB which is D. By analogy the same goes for the perpendicular bisector of BC. This means that the two intersect at D. Let k be the circle with diameter O_1O_2. Denote its centre by O. We have that

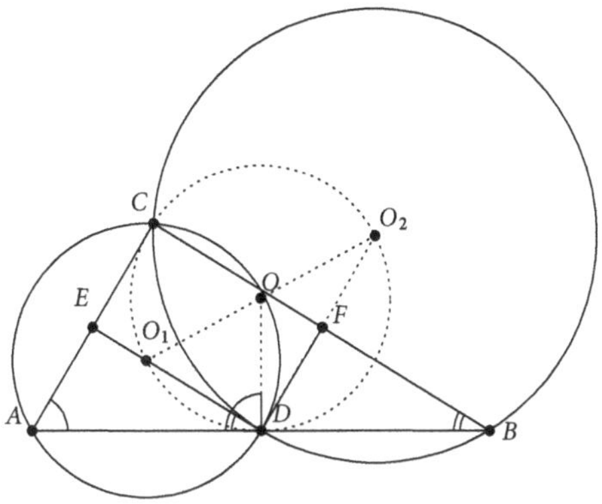

Figure 15.24 Problem 64.

$\angle O_1DO_2$ is a right angle, since the perpendicular bisectors are parallel to AC and BC respectively. This means that D lies on k. We have $\angle BAC = \angle ACD = \angle DO_1O = \angle O_1DO$ (since angles at the base of an isosceles triangle are equal and since $O_1D \perp AC$, $O_1O \perp CD$) and $\angle ADO_1 = \angle ABC$ (since $O_1D \parallel BC$). Hence we get $\angle ADO = \angle BAC + \angle ABC = \frac{\pi}{2}$, which means that K is tangent to AB at D.

Solution 2. Denote by k the circle that is tangent to the side AB in the point D and that passes through C. Let O be the centre of k. Then the points O, O_1 and O_2 are collinear as they all lie on the perpendicular bisector of the segment CD. Denote $\angle BAC = \alpha$. The centre angle $\angle DO_1C$ is then equal to 2α. The triangle ADC is isosceles, so that $\angle BDC = 2\alpha$ and therefore O_1 lies on the circle k (DC is a chord and AB is tangent to k, hence $\angle BDC = \frac{1}{2}\angle DOC$; it is possible that $\angle DOC > \pi$). Similarly, the point O_2 lies on the circle k. As we have already proved that the points O, O_1 and O_2 are collinear, the segment O_1O_2 is in fact the diameter of the circle k.

65. Consider a subset not containing any of the described pairs. If it contains elements that are less than or equal to $n/2$, take s to be the smallest of those. Then $2s$ cannot belong to the subset. Replace s with $2s$ in the subset; denote the new set by $\tilde{S}$. We need to make sure that $\tilde{S}$ still satisfies the requirements. It is obvious that if m and s are not relatively prime, then the the same is true

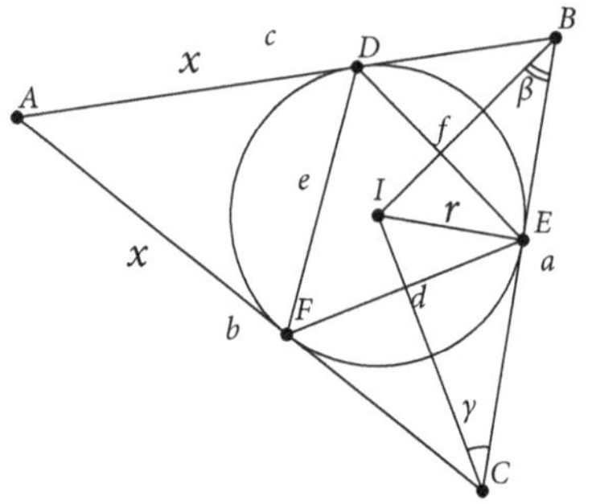

Figure 15.25 Problem 66

for m and $2s$. Also, if no element of the subset is divisible by s, then none will be divisible by $2s$; it remains to show that $2s$ will not be divisible by any other element of $\tilde{S}$. Assume that it is. Since s was the smallest element in S, this means that $2s$ is divisible by $2t \in S$, where t is a proper divisor of s. But then $\min\{S\} = s \geq 2t \in S$, a contradiction. Repeat the above procedure until all elements become greater than $n/2$. Since any two consecutive integers are relatively prime, the subset can contain at most one element from each such pair; hence it will have at most $[(n+2)/4]$ elements, except the special case $n = 4k + 1$. Then $[(n+2)/4] = k$, and we could possibly have $k + 1$ elements, but only when $2k + 1$ and $4k + 1$ belong to the subset. But they are relatively prime which is forbidden. The result holds for all n and the maximum is thus $[(n+2)/4]$.

66. Let $a = BC, b = CA, c = AB$ and $d = EF, e = FD, f = DE$. By AM–GM inequality

$$\frac{a}{f} + \frac{b}{d} + \frac{c}{e} \geq 3\sqrt[3]{\frac{abc}{def}}$$

thus it is sufficient to prove that

$$\sqrt[3]{\frac{abc}{def}} \geq 2 \quad \Leftrightarrow \quad abc \geq 8def.$$

Let $x = (p-a) = AD = AF$, $y = p-b$ and $z = p-c$. An application of the cosine rule to $\triangle ADF$ yields

$$e^2 = 2x^2(1-\cos A) = 2x^2\left(1 - \frac{b^2+c^2-a^2}{2bc}\right)$$
$$= x^2 \cdot \frac{a^2-(b-c)^2}{bc} = x^2 \cdot \frac{(y+z)^2-(y-z)^2}{bc} = \frac{4x^2yz}{bc}.$$

By the symmetry we have

$$d^2e^2f^2 = \frac{4^3x^4y^4z^4}{a^2b^2c^2}$$

and therefore need only to prove that

$$a^2b^2c^2 \geq 8\cdot 2^3x^2y^2z^2 \quad \Leftrightarrow \quad (x+y)(x+z)(y+z) \geq 8xyz.$$

But the last inequality follows if we multiply three AM–GM inequalities.

Solution 2. Let $\angle A = 2\alpha$, $\angle B = 2\beta$, $\angle C = 2\gamma$. Because the Gergonne triangle DEF has inscribed circle as its circumcircle and has angles $\frac{\pi}{2}-\alpha, \frac{\pi}{2}-\beta, \frac{\pi}{2}-\gamma$ we have $e = 2r\sin(\frac{\pi}{2}-\alpha) = 2r\cos\alpha$ and similarly $f = 2r\cos\beta$, $d = 2r\cos\gamma$.

On the other hand $a = (p-b)+(p-c) = r(\cot\beta+\cot\gamma)$ and similarly $b = r(\cot\alpha+\cot\gamma)$, $c = r(\cot\alpha+\cot\beta)$. Thus we need to prove that

$$\frac{\cot\beta}{2\cos\alpha}+\frac{\cot\gamma}{2\cos\alpha}+\frac{\cot\alpha}{2\cos\beta}+\frac{\cot\gamma}{2\cos\beta}+\frac{\cot\alpha}{2\cos\gamma}+\frac{\cot\beta}{2\cos\gamma} \geq 6.$$

The rearrangement inequality shows that it is sufficient to prove

$$2\left(\frac{\cot\alpha}{2\cos\alpha}+\frac{\cot\beta}{2\cos\beta}+\frac{\cot\gamma}{2\cos\gamma}\right) \geq 6 \quad \Leftrightarrow \quad \frac{1}{\sin\alpha}+\frac{1}{\sin\beta}+\frac{1}{\sin\gamma} \geq 6,$$

which follows from Jensen's or Karamata's inequality.

15.12 Solutions to Paper 12

67. Since

$$\frac{a_k}{s-a_k} = -1 + \frac{s}{s-a_k},$$

the inequality is equivalent to

$$s\left(\frac{1}{s-a_1}+\frac{1}{s-a_2}+\cdots+\frac{1}{s-a_n}\right) \geq n + \frac{n}{n-1} = \frac{n^2}{n-1}.$$

By the inequality between the harmonic and the arithmetic means we have

$$\frac{n}{\frac{1}{s-a_1}+\frac{1}{s-a_2}+\cdots+\frac{1}{s-a_n}} \le \frac{s-a_1+s-a_2+\cdots+s-a_n}{n} = \frac{ns-s}{n},$$

and hence

$$s\left(\frac{1}{s-a_1}+\frac{1}{s-a_2}+\cdots+\frac{1}{s-a_n}\right) \ge \frac{n^2}{n-1}.$$

It is obvious from the proof above that equality holds if and only if $a_1 = a_2 = \cdots = a_n$.

Solution 2. WLOG $a_1 \ge \cdots \ge a_n$. Then $\frac{1}{s-a_1} \ge \cdots \ge \frac{1}{s-a_n}$ as well. Applying Chebyshev's and AM–HM inequalities we get:

$$\begin{aligned}
&\frac{a_1}{s-a_1}+\frac{a_2}{s-a_2}+\cdots+\frac{a_n}{s-a_n}\\
&\quad\ge \frac{1}{n}(a_1+a_2+\cdots+a_n)\left(\frac{1}{s-a_1}+\frac{1}{s-a_2}+\cdots+\frac{1}{s-a_n}\right)\\
&\quad= \frac{s}{n}\left(\frac{1}{s-a_1}+\frac{1}{s-a_2}+\cdots+\frac{1}{s-a_n}\right)\\
&\quad\ge \frac{sn}{\left(\frac{1}{s-a_1}\right)^{-1}+\cdots+\left(\frac{1}{s-a_n}\right)^{-1}} = \frac{sn}{ns-s} = \frac{n}{n-1}.
\end{aligned}$$

Solution 3. If $a+b = c+d = S$ and $a \ge c \ge d \ge b \ge 0$ then it is easy to check (e.g. by Karamata) that

$$\frac{a}{S-a}+\frac{b}{S-b} \ge \frac{c}{S-c}+\frac{d}{S-d}.$$

If not all a_i are equal we can choose $a = a_i > \frac{s}{n}, b = a_j < \frac{s}{n}, c = \frac{s}{n}, d = a+b-c \ge 0$ and increase the number of $a_i = \frac{s}{n}$. When all be equal the problem be done.

Solution 4. WLOG $a_1 \ge \cdots \ge a_n$. Because $f(x) = \frac{x}{s-x}$ is convex on $(0,s)$ we can apply Karamata's or Jensen's inequality to get $nf\left(\frac{s}{n}\right) = \frac{n}{n-1}$ in the right-hand side.

68. Consider two ants who collide with each other. If we don't distinguish between ants, after the collision the situation will be identical to a situation where the ants "pass through" each other and continue as if nothing happened. Since there is an ant in the middle to begin with, and since one minute is exactly the time needed to crawl from the middle of the pole to one end and back to the middle, it means that there will certainly be an ant in the middle after one minute. We now need to determine the probability of this ant being Alice.

For each ant on the left half of the pole at time 0, there will be an ant on the right half after one minute, since, if we don't distinguish between ants and assume they pass through one another, an ant crawling towards the close end of the pole will have time to reach it and crawl back more than half the length of the pole, whereas an ant crawling towards the far end will not have time to reach the middle of the pole when crawling back. Hence, the number of ants on the left half at time 0 will be equal to the number of ants on the right half after one minute.

Observe now that the rules of collision imply that the order of the ants on the pole will remain the same regardless of the time elapsed since the beginning of the process. Denote the number of ants on the left half of the pole at time 0 by n, the number of ants on the right half at time 0 will then be $100 - n$. After one minute there will be $100 - n$ on the left half and n on the right one. Suppose that Alice is in the middle after one minute; the number of ants on her left side will then be $n = 100 - n$, since the order of the ants does not change. Thus Alice necessarily has to be the 51st ant from either end of the pole. Obviously the above condition is also sufficient.

The above reasoning actually means that we need to determine the probability of Alice being the 51st ant from either end, i.e. that there are exactly 50 ants on the left half of the pole and 50 on the right one at time 0. For any given ant the probability of it being placed on the left half is $\frac{1}{2}$ and is equal to the probability of it being placed on the right one. Hence, given 50 ants, the probability of them being placed to the left of the middle and the remaining 50 to the right is $\frac{1}{2^{50}} \cdot \frac{1}{2^{50}} = \frac{1}{2^{100}}$. Since there are $\binom{100}{50}$ ways of choosing 50 ants from the given 100, the probability of Alice being in the middle after one minute is

$$\binom{100}{50}\frac{1}{2^{100}}.$$

69. We start with two lemmas that are useful by themselves.

Lemma 15.8 *Consider the function $f(x) = x(S - x)$ where $m \leq x \leq M$. If $m + M \geq S$ then the minimum of f is $f(M)$ otherwise it is $f(m)$.*

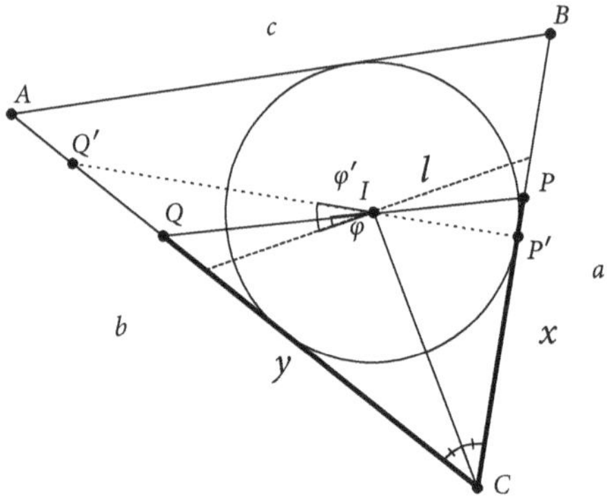

Figure 15.26 Problem 69

Proof. Rewriting $f(x)$ as $\frac{S^2}{4} - \left(x - \frac{S}{2}\right)^2$ we see that the minimum is attained either at $x = m$ or at $x = M$. Now the (sign of the) difference

$$f(M) - f(m) = S(M - m) - (M^2 - m^2) = (M - m)(S - M - m)$$

shows which of them we need to choose. □

Lemma 15.9 *If a, b, c are sides of a triangle then* $p^2 > b(a + c)$.

Proof. We see directly that

$$\begin{aligned} 4p^2 - 4b(a + c) &= (a + b + c)^2 - 4ab - 4bc \\ &= a^2 + b^2 + c^2 + 2ac - 2ab - 2bc = (a + c - b)^2. \end{aligned}$$ □

We return to our problem. Each line passing through the centre I of the inscribed circle will divide the given triangle ABC into one triangle and one quadrilateral or, in case the line coincides with one of the angle bisectors, into two triangles. For any vertex of the given triangle, consider all lines through I such that the part containing the vertex is a triangle (such lines will vary "between" the bisectors of the two angles at the two remaining vertices of the triangle). Denote the area of the small triangle at the vertex by T, the area of the given triangle by S. We need to find a line through I such that the function $f(T) = T(S - T)$ achieves its minimal value.

Lemma 15.8 shows that the function f will achieve its minimal value either when T is minimal or when T is maximal. This means we need first to find the positions of the line l through I for which the area of the cut-off triangle is minimal/maximal.

Suppose that a line l cuts the segments $x = CP$ and $y = CQ$ from the sides a and b of the triangle. According to Theorem 4.24

$$\frac{1}{x} + \frac{1}{y} = \frac{a+b+c}{ab} = \frac{2p}{ab}$$

thus the GM–HM inequality shows that xy has its minimal value when $x = y = \frac{ab}{p}$. Therefore the minimal area $xy \sin C$ of the cut-off-triangle with a vertex in C is equal to

$$m = \frac{a^2b^2}{2p^2}\sin C = \frac{abS}{p^2}$$

and is achieved when l is orthogonal to the angle bisector of the angle at C (show that l intersects both CA and BC in inner points!). Moreover if $a \le b \le c$ then similar calculations for vertices A and B shows that m is a global minimum.

Now we need to find a maximum. Consider the line l through I which is orthogonal to the angle bisector at C. Rotate l an angle φ around I (in either direction); regard the area of the cut-off-triangle as a function of φ, $T = T(\varphi)$. As long as the line cuts off a triangle with one vertex at C, the area $T(\varphi)$ will be an increasing function of φ. Indeed, in our picture, if $\varphi' > \varphi$ then from the angle bisector theorem it follows that

$$\frac{Q'I \cdot QI}{P'I \cdot PI} = \frac{Q'C}{P'C} \cdot \frac{QC}{PC} > 1.$$

Hence, the maximal area will be achieved when the line coincides with the angle bisector of A or the angle bisector of B. Thus the maximum area we find from the parts in which angle bisectors divide the triangle.

The angle bisector at C divides c and therefore the area S in relation $a : b$, thus if $a \le b \le c$ we need to choose between $\frac{b}{a+b}S$, $\frac{c}{a+c}S$, $\frac{c}{b+c}S$. Evidently the middle term $M = \frac{c}{a+c}S$ is the largest one.

Now to apply Lemma 15.8, we want to show that $m + M < S$. Indeed,

$$\frac{abS}{p^2} + \frac{c}{a+c}S < S \quad \Leftrightarrow \quad ab(a+c) + cp^2 < (a+c)p^2$$
$$\Leftrightarrow \quad ab(a+c) < ap^2$$

and we can use Lemma 15.9.

Now Lemma 15.8 shows that $f(T)$ has attains its minimum in m.

Thus, we have proved that the line with the required property is the line through I, orthogonal to the angle bisector of the largest angle of the triangle.

Remark Note that the inequality $m < S - M$ is, in fact, evident geometrically. It follows from the monotonicity because $S - M$ is the area of the triangle that the bisector at B cuts, but we wanted to mention a useful lemma.

70. Let $p = 2^{m+1} + 1$. Then from the given condition $3^{2^m} \equiv -1 \pmod{p}$ it follows that $(3, p) = 1$. Squaring both sides we obtain $3^{2^{m+1}} \equiv 1 \pmod{p}$, which implies that the order $\text{ord}_p(3)$ is a divisor of $2^{m+1} = p - 1$. So, $\text{ord}_p(3)$ is of the form 2^r, for some positive integer $r \leq m + 1$. If we suppose $r \leq m$, then $3^{2^m} \equiv 1 \pmod{p}$, which is a contradiction. Therefore, $r = m + 1$. On the other hand, by Euler's theorem, $\text{ord}_p(3)$ is a divisor of $\varphi(p)$. So, $2^{m+1} = p - 1$ divides $\varphi(p)$. Since $\varphi(p) \leq p - 1$, we have $\varphi(p) = p - 1$, i.e. p is a prime as desired.

Let now $p = 2^{m+1} + 1$ be a prime. Obviously $p > 3$ and $m + 1$ is an even number, otherwise $2 + 1$ is a divisor. Thus $p \equiv 2 \pmod{3}$.

Since $p \equiv 1 \pmod{4}$ by theorem 8.4 we get

$$\left(\frac{3}{p}\right) = (-1)^{\frac{(3-1)(p-1)}{4}} \left(\frac{p}{3}\right) = \left(\frac{2}{3}\right) = -1.$$

Thus,

$$-1 = \left(\frac{3}{p}\right) \equiv 3^{\frac{p-1}{2}} = 3^{2^m} \pmod{p},$$

and hence $p \mid (3^{2^m} + 1)$, which proves the converse of the first statement. Notice how efficient the Legendre symbols are!

71. No. We shall prove a general statement.

Theorem 15.10 *Let $g : \mathbb{R} \to \mathbb{R}$ be a function such that $g \circ g$ has exactly two (different) fixed points in $\mathbb{R}$, that are not fixed points of g itself. Then there exists no function $f : \mathbb{R} \to \mathbb{R}$ such that $f(f(x)) = g(x)$.*

Proof. Denote the fixed points of $g \circ g$, that are not fixed points of g, by y_1, y_2, where $y_1 \neq y_2$. Assume there exists a function f, satisfying the functional equation. We have

$$g(f(x)) = f(f(f(x))) = f(g(x)),$$

so that f and g commute, i.e. $f(g(x)) = g(f(x))$. Furthermore,

$$g(g(g(y_k))) = g(y_k), \qquad k = 1, 2,$$

which means that $g(y_k)$ is a fixed point of $g \circ g$. We have $g(y_k) \neq y_k$, as y_k is not a fixed point of g. Also, $g(y_k)$ cannot be a fixed point of g itself, since that would imply $y_k = g(g(y_k)) = g(y_k)$. The only remaining possibility is

$$g(y_1) = y_2, \qquad g(y_2) = y_1,$$

and it immediately follows that

$$f(f(y_1)) = g(y_1) = y_2, \qquad f(f(y_2)) = g(y_2) = y_1.$$

Also, since f and g commute,

$$\begin{aligned} g(g(f(y_{1(2)}))) &= g(f(g(y_{1(2)}))) \\ &= g(f(y_{2(1)})) = f(g(y_{2(1)})) = f(y_{1(2)}), \end{aligned}$$

which means that $f(y_{1(2)})$ is a fixed point of $g \circ g$.

If we assume that e.g. $f(y_1)$ is a fixed point of g itself, it would follow that

$$f(y_1) = g(f(y_1)) = f(g(y_1)) = f(y_2),$$

and thus $y_2 = g(y_1) = f(f(y_1)) = f(f(y_2)) = g(y_2) = y_1$, which is a contradiction. Hence, $f : \{y_1, y_2\} \mapsto \{y_1, y_2\}$.

Assume that $f(y_1) = y_1$; it would then follow that $f(f(y_1)) = f(y_1) = y_1$, a contradiction with $f(f(y_1)) = g(y_1) = y_2$. Hence $f(y_1) = y_2$, and, similarly, $f(y_2) = y_1$. But, this means that

$$f(f(y_1)) = f(y_2) = y_1,$$

which also contradicts $f(f(y_1)) = g(y_1) = y_2$.

The contradiction is due to the assumption that there exists a function f such that $f(f(x)) = g(x)$. Hence, such a function does not exist. □

It remains only to check that the function $g(x) = x^2 - 2$ satisfies the fixed point conditions of the theorem, which is easily done.

72. Identify each cell with its centre; the centres of all cells will then be the nodes in a lattice. Since the prince only moves one step at the time and does not revisit any cell, he will describe a curve without self-intersection and, hence, during his journey he will complete a simple closed lattice polygon C. Denote by D the bounded domain with boundary $\partial D = C$. The area of D is, according to Pick's formula (see Theorem 6.13), $A = 31$ ($a = 0, b = 64$). The vertical lines of the lattice cut D into rectangles of width 1. Each such rectangle is bounded from above and below by horizontal segments of the path.

The area of each rectangle is an integer, which is odd if and only if the bounding horizontal segments lie in horizontal rows of the board of different parity (e.g., if they lie in the third and in the sixth rows, the area will be 3). Since the total area is an odd integer, there has to be an odd number of rectangles of odd area. This means that the number n of moves in odd horizontal rows has to be odd. Indeed, such moves are included one at the time in rectangles of odd area and two at the time in rectangles of even area.

Colour the chessboard in "stripes": the odd horizontal rows in white and the even ones in black. There are 32 moves coming out of the 32 white cells; n of those (all horizontal ones) end in white cells, whereas $32 - n$ (all vertical ones) end in black cells. Since the prince ends his path in the initial cell (i.e., a cell of the same colour as at the start), the number of moves from black onto white cells has to be equal to the number of moves from white cells onto black ones. Thus, the number of all vertical moves has to be $64 - 2n$ and the number of horizontal moves has to be $2n$. The integer n being odd, the number of all horizontal moves cannot be equal to 32.

Note that in the original formulation the "prince" was called "lame rook." Much more about the problem can be found at [70].

15.13 Solutions to Paper 13

73. Let the number of players be n. The number of matches won by each player is an integer between 1 and $n - 1$, so these cannot be distinct as there are fewer than n possibilities. Hence two players A and B won the same number k of matches. Suppose without loss of generality that A beat B. Then even if B beat the remaining $k - 1$ players who lost to A, there must be a player C who beat A but was beaten by B. So A beat B, B beat C and C beat A.

74. Using the Cauchy–Schwarz inequality (or simply the inequality $\sqrt{ab}+\sqrt{cd} \le \sqrt{(a+c)(b+d)}$, proved on page 133 for $a, b, c, d \ge 0$) we get:

$$\begin{aligned}
&\sqrt{2}\left(\sqrt{a(a+b)^3} + b\sqrt{a^2+b^2}\right) \\
&\quad = \sqrt{2a(a+b)}\sqrt{(a+b)^2} + \sqrt{2b^2}\sqrt{a^2+b^2} \\
&\quad \le \sqrt{2a(a+b)+2b^2}\sqrt{(a+b)^2+a^2+b^2} \\
&\quad = \sqrt{2(a^2+ab+b^2)}\sqrt{2(a^2+ab+b^2)} \\
&\quad = 2(a^2+ab+b^2) = 2a^2+2b^2+2ab \\
&\quad \le 2a^2+2b^2+(a^2+b^2) \le 3(a^2+b^2),
\end{aligned}$$

with equality throughout iff $a = b$.

Solution 2. (basically the same) The inequality clearly holds if either a or b is zero. So, we may suppose that $a, b > 0$. Writing $s = a/b$, we must show that

$$\sqrt{2}\left(\sqrt{s(s+1)^3} + \sqrt{s^2+1}\right) \le 3(s^2+1).$$

To see that this is so, note that by AM–GM inequality

$$\begin{aligned}
\sqrt{2}\left(\sqrt{s(s+1)^3} + \sqrt{s^2+1}\right) &= (s+1)\sqrt{(2s)(s+1)} + \sqrt{2(s^2+1)} \\
&\le (s+1)\frac{2s+s+1}{2} + \frac{2+s^2+1}{2} \\
&= 2s^2+2s+2 \le 3(s^2+1).
\end{aligned}$$

with equality holding iff $s = 1$.

Solution 3. After squaring and simplifying, we are required to show that

$$6a^3b + 2ab^3 + 4b\sqrt{a(a+b)^3(a^2+b^2)} \le 7a^4 + 10a^2b^2 + 7b^4.$$

Note that $3a^4 + 3a^2b^2 \ge 6a^3b$, $\quad b^4 + a^2b^2 \ge 2ab^3$, with equality in both iff $a = b$. So, is

$$2a^4 + 3a^2b^2 + 3b^4 \ge 2b\sqrt{a(a+b)^3(a^2+b^2)}\,?$$

How about

$$4a^8 + 9b^8 + 12a^6b^2 + 21a^4b^4 + 18a^2b^6 \ge 4b^2a(a+b)^3(a^2+b^2)?$$

Expand the right-hand side to get

$$\begin{aligned}
&4b^2a(a+b)^3(a^2+b^2) \\
&\quad = 4a^6b^2 + 12a^5b^3 + 16a^4b^4 + 16a^3b^5 + 12a^2b^6 + 4ab^7.
\end{aligned}$$

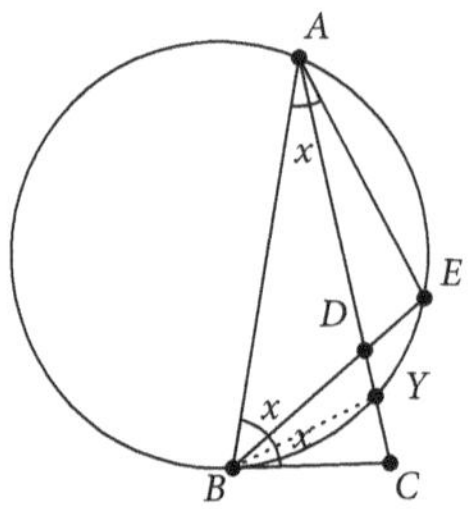

Figure 15.27 Problem 75.

So now we must prove that

$$4a^8 + 8a^6b^2 + 5a^4b^4 + 6a^2b^6 + 9b^8 \geq 12a^5b^3 + 16a^3b^5 + 4ab^7.$$

But $2b^8 + 2a^2b^6 \geq 4ab^7$, $5a^6b^2 + 5a^4b^4 \geq 10a^5b^3$, $a^8 + a^2b^6 \geq 2a^5b^3$, and

$$3a^8 + 3b^6b^2 + 3a^2b^6 + 7b^8 \geq 16\sqrt[16]{a^{48}b^{80}} = 16a^3b^5,$$

with equality in each of these iff $a = b$. This completes the proof.

75. Let the bisector of $\angle ABC$ meet AC at D and the circle at E (see Figure 15.27). Join AE. Let $\angle ABD = x = \angle DBC$. Then $\angle ACB = 2x$ since $AC = AB$. Hence $\angle BAC = \pi - 4x$. But since BC is tangent to T we have $\angle BAE = \angle CBE = x$, which implies $\angle DAE = \angle BAE - \angle BAC = x - (\pi - 4x) = 5x - \pi$. Since D is inside the circle $5x - \pi > 0$. Hence $\angle ABC = 2x > 2\pi/5 = 72°$.

Solution 2. Let Y be the point where AC meets T. Because $\angle YBC = \angle BAC$ we have

$$\angle DBC > \angle BAC,$$

which implies $\angle ABC > 2\angle BAC$. Now,

$$\begin{aligned}180° &= \angle ABC + \angle BAC + \angle ACB \\ &< \angle ABC + \frac{1}{2}\angle ABC + \angle ABC = \frac{5}{2}\angle ABC\end{aligned}$$

as required.

Solution 3. Let D be the point where the bisector of $\angle ABC$ meets AC.

We know that

$$BC^2 < CD \cdot AC = CD \cdot AB$$

Therefore

$$\frac{AB}{BC} > \frac{BC}{CD}$$

and since $\angle ABC = \angle BCD$ we deduce that $\angle BAC < \angle CBD = \frac{1}{2}\angle ABC$. (Equality would have meant the triangles are similar and corresponding angles are equal; the inequality means we have equality if we replace CD by CD', where D' is between C and D. We have $\triangle CBD' \sim \triangle BAC$; since its angle at B is obviously smaller than $\angle CBD$, the assertion follows.) Therefore

$$2\angle ABC + \frac{1}{2}\angle ABC > \angle ABC + \angle ACB + \angle BAC = 180°$$

as required.

76. Denote by A the set of all solutions of the equation. Partition A into subsets as follows:

For $(x_1, x_2, \dots, x_q) \in A$, define $A(x_1, x_2, \dots, x_q)$ as the set of all $(y_1, y_2, \dots, y_q)$, where $(y_1, y_2, \dots, y_q)$ is a permutation of $(x_1, x_2, \dots, x_q)$. Suppose $a_i \in S, 1 \le i \le n$, are distinct and $(x_1, x_2, \dots, x_q)$ is an arrangement of k_1 a_1's, k_2 a_2's, ..., k_n a_n's with $k_i \ge 1$. The set $A(x_1, x_2, \dots, x_q)$ has

$$\frac{q!}{k_1!k_2!\cdots k_n!}$$

elements. Hence $\#A(x_1, x_2, \dots, x_q)$ is a multiple of q unless q is cancelled ($\#A$ denotes the number of elements of A). Since q is a prime this only happens for $n = 1$ and $k_1 = q$. However, in this case we have

$$qa_1 \equiv 0 \pmod p \quad \Leftrightarrow \quad a_1 \equiv 0 \pmod p.$$

But S does not contain a multiple of p, and so this case does not arise. The result follows.

77. Let us divide the group E_n into two parts $(EA)_n$ and $(EB)_n$ according to whether the word terminates by the letter A or B. Similarly, divide the group O_n into two parts $(OA)_n$ and $(OB)_n$ according to whether the word terminates by the letter A or B. Denote further by e_n, o_n, $(eA)_n$, $(eB)_n$, $(oA)_n$,

$(oB)_n$ in turn the number of elements of E_n, O_n, $(EA)_n$, $(EB)_n$, $(OA)_n$, and $(OB)_n$, respectively. For any natural number n,

$$e_n = (eA)_n + (eB)_n,$$
$$o_n = (oA)_n + (oB)_n. \tag{15.9}$$

Any word from the group $(EA)_{n+1}$ arises by appending the letter A either to a word from $(EA)_n$, or to a word from $(OB)_n$. Consequently,

$$(eA)_{n+1} = (eA)_n + (oB)_n.$$

Similarly,

$$(eB)_{n+1} = (eA)_n + (eB)_n,$$
$$(oA)_{n+1} = (eB)_n + (oA)_n,$$
$$(oB)_{n+1} = (oA)_n + (oB)_n. \tag{15.10}$$

For $n = 1$, the groups have the following forms

$$(EA)_1 = \{A\}, \quad (EB)_1 = \{B\}, \quad (OA)_1 = \varnothing, \quad (OB)_1 = \varnothing,$$

implying that $(eA)_1 = (eB)_1 = 1$ and $(oA)_1 = (oB)_1 = 0$.

Let us assume that for some natural number k the groups $(EA)_k$ and $(EB)_k$ contain the same number of elements, which we denote by p, and simultaneously the groups $(OA)_k$ and $(OB)_k$ have the same number of elements, which we denote by q. Assume further that $p \neq q$, as is the case for $k = 1$, when $p = 1$ and $q = 0$. Let us write down the numbers of elements of the various groups for $n = k, k+1, k+2, k+3, k+4$, using the formulas (15.9) and (15.10).

n	k	$k+1$	$k+2$	$k+3$	$k+4$
$(eA)_n$	p	$p+q$	$p+3q$	$2p+6q$	$6p+10q$
$(eB)_n$	p	$2p$	$3p+q$	$4p+4q$	$6p+10q$
$(oA)_n$	q	$p+q$	$3p+q$	$6p+2q$	$10p+6q$
$(oB)_n$	q	$2q$	$p+3q$	$4p+4q$	$10p+6q$
e_n	$2p$	$3p+q$	$4p+4q$	$6p+10q$	$12p+20q$
o_n	$2q$	$p+3q$	$4p+4q$	$10p+6q$	$20p+12q$

From the table, a number of facts can be gleaned. Since $p \neq q$, we also have $2p \neq 2q$, $3p+q \neq p+3q$ and $6p+10q \neq 10p+6q$. Thus we see that

$e_k \neq o_k$, $e_{k+1} \neq o_{k+1}$, $e_{k+2} = o_{k+2}$, $e_{k+3} \neq a_{k+3}$, and that the groups $(EA)_{k+4}$ and $(EB)_{k+4}$ contain again the same number of elements, as do the groups $(OA)_{k+4}$ and $(OB)_{k+4}$, and these two numbers are again distinct.

Using induction, it follows that the table has the same properties as mentioned above for any $k = 4m + 1$, where m is a nonnegative integer; hence, the equality $e_n = o_n$ holds if and only if $n = k + 2 = 4m + 3$.

Conclusion: The groups E_n and O_n have the same number of elements if and only if $n = 4m + 3$, where m is a nonnegative integer.

Solution 2. From the relations (15.9) and (15.10) we get for each $n \geq 2$ the equalities

$$\begin{aligned} e_{n+1} &= (eA)_{n+1} + (eB)_{n+1} = (eA)_n + (oB)_n + e_n \\ &= 2e_n + (oB)_n - (eB)_n = 2e_n + o_{n-1} - e_{n-1} \\ &= 2e_n - 2e_{n-1} + \underbrace{o_{n-1} + e_{n-1}}. \end{aligned}$$

The underbraced sum is equal to the number of all words of length $n - 1$, i.e. to 2^{n-1}. This means that the sequence $\{e_n\}$ satisfies the recurrence relation

$$e_{n+1} = 2e_n - 2e_{n-1} + 2^{n-1} \quad (n = 2, 3, \ldots), \tag{15.11}$$

which (in combination with the initial values $e_1 = 0$, $e_2 = 1$) makes it possible to successively determine all the numbers e_n. Using the theory of difference equations (see section 11.3), a solution to such a problem can be found in the explicit form

$$e_n = 2^{n-1} - (\sqrt{2})^{n-1} \cos \frac{\pi(n-1)}{4},$$

which implies that the sought equality $e_n = 2^{n-1}$ occurs precisely for those numbers n which are of the form $4m + 3$. We therefore try to find the relationship between the numbers e_{n+4} and e_n. Using (15.11) we successively determine

$$\begin{aligned}
e_{n+2} &= 2e_{n+1} - 2e_n + 2^n,\\
e_{n+3} &= 2e_{n+2} - 2e_{n+1} + 2^{n+1} = 2(2e_{n+1} - 2e_n + 2^n) - 2e_{n+1} + 2^{n+1}\\
&= 2e_{n+1} - 4e_n + 2^{n+2},\\
e_{n+4} &= 2e_{n+3} - 2e_{n+2} + 2^{n+2}\\
&= 2(2e_{n+1} - 4e_n + 2^{n+2}) - 2(2e_{n+1} - 2e_n + 2^n) + 2^{n+2}\\
&= -4e_n + 2^{n+3} - 2^{n+1} + 2^{n+2} = -4e_n + 2^{n+3} + 2^{n+1},
\end{aligned}$$

or

$$e_{n+4} - 2^{n+3} = -4(e_n - 2^{n-1}).$$

The last relation immediately implies that the equality $e_{n+4} = 2^{n+3}$ holds if and only if $e_n = 2^{n-1}$. Since among the numbers $n = 1, 2, 3, 4$ the last equality occurs only for $n = 3$, the conjecture that the sought n are precisely those of the form $n = 4m + 3$ is thus proved.

78. The length of the radius is clearly irrelevant. We shall therefore assume that $R = 1$. Further, we can interpret all points as complex numbers. WLOG we can choose

$$A = 1, \quad B = e^{i\frac{\pi}{3}}, \quad C = e^{i\alpha}, \quad D = e^{i(\alpha+\frac{\pi}{3})}, \quad E = e^{i\beta}, \quad F = e^{i(\beta+\frac{\pi}{3})},$$

for some real α and β. We then have

$$M_1 = \frac{1}{2}(e^{i\frac{\pi}{3}} + e^{i\alpha}),\ M_2 = \frac{1}{2}(e^{i(\alpha+\frac{\pi}{3})} + e^{i\beta}),\ M_3 = \frac{1}{2}(e^{i(\beta+\frac{\pi}{3})} + 1),$$

and hence

$$\begin{aligned}
M_2 - M_3 &= \frac{1}{2}(e^{i\alpha}e^{i\frac{\pi}{3}} + e^{i\beta}(1 - e^{i\frac{\pi}{3}}) - 1) = \frac{1}{2}(e^{i\alpha}e^{i\frac{\pi}{3}} + e^{i\beta}e^{-i\frac{\pi}{3}} - 1),\\
M_1 - M_3 &= \frac{1}{2}(e^{i\frac{\pi}{3}} + e^{i\alpha} - e^{i\beta}e^{i\frac{\pi}{3}} - 1) = \frac{1}{2}(e^{i\alpha} - e^{i\beta}e^{i\frac{\pi}{3}} - e^{-i\frac{\pi}{3}})\\
&= \frac{1}{2}e^{-i\frac{\pi}{3}}(e^{i\alpha}e^{i\frac{\pi}{3}} - e^{i\beta}e^{2i\frac{\pi}{3}} - 1)\\
&= \frac{1}{2}e^{-i\frac{\pi}{3}}(e^{i\alpha}e^{i\frac{\pi}{3}} + e^{i\beta}e^{-i\frac{\pi}{3}} - 1).
\end{aligned}$$

This means that $M_2 - M_3 = e^{i\frac{\pi}{3}}(M_1 - M_3)$. Thus the triangle $M_1M_2M_3$ is equilateral.

15.14 Solutions to Paper 14

79. For $m = 0$ we get that $2f(n) = f(kn)$. Furthermore, $m = 1$ gives

$$f(n+1) + f(n-1) = f(kn) = 2f(n),$$

which is a second order linear recurrence equation. Its characteristic equation is $r^2 - 2r + 1 = 0$, with the solutions $r_{1,2} = 1$. All possible solutions are given by $f(n) = an+b$, where $a, b \in \mathbb{R}$. We need to pick out those among the solutions that solve the given equation. We have that $a(n+m)+a(n-m)+b = a(kn)+b$ and hence $a(k-2)n - b = 0$ for all n. This means that $b = 0$ and $a = 0$ for $k \neq 2$, so that for $k \neq 2$ we only have the so called *trivial solution*, $f \equiv 0$. For $k = 2$ the coefficient a can be an arbitrary real number, so that for $k = 2$ we have the solutions $f(n) = an$, $a \in \mathbb{R}$.

Remark Rewriting the recursive equation as $f(n) = \frac{f(n+1)+f(n-1)}{2}$ we could see that every term in the sequence $x_n = f(n)$ is the arithmetic mean of its neighbours which is a characteristic property of an arithmetic sequence, thus we could write $f(n) = an + b$ directly.

Solution 2. We have $2f(n) = f(kn)$, which implies $f(0) = 0$. Also, choosing $m = n$ we get $f(2n) = f(kn) = 2f(n)$. Set $f(1) = a$. We then have $f(2) = 2f(1) = 2a$, $f(3) + f(1) = 2f(2)$, which gives $f(3) = 4a - a = 3a$. By induction $f(n) = na$. Hence $f(k) = ka$, on the other hand $f(k) = 2f(1) = 2a$. This means that $(k-2)a = 0$, so for $k = 2$ we have the solutions $f(n) = an$, $a \in \mathbb{R}$, whereas for $k \neq 2$ the trivial solution is unique.

Remark Another way to induction was to show first that $f(2^m) = 2^m a$ and then using a binary representations to write an arbitrary n as a sum of powers of two.

80. If $ab + bc + ca \leq a + b + c$, then the Cauchy-Schwarz inequality solves the problem (using, e.g. $a^2c = (a\sqrt{c})^2$, $\frac{1}{c} = \left(\frac{1}{\sqrt{c}}\right)^2$):

$$\frac{a}{b} + \frac{b}{c} + \frac{c}{a} = \frac{a^2c + b^2a + c^2b}{abc} \geq \frac{\frac{(a+b+c)^2}{\frac{1}{a}+\frac{1}{b}+\frac{1}{c}}}{abc} \geq a + b + c.$$

Otherwise, the same inequality gives (now $a^2c = \left(\frac{ac}{\sqrt{c}}\right)^2, c = (\sqrt{c})^2$):

$$\frac{a}{b} + \frac{b}{c} + \frac{c}{a} = \frac{a^2c + b^2a + c^2b}{abc} \geq \frac{\frac{(ab+bc+ca)^2}{a+b+c}}{abc} \geq a + b + c.$$

(Here we have used the fact that $abc \leq 1$.)

Solution 2. Replacing a, b, c by ta, tb, tc with $t = \frac{1}{\sqrt[3]{abc}}$ preserves the value of the quantity in the left-hand side of the inequality and increases the value of the right-hand side and makes $at \cdot bt \cdot ct = abct^3 = 1$. Hence we may assume WLOG that $abc = 1$. Then there exist positive real numbers x, y, z such that $a = \frac{y}{x}, b = \frac{z}{y}, c = \frac{x}{z}$. The rearrangement inequality gives

$$x^3 + y^3 + z^3 \geq x^2y + y^2z + z^2x.$$

Thus

$$\frac{a}{b} + \frac{b}{c} + \frac{c}{a} = \frac{x^3 + y^3 + z^3}{xyz} \geq \frac{x^2y + y^2x + z^2x}{xyz} = a + b + c,$$

as desired.

Solution 3. Using the AM–GM inequality, we deduce that

$$\frac{2a}{b} + \frac{b}{c} \geq 3\sqrt[3]{\frac{a^2}{bc}} \geq 3a.$$

Similarly, $\frac{2b}{c} + \frac{c}{a} \geq 3b$ and $\frac{2c}{a} + \frac{a}{b} \geq 3c$. Adding these three inequalities, the conclusion is immediate.

Solution 4. Let $x = \sqrt[9]{\frac{ab^4}{c^2}}$, $y = \sqrt[9]{\frac{ca^4}{b^2}}$, $z = \sqrt[9]{\frac{bc^4}{a^2}}$. Consequently, $a = xy^2, b = zx^2, c = yz^2$, and also $xyz \leq 1$. Thus, using the rearrangement inequality, we find that

$$\sum_{\text{cyclic}} \frac{a}{b} = \sum_{\text{cyclic}} \frac{x^2}{yz} \geq xyz \sum_{\text{cyclic}} \frac{x^2}{yz} = \sum_{\text{cyclic}} x^3 \geq \sum_{\text{cyclic}} xy^2 = \sum_{\text{cyclic}} a.$$

Solution 5. Using the weighted AM–GM inequality $\frac{2}{3}x + \frac{1}{3}y \geq x^{\frac{2}{3}}y^{\frac{1}{3}}$ we get

$$\frac{2}{3}a^2c + \frac{1}{3}ab^2 \geq (a^2c)^{\frac{2}{3}}(ab^2)^{\frac{1}{3}} = a^{\frac{5}{3}}b^{\frac{2}{3}}c^{\frac{2}{3}}.$$

Taking a cyclic sum of three such inequalities we get

$$a^2c + ab^2 + bc^2 \geq (a + b + c)a^{\frac{2}{3}}b^{\frac{2}{3}}c^{\frac{2}{3}} \geq (a + b + c)abc$$

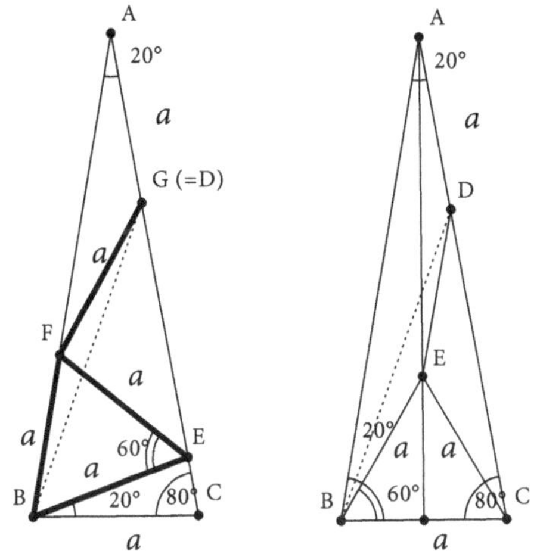

Figure 15.28 Problem 81

and it remains to divide both sides with abc.

81. Since the triangle is isosceles, we have that $\angle B = \angle C = 80°$. Because $80° - 20° = 60°$ the idea to use an equilateral triangle seems very natural. We will consider two different approaches.

Denote $AD = BC = a$ (see the first picture on figure 15.28). Choose the point E on AC so that $\angle CBE = 20°$. It follows that $\angle BEC = 80°$ and hence $BE = BC = a$. Choose now F on AB so that $\angle FEB = 60°$. This means that $\triangle FEB$ is equilateral and $EF = BF = a$. Now, choose G on AC so that $\angle FGE = 40°$. Since $\angle GEF = 180° - 80° - 60° = 40°$, it follows that $GF = EF = a$. The angle FGE is an exterior angle of $\triangle AFG$. This implies $\angle AFG = 40° - 20° = 20°$ and hence $AG = GF = a$. This is possible only if the points D and G coincide. We also have that $\angle DFA$ is an exterior angle of $\triangle FBD$, which is isosceles. It follows that $\angle FDB = 10°$ and hence $\angle BDC = 30°$.

Solution 2. Let E be a point on the altitude such that $EB = BC = a$ (see the second picture in Figure 15.28). Then EBC is equilateral and

$$\angle EBA = \angle CBA - \angle CBE = 20° = \angle BAC.$$

Thus in the quadrilateral $EBAD$ we have that $\angle EBA = \angle BAD$ and $BE = AD = a$, which implies (show!) that $EBAD$ is an isosceles trapezoid. Then

$$\angle BDC = 180° - \angle BCD - \angle CBD = 180° - 80° - (60° + 10°) = 30°.$$

Solution 3. We need two parameters and choose $\alpha = \angle BDC$ and $a = AD = BC$. The sine rule in the triangles ABD and BDC gives:

$$\frac{\sin(\alpha - 20°)}{a} = \frac{\sin 20°}{BD}; \qquad \frac{\sin\alpha}{a} = \frac{\sin 80°}{BD}.$$

Hence,

$$\frac{\sin(\alpha - 20°)}{\sin\alpha} = \frac{\sin 20°}{\sin 80°} \quad \Rightarrow \quad \frac{\sin\alpha\cos 20° - \cos\alpha\sin 20°}{\sin\alpha \cdot \sin 20°} = \frac{1}{\sin 80°}$$
$$\Rightarrow \quad \cot\alpha - \cot 20° = \frac{1}{\sin 80°}.$$

This equation already gives the formal answer $\alpha = \operatorname{arccot}\left(\cot 20° - \frac{1}{\sin 80°}\right)$, but to get it as 30° we need more trigonometry:

$$\begin{aligned}
\cot 20° - \frac{1}{\sin 80°} &= \frac{\cos 20°}{\sin 20°} - \frac{1}{\cos 10°} = \frac{\cos 20°}{\sin 20°} - \frac{2\sin 10°}{\sin 20°} \\
&= \frac{\cos 20° - 2\sin 10°}{\sin 20°} = \frac{\sin 70° - \sin 10° - \sin 10°}{\sin 20°} \\
&= \frac{2\cos\frac{80°}{2}\sin\frac{60°}{2} - \sin 10°}{\sin 20°} = \frac{\cos 40° - \sin 10°}{\sin 20°} \\
&= \frac{\sin 50° - \sin 10°}{\sin 20°} = \frac{2\cos\frac{60°}{2}\sin\frac{40°}{2}}{\sin 20°} = 2\cos 30° = \sqrt{3}.
\end{aligned}$$

Direct solutions are not always the best!

82. First let us examine what such a cyclic quadrangle $ABCD$ with a sixty-degrees angle at the vertex B and congruent sides BC and CD can look like. Let k denote the circumcircle of the quadrangle $ABCD$. The equality $\angle ABC = 60°$ already determines the length of the diagonal AC, since it is a chord subtended at the circumference by an angle of 60°. The vertex D then has to be an interior point of the shorter arc AC of the circle k (in the half-plane opposite to ACB) and the vertex B is the image of D under the symmetry with respect to the line SC, where S is the centre of the circle k.

Since $BC = CD$, by hypothesis, the angles BAC and CAD (subtending the same chords) are congruent. We therefore see that the half-lines AD and

AB are symmetric with respect to the axis AC. Denote by X the image of the point D under this symmetry. The point X clearly lies in the interior of the side AB (the image of the shorter arc AC lies wholly in the interior domain of the circle k), and since $CX = CD = BC$, the triangle XBC is isosceles. The triangle XBC is even equilateral, since the magnitude of its angle at the vertex B is 60°. Thus $BX = BC = CD$. By symmetry also $DA = XA$, so $CD + DA = BX + XA = AB$, which is the desired equality.

It is easy to see that the converse does not hold. It is enough to consider a quadrangle $ABCD$, possessing the properties described in the statement of the problem, which in addition satisfies $CD \neq DA$. (That such a quadrangle exists follows from our analysis at the beginning of this solution.) Now if we interchange the sides CD and DA, i.e., replace the vertex D by the vertex D' axially symmetric to D with respect to the perpendicular bisector of the diagonal AC, we obtain a cyclic quadrangle $ABCD'$, with a sixty degrees angle at the vertex B, which still satisfies the equality $CD' + D'A = DA + CD = AB$, but at the same time $BC = CD = D'A \neq D'C$.

Solution 2. If we now denote by φ the angle subtended at the circumference by the congruent chords BC and CD ($0° < \varphi < 60°$), it is easily seen that the angle corresponding to the chord DA is $60° - \varphi$, and to the chord AB, $120° - \varphi$. If R is the radius of the circumcircle the equality to be proved is equivalent to

$$2R \sin\varphi + 2R\sin(60° - \varphi) = 2R\sin(120° - \varphi).$$

Since $\sin(120° - \varphi) = \sin(60° + \varphi)$, the last equality is seen (upon a simple manipulation) to be in turn equivalent to

$$\sin\varphi = 2\cos 60° \sin\varphi,$$

which is trivially fulfilled.

As in the previous solution we observe that the equality $CD + DA = AB$ remains in force if the sides CD and DA of the given quadrangle are interchanged. The new quadrangle will still be cyclic, the magnitude of the interior angle at the vertex B is unchanged, but the equality $BC = CD$ turns into $BC = DA$.

Solution 3. Denote the lengths of sides of the quadrangle $ABCD$, satisfying the hypothesis of the problem, in the usual way by a, b, c, d. Since the interior angles at the vertices B and D have the magnitudes of 60° and 120°, respectively, the cosine law for the triangles ABC and CDA implies (upon expressing the quantity AC^2 in two ways) that

$$a^2 + b^2 - ab = c^2 + d^2 + cd. \tag{15.12}$$

a) Since $b = c$, the equality (15.12) in turn implies

$$\begin{aligned} a^2 + c^2 - ac &= c^2 + d^2 + cd, \\ a^2 - d^2 &= ac + cd, \\ (a-d)(a+d) &= c(a+d), \\ a - d &= c. \end{aligned}$$

The equality $a = c + d$, which we wanted to prove, is thus seen to hold.

b) If $a = c + d$, then substituting for a into (15.12) yields

$$(c+d)^2 + b^2 - (c+d)b = c^2 + d^2 + cd.$$

Upon an easy manipulation this gives $(b-c)(b-d) = 0$, which implies that either $b = c$ or $b = d$. The converse implication thus does not hold in general.

83. The positive numbers a_1, a_2, a_3, a_4 can be the lengths of the sides of a quadrilateral iff $s - a_k > 0$, $k = 1, 2, 3, 4$. Denote $x_i = s - a_i > 0$. Note that $\sum x_i = 4s - \sum a_i = 2s$, thus $s + a_i = 2s - x_i = \sum_{j \neq i} x_j$. Thus we need to prove that

$$\begin{aligned} \sum_i \frac{1}{s + a_i} &= \frac{1}{x_1 + x_2 + x_3} + \frac{1}{x_1 + x_3 + x_4} + \frac{1}{x_1 + x_2 + x_4} + \frac{1}{x_2 + x_3 + x_4} \\ &\leq \frac{2}{9} \sum_{1 \leq i < j \leq 4} \frac{1}{\sqrt{x_i x_j}} = \frac{1}{9} \sum_{i \neq j} \frac{1}{\sqrt{x_i x_j}}, \end{aligned}$$

for all $x_i > 0$, $i = 1, 2, 3, 4$, $(j = 1, 2, 3, 4)$. It is enough to prove that

$$\frac{1}{x_1 + x_2 + x_3} \leq \frac{1}{9} \left(\frac{1}{\sqrt{x_1 x_2}} + \frac{1}{\sqrt{x_2 x_3}} + \frac{1}{\sqrt{x_3 x_1}} \right)$$

for all $x_1, x_2, x_3 > 0$. The AM–HM inequality and the rearrangement (or Cauchy–Schwarz) inequality give

$$\frac{3}{\frac{1}{\sqrt{x_1 x_2}} + \frac{1}{\sqrt{x_2 x_3}} + \frac{1}{\sqrt{x_3 x_1}}} \leq \frac{\sqrt{x_1 x_2} + \sqrt{x_2 x_3} + \sqrt{x_3 x_1}}{3} \leq \frac{x_1 + x_2 + x_3}{3},$$

and the inequality follows. Equality occurs iff $\sqrt{x_1 x_2} = \sqrt{x_2 x_3} = \sqrt{x_3 x_1} = \cdots$ i.e. iff $x_1 = x_2 = x_3 = x_4$, which is equivalent to $a_1 = a_2 = a_3 = a_4$.

Solution 2. The numbers $s - a_i$ are positive, and the AM–GM inequality gives

$$\frac{1}{3}\sum_{j\neq i}(s-a_j) \geq \Big(\prod_{j\neq i}(s-a_j)\Big)^{1/3}, \tag{15.13}$$

for every $i = 1,2,3,4$. Since $\sum_{j\neq i}(s-a_j) = s + a_i$, relation (15.13) leads to

$$\frac{3}{s+a_i} \leq \Big(\prod_{j\neq i}\frac{1}{\sqrt{s-a_j}}\Big)^{2/3}. \tag{15.14}$$

On the other hand, the AM–GM inequality gives

$$\begin{aligned}\frac{1}{3}\sum_{j<k,j,k\neq i}\frac{1}{\sqrt{(s-a_j)(s-a_k)}} &\geq \Big(\prod_{\substack{j<k,\\ j,k\neq i}}\frac{1}{\sqrt{(s-a_j)(s-a_k)}}\Big)^{1/3}\\ &= \Big(\prod_{j\neq i}\frac{1}{\sqrt{s-a_j}}\Big)^{2/3}.\end{aligned} \tag{15.15}$$

Combining (15.14) and (15.15) and adding we get the required inequality. The equality takes place if and only if the quadrilateral is a rhombus.

Solution 3. We can write, using AM–GM–HM inequalities

$$\begin{aligned}\frac{2}{9}\sum_{i<j}\frac{1}{\sqrt{(s-a_i)(s-a_j)}} &\geq \frac{4}{9}\sum_{i<j}\frac{1}{a_i+a_j}\\ &= \frac{2}{9}\sum_{i=1}^{4}\Big(\sum_{i\neq j}\frac{1}{a_i+a_j}\Big) \geq \frac{2}{9}\sum_{i=1}^{4}\frac{9}{2a_i+2s} = \sum_{i=1}^{4}\frac{1}{s+a_i}.\end{aligned}$$

84. Assume that there exists such an n and denote by p the smallest prime divisor of n. Obviously $p > 3$. It follows that n and $p-1$ are relatively prime and hence there exist integers a and b such that $a(p-1) + bn = 1$. Using Fermat's little theorem we then get (modulo p):

$$2 \equiv 2^{a(p-1)}\cdot 2^{bn} \equiv 1\cdot 3^{bn} \equiv 3^{a(p-1)}\cdot 3^{bn} \equiv 3,$$

which is a contradiction.

15.15 Solutions to Paper 15

85. We shall prove the inequality

$$\frac{a_1}{1+xa_1} + \frac{a_2}{1+xa_2} + \cdots + \frac{a_n}{1+xa_n} \leq \frac{n}{n+x},$$

for all $x \geq 0$. It is obvious that equality holds for $x = 0$. Let $x > 0$. We have

$$\begin{aligned}
&\frac{a_1}{1+xa_1} + \frac{a_2}{1+xa_2} + \cdots + \frac{a_n}{1+xa_n} \\
&\quad = \frac{1}{x}\left(\frac{1+xa_1-1}{1+xa_1} + \frac{1+xa_2-1}{1+xa_2} + \cdots + \frac{1+xa_n-1}{1+xa_n}\right) \\
&\quad = \frac{1}{x}\left(n - \frac{1}{1+xa_1} - \frac{1}{1+xa_2} - \cdots - \frac{1}{1+xa_n}\right).
\end{aligned}$$

What remains to prove is that

$$\frac{1}{1+xa_1} + \frac{1}{1+xa_2} + \cdots + \frac{1}{1+xa_n} \geq \frac{n^2}{n+x}.$$

The inequality between harmonic and arithmetic means gives

$$\frac{1}{1+xa_1} + \frac{1}{1+xa_2} + \cdots + \frac{1}{1+xa_n} \geq \frac{n^2}{1+xa_1+xa_2+\cdots+xa_n} = \frac{n^2}{n+x}.$$

For $x \neq 0$ equality occurs iff all terms in the sum above are equal, i.e. iff $a_1 = a_2 = \cdots = a_n = \frac{1}{n}$.

Solution 2. Let $S = \sum \frac{a_i}{1+xa_i}$. WLOG $a_1 \geq \cdots \geq a_n$. Then

$$\frac{a_1}{1+xa_1} \leq \frac{a_2}{1+xa_2} \leq \cdots \leq \frac{a_n}{1+xa_n}$$

and according to Chebyshev's inequality we have

$$S \leq \frac{1}{n}(a_1 + \cdots + a_n)\left(\frac{1}{1+xa_1} + \frac{1}{1+xa_2} + \cdots + \frac{1}{1+xa_n}\right)$$

Hence

$$S \leq \frac{1}{n}\left(1 - \frac{a_1x}{1+xa_1} + 1 - \frac{a_2x}{1+xa_2} + \cdots + 1 - \frac{a_nx}{1+xa_n}\right) = \frac{1}{n}(n - xS),$$

which gives $S \leq \frac{1}{1+\frac{x}{n}} = \frac{n}{n+x}$.

Solution 3. Another way to use Chebyshev's inequality is:

$$\frac{n+x}{n}\left(\sum_{i=1}^{n} \frac{a_i}{1+xa_i}\right) = \frac{1}{n}\left(\sum_{i=1}^{n}(1+a_ix)\right)\left(\sum_{i=1}^{n} \frac{a_i}{1+xa_i}\right) \leq \sum_{i=1}^{n} a_i = 1.$$

Solution 4. $f(t) = \frac{t}{1+tx}$ is concave and Jensen's or Karamata's inequality gives

$$\sum_{i=1}^{n} f(a_i) \leq nf\left(\frac{\sum a_i}{n}\right) = nf\left(\frac{1}{n}\right) = \frac{n}{n+x}.$$

86. If (x, y, z) is a solution of the given system, then clearly $xyz \neq 0$. We can therefore multiply the three equations by the factors yz, zx and xy respectively, and look for solutions of the equivalent system

$$x^2yz = y + z, \quad xy^2z = x + z, \quad xyz^2 = x + y \tag{15.16}$$

in the domain of all nonzero real numbers. Adding up the left-hand and the right-hand sides of this system yields, after a simple manipulation, the equation

$$(xyz - 2)(z + y + z) = 0.$$

We thus see that either $xyz = 2$, or $x + y + z = 0$.

Case 1 Let $xyz = 2$. Upon substituting for xyz into the system (15.16) we obtain

$$2x = y + z, \quad 2y = x + z, \quad 2z = x + y,$$

which is equivalent to the system

$$3x = x + y + z, \quad 3y = x + y + z, \quad 3z = x + y + z.$$

It follows that $x = y = z$. In view of the condition $xyz = 2$, we thus get $x = y = z = \sqrt[3]{2}$. Straightforward verification shows that the triple $(\sqrt[3]{2}, \sqrt[3]{2}, \sqrt[3]{2})$ is indeed a solution of the system (15.16), hence also of the original system of equations.

Case 2 Let $x + y + z = 0$. From the first equation in (15.16) we get $x^2yz = -x$, which in view of the condition $x \neq 0$ implies $xyz = -1$. Let us check that, conversely, any triple of nonzero real numbers (x, y, z) satisfying the system of two equations

$$x + y + z = 0, \quad xyz = -1, \tag{15.17}$$

is a solution of the original system: indeed, from the equalities (15.17) it follows that

$$\frac{1}{y} + \frac{1}{z} = \frac{y + z}{yz} = \frac{-x}{-1/x} = x^2.$$

(Owing to the symmetry of the given system it is enough to check only one equation.)

The system (15.17) has infinitely many solutions in the domain of nonzero real numbers, which can be described, for instance, by choosing one of the variables (e.g. z) as a parameter. This yields the system

$$x + y = -z, \quad xy = -\frac{1}{z}.$$

Substituting for x from the first equation into the second gives

$$(y + z)y = \frac{1}{z},$$

that is,

$$y^2 + yz - \frac{1}{z} = 0. \tag{15.18}$$

This is a quadratic equation with unknown y and parameter z. Its discriminant is equal to $D = z^2 + 4/z$. The necessary and sufficient condition for the equation to have real roots is the inequality $D \geq 0$. Solving the inequality $(z^3 + 4)/z \geq 0$ leads to the following condition for the parameter z

$$z \in (-\infty, -\sqrt[3]{4}) \cup (0, \infty). \tag{15.19}$$

Under the condition (15.19), the quadratic equation (15.18) has the roots

$$y_1 = \frac{-z + \sqrt{z^2 + 4/z}}{2} \quad \text{and} \quad y_2 = \frac{-z - \sqrt{z^2 + 4/z}}{2},$$

which under the relation $x = -y - z$ correspond to the values

$$x_1 = \frac{-z - \sqrt{z^2 + 4/z}}{2} \quad \text{and} \quad x_2 = \frac{-z + \sqrt{x^2 + 4/z}}{2};$$

further, the equality $(x_1, y_1) = (x_2, y_2)$ holds only for $z = -\sqrt[3]{4}$.

Conclusion: The given system has a solution $x = y = z = \sqrt[3]{2}$. All other solutions are the triples (x, y, z) of the form

$$(x, y, z) = \left(\frac{-z \pm \sqrt{z^2 + 4/z}}{2}, \frac{-z \mp \sqrt{z^2 + 4/z}}{2}, z\right),$$

where z is any number satisfying (15.19).

Remark Note that

$$xyz = \frac{y + z}{x} = \frac{z + x}{y} = \frac{x + y}{z} \quad \Rightarrow \quad xyz = \frac{y + z + z + x + x + y}{x + y + z} = 2.$$

But this nice argument works only when $x+y+z \neq 0$, thus the case $x+y+z = 0$ should be considered separately!

87. Substituting $x = 0$ in the equation yields $f(f(y)) = (f(0))^2 + y$, whence f is surjective. In particularly there exists a solution a for the equation $f(x) = 0$. Substituting a for x, we get $f(f(y)) = y$. Using this and the surjectivity, the substitution of $f(x)$ for x in the functional equation gives

$$f(xf(x) + f(y)) = x^2 + y.$$

Comparing this with the original equation we have $(f(x))^2 = x^2$. Hence, $f(x) = \pm x$ for each x.

If for some nonzero reals x and y we have $f(x) = -x$ and $f(y) = y$, then the main equation implies $f(-x^2 + y) = x^2 + y$, which contradicts $f(-x^2 + y) = \pm(-x^2 + y)$. Thus we conclude that either $f(x) = x$ for all x, or $f(x) = -x$ for all x. These functions obviously satisfy the functional equation.

Remark If we put $a = 1 \cdot f(1) + f(-(f(1)^2)$ we get $f(a) = 0$ directly from the functional equation. This shows how creative one can be solving a functional equation.

88. a) If the triangle ABC is acute then by the cosine theorem for $\triangle AHB$ we find

$$AB^2 = AH^2 + BH^2 - 2AH \cdot BH \cos(\pi - \gamma).$$

Note that if AF is the altitude then $\angle FHC = \beta$ and

$$CH = \frac{CF}{\sin\beta} = \frac{AC\cos\gamma}{\sin\beta} = 2R\cos\gamma$$

Since $AB = 2R\sin\gamma$ we obtain a useful fact: $AB^2 + CH^2 = 4R^2$. Therefore

$$4R^2 = AH^2 + BH^2 + CH^2 + \frac{AH \cdot BH \cdot CH}{R}.$$

Then $4R^3 = 7R + 3$, i.e. $(R + 1)(2R + 1)(2R - 3) = 0$, whence $R = \frac{3}{2}$.

If the triangle ABC is obtuse, we find analogously

$$4R^2 = AH^2 + BH^2 + CH^2 - \frac{AH \cdot BH \cdot CH}{R}$$

and therefore $4R^3 = 7R - 3$, i.e. $(R - 1)(2R - 1)(2R + 3) = 0$. Since $3 = AH \cdot BH \cdot CH < (2R)^3$, we conclude that $R = 1$.

The existence of $\triangle ABC$ with $R = \frac{3}{2}$ and $R = 1$ follows from (b).

b) Since $S = \frac{AB \cdot BC \cdot CA}{4R}$, we have

$$S^2 = \frac{(4R^2 - AH^2)(4R^2 - BH^2)(4R^2 - CH^2)}{16R^2}.$$

Setting $AH^2 = x$, $BH^2 = y$, $CH^2 = z$ and $4R^2 = t$, we obtain

$$S^2 = \frac{t^3 - 7t^2 + t(xy + yz + zx) - 9}{4t}.$$

Without loss of generality we may assume that $x \geq y \geq z$. Then $x \geq \frac{7}{3}$ and therefore

$$xy + yz + zx = \frac{9}{x} + x(7 - x) = 15 - \frac{(x-3)^2(x-1)}{x} \leq 15,$$

where equality holds iff $x = 3$. Hence

$$S^2 \leq \frac{t^3 - 7t^2 + 15t - 9}{4t}.$$

Since $R = \frac{3}{2}$ or $R = 1$, we conclude that $S_{\max} = \sqrt{8}$ and it is obtained for an acute $\triangle ABC$ with $R = \frac{3}{2}$, $AH = BH = \sqrt{3}$ and $CH = 1$. The sides of this triangle are $\sqrt{6}, \sqrt{6}$ and $\sqrt{8}$.

It remains to prove that there exists a triangle with $R = 1$; a simple calculation shows that the triangle with $AH = BH = \sqrt{3}$, $CH = 1$ and $AB = \sqrt{3}$, $BC = CA = 1$ has the required properties.

Alternative solution (sketch). Regard the triangle as imbedded in the complex plane with the circumcentre at zero and A at R. We then have $B = Re^{2i\gamma}$, $C = Re^{2i(\alpha+\gamma)}$ and $H = A + B + C$. It follows that

$$\begin{aligned} AH^2 = R^2|e^{2i\gamma}|^2|1 + e^{2i\alpha}|^2 &= R^2((1 + \cos 2\alpha)^2 + \sin^2 2\alpha) \\ &= 2R^2(1 + \cos 2\alpha), \end{aligned}$$

and similarly for BH, CH (α, β, γ are the angles of the triangle at A, B, C respectively). This leads to the equality

$$7 = AH^2 + BH^2 + CH^2 = 2R^2(3 + \cos 2\alpha + \cos 2\beta + \cos 2\gamma).$$

Similar calculations yield

$$3 = AH \cdot BH \cdot CH = 8R^3|\cos\alpha\cos\beta\cos\gamma|.$$

The trigonometric identity (prove it!)

$$\cos 2\alpha + \cos 2\beta + \cos 2\gamma = -1 - 4\cos\alpha\cos\beta\cos\gamma \quad \text{for} \quad \alpha+\beta+\gamma=\pi,$$

then gives $4R^3 = 7R \pm 3$, the sign depending on whether the triangle is acute or obtuse.

89. Let $x = 2^n\alpha$ for some $n \in \mathbb{N}_0 = \mathbb{N} \cup \{0\}$. From

$$\cos 2x = 2\cos^2 x - 1,$$
$$\cos 4x = 2(2\cos^2 x - 1)^2 - 1 = 8\cos^4 x - 8\cos^2 x + 1 < 0$$

and the condition $\cos 4x = \cos 2^{n+2}\alpha < 0$, it follows

$$\frac{2-\sqrt{2}}{4} < \cos^2 x < \frac{2+\sqrt{2}}{4},$$

i.e.

$$\cos x \in \left(-\frac{\sqrt{2+\sqrt{2}}}{2}, -\frac{\sqrt{2-\sqrt{2}}}{2}\right) \cup \left(\frac{\sqrt{2-\sqrt{2}}}{2}, \frac{\sqrt{2+\sqrt{2}}}{2}\right).$$

But $\cos x = \cos 2^n\alpha < 0 \Rightarrow \cos 2^n\alpha < -\frac{\sqrt{2-\sqrt{2}}}{2} < -\frac{1}{3}$ for every $n \in \mathbb{N}_0$.

Now let us prove, using mathematical induction, that

$$\left|\cos 2^n\alpha + \frac{1}{2}\right| \geq \left(\frac{5}{3}\right)^n \left|\cos\alpha + \frac{1}{2}\right|, \quad \forall n \in \mathbb{N}_0.$$

For $n = 0$ it is obvious. Suppose $n > 0$ and the inequality holds for $n-1$. Then

$$\begin{aligned}\left|\cos 2^n\alpha + \frac{1}{2}\right| &= 2\cdot\left|\cos^2 2^{n-1}\alpha - \frac{1}{4}\right| \\ &= 2\cdot\left|\cos 2^{n-1}\alpha - \frac{1}{2}\right|\cdot\left|\cos 2^{n-1}\alpha + \frac{1}{2}\right| \\ &\geq 2\cdot\left(\frac{1}{3}+\frac{1}{2}\right)\cdot\left(\frac{5}{3}\right)^{n-1}\cdot\left|\cos\alpha + \frac{1}{2}\right| \\ &= \left(\frac{5}{3}\right)^n\cdot\left|\cos\alpha + \frac{1}{2}\right|\end{aligned}$$

proves the inequality.

On the other hand $|\cos 2^n\alpha + \frac{1}{2}| \leq 1$ because $-1 \leq \cos 2^n\alpha < 0$. Therefore

$$0 \leq \left|\cos\alpha + \frac{1}{2}\right| \leq \left(\frac{3}{5}\right)^n, \quad \forall n \in \mathbb{N}_0.$$

Taking the limit $n \to \infty$ implies $\cos\alpha + \frac{1}{2} = 0$, i.e.

$$\alpha = \pm\frac{2\pi}{3} + 2k\pi, \quad k \in \mathbb{Z}.$$

Solution 2. Because of the 2π-periodicity it is sufficient to find all $\alpha \in [0, 2\pi)$, which satisfy the conditions of the problem.

The number $\frac{\alpha}{2\pi}$ cannot be a diadic fraction, i.e. a number of the form $\frac{\alpha}{2\pi} = \frac{m}{2^n}$ for some $m, n \in \mathbb{N}_0$, because it would mean that $\cos(2^n\alpha) = \cos(2m\pi) = 1 > 0$.

Every number from $[0, 1)$, which is not a diadic fraction, can be represented in a unique way using the binary number system. Let

$$\frac{\alpha}{2\pi} = 0, b_1 b_2 b_2 \dots b_n b_{n+1} \dots \tag{15.20}$$

be the binary form of the number $\frac{\alpha}{2\pi}$. Then

$$\frac{2^n\alpha}{2\pi} = b_1 b_2 b_3 \dots b_n, b_{n+1} b_{n+2} \dots .$$

However,

$$\begin{aligned} \cos(2^n\alpha) < 0 \quad &\Leftrightarrow \quad 2^n\alpha \in \bigcup_{k\in\mathbb{Z}} \left(2k\pi + \frac{\pi}{2}, 2k\pi + \frac{3\pi}{2}\right) \\ &\Leftrightarrow \quad \frac{2^n\alpha}{2\pi} \in \bigcup_{k\in\mathbb{Z}} \left(k + \frac{1}{4}, k + \frac{3}{4}\right). \end{aligned}$$

Consider two cases:

1) If $b_{n+1} = 0$ then $0, 0b_{n+2}b_{n+3}\dots$ is greater than $\frac{1}{4}$ only if $b_{n+2} = 1$;

1) If $b_{n+1} = 1$ then $0, 1b_{n+2}b_{n+3}\dots$ is less than $\frac{3}{4}$ only if $b_{n+2} = 0$.

Hence, in the representation (15.20) every two neighbouring digits are different. Finally,

$$\frac{\alpha}{2\pi} = 0, 010101\dots \quad \text{or} \quad \frac{\alpha}{2\pi} = 0, 101010\dots,$$

i.e.

$$\frac{\alpha}{2\pi} = \frac{\frac{1}{4}}{1 - \frac{1}{4}} = \frac{1}{3} \quad \text{or} \quad \frac{\alpha}{2\pi} = \frac{\frac{1}{2}}{1 - \frac{1}{4}} = \frac{2}{3}$$

and the solutions up to 2π-periodicity are $\alpha = \frac{2\pi}{3}$ and $\alpha = \frac{4\pi}{3}$.

90. Using that p_n is odd for all $n > 1$, it is easy to prove that $p_n > 2n + 1$ for all $n \geq 7$. Denote $s_n = p_1 + p_2 + \dots + p_n$.

We have $s_7 = 2 + 3 + 5 + 7 + 11 + 13 + 17 = 58 > 49 = 7^2$. We shall use induction to prove that $s_n > n^2$ for all $n \geq 7$. Indeed, $s_{k+1} = s_k + p_{k+1} > k^2 + 2k + 1 = (k + 1)^2$ for $k \geq 7$. Also, it is easily checked that the same inequality holds for $1 \leq n \leq 6$. Thus $s_n > n^2$ for all $n \in \mathbb{N}$. Denote by m^2 the largest square which does not exceed s_n. According to what was proved above, we have $m \geq n$, i.e. $m = n + k,\ k \geq 0$. Thus we have proved that for all n there exists a $k \geq 0$ such that

$$(n + k)^2 \leq s_n \leq (n + k + 1)^2.$$

We shall prove that this implies $p_{n+1} > 2(n + k) + 1$. Assume the opposite, i.e. that $p_{n+1} \leq 2(n + k) + 1$. For all $n \geq 2$ we have $p_{n+1} \geq p_n + 2$. It follows that

$$\begin{aligned}
p_n &\leq 2(n + k) + 1 - 2 = 2(n + k) - 1,\\
p_{n-1} &\leq 2(n + k) + 1 - 4 = 2(n + k) - 3,\\
&\vdots\\
p_{n-j} &\leq 2(n + k) + 1 - 2(j + 1) = 2(n + k) - (2j + 1)\\
&\vdots\\
3 = p_2 &\leq 2(n + k) + 1 - 2(n - 1) = 2(n + k) - (2n - 3)\\
2 = p_1 &\leq 2(n + k) + 1 - 2n + 1 = 2(n + k) - (2n - 1) + 1.
\end{aligned}$$

Adding all the above inequalities we get

$$\begin{aligned}
s_n &= p_1 + p_2 + \cdots + p_n \leq 2n(n + k) - (1 + 3 + \cdots + (2n - 1)) + 1\\
&= 2n(n + k) - n^2 + 1 = n^2 + 2nk + k^2 - k^2 + 1 = (n + k)^2 - (k^2 - 1),
\end{aligned}$$

i.e. $s_n \leq (n + k)^2 - (k^2 - 1)$. In fact, because $p_5 - p_4 = 11 - 7$ we should get less and in any case $s_n \leq (n + k)^2 - 1$. But by assumption we have $s_n \geq (n + k)^2$. The contradiction shows that $p_{n+1} > 2(n + k) + 1$. Hence

$$s_{n+1} = s_n + p_{n+1} > (n + k)^2 + 2(n + k) + 1 = (n + k + 1)^2 > s_n,$$

i.e.

$$s_n < (n + k + 1)^2 < s_{n+1},$$

and the statement follows.

Solution 2. We show by induction that every interval between s_{n-1} and s_n contains a square a_n^2 such that $(a_n + 1)^2 > s_n$ and $2a_n - 1 \leq p_n$.

Note that $2a_n + 1 \le p_n + 2 \le p_{n+1}$, thus

$$(a_n + 1)^2 = a_n^2 + 2a_n + 1 < s_n + p_{n+1} = s_{n+1}$$

and we can take $a_{n+1} = a_n + 1$ if this is the only square in the interval (s_n, s_{n+1}).

Otherwise we consider the last square m^2 from this interval. Because the previous square $(m-1)^2$ also belongs to this interval their difference $2m-1$ is less than p_{n+1} and we can take $a_{n+1} = m$.

15.16 Solutions to Paper 16

91. *Solution 1.* Let f be any function with the required property. Denote $f(1) = p$; then clearly $p > 0$.

Taking $x = 1$ and $y = 1$ in the given functional equation, we get after a small manipulation

$$p = f(p). \tag{15.21}$$

Taking further $x = p, y = 1$ yields

$$p^2(f(p) + p) = (p+1)f(f(p)),$$

so by (15.21)

$$2p^3 = (p+1)p.$$

This algebraic equation has three real roots $-\frac{1}{2}, 0, 1$. The only root satisfying the condition $p > 0$ is $p = 1$; thus

$$f(1) = 1.$$

Let finally t be an arbitrary positive real number. Taking $x = 1$ and $y = t$ in the given equation, we obtain

$$1 + f(t) = (1+t)f(t).$$

Solving for $f(t)$ gives

$$f(t) = \frac{1}{t}.$$

An easy verification shows that the function $f(t) = 1/t$ indeed satisfies the required functional equation.

Solution 2. If we put $y = x$ in the equation we find that $xf(x)$ is a fix point for each x. If $x = f(x)$ is a fix point and $y = 1$ we obtain an equation

$$x^2(x + f(1)) = (x+1)x \quad \Rightarrow \quad x^2 + x(f(1) - 1) - 1 = 0,$$

that have only one positive solution c. Thus for arbitrary x

$$xf(x) = c \quad \Rightarrow \quad f(x) = \frac{c}{x}$$

and we need only to determine the constant c which is trivial.

92. *Solution 1.* For positive real numbers $\alpha, \beta, \gamma, x, y, z$ with $\alpha\beta\gamma = xyz = 1$ and $x, y, z \in [\alpha, \gamma]$, we have $x + y + z \leq \alpha + \beta + \gamma$. This is because

$$\begin{aligned}&(\alpha + \beta + \gamma) - (x + y + z)\\ &\quad = \begin{cases} \beta\gamma(x-\alpha)(z-\alpha) + xz(\gamma - y)(y - \beta) \geq 0, & \text{if } y \geq \beta \\ \alpha\beta(\gamma - x)(\gamma - z) + xz(y - \alpha)(\beta - y) \geq 0, & \text{if} y \leq \beta. \end{cases}\end{aligned}$$

We are going to apply this observation to α, β, γ defined by

$$\{\alpha, \beta, \gamma\} = \left\{\frac{b}{a}, \frac{c}{b}, \frac{a}{c}\right\}. \quad \alpha \leq \beta \leq \gamma.$$

Since $(c{+}a)/(b{+}c) \in [\min\{c/b, a/c\}, \max\{c/b, a/c\}] \subseteq [\alpha, \gamma]$ and similarly for $(a+b)/(c+a)$ and $(b+c)/(a+b)$, we have

$$\frac{c+a}{b+c} + \frac{a+b}{c+a} + \frac{b+c}{a+b} \leq \alpha + \beta + \gamma = \frac{b}{a} + \frac{c}{b} + \frac{a}{c}.$$

Since $(b+a)/(b+c) \in [\min\{1, a/c\}, \max\{1, a/c\}] \subseteq [\alpha, \gamma]$ and similarly for $(c+b)/(c+a)$ and $(a+c)/(a+b)$, we have

$$\frac{b+a}{b+c} + \frac{c+b}{c+a} + \frac{a+c}{a+b} \leq \alpha + \beta + \gamma = \frac{b}{a} + \frac{c}{b} + \frac{a}{c}.$$

Adding these, we get the desired inequality by using $a + b + c = 1$ for the left-hand side.

*Solution 2.*The solution above is hardly possible to find. A more realistic solution is to use some standard inequality, but the difficulty is that all of them leads us to wrong direction. Besides that we have no complete symmetry, only a cyclic one. In any case the first step is to transform the inequality.

$$\sum_{\text{cyclic}} \frac{2a+b+c}{b+c} \le 2\sum_{\text{cyclic}} \frac{a}{c} \quad \Leftrightarrow \quad 3+\sum_{\text{cyclic}} \frac{2a}{b+c} \le 2\sum_{\text{cyclic}} \frac{a}{c}$$

$$\Leftrightarrow \quad 3 \le 2\sum_{\text{cyclic}} \left(\frac{a}{c} - \frac{a}{b+c}\right) = 2\sum_{\text{cyclic}} \left(\frac{ab}{c(b+c)}\right).$$

Now we are able to apply Cauchy-Schwarz inequality:

$$2\sum_{\text{cyclic}} \left(\frac{ab}{c(b+c)}\right) = \sum_{\text{cyclic}} (b+c) \sum_{\text{cyclic}} \left(\frac{ab}{c(b+c)}\right) \ge \left(\sum_{\text{cyclic}} \sqrt{\frac{ab}{c}}\right)^2$$

and we need only to prove that

$$\left(\sum_{\text{cyclic}} \sqrt{\frac{ab}{c}}\right)^2 \ge 3 = 3(a+b+c).$$

If $x = \sqrt{\frac{bc}{a}}$, $y = \sqrt{\frac{ca}{b}}$, $z = \sqrt{\frac{ab}{c}}$ this looks like

$$(x+y+z)^2 \ge 3(xy+yz+xz)$$

and follows for example from Muirhead inequality (see page 125) or the rearrangement inequality.

93. *Solution 1.* Let p be an arbitrary prime, greater than n. Let N be a natural number such that $N > p+(n-1)n!$. Set $a_0 = N!+p$, $a_1 = N!+p+n!$, $a_2 = N!+p+2(n!), \dots, a_{n-1} = N!+p+(n-1)(n!)$. It is obvious that the numbers $a_0, a_1, a_2, \dots, a_{n-1}$ form an arithmetic progression with first term $N!+p$ and difference $n!$. We shall now prove that the numbers $a_0, a_1, a_2, \dots, a_{n-1}$ are composite numbers. For all $0 \le k < n$ we have $p+k(n!) \le N$, which means that the number $p+k(n!)$ is a divisor of $N!$. It now follows that

$$a_k = N!+p+k(n!) = (p+k(n!))c,$$

where c is an integer. Since obviously $c > 1$ it follows that a_k is a composite number.

Next we shall prove that for $0 \le k < m \le n$ the numbers a_k and a_m are relatively prime. Assume the opposite, i.e. assume that there exists a prime $q(>1)$ such that q is a divisor of both a_k and a_m. Then q is also a divisor of $a_m - a_k$. We have

$$a_m - a_k = [N!+p+m(n!)] - [N!+p+k(n!)] = (m-k)(n!),$$

i.e. q is a divisor of $(m-k)(n!)$. Since $1 \le m-k < n$, we have $(m-k)(n!) = 1 \cdot 2 \cdot 3 \cdot \dots \cdot (m-k-1) \cdot (m-k)^2 \cdot (m-k+1) \cdot \ldots \cdot n$. The number q being a prime it divides $(m-k)^2$ if and only if it divides $m-k$. Hence q is a divisor of $n!$. Since $N > n$ and $n!$ divides $N!$, it follows that q is a divisor of $N!$. Hence q divides $p = a_m - N! - m(n!)$ and, as both are primes, it follows that $q = p$. But, we chose p to be greater than n, which contradicts $p \mid n!$. The contradiction means that our assumption was false to begin with and hence, that the terms of the arithmetic progression are relatively prime.

Solution 2. Let $p_1, \ldots, p_m$ be all primes less than n and d be their product. Let $q_0, \ldots, q_{n-1}$ be different primes greater than n. Consider the following system of congruences:

$$x = 1 \pmod{d}; \quad x + d(i-1) = 0 \pmod{q_i}, \quad i = 0, \ldots, n-1.$$

By the Chinese remainder theorem the system has infinitely many solutions and we choose one of them greater than all q_i. Then the numbers $x + d(i-1)$ being divisible by q_i form an arithmetic progression consisting of composite numbers and we need only to prove that they are relatively prime.

Suppose that on contrary some prime q divides $x+d(i-1)$ and $x+d(j-1)$ for $0 \le i < j < n$. Then q divides their difference $d(j-i)$ and therefore should be one of p_k. This means that q divides d and therefore divides $x = (x+d(i-1)) - d(i-1)$. But $x = 1 \pmod{d}$ implies $x = 1 \pmod{q = p_k}$, which is a contradiction.

94. We use complex numbers. Choose the coordinate system such that $A = 0, B = 1, C = c$. If the triangle XYZ is a positively oriented equilateral triangle, then the coordinates x, y, z of its vertices satisfy

$$z = x + \omega(y-x), \quad \text{where} \quad \omega = e^{\frac{i\pi}{3}} = \frac{1}{2} + \frac{i\sqrt{3}}{2}.$$

Observe that $\omega^3 = -1, \omega \ne -1$, and hence $\omega^2 - \omega + 1 = 0$, i.e., $\omega^2 = \omega - 1$.

Now, the coordinates d, e, f, g, h, i of the points D, E, F, G, H, I equal:

$$\begin{aligned}
d &= 0 + \omega(1-0) = \omega,\\
e &= 1 + \omega(0-1) = 1-\omega,\\
f &= c + \omega(0-c) = c - \omega c,\\
g &= d + \omega(f-d) = \omega + \omega(c - \omega c - \omega)\\
&= \omega + \omega c - \omega^2 c - \omega^2 = \omega + \omega c - (\omega-1)c - (\omega-1)\\
&= c+1,\\
h &= e + \omega(c-e) = 1-\omega+\omega(c-1+\omega)\\
&= 1-\omega+\omega c - \omega + \omega - 1 = \omega c - \omega,\\
i &= g + \omega(h-g) = c+1+\omega(\omega c - \omega - c - 1)\\
&= c+1+(\omega-1)c - (\omega-1) - \omega c - \omega = 2(1-\omega).
\end{aligned}$$

The claim of the problem now follows from $A = 0, E = 1-\omega, I = 2(1-\omega)$.

95. We start from an estimate for the length of the angle bisector l_c. According to formula on page 150 we have $l_c^2 = ab - xy$, where x and y are the parts into which the angle bisector l_c divides c; x adjacent to a and y adjacent to b. By the angle bisector theorem we have

$$\frac{x}{a} = \frac{y}{b}.$$

Since $x + y = c$, it is easy to deduce that

$$x = \frac{ac}{a+b}, \qquad y = \frac{bc}{a+b}.$$

Hence

$$l_c^2 = ab - \frac{abc^2}{(a+b)^2} = \frac{ab}{(a+b)^2}\left((a+b)^2 - c^2\right).$$

Using AM-GM inequality we get:

$$l_c^2 \le \frac{(a+b)^2 - c^2}{4} = \frac{ab}{2} + \frac{a^2+b^2-c^2}{4} = \frac{ab}{2} + \frac{ab}{2}\cos C.$$

We also have

$$\frac{ab}{2} = \frac{a \cdot 2R\sin B}{2c} = \frac{Ra\sin B}{c} = Rh_c.$$

Hence

$$l_c^2 \le Rh_c + Rh_c\cos C,$$

and similarly for the other angle bisectors. It follows that

$$\frac{l_a^2}{h_a} + \frac{l_b^2}{h_b} + \frac{l_c^2}{h_c} \leq 3R + R(\cos A + \cos B + \cos C).$$

We now use formula for the cosine sum on page 299 to deduce

$$\frac{l_a^2}{h_a} + \frac{l_b^2}{h_b} + \frac{l_c^2}{h_c} \leq 3R + R\left(\frac{R+r}{R}\right) = 4R + r.$$

96. *Solution 1.* a) Let 29 contestants solve all the problems except the 8th one, while the 30th contestant solves the 8th problem only. Then he solves fewer problems than any other contestant and scores 29 points while any of the others scores 7 points only.

b) Suppose contestant A scores less points than any of the others, and at the same time he solves more problems than any of the other contestants. Let, for definiteness, A solve the first k of the eight problems proposed. Let also a_i be the numbers of the contestants which solve the i–th problem. respectively. Then the total number of points scored by A is $(30 - a_1) + \cdots + (30 - a_k) = 30k - (a_1 + \cdots + a_k)$.

It is not difficult to show that this number does not exceed 58 (show it or look at the second solution!). Hence,

$$a_1 + \cdots + a_k \geq 30k - 58. \tag{15.22}$$

Further, all 29 contestants (except A) together solve $(a_1 - 1) + \cdots + (a_k - 1)$ from the first k problems. If $(a_1 - 1) + \cdots + (a_k - 1) \geq (k-2) \cdot 29 + 1$, then there exists at least one of these 29 contestants, say, B, who solves at least $(k-2) + 1/29$, that is at least $k - 1$ problems. By condition, B solves less problems than A, hence, he solves exactly these $k-1$ problems. But A solves all these $k-1$ problems and one more. So A scores more points than B, a contradiction. Therefore, $(a_1 - 1) + \cdots + (a_k - 1) \leq (k-2) \cdot 29$, so that

$$a_1 + \ldots + a_k \leq (k-2) \cdot 29 + k = 30k - 58. \tag{15.23}$$

From (15.22) and (15.23) we have $a_1 + \ldots + a_k = 30k - 58$. Then A scores exactly $30k - (a_1 + \ldots + a_k) = 58$ points. Thus, the total number S of points scored by all contestants satisfies the inequality $S \geq 58 + 29 \cdot 59 = 30 \cdot 59 - 1$.

On the other hand,

$$\begin{aligned}
S &= a_1(30-a_1)+\cdots+a_k(30-a_k)+a_{k+1}(30-a_{k+1})+\cdots+a_8(30-a_8)\\
&\le k\cdot\left(30-\frac{a_1+\cdots+a_k}{k}\right)\cdot\frac{a_1+\cdots+a_k}{k}+(8-k)\cdot 15^2\\
&= k\cdot\left(30-\frac{30k-58}{k}\right)\cdot\frac{30k-58}{k}+(8-k)\cdot 225\\
&= \frac{58}{k}\cdot(30k-58)+225(8-k)=58\cdot 30+8\cdot 225-\left(\frac{58^2}{k}+15^2k\right).
\end{aligned}$$

(Here we use Jensen's inequality for the concave function $y = x(30-x)$ to estimate first k terms along with the fact that the maximum of the function is attained for $x = 15$ to estimate the rest.) Hence,

$$30\cdot 59-1\le 58\cdot 30+8\cdot 225-\left(\frac{58^2}{k}+15^2k\right).$$

Then

$$\frac{58^2}{k}+15^2k\le 30(58-59)+1+8\cdot 225=1771.$$

Note that the minimum of the expression $\frac{58^2}{k}+15^2k$ is attained for $\frac{58^2}{k}=15^2k$, i.e., for $k_0=\frac{58}{15}=4-\frac{2}{15}$. For $k\ge k_0$ the function $\frac{58^2}{k}+15^2k$ increases, so for $k\ge 5$ we have $\frac{58^2}{k}+15^2k\ge\frac{58^2}{5}+15^2\cdot 5$. Thus $1771\ge\frac{58^2}{5}+15^2\cdot 5$, which is not true.

Solution 2. We consider only non-trivial part (b) here. Let a_i be the number of students solved problem i. Each problem can give

$$a_i(30-a_i)\le\left(\frac{a_i+(30-a_i)}{2}\right)^2=15^2$$

points (we have used AM–GM inequality here). Totally we have at most $8\cdot 15^2$ points and a mean value is not exceed $\frac{8\cdot 15^2}{30}=60$ points. It means that A get at most 59 points. If s|he solved k first problems then others solved at least one of the last $8-k$ problems and at most $k-2$ of the first k (because if somebody of them solved only $k-1$ first problems s|he would have less points than A). Thus

$$\sum_{i=1}^{k}a_i\le k+29(k-2),$$

and A get

$$\sum_{i=1}^{k}(30-a_i)\ge 30k-(k+29(k-2))=58 \qquad (15.24)$$

which means that s|he get $y = 59$ or $y = 58$ points. It is important that the inequalities give rather close estimates. To analyze them easier we put $a_i = 15 + b_i$.

The total number of points S is at least $y + 29(y+1)$ and is equal to

$$\sum_{i=1}^{8} a_i(30 - a_i) = \sum_{i=1}^{8}(15 + b_i)(15 - b_i) = 1800 - \sum_{i=1}^{8} b_i^2.$$

If $y = 59$ then

$$\sum_{i=1}^{8} b_i^2 \le 1800 - 30 \cdot 59 + 29 = 30(60 - 59) - 29 = 1.$$

Almost all $b_i = 0$ and at most one of them is ±1. Because A get 59 points and almost all $a_i = 15$ it means that $k = 4$, one of the problem (that A solved) gives 14 points and other – 15 points. But in this case three solved problems would give at most 45 points which is less than 59 and we get a contradiction.

Thus $y = 58$ and

$$\sum_{i=1}^{8} b_i^2 \le 1800 - 30 \cdot 58 + 29 = 30 \cdot (60 - 58) - 29 = 31.$$

Besides that the inequality in (15.24) become equality thus everybody except A solved $k - 2$ problems from the first k and one in the rest. This means that

$$a_1 + \cdots + a_k = k + 29(k-2) = 30k - 58 \quad \Leftrightarrow$$
$$b_1 + \cdots + b_k = 15 \cdot k - 58$$

and

$$a_{k+1} + \cdots + a_8 = 29 \quad \Rightarrow \quad b_{k+1} + \cdots + b_8 = 15 \cdot (8 - k) - 29.$$

Thus we have

$$b_1 + \cdots + b_8 = 15 \cdot k - 58 + 15 \cdot (8 - k) - 29 = 120 - 58 - 29 = 33 > 31$$

a contradiction because $b_i^2 \ge b_i$.

Note how the inequalities helped each other to improve their estimates!

15.17 Solutions to Paper 17

97. The inequality between harmonic and arithmetic means and the Cauchy-Schwarz inequality give

$$\frac{1}{1+2ab}+\frac{1}{1+2bc}+\frac{1}{1+2ca} \geq \frac{9}{3+2(ab+bc+ca)} \geq \frac{9}{3+2(a^2+b^2+c^2)} = 1.$$

In the first inequality equality occurs iff $ab = bc = ca$, which is equivalent to $a = b = c$ (remember that $a, b, c > 0$). The same condition suffices to guarantee equality in the second inequality; hence, equality occurs iff $a = b = c = 1$.

Solution 2. Applying the AM–GM inequality to each denominator, one obtains

$$\frac{1}{1+2ab}+\frac{1}{1+2bc}+\frac{1}{1+2ca} \geq \frac{1}{1+a^2+b^2}+\frac{1}{1+b^2+c^2}+\frac{1}{1+c^2+a^2}.$$

Applying now the AM–HM inequality to the whole expression, one obtains

$$\frac{1}{1+a^2+b^2}+\frac{1}{1+b^2+c^2}+\frac{1}{1+c^2+a^2} \geq 3\cdot\frac{3}{(1+a^2+b^2)+(1+b^2+c^2)+(1+c^2+a^2)} = \frac{9}{3+2(a^2+b^2+c^2)} = \frac{9}{3+2\cdot 3} = 1.$$

Solution 3. Applying the AM-.HM inequality, one obtains

$$\frac{1}{1+2ab}+\frac{1}{1+2bc}+\frac{1}{1+2ca} \geq \frac{9}{3+2ab+2bc+2ca} = \frac{9}{a^2+b^2+c^2+2ab+2bc+2ca} = \frac{9}{(a+b+c)^2}.$$

Jensen's inequality for the square function (or the inequality between arithmetic and quadratic means) establishes

$$\frac{(a+b+c)^2}{9} \leq \frac{a^2+b^2+c^2}{3} = \frac{2}{3} = 1.$$

Thus

$$\frac{1}{1+2ab}+\frac{1}{1+2bc}+\frac{1}{1+2ca} \geq \frac{9}{(a+b+c)^2} \geq 1.$$

Solution 4. Multiplying both sides of the desired inequality by

$$\frac{1}{8}(1+2ab)(1+2bc)(1+2ca),$$

one gets the equivalent inequality

$$\frac{1+ab+bc+ac}{4} \geq a^2b^2c^2.$$

Note that

$$1 = \frac{3}{3} = \frac{a^2+b^2+c^2}{3} \geq \sqrt[3]{a^2b^2c^2},$$

giving $abc \leq 1$; thus

$$a^2b^2c^2 = (abc)^2 \leq \sqrt{abc}.$$

So indeed one has

$$\frac{1+ab+bc+ac}{4} \geq \sqrt[4]{a^2b^2c^2} = \sqrt{abc} \geq a^2b^2c^2.$$

98. a) Define the *value* of a word w as the number $b-a$ where b and a are the numbers of B's and A's, respectively, in w. Note that any allowed operation preserves the value of a word *modulo* 3. The value of the known BAU word is $7-8=-1$ and the value of the word under consideration is $9-8=1$, so the latter cannot belong to BAU.

b) To begin with, note that the allowed operations make it possible to interchange any two consecutive letters of a word. Indeed, if these letters are BA, AU or UB then replace them with UU, BB or AA, respectively, then invert the order of letters, then replace the letters UU, BB or AA back, and finally invert the order once more. If the letters to be interchanged are AB, UA or BU then perform all these steps in reverse order.

Iteration of the interchanging operation enables us to interchange any two letters (move the second letter side by side with the first, interchange them, and finally move the first letter to the original place of the second). Note that the known BAU word contains 7 letters B, 8 letters A and 10 letters U, the word under consideration has 8 letters B, 9 letters A and 8 letters U. Thus we can replace some consecutive letters UU with BA and then reorder the letters in the word, obtaining the required word.

Solution 2. a) Part b) of the previous solution implies that a word w belongs to BAU if and only if there is a word w' in BAU containing any letter the same number of times as w. Suppose the right numbers of letters can be obtained from the known BAU word by a sequence of the allowed operation. Let x be the difference between the number of performed replacements $BA \to UU$ and the number of performed replacements $UU \to BA$; analogously define y and z for the rules $AU \to BB$ and $UB \to AA$, respectively. The known word contains 7 letters B, 8 letters A and 10 letters U, the word under consideration has 9 letters B, 8 letters A and 8 letters U. This leads to the system of equations

$$\begin{cases} 7 - x + 2y - z = 9 \\ 8 - x - y + 2z = 8 \\ 10 + 2x - y - z = 8 \end{cases}$$

(each equation describes the number of occurrences of one of the letters). This system has no integer solutions, so the answer must be "no".

b) Here the word under consideration has 8 letters B, 9 letters A and 8 letters U. The corresponding system of equations is

$$\begin{cases} 7 - x + 2y - z = 8 \\ 8 - x - y + 2z = 9 \\ 10 + 2x - y - z = 8 \end{cases}$$

One of its solutions is $x = 0,\ y = z = 1$. Hence one can construct the word under consideration from the known word, performing both the replacements $AU \to BB$ and $UB \to AA$ once and then reordering the letters.

99. Consider Figure 15.29. The diagonal AC is a diameter of the circumcircle of the square $ABCD$, so by the Thales theorem the angle ALC is right. The point K is therefore the intersection of the altitudes CD and AL of the triangle ACM, hence also the line MK is perpendicular to AC and intersects the side BC of the given square at an interior point N, since $MK \parallel DB$.

The desired assertion can now be proved in several ways.

1. The quadrangles $BCLD$ and $KLMD$ are cyclic, hence

$$\angle NBL = \angle CBL = \angle CDL = \angle KDL = \angle KML = \angle NML.$$

As the points B and M lie in the same half-plane determined by the line NL, the points B, L, M and N are concyclic.

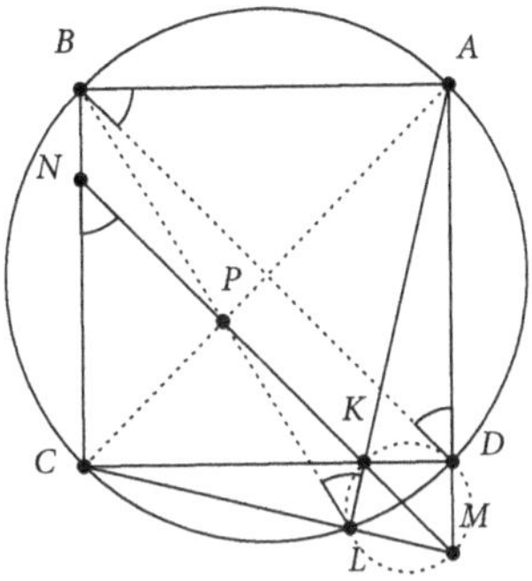

Figure 15.29 Problem 99.

2. Since $MN \parallel DB$, we have $\angle MNC = 45°$, so also the angle BLA subtended by the chord BA of the circle k has a magnitude of 45°, where $\angle BLM = \angle BNM = 135°$. The points L and N lie in the same half-plane determined by the line MB, hence the points B, L, M and N are concyclic.

3. Denote by P the foot of the altitude from the vertex M onto the side AC, and consider the quadrangles $ABNP$, $APKD$ and $DKLM$. By the Thales theorem, all these quadrangles are cyclic. The vertex C of the given square $ABCD$ lies in the exterior of each of the three circumcircles of these cyclic quadrangles, so using the theorem about the power of the point C with respect to the circumcircle of the quadrangles $ABNP$, $APKD$ and $DKLM$, respectively, we obtain the following three equalities

$$CN \cdot CB = CP \cdot CA,$$
$$CP \cdot CA = CK \cdot CD,$$
$$CK \cdot CD = CL \cdot CM,$$

which immediately imply that

$$CN \cdot CB = CL \cdot CM.$$

It again follows that the points B, L, M and N are concyclic.

100. We present numbers ended with a as $10^k b + a$. It is easy to see that the ordered set a is stable if and only if a^2 is ended with a, i.e. $a^2 - a = a(a-1)$ is divisible by $10^k = 2^k 5^k$ (if a is started with $0'$s we identify a as the corresponding positive integer, and if all digits of a are $0'$s we identify a as 0). Since a and $a-1$ are coprime, it follows that exactly four different cases are possible:

1) $10^k \mid a$, 2) $10^k \mid (a-1)$, 3) $2^k \mid a$ and $5^k \mid (a-1)$, 4) $5^k \mid a$ and $2^k \mid a-1$.

Consider these cases.

Case 1 Since $0 \le a < 10^k$, there is exactly one collection $a = 0 \dots 0$.

Case 2 Since $-1 \le a - 1 < 10^k$, there is exactly one collection $a - 1 = 0 \dots 0$, i.e. $a = 0 \dots 01$.

Case 3 In this case $a \equiv 0 \pmod{2^k}$ and where $a \equiv 1 \pmod{5^k}$. According to Chinese remainder theorem there exists exactly one a with those conditions in the interval $[0, 10^k - 1]$. If a has less then k digits, we write additionally corresponding number of 0's at its left.

Case 4 Similar to case 3.

101. An inversion with centre O transforms the circle C into a straight line d. Points P, Q, R become the intersections of d with the straight lines OA, OB, OC and K, L, M become the common points of d with BC, CA, AB respectively.

Thus the conclusion reduces to proving that the circles $(O, P, K), (O, Q, L)$, and (O, R, M) have a second common point. In order to do this, it is enough to show that they have a common radical axis.

Let X be the intersection of the radical axis of the circles $(O, Q, L), (O, R, M)$ with d. Then, considering oriented segments, we get $\overline{XQ} \cdot \overline{XL} = \overline{XR} \cdot \overline{XM}$ and it is enough to prove that $\overline{XP} \cdot \overline{XK} = \overline{XQ} \cdot \overline{XL}$.

By Menelaus theorem for the triangles OPQ, OQR, ORP and straight lines AB, BC, CA respectively, we get

$$\frac{\overline{OA}}{\overline{PA}} \cdot \frac{\overline{PM}}{\overline{QM}} \cdot \frac{\overline{QB}}{\overline{OB}} = 1, \qquad \frac{\overline{OB}}{\overline{QB}} \cdot \frac{\overline{QK}}{\overline{RK}} \cdot \frac{\overline{RC}}{\overline{OC}} = 1, \qquad \frac{\overline{OC}}{\overline{RC}} \cdot \frac{\overline{RL}}{\overline{PL}} \cdot \frac{\overline{PA}}{\overline{OA}} = 1$$

whence $\overline{PM} \cdot \overline{QK} \cdot \overline{RL} = \overline{QM} \cdot \overline{RK} \cdot \overline{PL}$.

Take now coordinates on d, with X as origin (denoted with same but small letters). Then $(m-p)(k-q)(l-r) = (m-q)(k-r)(l-p)$ and $ql = mr$, which leads to $(pk - ql)(r + m - q - l) = 0$.

If $r + m = q + l$ we get (using the equal products) $\{M, R\} = \{Q, L\}$, in which case the conclusion is obvious; otherwise $pk = ql$, whence $\overline{XP} \cdot \overline{XK} = \overline{XQ} \cdot \overline{XL}$.

102. Assume that $x = P(f(x), g(x))$, $\deg f = 2$, $\deg g = 3$. Slightly changing P if necessary, we can assume WLOG that $f(x) = x^2 + ax + b$. Since we can substitute $x' = x + \frac{a}{2}$ and add $\frac{a}{2}$ to P, we can further assume that $a = 0$. Again, after a slight change of P we can also assume that $b = 0$, i.e. WLOG $f(x) = x^2$. Similarly, we can assume WLOG that $g(x) = x^3 + cx^2 + dx$. Finally, by regarding $g(x) - cf(x)$ instead and after slightly changing P, we can assume that $c = 0$, so that g consists of odd powers of x. If now (after all changes and substitutions) $P(s, t) = \sum a_{ij} s^i t^j$, it follows that for even j the corresponding term is an even power of x, whereas for odd j it is an odd power of x. Since x is odd, this means that all even powers cancel out, so that only the terms with j odd are left. But this implies that $g(x)$ divides $P(f(x), g(x)) = x$, which is impossible. Hence, there exists no polynomial $P = P(s, t)$ such that $x = P(f(x), g(x))$.

As to the second question, f, g, P can be chosen so that $x = P(f(x), g(x))$, e.g. take $f(x) = x^2$, $g(x) = x^4 + x$, $P(s, t) = t - s^2$.

Remark In fact one can prove more: if $\deg f(x) = m \le \deg g(x) = n$ and $(m, n) = 1$ then no one polynomial of degree less than m can be represented as a polynomial in f and g, but this is more advanced result (see [43]).

15.18 Solutions to Paper 18

103. For $n = 1$ the given inequality assumes the form

$$\sqrt{2} \le \frac{1}{2}(p + 1), \quad \text{that is,} \quad p \ge 2\sqrt{2} - 1.$$

Denote $p_1 = 2\sqrt{2} - 1$. We have found that no number p smaller than p_1 has the required property. Hence, the sought number is p_1, provided we can show that for any $n \ge 1$

$$F(n) = \sqrt{1^2 + 1} + \sqrt{2^2 + 1} + \sqrt{3^2 + 1} + \cdots + \sqrt{n^2 + 1} \le \frac{1}{2} n(n + p_1).$$

We shall prove this by induction. For $n = 1$ the inequality is fulfilled.

Assume that the inequality holds for some natural number n, and let us show that it then also holds for $n+1$. Since

$$F(n+1) = F(n) + \sqrt{(n+1)^2+1},$$

it follows from the induction hypothesis and the definition of p_1 that

$$F(n+1) \le \frac{1}{2}n(n+2\sqrt{2}-1) + \sqrt{(n+1)^2+1}.$$

All we now need to show is that

$$\frac{1}{2}n(n+2\sqrt{2}-1) + \sqrt{(n+1)^2+1} \le \frac{1}{2}(n+1)(n+1+2\sqrt{2}-1).$$

which (after small manipulation) is equivalent to the inequality

$$\sqrt{(n+1)^2+1} \le n + \sqrt{2},$$

the validity of which is easily checked by taking squares on both sides:

$$(n+\sqrt{2})^2 = n^2 + 2\sqrt{2}n + 2 \ge n^2 + 2n + 2 = (n+1)^2 + 1.$$

Conclusion: the sought real number is $p = 2\sqrt{2} - 1$.

104. Construct an oriented graph with vertices at the cities and edges along the roads in such a way that the one-way roads correspond to oriented edges and the rest of the roads correspond to non-oriented edges. We shall show that at any moment a non-oriented edge can be oriented in such a way that the connectedness of the graph is preserved. This obviously implies that if both ministers play correctly the result will be a draw, so that both ministers will keep their posts.

Assume now that the edge connecting u and v has not yet been oriented. If there is a road connecting the two cities along some other edges (without loss of generality we can assume that it goes from u to v), we can orient the chosen edge from v to u. Then in any road containing the chosen edge in the opposite direction we can replace this edge by the existing road from u to v and the connectedness will not be lost.

Thus it remains to consider the situation when there is no road connecting u and v that does not contain the chosen edge (u, v). Consider the set U of all vertices that can be reached from u without passing (u, v) (including u itself). Since the graph is connected, for each element of U there exists a road from that element to v. Here we only assumed that any such road contains (u, v).

Hence for each element of U we can reach u without passing (u,v) (since we have to be at u in order to pass (u,v)). This implies that any vertex in U can be reached from any other vertex in U, without passing the edge (u,v).

Denote by V the analogous set for the vertex v. It is not difficult to prove that it has the same property, i.e. any vertex in V can be reached from any other vertex in V, without passing the edge (u,v).

Since the graph is connected, any of its vertices has to belong to exactly one of the sets U and V. Also, since the original graph preserved its connectedness after removing (u,v), there has to exist at least one more edge between U and V except (u,v). Suppose this edge connects $u_1 \in U$ and $v_1 \in V$. At the particular moment of the game this edge may be oriented or not, but it can certainly be passed at least in one of the two possible directions, say from u_1 to v_1. But, this means there is a road connecting u and v that does not contain the chosen edge (u,v), namely: $u \Rightarrow u_1 \Rightarrow v_1 \Rightarrow v$. This is a contradiction to what was assumed earlier and the statement is thus proved.

105. Note that

$$
\begin{aligned}
&S(BOC)\cdot S(DOE)\cdot S(FOA) = \\
&\quad = \left(\frac{1}{2}BO\cdot CO\sin\beta\right)\cdot\left(\frac{1}{2}DO\cdot EO\sin\alpha\right)\cdot\left(\frac{1}{2}FO\cdot AO\sin\gamma\right) = \\
&\quad = \left(\frac{1}{2}AO\cdot BO\sin\alpha\right)\cdot\left(\frac{1}{2}CO\cdot DO\sin\gamma\right)\cdot\left(\frac{1}{2}EO\cdot FO\sin\beta\right) = \\
&\quad = S(AOB)\cdot S(COD)\cdot S(EOF) = 4\cdot 6\cdot 9 = 216.
\end{aligned}
$$

From the AM–GM inequality it follows that

$$
\begin{aligned}
S(ABCDEF) &= S(AOB) + S(COD) + S(EOF) \\
&\qquad + S(BOC) + S(DOE) + B(FOA) \\
&= 4 + 6 + 9 + S(BOC) + S(DOE) + S(FOA) \\
&\geq 19 + 3\sqrt[3]{S(BOC)\cdot S(DOE)\cdot S(FOA)} \\
&= 19 + 3\sqrt[3]{216} = 37.
\end{aligned}
$$

It suffices to give an example of the hexagon with area equal to 37, but this is achieved by taking

$$AO = 2\sqrt{3}, \quad BO = DO = \frac{8}{3}, \quad CO = EO = 3\sqrt{3},$$
$$FO = 4, \quad \alpha = \beta = \gamma = 60°.$$

It is easy to verify that this hexagon is convex.

106. Let

$$k = \frac{a^2 + ab + b^2}{ab - 1}, \; k \in \mathbb{N}.$$

In the case $a = b$, the number $k = 3 + \frac{3}{a^2-1}$ is a positive integer only if $a = 0$ or $a = 2$, whence $k = 0$ or $k = 4$.

The case $b = 0$ leads to $k = -a^2$, hence $a = 0$ and $k = 0$.

We will now look for solutions (a, b) such that $a > b > 0$. The given relation is equivalent to

$$a^2 - (k-1)ab + b^2 + k = 0$$

and a classical approach is an attempt to create a new solution (a', b') of the same equation with $a' = b$. Let us see how we can choose $b' = c$ in this case. The equation will be

$$b^2 - (k-1)bc + c^2 + k = 0$$

and if we will consider the difference of two equations we get

$$c^2 - (k-1)bc + (k-1)ab - a^2 = 0.$$

We can solve it, but it is more elegant to note that $c = a$ is one evident solution, thus another one we get by Vieta's theorem $c = (k-1)b - a$. Thus if (a, b) is a solution, then $(a', b') = (b, (k-1)b - a)$ is also a solution. But do we have the conditions $a' > b' > 0 \Leftrightarrow b > (k-1)b - a > 0$?

The inequality $(k-1)b - a > 0$ is true in all cases because it becomes successively

$$k > \frac{a+b}{b}, \quad \frac{a^2 + ab + b^2}{ab - 1} > \frac{a+b}{b}, \quad b^3 > -a - b.$$

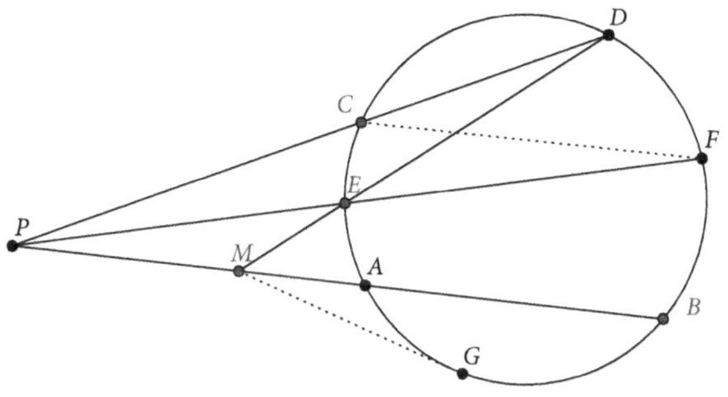

Figure 15.30 Problem 107.

In the same way, the inequality $b > (k-1)b - a$ is successively equivalent to

$$k < \frac{2b+a}{b}, \qquad \frac{a^2+ab+b^2}{ab-1} < \frac{2b+a}{b},$$

$$b^3 - 2b < a(b^2-1), \qquad a > b + \frac{3b}{b^2-1}$$

for $b > 1$. If $3b < b^2 - 1$, that is $b \geq 4$, the previous inequality is true, therefore for each solution (a, b) with $a > b \geq 4$ we find a solution (b, c) with $b > c > 0$. Thus, the descent method shows that each solution corresponds to a solution with $b \leq 3$, but the same k.

For $b = 1$ we get $k = a + 2 - \frac{3}{a-1}$, $a = 4$ or $a = 2$, $k = 7$.

For $b = 2$ we get $4k = 2a + 5 + \frac{21}{2a-1}$, $a = 4$ or $a = 11$, $k = 4$ or $k = 7$.

For $b = 3$ we get $9k = 3a + 10 + \frac{91}{3a-1}$ which does not yield any solution.

Thus, the answer is $k \in \{0, 4, 7\}$.

107. ("if") Assume that CF is parallel to PB. It then follows that $\angle CFP = \angle EPM$. Since $\angle CFP = \angle CDE$, it follows that $\angle PDM = \angle EPM$. We can now conclude that the triangles $\triangle PME$ and $\triangle DMP$ are similar and, accordingly,

$$\frac{PM}{DM} = \frac{ME}{PM}, \quad \text{i.e.} \quad PM^2 = ME \cdot MD.$$

It remains to show that there is a tangent to the circle passing through M, i.e. that M lies outside the circle. Since CF is parallel to PB, the point F must lie on the arch BD that does not contain the point A. The point F must then be an interior point for $\angle BPD$. It follows that PF lies inside $\angle BPD$ and, consequently, E belongs to the arch AC that does not contain B. Thus the points E and D lie on the same of the two arches into which AB divides the circle and the chords AB and ED cannot have a common point. This means that the lines DE and PB intersect outside the circle.

("only if") Let it now be given that there is a tangent to k, passing through M and that the square of the length of this tangent is equal to $PM^2 = ME \cdot MD$. The point M being an exterior point for the circle, it cannot lie on the chord AB.

Assume to begin with that M lies between P and A. The assertion then follows from the similarity of $\triangle PME$ and $\triangle DMP$, since the above similarity implies $\angle EPM = \angle PDM = \angle CFP$, and we can deduce that CF is parallel to PB.

It remains to show that M has to lie between P and A. It follows easily from the fact that the length of the tangent through M (= PM) has to be smaller than the total length of any secant line from M and larger than its exterior part.

108. We will show by induction that

$$f_n(x) = x(f_{n-1}(x) - f_{n-2}(x)).$$

For $n = 2$ the statement is trivial: $x^2 - x = x(x-1)$. Assume that the assertion holds for some n. We then have

$$\begin{aligned} f_{n+1} &= \frac{f_n^2(x) - x^n}{f_{n-1}(x)} = \frac{f_n(x) \cdot x(f_{n-1}(x) - f_{n-2}(x)) - x^n}{f_{n-1}(x)} \\ &= x f_n(x) - x \cdot \frac{f_n(x) f_{n-2}(x) + x^{n-1}}{f_{n-1}(x)} \\ &= x f_n(x) - x \cdot \frac{\left(\frac{f_{n-1}^2(x) - x^{n-1}}{f_{n-2}(x)}\right) \cdot f_{n-2}(x) + x^{n-1}}{f_{n-1}(x)} \\ &= x f_n(x) - x f_{n-1}(x), \end{aligned}$$

which proves the induction step. It is now easy to show, again using induction, that all $f_n(x)$ are polynomials with integer coefficients.

15.19 Solutions to Paper 19

109. We have

$$\frac{1}{1+x_1+x_1x_2}+\frac{1}{1+x_2+x_2x_3}+\cdots+\frac{1}{1+x_n+x_nx_1}$$
$$>\frac{1}{1+x_1+x_1x_2+\cdots+x_1x_2\cdots x_{n-1}}$$
$$+\frac{1}{1+x_2+x_2x_3+\cdots+x_2x_3\cdots x_n}+\cdots$$
$$+\frac{1}{1+x_n+x_nx_1+\cdots+x_nx_1x_2\cdots x_{n-2}}.$$

We further multiply the nominator and the denominator of the first term by x_n, of the second term by $x_nx_1, \ldots$, of the n-th term by $x_nx_1x_2\ldots x_{n-1}$, and, keeping in mind that $x_1x_2\ldots x_n = 1$, we get

$$\frac{x_n}{x_n+x_nx_1+\cdots+x_nx_1x_2\cdots x_{n-1}}$$
$$+\frac{x_nx_1}{x_nx_1+x_nx_1x_2+\cdots+x_nx_1\cdots x_{n-1}+x_n}$$
$$+\cdots+\frac{x_nx_1x_2\cdots x_{n-1}}{x_nx_1x_2\cdots x_{n-1}+x_n+x_nx_1+\cdots+x_nx_1x_2\cdots x_{n-2}}=1.$$

110. It is clear that if we prove that $f(n) \le n$ for any non-negative n we shall get by easy induction that $f(n) = n$ for any n. Indeed, from $f(0) \le 0$ we get $f(0) = 0$ and if we suppose $f(k) = k$ for any $k \le n$, from $f(n+1) \le n+1$ and the injectivity condition, we get $f(n+1) = n+1$.

Suppose, by way of contradiction, that $f(n) > n$ for some non-negative n and denote $f^k = f \circ f \circ \ldots \circ f$, the k-composition of f with itself, $k = 2, 3, \ldots$. We easily obtain by induction that $f^k(n) < f(n)$ for any $k = 2, 3, \ldots$. Indeed,

$$f(f(n)) \le \frac{n+f(n)}{2} < \frac{f(n)+f(n)}{2} = f(n)$$

and if $f^p(n) < f(n)$ for $p = 2, \ldots, k$ we have

$$f^{k+1}(n) = f(f(f^{(k-1)}(n))) \le \frac{f^{k-1}(n)+f^k(n)}{2} < f(n).$$

Thus the sequence $(f^k(n))_k$ is a bounded sequence of non-negative integers and so $f^p(n) = f^q(n)$ for some $p < q$. Since f is injective, it follows that $f^{q-p}(n) = n$, therefore $f^{q-p+1}(n) = f(n)$, a contradiction.

It is obvious that the identity function verifies the given relation and is injective.

Remark There are many other ways of proving the result and maybe the simplest is to make connections with a well-known fact on limits of sequences: if a sequence of non-negative numbers, say $(a_n)_n$ is sub-additive, that is $a_{n+1} \leq \frac{a_n + a_{n-1}}{2}$ for any n, then the sequence has a finite limit.

In our case, for any natural n the sequence $(f^k(n))_k$ is sub-additive, and being a convergent sequence of integers is constant beginning from some term. This contradicts the injectivity, as above, if f is not the identity function.

111. Note that

$$\frac{OA_1}{AA_1} = \frac{\text{area}(OBC)}{\text{area}(ABC)} = \frac{OB \cdot OC \cdot BC}{4R_1} \cdot \frac{4R}{AB \cdot AC \cdot BC},$$

so we have to prove that

$$\sum_{\text{cyclic}} OB \cdot OC \cdot BC \geq AB \cdot AC \cdot BC.$$

Consider the complex coordinates $O(0), A(a), B(b), C(c)$. The wanted conclusion reads

$$\sum_{cyclic} |b| \cdot |c| \cdot |b - c| \geq |a - b| \cdot |b - c| \cdot |c - a|,$$

that is

$$\sum_{\text{cyclic}} |b^2 c - c^2 b| \geq |ab^2 + bc^2 + ca^2 + ca^2 - a^2 b - b^2 c - c^2 a|,$$

which is obvious.

112. Denote by M the point of intersection of O_1O_2 and AC (see figure 15.31), and apply Menelaus' theorem to the triangle ABC and the straight line O_1O_2M

$$\frac{AM}{MC} \cdot \frac{CO_2}{O_2B} \cdot \frac{BO_1}{O_1A} = 1.$$

Since the triangles COB and BOA are right triangles with altitudes OO_2 and OO_1, by theorem 4.3 we get

$$\frac{CO_2}{O_2B} = \frac{OC^2}{OB^2} \quad \text{and} \quad \frac{BO_1}{O_1A} = \frac{OB^2}{OA^2}.$$

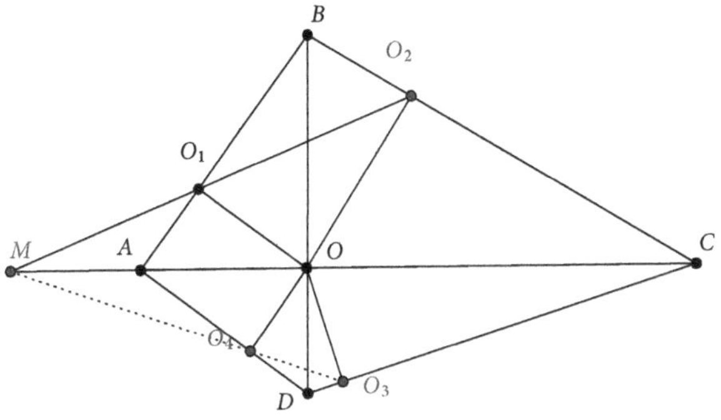

Figure 15.31 Problem 112.

It now follows that

$$\frac{AM}{MC} = \frac{AO^2}{OC^2}.$$

If we denote by N the point of intersection of O_3O_4 and AC, in a similar way we get

$$\frac{AN}{NC} = \frac{AO^2}{OC^2}.$$

Both M and N are external points for the segment AC and divide it in the same ratio taking A as starting point (in other words, they have the same barycentric coordinates relative A and C). Hence $M = N$ and the lines O_1O_2, O_3O_4 and AC are concurrent.

113. We use induction with respect to n. For $n = 1, 2, 3$, we see directly that the number of edges is not greater than $n^2/4$. Assume that the statement holds for all graphs with less than n vertices and no triangles. Now, given a graph with n vertices (and no triangles), take an arbitrary edge l, connecting A and B, and consider the subgraph without A and B (and all the edges from them). According to our assumption, it has no more than $(n-2)^2/4$ edges. All the other edges are either connected with A or with B. If C is a vertex different from A, B, it cannot be connected with both of them. This means that we have at most $n-2$ such edges and l itself and hence the graph has at most $(n-2)^2/4 + (n-1) = n^2/4$ edges.

114. Let $x = \sin^2 t$. Then

$$P_{i,n+1} = (1-x)P_{i,n} + xP_{i+1,n}, \qquad P_{0,2005} = 1-2x.$$

Let $B(n,k) = \binom{n}{k}x^k(1-x)^{n-k}$ be the so-called **Bernstein polynomial**. Note that it is equal to zero if $k < 0$ or $k > n$. Note also that

$$(1-x)B(n,k) + xB(n,k-1) = B(n,k+1).$$

We shall first show by induction that

$$P_{i,n} = \sum_{k=0}^{n} B(n,k)P_{i+k}.$$

This is obvious for $n = 0$ and

$$P_{i,n+1} = (1-x)P_{i,n} + xP_{i+1,n}$$
$$= \sum_{k=0}^{n}(1-x)B(n,k)P_{i+k} + \sum_{k=0}^{n} xB(n,k)P_{i+k+1}$$

(adding some zero terms)

$$= \sum_{k=0}^{n+1}(1-x)B(n,k)P_{i+k} + \sum_{k=-1}^{n} xB(n,k)P_{i+k+1}$$

(replacing $k+1$ by k in the second sum)

$$= \sum_{k=0}^{n+1}(1-x)B(n,k)P_{i+k} + \sum_{k=0}^{n+1} xB(n,k-1)P_{i+k}$$
$$= \sum_{k=0}^{n+1}\left((1-x)B(n,k) + xB(n,k-1)\right)P_{i+k} = \sum_{k=0}^{n+1} B(n+1,k)P_{i+k}.$$

Now we shall prove, using the same method, the following formula for the derivative:

$$P'_{0,n} = n\sum_{k=0}^{n-1} B(n-1,k)\Delta P_k,$$

where $\Delta P_k = P_{k+1} - P_k$. To do this we check directly that

$$B(n,k)' = n\left[B(n-1,k-1) - B(n-1,k)\right]$$

and get

$$P'_{0,n} = n\left(\sum_{k=0}^{n} B(n-1,k-1)P_k - \sum_{k=0}^{n} B(n-1,k)P_k\right)$$

(removing some zero terms)

$$= n\left(\sum_{k=1}^{n} B(n-1,k-1)P_k - \sum_{k=0}^{n-1} B(n-1,k)P_k\right)$$

(and replacing $k-1$ in the first sum by k)

$$= n\left(\sum_{k=0}^{n-1} B(n-1,k)P_{k+1} - \sum_{k=0}^{n-1} B(n-1,k)P_k\right) = n\sum_{k=0}^{n-1} B(n-1,k)(P_{k+1}-P_k).$$

If we put $x = 0$, we get

$$P'_{0,n}(0) = n\Delta P_0.$$

Easy induction now gives

$$P''_{0,n}(0) = n(n-1)\Delta^2 P_0, \qquad P'''_{0,n}(0) = n(n-1)(n-2)\Delta^3 P_0, \qquad \ldots.$$

In our example we have

$$P'_{0,2005} = -2, \qquad P''_{0,2005} = 0, \qquad P'''_{0,2005} = 0,$$

which means that

$$2005(P_1 - P_0) = -2; \qquad \Delta^2 P_0 = \Delta(P_1 - P_0) = 0,$$

and hence

$$P_2 - P_1 = P_1 - P_0 = -2/2005;$$
$$\Delta^3 P_0 = \Delta^2(P_1 - P_0) = 0, \qquad \Delta^2 P_1 = \Delta^2 P_0 = 0,$$

which implies

$$P_3 - P_2 = P_2 - P_1 = -2/2005, \qquad \ldots,$$

and so on. We see that $P_{i+1} - P_i = -2/2005$ for all i. Since $P_0 = P_{0,2005}(0) = 1$, we get $P_i = 1 - 2i/2005$.

Solution 2. We shall prove by induction that

$$P_{i,n} = \sum \binom{n}{k} \cos^{2(n-k)} t \cdot \sin^{2k} t \cdot P_{k+i},$$

where the sum is taken over all k; we assume that $\binom{n}{k} = 0$ for $k < 0$ and for $k > n$. The statement is true for $n = 0$, since $P_{i,0} = P_i$. Assume that the statement holds for a certain n and for all i. We have

$$\begin{aligned}
P_{i,n+1} &= (\cos^2 t)P_{i,n} + (\sin^2 t)P_{i+1,n} \\
&= \cos^2 t \sum \binom{n}{k} \cos^{2(n-k)} t \cdot \sin^{2k} t \cdot P_{k+i} \\
&\quad + \sin^2 t \sum \binom{n}{k} \cos^{2(n-k)} t \cdot \sin^{2k} t \cdot P_{k+i+1} \\
&= \sum \binom{n}{k} \cos^{2(n-k+1)} t \cdot \sin^{2k} t \cdot P_{k+i} \\
&\quad + \sum \binom{n}{k} \cos^{2(n-k)} t \cdot \sin^{2k+2} t \cdot P_{k+i} \\
&= \sum \binom{n}{k} \cos^{2((n+1)-k)} t \cdot \sin^{2k} t \cdot P_{k+i} \\
&\quad + \sum \binom{n}{k-1} \cos^{2((n+1)-k)} t \cdot \sin^{2k} t \cdot P_{k+i} \\
&= \sum \left[\binom{n}{k} + \binom{n}{k-1}\right] \cos^{2((n+1)-k)} t \cdot \sin^{2k} t \cdot P_{k+i} \\
&= \sum \binom{n+1}{k} \cos^{2((n+1)-k)} t \cdot \sin^{2k} t \cdot P_{k+i}.
\end{aligned}$$

Hence the statement is true for all n.

Let $a = \sin^2 t$, thus

$$P_{0,n} = \sum \binom{n}{k} a^k (1-a)^{n-k} P_k.$$

The coefficient for a^i in this sum is

$$\sum_{k=0}^{i} \binom{n}{k}\binom{n-k}{i-k}(-1)^{i-k} P_k.$$

Since $P_{0,2005} = \cos 2t = 1 - 2a$, a comparison of the coefficients for a^i in the two formulas gives $P_0 = 1$; $-2005P_0 + 2005P_1 = -2$, which implies $P_1 = \frac{2003}{2005}$. Further, for $i \geq 2$,

$$\sum_{k=0}^{i} \binom{2005}{k}\binom{2005-k}{i-k}(-1)^{i-k} P_k = 0,$$

and hence we get a formula for P_i in terms of $P_0, P_1, \ldots, P_{i-1}$. The explicit formula $P_i = 1 - 2i/2005$ follows by induction.

15.20 Solutions to Paper 20

115. *Solution 1.* The given inequality is equivalent to

$$\frac{a}{b}+\frac{b}{a}+4\geq 3\sqrt[3]{\frac{a}{b}}+3\sqrt[3]{\frac{b}{a}},$$

(take both sides to the power 3). The inequality between the arithmetic and the geometric means gives

$$\frac{a}{b}+1+1\geq 3\sqrt[3]{\frac{a}{b}},$$

and also

$$\frac{b}{a}+1+1\geq 3\sqrt[3]{\frac{b}{a}}.$$

In both inequalities equality will hold iff $a = b$. Adding the two gives

$$\frac{a}{b}+\frac{b}{a}+4\geq 3\sqrt[3]{\frac{a}{b}}+3\sqrt[3]{\frac{b}{a}}.$$

Solution 2. We shall use the inequality between power means of order $\frac{1}{3}$ and $\frac{1}{2}$

$$\left(\frac{\sqrt[3]{a/b}+\sqrt[3]{b/a}}{2}\right)^3\leq\left(\frac{\sqrt{a/b}+\sqrt{b/a}}{2}\right)^2,$$

with equality iff $\frac{a}{b} = \frac{b}{a}$, i.e. iff $a = b$. Since

$$\left(\sqrt{a/b}+\sqrt{b/a}\right)^2=(a+b)\left(\frac{1}{a}+\frac{1}{b}\right),$$

we get the desired inequality with equality iff $a = b$.

Solution 3. According to Jensen's inequality applied to concave function $f(x) = x^{\frac{2}{3}}$ we have

$$\frac{a^{\frac{2}{3}}+b^{\frac{2}{3}}}{2}\leq\left(\frac{a+b}{2}\right)^{\frac{2}{3}}\quad\Leftrightarrow\quad\left(a^{\frac{2}{3}}+b^{\frac{2}{3}}\right)^3\leq 2(a+b)^2,$$

which is equivalent to our statement (divide by ab to see that).

Solution 4. Put $x = \sqrt[3]{\frac{a}{b}}$. Then we need to prove that

$$3\left(x+\frac{1}{x}\right)\leq 4+x^3+\frac{1}{x^3}.$$

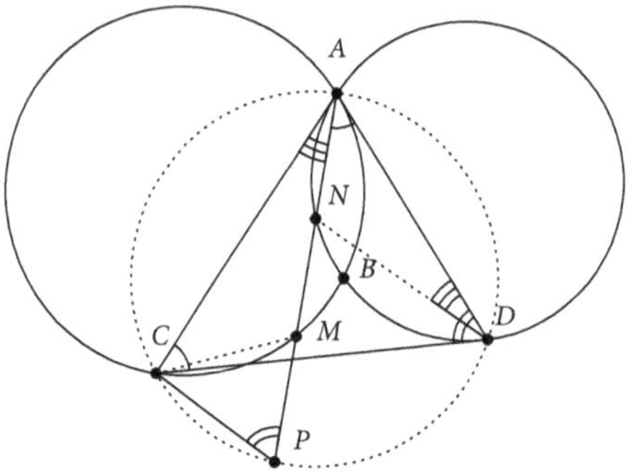

Figure 15.32 Problem 116. Solution 1.

Now put $y = x + \frac{1}{x} \geq 2$ and it remains to prove that

$$3y \leq 4 + y^3 - 3y \quad \Leftrightarrow \quad 6y \leq 4 + y^3.$$

We have the equality for $y = 2$ and obvious inequality $6 \leq 3y^2$ for the derivatives when $y \geq 2$.

116. *Solution 1.* All we need is angle chasing. WLOG we can consider the case when the ray is interior to the angle $\angle BAC$. Then

$$\angle CMP = \angle MCA + \angle CAM = \angle MAD + \angle CAM = \angle CAD.$$

Also, $\angle CPM = \angle CDA$, since both subtend the cord AC in the circumcircle ACD. Therefore, the triangles $\triangle ACD$ and $\triangle MCP$ are similar, and thus

$$\frac{MC}{AC} = \frac{MP}{AD}.$$

Furthermore, $\angle ACM = \angle NAD$ and $\angle CAM = \angle ADN$, so $\triangle ACM \sim \triangle DAN$. Hence

$$\frac{AN}{AD} = \frac{MC}{AC}.$$

Thus $MP = AN$ and $AM = NP$, as claimed.

Solution 2. Let Q be the intersection point of the line DN and the circumcircle. Because $\angle CAP = \angle ADN = \angle ADQ$ we get that $|AQ| = |CP|$ and therefore $AQCP$ is isosceles trapezoid (see theorem 4.9) and $\angle QAN = \angle CPM$. Besides that $\angle CMP = \angle ACM + \angle CAM = \angle NAD + \angle NDA = \angle ANQ$.

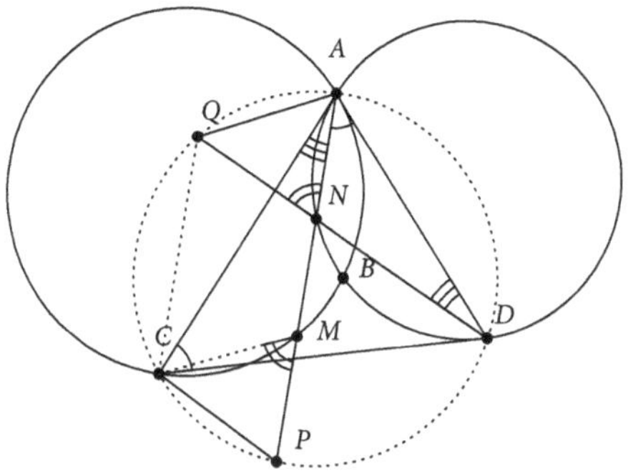

Figure 15.33 Problem 116. Solution 2.

Therefore the triangles CPM and AQN are similar and (having the equal sides) congruent, which proves that $MP = AN \Leftrightarrow AM = MP$.

117. Such a line will divide the triangle into a triangle with area S_1 and a quadrilateral (possibly another triangle) with area $S_2 = S - S_1$, where S denotes the area of the original triangle. Since $S_1, S_2 > 0$ and $S_1 + S_2 = S$ whatever the position of the line, the product $S_1 S_2$ will be maximal for $S_1 = S_2$. The problem is thus reduced to: find a line through the incentre which divides the area of the triangle into two equal parts.

Denote by C the vertex of the original triangle $\triangle ABC$ which is also a vertex in the smaller cut-off triangle $\triangle A_1B_1C$. Let $BC = a$, $CA = b$, $AB = c$, and $x = B_1C$, $y = CA_1$. Theorem 4.24 gives $\frac{1}{x} + \frac{1}{y} = \frac{a+b+c}{ab} = \frac{2p}{ab}$. We now want to find x and y such that $S_1 = \frac{1}{2}xy \sin C = \frac{1}{2}S = \frac{1}{4}ab \sin C$, or, equivalently, such that $xy = \frac{1}{2}ab$. We get the system of equations

$$x + y = p, \quad xy = \frac{1}{2}ab,$$

which has the solutions

$$x, y = \frac{p \pm \sqrt{p^2 - 2ab}}{2}$$

and the problem is thus (formally) solved. Observe that $x, y > 0$. Obviously there exist triangles such that C cannot be chosen to be any vertex of the original triangle (e.g., take a isosceles triangle with a very short third side compared to the two equal ones). We shall show that there exists a line as required provided C is the vertex between the longest and the shortest sides of the triangle. Assume therefore that $b \le c \le a$. If we start with the position when the line coincides with the angle bisector from the vertex A and rotate it continuously till it coincides with the angle bisector from B, the area of the cut-off triangle will change continuously from $S_A \le \frac{1}{2}S$ to $S_B \ge \frac{1}{2}S$, so the existence of the line with the required properties follows by continuity. Let us show that the same result follows by the formulas for x and y. We have to show that $p^2 - 2ab \ge 0$, $\frac{p+\sqrt{p^2-2ab}}{2} \le a$ and that $\frac{p-\sqrt{p^2-2ab}}{2} \le b$. We have

$$\begin{aligned} 4(p^2 - 2ab) &= (a+b+c)^2 - 8ab \ge (a+2b)^2 - 8ab \\ &= a^2 + 4ab + 4b^2 - 8ab = (a-2b)^2 \ge 0. \end{aligned}$$

To show that

$$p + \sqrt{p^2 - 2ab} \le 2a,$$

let us note that $2a - p = \frac{1}{2}(4a - a - b - c) = \frac{1}{2}(3a - b - c) > 0$, so by squaring we get the equivalent inequality

$$p^2 - 2ab \le 4a^2 - 4ap + p^2,$$

which is equivalent to

$$2p \le 2a + b.$$

Since $c \le a$, the last inequality is obvious. To show that

$$p - \sqrt{p^2 - 2ab} \le 2b,$$

we first note that the inequality is trivial for $p - 2b < 0$. For $p - 2b \ge 0$ squaring gives the equivalent inequality $2p \ge a + 2b$, which follows from $c \ge b$.

118. *Solution 1.* Let $A = \{a_1, a_2, a_3, a_4, \ldots, a_k\}$, where $a_i \in \mathbb{N}$, $i = 1, 2, \ldots, k$, and $a_1 < a_2 < a_3 < a_4 < \cdots < a_k$.

If $A + A$ has $2k - 1$ distinct elements, then the elements of $A + A$ are given by $a_i + a_j$, $1 \le i \le j \le k$. These are represented as follows:

$$\begin{gathered} 2a_1, a_1 + a_2, a_1 + a_3, \ldots, a_1 + a_k, 2a_2, a_2 + a_3, \ldots, \\ a_2 + a_k, \ldots, 2a_{k-1}, a_{k-1} + a_k, 2a_k. \end{gathered}$$

Since $a_1 < a_2 < a_3 < a_4 < \cdots < a_k$, we have

$$2a_1 < a_1 + a_2 < 2a_2 < a_2 + a_3 < 2a_3 < \cdots < 2a_{k-1} < a_{k-1} + a_k < 2a_k.$$

These numbers, therefore, make up the $2k-1$ distinct elements of $A+A$ which can be represented either as $2a_i$, $i = 1, 2, \ldots, k$, or as $a_j + a_{j+1}$, $j = 1, 2, \ldots, k-1$. We also know that for $i = 1, 2, \ldots, k-2$, $a_i + a_{i+1} < a_i + a_{i+2} < a_{i+1} + a_{i+2}$. But $a_i + a_{i+1} < 2a_{i+1} < a_{i+1} + a_{i+2}$. As there are only $2k-1$ distinct elements in $A + A$, we have $2a_{i+1} = a_i + a_{i+2}$ for $i = 1, 2, \ldots, k-2$. Hence, the k elements in A form an arithmetic progression of length k.

Conversely, if the elements of A form an arithmetic progression of length k, then $a_i = a_1 + (i-1)d$, where d is a positive integer, $i = 1, 2, \ldots, k$. Thus, each element in $A + A$ is of the form $2a_1 + (i + j - 2)d$, where $1 \le i \le j \le k$. But $2 \le i + j \le 2k$. Hence, $A + A$ has exactly $2k - 1$ distinct elements.

Solution 2. Set each some $a_i + a_j$ on the point (i, j) of a square. Consider the sums on the diagonal $i + j = n$ for some $n = 3 \ldots, 2k$. For each of them there exists a smaller sum on the previous diagonal $i + j = n - 1$ (either below or on the left). Thus for any sum $a_i + a_j$ there exists a path from $(1, 1)$ to (k, k) through (i, j) containing different elements from each diagonal and containing $a_i + a_j$. Because we have $2k - 1$ such diagonals the elements on the diagonals should be equal. In particularly for $i < j$ we have $a_i + a_j = a_{i+1} + a_{j-1} \Leftrightarrow a_{i+1} - a_i = a_{j+1} - a_j$. Thus the differences between the consequent numbers is a constant which implies that the sequence is an arithmetic progression.

Vise versa, for an arithmetic progression the sums on the diagonals are equal thus $A + A$ has exactly 2k-1 elements.

119. Choose $k \in \mathbb{N}$. Consider the sequence $\{r_n\}_{n\ge0}$, where r_n is the remainder when dividing a_n by k. We have

$$r_0 = r_1 = r_2 = 1, \qquad r_{n+3} \equiv r_{n+2} r_{n+1} + r_n \pmod{k}.$$

Since there are finitely many values of r_n, the sequence has to be periodic, starting from some point. Similarly, we can use the equality to determine r_n given $r_{n+3}, r_{n+2}, r_{n+1}$ and by the same argument as above we get that the sequence $\{r_n\}_{n\in\mathbb{Z}}$ (starting at an arbitrary place) is periodic backwards. We have $r_{-1} = r_2 - r_1 r_0 = 1 - 1 = 0$ and hence there exists an index $n_k > 0$ such that $r_{n_k} = 0$.

120. We will consider the case when area is equal to 4 first. Reduce first every square in such a way that its area becomes equal to 4^{-n} for the smallest possible integer n. To do this simply write its side in the binary system and leave only the first nonzero digit (e.g., if its side is equal to 0.0101101..., we reduce it to 0.01). Note that the total area is thus reduced but not more then 4 times and it is sufficient to prove the following:

If every square has area 4^{-n} for some integer n and their total area is at least 1 then it is possible to cover the unite square.

We use induction on the number of squares. The base (if there is only one square) is trivial. Assume that the statement holds for a certain number of squares. Increase this number by one and consider the smallest square with area 4^{-k}. If there are four such squares we can replace them by one with area 4^{1-k} and use the assumption. If there are $m \le 3$ such squares, then writing the total area S in the base 4 we see that the last digit is exactly m (because they are the smallest ones and every square has area 4^{-n}). This means we can simply throw away those m squares. The total area will still be greater than or equal to 1 and again we can use the assumption.

Here is an algorithmic description of the same idea. We have black squares that we need to cover (one in the beginning) with the help of white squares. Is some white square can cover a black one we do this and consider the remaining squares. If not we divide each black square in four equal squares and continue the process. We stop when we cover the last black square. Why this happens? First, the area of the covering white square is not more than four times larger than the area of a covered black square (otherwise we could cover a larger square in the previous step). Thus the total area of white squares will be always at least four times large then total area of black squares. But sooner or later the size of the black squares became less than the smallest white square.

Now let area be equal to 3. It is still sufficient. Here is a nice proof suggested by A.Chapovalov. WLOG we can suppose that all the squares we use for covering have sizes less then 1. We cover our square strip by strip. We start from the largest square with the size a_0 and put it at the left corner of the lower side. Then we attach the square of the next size to the right and continue until we achieve the right corner (and we achieve it because the sum of the lengths is larger than the sum of squares which is larger than one). Suppose that the last square we used here have size a_1. Now we have covered a strip of the size $1 \times a_1$ and have used not more than $a_0(1 + a_1)$ of area of squares. We cut the strip, through away used squares and repeat the procedure with the remaining squares (ignoring the fact that some part of our area was already covered).

If we finish the next strip with the last square of the size a_2 then we cut the strip again and continue until either we cover all the strips (and our square as well) or we have no more squares and finish on the strip number n with the last square of the size a_n, and an uncompleted (maybe empty) next strip. We want to show that the second alternative is in fact impossible. Suppose that nevertheless this was the case. First of all $a_1 + \dots + a_n < 1$ otherwise the square would be covered.

Let us estimate how much of area we have used. To cover the strip number k we have used less than $a_{k-1}(1 + a_k)$ of our area of squares, because the size of the largest (first) square was less than a_{k-1}. For the uncovered strip we have used less then $a_n \cdot 1$ of area. Because $a_{k-1}(1+a_k) = a_{k-1} + a_{k-1}a_k < a_{k-1} + a_k$ together we have used less than

$$(a_0 + a_1) + (a_1 + a_2) + \dots + (a_{n-1} + a_n) + a_n$$
$$< 1 + 2(a_1 + a_2 + \dots + a_n)$$

of area. But we have used all the squares with the total area 3 thus $a_1 + a_2 + \dots + a_n > 1$ which contradicts to the inequality above.

CHAPTER 16

Sources

Most of the problems have been taken from various materials distributed at IMO ([1]–[25]). We are very thankful for those collections of problems. Unfortunately, we were unable to point out the author of every problem so we cite only the source where *we* have found the problem or its idea, though it is not necessarily the origin of the problem. We prefer to cite the books even if we know that the problems here are taken from different national competitions (e.g. Russian) or are theorems first proved by famous scientists like P. Erdős. Sometimes we do not know the actual origin of problem and skip the reference, marking them by F. For problems that we consider as original we write O.

Table 16.1 Sources for the problems from the 20 papers.

Paper	Prob 1	Prob2	Prob 3	Prob 4	Prob 5	Prob 6
1	[17]	[10]	[17]	[17]	[17]	[18]
2	[31]	[31]	[31]	[42]	[31]	[35]
3	[45]	[1]	[7]	[7]	[42]	[4]
4	[17]	[55]	[17]	[17]	[17]	[53]
5	[27]	[27]	[27]	[27]	[27]	[27]
6	[9]	[9]	[9]	[5]	[3]	[8]
7	[24]	[24]	[12]	[45]	[48]	[46]
8	F	[18]	[10]	F	F	F
9	[23]	[17]	[17]	[3]	[17]	[13]
10	[2]	F	[65]	[25]	[19]	[35]
11	F	[11]	[22]	[22]	[17]	[16]
12	[35]	[69]	[55]	[16]	[27]	[44]
13	[14]	[14]	[14]	[14]	[8]	F
14	F	[28]	[47]	[8]	[17]	[41]
15	F	[8]	[15]	[6]	[20]	[50]
16	[8]	[15]	[41]	[7]	[55]	[3]
17	[9]	[9]	[8]	F	[17]	O
18	[8]	F	[3]	[17]	[56]	O
19	[28]	[17]	[17]	F	[26]	O
20	[8]	[17]	O	[38]	F	O

BIBLIOGRAPHY

[1] 22nd Albanian Mathematical Olympiad for High Schools.

[2] Problems of the 34th Austian Mathematical Olympiad 2003.

[3] Belorussian Mathematical Olympiad, 2002–2004.

[4] The British Mathematical Olympiad, 2001.

[5] The British Mathematical Olympiad, 2002.

[6] Bulgarian Mathematical Competitions, 2004.

[7] Mathematical Competitions in Croatia, Zagreb, 2001–2004.

[8] Czech and Slovak Mathematical Olymiad, 2000–2003.

[9] Estonian Math Competitions, 2001–2004.

[10] Hellenic Mathematical Competitions, 2003.

[11] Kürschák Competition, 1975.

[12] IMO Training Camp, 2002, India.

[13] 20th Iranian Mathematical Olympiad, 2003.

[14] 17th Irish Mathematical Olympiad, 2004.

[15] The 14th Japanese Mathematical Olympiad, 2004.

[16] The 16th Korean Mathematical Olympiad, 2003.

[17] Romanian Mathematical Competition, 2002–2005.

[18] Н.Х.Агаханов, О.К.Подлипский, *Математические олимпиады Московской области 1993–2002*, 2003, МФТИ, Москва, ISBN 5-89155-100-0

[19] С.Л.Берлов, С.В.Иванов, *Петербургские математические олимпиады*, 1998, Лань, СПб, ISBN 5-8114-0083-7.

[20] Н.Б.Васильев, А.А.Егоров, *Задачи Всесоюзных математических олимпиад*, 1988, Наука, Москва, ISBN 5-02-013730-8.

[21] Deveto Republičko Takmičeje Iz Matematike Učenika Srednjih Škola Republike Srpske, Banja Luka, 27. IV, 2002.

[22] 47th Math Olympiad in Slovenia, 2003.

[23] Thai Mathematical Olympiad, 2002.

[24] The 2002 Vietnamese Mathematical Olympiad, Hanoi, 2002.

[25] The 2003 Vietnamese Mathematical Olympiad, Hanoi, 2003.

[26] M.Aigner, G.M. Ziegel, *Proofs from THE BOOK*, Springer, ISBN 3-540-67865-4

[27] American Mathematical Monthly.

[28] T.Andeescu, V.Cîrtoaje, G.Dospinescu, M.Lascu, *Old and New Inequalities*, GIL, 2004, ISBN 973-9417-35-3.

[29] A.H. Beiler, *Recreations in the Theory of Numbers*, Dover, 1966.

[30] Béla Bolabás, *The Art of Mathematics. Coffee Time in Memphis*, 2006, Cambridge University Press, ISBN 978-0-521-6935-0

[31] *Crux Mathematicorum with Mathematical Mayhem*, 28: N1, February 2002.

[32] Ronald L. Graham, *Rudiments of Ramsey Theory*, 1981, AMS, Providence, Rhode Island.

[33] Marshall Hall, Jr., *Combinatorial theory*, 1967, Braisdell Publishing Company.

[34] Halmos Paul. R., *I want to be a mathematician*, 1985, Springer, 421p.

[35] D.Djukić, V. Janković, I.Matić, N.Petrović, *The IMO Compendium*, Springer, ISBN 978-0387-24299-6.

[36] Johnson, Roger A., *Modern Geometry*, 1929, Houghton Mifflin, Boston (republished facsimile by Dover 1960, 2007 as *Advanced Euclidean Geometry*)

[37] Kolja Knauer, Piotr Micek, Torsten Ueckerdt, *How to eat* 4/9 *of a pizza*, Discrete Mathematics (311), 2011, p. 1635–1645.

[38] Mathematical Medley, 2004, vol. 31, No 2.

[39] Murty, Ram, *Prime Numbers and Irreducible Polynomials*, The American Mathematical Monthly, Vol. 109, No. 5, p. 452-458, 2002.

[40] Stanley, Richard P., *Catalan addendum to Enumerative Combinatorics*, Volume 2, 1998, www-math.mit.edu/~rstan/ec/catadd.pdf

[41] W.Serpinski, 200 *Zadan z Elementarnej Teorii Liczb*, PZWS, Warsazawa.

[42] The Mathematical Gazette, vol. 86, N 506, July 2002.

[43] Anna Torstensson, Victor Ufnarovski, Hans Öfverbeck, *Using resultants for SAGBI basis verification in the univariate polynomial ring*, Journal of Symbolic Computation, Vol. 40, No. 3, p. 1087–1105, 2005.

[44] Л.Э. Медников, А.В. Шаповалов, *Турнир городов: мир математики в задачах*, МЦНМО, 2012, Москва, ISBN 978-5-4439-0303-3.

[45] В.А.Садовничий, А.С.Подколозин, *Задачи студенчиских олимпиад по математике*, 1978, Наука.

[46] В.А.Уфнаровский, *Математический аквариум*, Штиинца, 2003, 255p.

[47] И.Ф. Шарыгин, *Геометрия. Планиметрия, 9-11 классы*, 2001 , Дрофа, Москва, ISBN 5-7107-3283-4.

[48] М.И.Башмаков, Б.М.Беккер, В.М.Гольховой, *Задачи по математике. Алгебра и Анализ*, 1982, Наука, Москва.

[49] Белов А., Сапир М., *И возвращается ветер... , или периодичность в математике*, Квант, 1990, no. 4.

[50] В.И.Берник, И.К.Жук, О.В.Мельников, *Сборник олимпиадных задач по математике*, 1980, Народная асвета, Минск.

[51] В.В.Прасолов, *Задачи по планиметрии*, 1991, Наука, Москва.

[52] В.В.Прасолов, *Задачи по алгебре, арифметике и анализу*, 2011, МЦНМО, Москва, ISBBN 978-5-94057-810-9.

[53] Александър Кючуков, Петър Недевски *Функционални и диференциални уравнения*, Академично издателство "Проф. Марин Дринов" София, 1995.

[54] В.П.Солтан, С.И.Мейдман, *Тождества и неравенства в треугольнике*, Штиинца, 1982, 60р.

[55] Списание *Математика*, София, 1972, 3–5.

[56] Списание *Математика*, София, 1974, 1.

[57] А.Я.Хинчин, *Цепные дроби*, 1978, Наука, Москва.

[58] Arseniy Akopyan, *Geometry in Figures*, 2011, www.mccme.ru/~akopyan/papers.htm.

[59] Keith Burns and Boris Hasselblatt, *The Sharkovsky theorem: a Natural direct proof*, math.arizona.edu/~dwang/BurnsHasselblattRevised-1.pdf

[60] www.geogebra.org

[61] mathematicalbuffet.com

[62] On-line Encyclopedia of Integer Sequences, oeis.org/

[63] en.wikipedia.org/wiki/Jensen's_inequality

[64] en.wikipedia.org/wiki/Karamata's_inequality

[65] en.wikipedia.org/wiki/Japanese_theorem_for_cyclic_polygons

[66] www.artofproblemsolving.com/wiki/index.php?title=Maclaurin

[67] www.artofproblemsolving.com/articles/files/MildorfInequalities.pdf

[68] www.math.ust.hk/excalibur/v11_n1.pdf

[69] www.msri.org/web/msri/about-msri/news/emissary-newsletter

[70] sasja.shap.homedns.org/Zadachi/Statnye.html

INDEX